OCEAN MIXING

The stratified ocean mixes episodically in small patches where energy is dissipated and density smoothed over scales of centimeters. The net effect of these countless events affects the shape of the ocean's thermocline, how heat is transported from the sea surface to the interior, and how dense bottom water is lifted into the global overturning circulation. This book explores the primary factors affecting mixing, beginning with the thermodynamics of seawater, how they vary in the ocean, and how they depend on the physical properties of seawater. Turbulence and double diffusion are then discussed, which determines how mixing evolves and the different impacts it has on velocity, temperature, and salinity. It reviews insights from both laboratory studies and numerical modeling, emphasizing the assumptions and limitations of these methods. This is an excellent reference for researchers and graduate students working to advance our understanding of mixing, including oceanographers, atmospheric scientists, and limnologists.

M. C. GREGG is an emeritus professor of oceanography at the University of Washington. He is a leading expert on small-scale mixing processes and turbulence in the ocean, and has devoted his career to understanding these processes and how they impact larger-scale ocean dynamics. He was awarded the Henry Stommel Research Medal by the American Meteorological Society for his work on mixing and turbulence and is a fellow of the American Geophysical Union (AGU), American Meteorological Society (AMS), and the American Association for the Advancement of Science (AAAS).

OCEAN MIXING

M. C. GREGG
University of Washington

CAMBRIDGE
UNIVERSITY PRESS

University Printing House, Cambridge CB2 8BS, United Kingdom

One Liberty Plaza, 20th Floor, New York, NY 10006, USA

477 Williamstown Road, Port Melbourne, VIC 3207, Australia

314–321, 3rd Floor, Plot 3, Splendor Forum, Jasola District Centre, New Delhi – 110025, India

79 Anson Road, #06–04/06, Singapore 079906

Cambridge University Press is part of the University of Cambridge.

It furthers the University's mission by disseminating knowledge in the pursuit of education, learning, and research at the highest international levels of excellence.

www.cambridge.org
Information on this title: www.cambridge.org/9781107173804
DOI: 10.1017/9781316795439

First published 2021

Printed in United Kingdom by TJ Books Limited, Padstow Cornwall

A catalogue record for this publication is available from the British Library.

Library of Congress Cataloging-in-Publication Data
Names: Gregg, M. C., 1939– author.
Title: Ocean mixing / M. C. Gregg, University of Washington.
Description: Cambridge, UK ; New York, NY : Cambridge University Press, 2021. |
Includes bibliographical references and index.
Identifiers: LCCN 2020042239 (print) | LCCN 2020042240 (ebook) |
ISBN 9781107173804 (hardback) | ISBN 9781316795439 (epub)
Subjects: LCSH: Oceanic mixing.
Classification: LCC GC299 .G74 2021 (print) | LCC GC299 (ebook) |
DDC 551.46/2–dc23
LC record available at https://lccn.loc.gov/2020042239
LC ebook record available at https://lccn.loc.gov/2020042240

ISBN 978-1-107-17380-4 Hardback

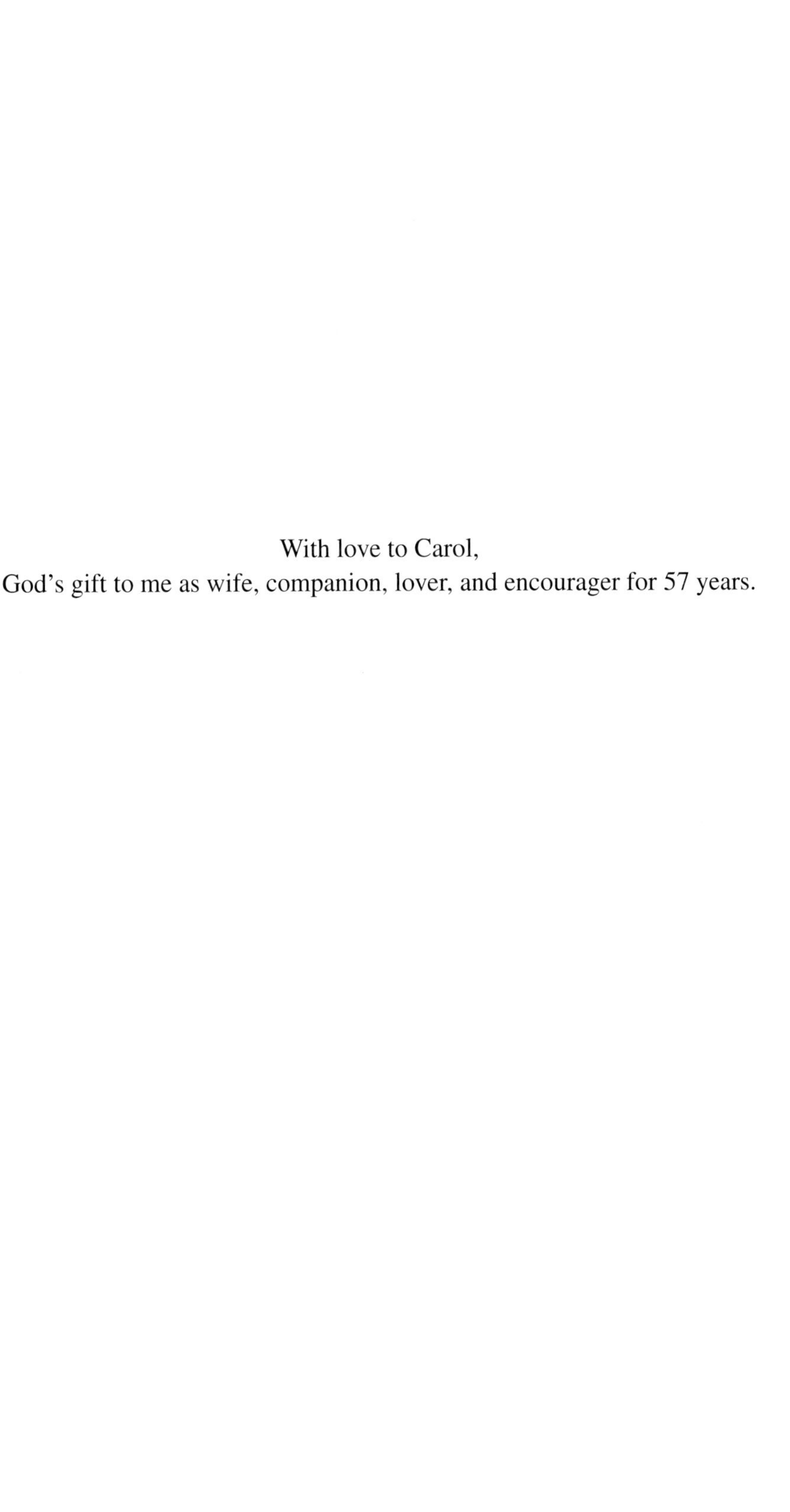

With love to Carol,
God's gift to me as wife, companion, lover, and encourager for 57 years.

Contents

Color plates can be found between pages 178 and 179

Preface

This book began over thirty years ago as a joint endeavor with Chris Garrett, but was repeatedly put off as the need for ever more detailed funding proposals increased. Taken from the name of a course I taught for many years at the University of Washington, the title reflects the focus on thermodynamic mixing, the final stage of the cascade of energy and scalar variance from the size of ocean basins to centimeters. In presenting results of decades of research as well as the most important outstanding issues, the book concentrates on the stratified ocean, the flywheel of Earth's climate.

The understanding of mixing described here stems from work begun in the late 1960s that has reached a first-order understanding by comparison with prior ignorance. This confidence comes from comparing aspects of mixing that we can measure with thickening rates of tracer clouds. We must be mindful, however, that the microstructure part of the edifice is based on a string of ad hoc approaches bridging gaps in our observational capability. Going forward, filling in these gaps should be a high priority. Both the gaps and possible future directions are presented, in part using short 'perspective' sections with my personal views.

Beginning with an overview of mixing, Chapter 1 proceeds to the role of mixing in the meridional overturning circulation and concludes with ocean-wide budgets of energy and scalar variances. Chapter 2 treats the physical properties of seawater affecting mixing and the application of thermodynamics to those properties. The resulting heat and salt conservation equations lead to equilibrium and well-mixed reference states for mixing, and later are used to infer turbulent diffusivities from measured dissipation rates.

Focusing on how microstructure is measured and interpreted, Chapter 3 is not intended as a stand-alone treatment of turbulence. The aim, rather, is to examine turbulent cascades of energy and scalar variance to ultimate dissipation and diffusion, with an eye to how well they can be estimated with one-dimensional measurements. Differential diffusion and horizontal cascades are also examined as topics of increasing interest. Concentration on dissipation-scale processes concludes with double diffusion in Chapter 4. After reviewing its theoretical basis and the conditions for its occurrence, the discussion shifts to diapycnal fluxes, staircases, and thermohaline intrusions.

With the spatial scales that must be measured having been established, Chapter 5 examines sensors and vehicles. Most direct mixing measurements have come from airfoils and thermistors; other probes show promise, but none are regularly producing useful data. Owing to the enormous span of dissipation rates in the ocean, all sensors have limited spatial resolution, sensitivity, or noise level when outside the range of weak-to-moderate turbulence produced by internal waves close to background levels. Owing to the importance of finescale (1–100 m) structures in generating and modulating mixing, these probes are also examined, particularly as they affect the vertical scales over which density overturns can be resolved. Because these probes are carried on many types of vehicles, the discussion examines how the capabilities and limitations of the platforms shape what we can observe.

Chapters 6 and 7 focus on internal waves and their interactions that drive mixing. Beginning with the basic equations, the discussion includes signatures of internal waves most often observed, as well as the degenerate solution termed the 'vortical mode'. An appendix, prepared with R.-C. Lien, develops expressions needed to apply the Garrett-Munk internal wave spectrum to observations. After reviewing production and propagation of internal wave energy, the emphasis is on energy transfer from large to small scales, the expressions used to quantify it, and how they compare with observations. The saturated range of length scales separating linear internal waves from turbulence is also considered.

The final chapter, 8, examines how the different aspects of mixing come together in the pycnocline of the open ocean. After considering the structure of the pycnocline, the nature of its finestructure is examined as the basis for the first global maps of mixing. After discussing patterns in these maps, mixing patches are characterized, including identification of which ones are produced by salt fingers. The chapter concludes with overviews of mixing in three important regions where mixing differs greatly from the open-ocean pycnocline: the Southern Ocean, the Arctic, and ocean ridges.

Because the subject is evolving rapidly and has much to do before embalming accumulated knowledge for display, I include background of how the topics came to their present state, hoping that this approach may aid readers to better see paths to explore. In addition to colleagues, graduate students, and postdocs who have educated me along the way, I am deeply indebted to Ren-Chieh Lien, Howard Stone, Rob Pinkel, Eric Kunze, Tom Sanford, and Matthew Alford for critiquing the chapters. They, however, could do only so much, and remaining problems are mine.

Since the book was completed, my close friend and colleague Tom Sanford died unexpectedly. His work is cited in many places throughout the text, but here I wish to acknowledge the important role he played in my research and the value of the encouragement he provided during the decades of our collaboration. His integrity, scientific acumen, and good sense helped me greatly, and I know that many others also benefited from knowing Tom.

1

Mixing and Its Role in the Ocean

1.1 Overview

Defined in dictionaries as the 'combining of things so the resulting substance is uniform in composition, whether or not the separate elements can be distinguished', 'mixing' is often a general term that depends on the length scales being considered. In the North Atlantic, 100 km-diameter eddies are viewed as mixing agents when they reduce variability across the Gulf Stream. Mixing, however, is not complete until molecular viscosity smoothes velocity fluctuations to remove significant shear. Consequently, the viscous dissipation rate,

$$\epsilon \equiv \nu\overline{(\nabla \mathbf{v}')^2} \quad \left[\mathrm{W\,kg^{-1}}\right], \tag{1.1}$$

is the fundamental parameter of turbulence, as it equals the flux of energy from large to small scales when turbulence is at steady state. In stratified fluids, fluctuations of temperature and salinity produced by turbulence are smoothed at rates parameterized by

$$\chi_{\mathrm{T}} \equiv 2\kappa_T\overline{(\nabla\Theta')^2} \quad \left[\mathrm{K^2\,s^{-1}}\right], \tag{1.2}$$

$$\chi_{\mathrm{S}} \equiv 2\kappa_S\overline{(\nabla S')^2} \quad \left[\mathrm{s^{-1}}\right], \tag{1.3}$$

where Θ is conservative temperature (Section 2.2.2), S is salinity in parts per thousand,[1] and primes denote fluctuations about means. In the ocean, these variances are contributed by gradients over tens of centimeters to fractions of a millimeter. How this happens in the stratified ocean is the focus of this book.

Storage and transport of heat in the ocean, the flywheel of global climate, is a major concern of climatologists, who in turn rely on oceanographers to

[1] In this form, χ_S is proportional to $(\mathrm{g_{ss}/kg_{sw}})^2\,\mathrm{s}^{-1}$ where $\mathrm{g_{ss}}$ is grams of sea salt and $\mathrm{kg_{sw}}$ is kilograms of seawater. If the salinity gradient were in concentration units, represented by s, the proportionality would be to $(\mathrm{kg_{ss}/kg_{sw}})^2\,\mathrm{s}^{-1}$.

estimate bulk mixing coefficients from measurements of centimeter-scale gradients contributing to ϵ, χ_T, and χ_S. The scales at which this occurs are determined by the intensity of the turbulence, ϵ, and the properties of seawater (Chapter 2). Termed 'microstructure', these length scales are the smallest of physical significance in the ocean. How the measurements are made (Chapter 5), the assumptions and uncertainties accompanying them (Chapter 3), and their relation to larger-scale processes producing microstructure are the core of mixing research.

Two processes, double diffusion (Chapter 4) and internal waves, produce nearly all microstructure mixing the pycnocline. Double diffusion is driven by the 100-fold contrast in molecular diffusivities of heat and salt. In addition to forming convecting layers when stable cool fresh water overlies warm saline water, double diffusion generates centimeter-diameter salt fingers when the vertical gradients are reversed. A few centimeters in diameter under oceanic conditions, salt fingers can form staircases when their fluxes are intense. One of the major issues, however, is the contribution of salt fingers to heat and salt fluxes over large areas of the pycnocline where fingering does not form staircases. In these situations fingers are difficult to detect in profiles but have been observed in horizontal measurements.

Internal waves, the second process, are mostly generated at the sea surface by wind and at the seafloor by low-frequency currents over rough bottoms (Chapter 6). As they propagate from the boundaries, internal waves transport energy vertically as well as horizontally, creating patches of turbulence as they break. Viscosity dissipates the turbulence and enhances heat and salt diffusion by straining their mean gradients until diffusion smoothes them.

Numerical models of ocean circulation or even regional models have little use for the dissipation rates that are the most direct results of microstructure measurements. Rather, assuming that vertical turbulent fluxes depend linearly on background gradients, small-scale mixing is represented in models by turbulent, a.k.a. eddy, diffusivities times the mean gradients, e.g. the vertical turbulent heat flux is

$$J_Q^z = -\rho c_p K_T \partial \overline{T}/\partial z \quad \left[\mathrm{W\,m^{-2}}\right], \tag{1.4}$$

where ρ is the density of seawater, c_p is the specific heat of seawater at constant pressure, and K_T is the eddy coefficient, or turbulent diffusivity, for heat. Unlike molecular diffusivities, which depend only on the temperature, salinity, and pressure of the water, eddy diffusivities also depend on the intensity and structure of the turbulence. Presently, eddy diffusivities are estimated from microstructure by assuming that the average rate of turbulent production balances the average rate of its dissipation by viscosity or by scalar diffusion. For example,

$$K_T \approx \chi_T / 2(\partial\overline{\theta}/\partial z)^2 \quad \left[\mathrm{m^2\,s^{-1}}\right], \tag{1.5}$$

is the eddy diffusivity for heat in terms of one microstructure parameter, χ_T, and the mean gradient of potential temperature (Section 3.11.2). The equivalent production–dissipation balance for turbulent kinetic energy leads to

$$K_\rho = \Gamma_{\text{mix}}\, \epsilon / N^2 \quad \left[\text{m}^2\ \text{s}^{-1}\right] \tag{1.6}$$

for the diapycnal eddy coefficient for density, K_ρ, with Γ_{mix} as the mixing coefficient, a measure of the efficiency of stratified turbulence. Owing to the ad hoc nature of this procedure, it has been tested multiple times by comparison with diffusivities inferred from the vertical thickening of artificial tracers injected into the pycnocline (Sections 3.12.1 and 5.8). Verification within a factor of two by these tests is the principal justification for accepting diffusivities from microstructure.

Although microstructure measurements are the basis for quantifying mixing, they are too limited to cover the expanse of the ocean. Fortunately, mixing produced by internal waves can be usefully estimated from finescale, loosely defined as 1–100 m, measurements of shear and strain using models of the rate at which wave–wave interactions transfer energy to small scales and ultimate dissipation (Chapter 7). Applying these models to large archives of data collected for other purposes has provided the first global maps of mixing intensity (Chapter 8). In addition to revealing variability with latitude, depth, season, and bottom roughness in the pycnocline of the open ocean, these maps reveal the uniqueness of regimes like the Southern Ocean and the Arctic.

Other than demonstrating that mixing in the pycnocline occurs episodically in patches, measurements to date provide limited insight into specifics of how the mixing evolves as internal waves break. Therefore, present understanding is also based on inferences from large-scale ocean variability, analytic and numerical models, finescale measurements, and process studies. Although much has been learned from these varied approaches, described in Section 1.2, one of the major mixing issues, the meridional overturning circulation, is not well understood. It is examined in Section 1.3, beginning with the formation of Antarctic Bottom Water (AABW) and continuing through a sequence of approaches taken to quantify its upwelling. Finally, the global levels of mixing required to uplift bottom water are constrained by the rates at which energy and scalar variability are produced in the ocean (Section 1.4).

1.2 How is Mixing Studied?

Mixing is studied by inferences, microstructure and tracer observations, proccess studies, finestructure observations, and integrated programs. Before the 1960s, microstructure could not be detected, and mixing was inferred from observations over larger scales. Although detailed measurements now anchor our understanding, they are not sufficient by themselves. Rather, what we know is a composite formed from these related endeavors, which in the end must converge to the same understanding of how the ocean is mixed.

1.2.1 Inferences

Where oceanic regimes are dominated by one-dimensional dynamics, simplified momentum or heat equations can be used to infer turbulent, a.k.a. eddy, coefficients. Ekman (1905) pioneered this approach when he realized that the near-surface spiral he calculated was much too compressed using molecular viscosity. He resolved the discrepancy by replacing molecular viscosity with an eddy coefficient large enough to obtain a reasonable length scale. Of equally enduring significance, Munk (1966) simplified the heat equation of the abyssal Pacific thermocline to a vertical balance between turbulent diffusion and steady upwelling of AABW (Figure 1.1, left). For the assumed rate of upwelling, fitting the exponential solution to the observed profile yielded a constant eddy diffusivity of $K_T \sim 10^{-4}$ m^2 s^{-1}, a value subsequently known as the 'canonical eddy diffusivity'. In the stratified ocean, K_T and the related scalar eddy diffusivities K_S and K_ρ operate on the background stratification and are thus are of primary importance to the general circulation.

Other inferences considered bottom water flowing through deep basins, where the principal interaction was mixing with overlying warmer water (Figure 1.1, middle). As an example, inferences from temperature changes through Romanche Fracture Zone gave $K_T \sim 10^{-1}$ m^2 s^{-1}, three orders of magnitude more intense than the canonical value (Ferron et al., 2003), but consistent with direct microstructure measurements. Finally, warming of the seasonal thermocline by vertical turbulent diffusion (Figure 1.1, right) has been modeled by equating the diffusive term with the time derivative of temperature. As computational capabilities have improved, elaborate inverse methods have largely replaced simple inferences (Wunsch,

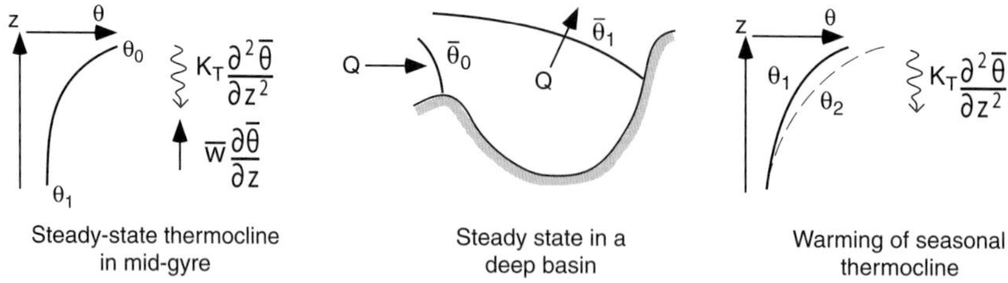

Figure 1.1 Inferences of thermal eddy coefficients for (left) a steady-state main thermocline with constant upwelling, (middle) changes in water flowing through a basin, and (right) warming of a seasonal thermocline.

2006), but the goal remains the same, to develop physical models consistent with observations.

1.2.2 Microstructure and Tracer Observations

The first successful turbulence measurements in seawater were made by towing hot-film velocity probes in a turbulent tidal channel (Grant et al., 1959). Profiling, however, soon dominated microstructure observations in the open ocean, first with untethered free-fall tubes and later with loosely tethered packages. The profiles revealed a wide range of mixing levels, with the most intense turbulence occurring in estuaries. Typical results showed ϵ and χ_T in relation to velocity and density (Figure 1.2). Now, microstructure probes are widely deployed on floats, gliders, autonomous vehicles, and moorings. Sensors have not evolved at a similar rate, but when deployed in clusters, measurements from multiple platforms are providing detailed information about mixing in mesoscale structures and in long time series.

In contrast to instantaneous mixing rates from microstructure, the spread of artificial tracers yields net mixing between observations. Artificial tracers have been injected into pycnoclines several times to test procedures for estimating eddy

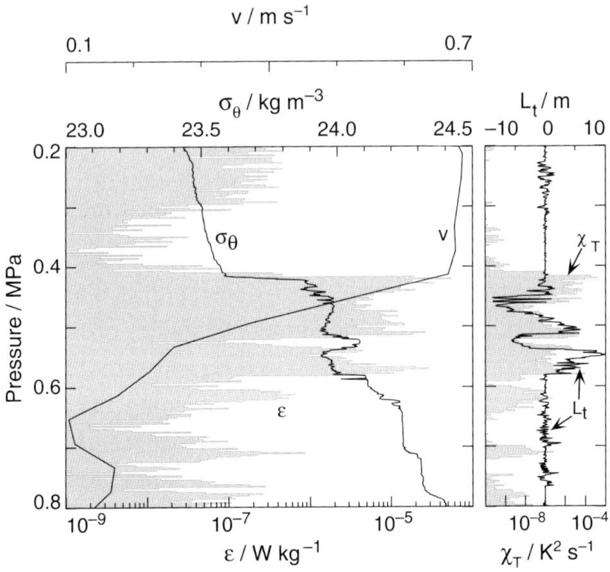

Figure 1.2 (Left) Along-channel velocity, v, potential density, σ_θ, and viscous dissipation rate, ϵ, through an overturn in profile 5854 in Figure 1.4 from Admiralty Inlet. (Right) Overturning displacements, L_t, and the rate of thermal dissipation, χ_T. Pressures 0.2–0.8 MPa correspond to depths of 20–80 m. (Data from Seim and Gregg, 1994)

diffusivities from microstructure (Section 5.8). Releases in coastal waters were observed for weeks, and those in the open ocean for months. In all cases, net diffusivities agreed within a factor of two with microstructure estimates. These comparisons are the principal justification for believing the ad hoc procedures for inferring diffusivity from dissipation rates.

1.2.3 Process Studies

Owing to the ease with which double diffusivity staircases were disrupted in laboratory tanks (Stern, 1960) and the belief that the pycnocline is moderately or strongly turbulent, double diffusion was initially treated as a curiosity rather than as an important oceanic process. Nonetheless, extensive laboratory studies (Turner, 1973) explored it and quantified buoyancy fluxes across staircase interfaces (Chapter 4). Subsequent observations, however, indicate that oceanic staircases formed by salt fingering differ in important respects from those in laboratories. Some analytic models address these issues, but a major need now is more detailed observations of staircases at sea. The more important unknown, though, is the role of fingering where it is does not form staircases. Some microstructure-based estimates suggest that fingers may contribute half the vertical flux in typical pycnoclines lacking staircases, and patches of fingers have been observed. Owing to the scale of the internal waves affecting the fingers, they cannot be simulated accurately in laboratories. Thus, laboratory experiments with double diffusion appear to have run their course, and the hope is for realistic simulations including fingers and the large-scale internal waves affecting them.

Visual identification of shallow internal waves breaking as Kelvin–Helmholtz shear instabilities (Woods, 1968) spawned a continuing sequence of laboratory and numerical studies seeking to understand mixing during these episodic events. For instance, Figure 1.3 illustrates where density instabilities develop in overturning billows. A major goal of laboratory and numerical studies of shear instabilities is to determine $\Gamma_{\rm mix}$ and its dependence on the strength and age of the instability. Other than by comparing microstructure and tracer spreading rates, mixing efficiency has not been determined in the ocean, as it requires simultaneous observations of microstructure and changes in potential energy. Numerical simulations can in principle do this, but those to date have concentrated on thin interfaces separating homogenous layers. The pycnocline, however, is irregularly steppy, with high gradients bounded by lesser gradients that nonetheless are well stratified and provide paths for small internal waves generated by the instabilities to propagate away, removing some of the energy.

The most successful mixing process studies to date modeled, analytically and numerically, the rate of energy transfer by interactions between internal waves

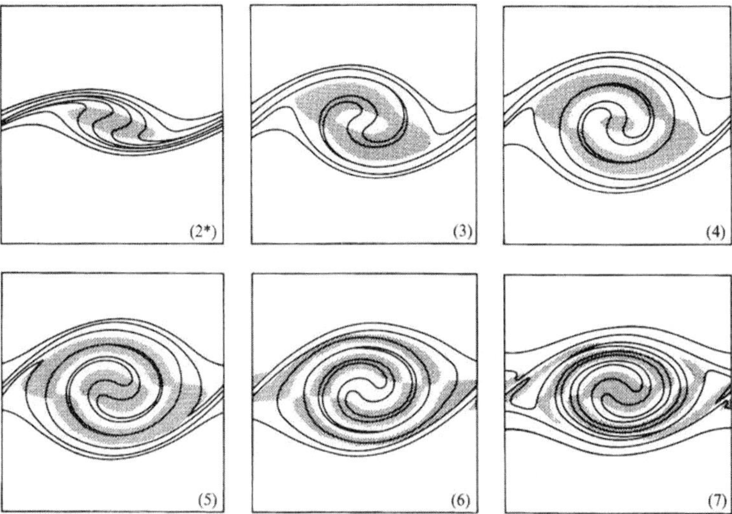

Figure 1.3 Simulation of a Kelvin–Helmholtz instability, shown initially when instabilities develop (2*), marked by shading. The Reynolds number was 500 at times of maximum (3) and minimum (6) Reynolds number stress, maximum kinetic energy (4), and zero Reynolds stress (5 & 7). (From Klaassen and Peltier, 1985).

(Section 7.5). These calculations demonstrated a net energy flux from large to small scales, which at steady state equals the turbulent dissipation rate, ϵ, produced by breaking internal waves. Formulation of the flux in terms of background stratification and internal wave shear and strain enables estimation of dissipation rates from finescale measurements.

1.2.4 Finestructure Observations

Finescale measurements relate mixing patches to the structures creating them, including the mean gradients needed for estimating eddy coefficients. They, however, do not usually resolve lateral variability well enough to identify mixing structures. Where strong stratification exists in estuaries and near coasts, unaliased images of strong events can be obtained using tow chains or backscatter from narrow beams of high-frequency acoustics. The example in Figure 1.4 shows a train of overturning billows similar to Kelvin–Helmholtz instabilities in laboratories, with the the profile in Figure 1.2 passing through an overturn. The image was essential in identifying the nature of the mixing patch.

In addition to being background for microstructure, finestructure contains the information needed to apply parameterizations from process studies to estimate mixing rates in most of the ocean. This allows mixing to be estimated within a factor of two from finescale profiles collected during large-scale surveys, as well as

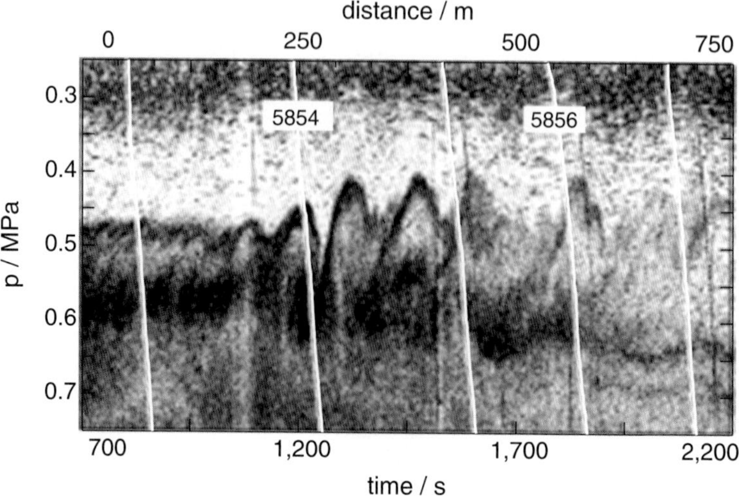

Figure 1.4 Intensity of 200 kHz acoustic backscatter from overturning billows in Admiralty Inlet. The track of profile 5854 from Figure 1.2 is labeled. By homogenizing the water, the intense mixing reduced backscatter amplitude by profile 5857. (Adapted from Seim and Gregg, 1994)

from the 4,000 Argo floats taking temperature and salinity profiles throughout the ocean every ten days. The results led to the first global estimates of mixing rates (Section 8.4).

1.2.5 Integrated Studies

Early mixing studies were often solo affairs, sampling microstructure and a few finescale variables from one ship during a few weeks in a small area. As measurements became more reliable and more accepted, microstructure was included in large-scale experiments, e.g. Polymode, but it was incidental to program goals. More recent programs include microstructure as an essential component. One example, the Hawaii Ocean Mixing Experiment (HOME) (Section 8.9.2), was inspired by satellite observations of coherent internal tides radiating from the islands. Subsequent modeling predicted the sources, and intensive observations verified the predictions, in addition to estimating accompanying local mixing. A successor program followed one of the northward internal tide beams to examine its effect on mixing far from the islands.

Owing to the importance of the Southern Ocean to heat and carbon transfer between atmosphere and ocean, several programs have addressed locations expected to mix intensely (Section 8.7). The Diapycnal and Isopycnal Mixing Experiment (DIMES) focused on Drake Passage, while the Southern Ocean

Finestructure Experiment (SOFine) worked around Kerguelen Plateau, two of the three sites providing the principal drag on the Antarctic Circumpolar Current (ACC), which flows unimpeded around the globe, driven by strong zonal winds. Seeking a comprehensive understanding, these programs included modeling, moorings, sections of fine and microstructure profiles, analysis of satellite data, and a tracer release. Observations and insights discussed in subsequent chapters have come from all of these types of programs.

1.3 The Meridional Overturning Circulation

Circling the globe between 1872 and 1876 on the first modern oceanographic expedition, *H.M.S. Challenger* traced the densest bottom water to Antarctica (Figure 1.5). Decreasing density of AABW as it flows north is evidence of mixing, but how and where it upwells remains a major issue. Present understanding of the overturning circulation includes northward bottom flows, upwelling in the three ocean basins, and shallow return flows. Mixing is central to forming the dense water, modifying it during descent, and subsequently lifting it to depths reached by wind-driven upwelling.

1.3.1 Formation of Antarctic Bottom Water (AABW)

Bottom water is formed at a few sites around Antarctica where polynyas, persistent ice-free bands, expose seawater on the continental shelf to winter winds off the continent (Figure 1.6). Owing to the difficulty in observing these locations,

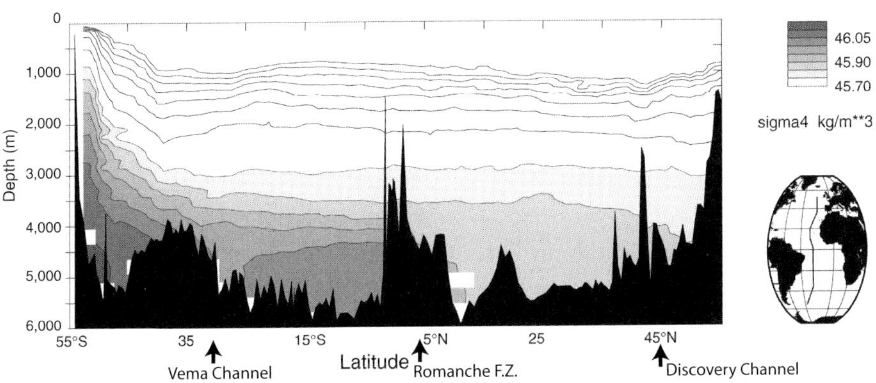

Figure 1.5 Potential density along a meridional section in the Atlantic. As AABW flows north, its density decreases abruptly in narrow gaps between basins. (From Bryden and Nurser, 2003. © American Meteorological Society. Used with permission)

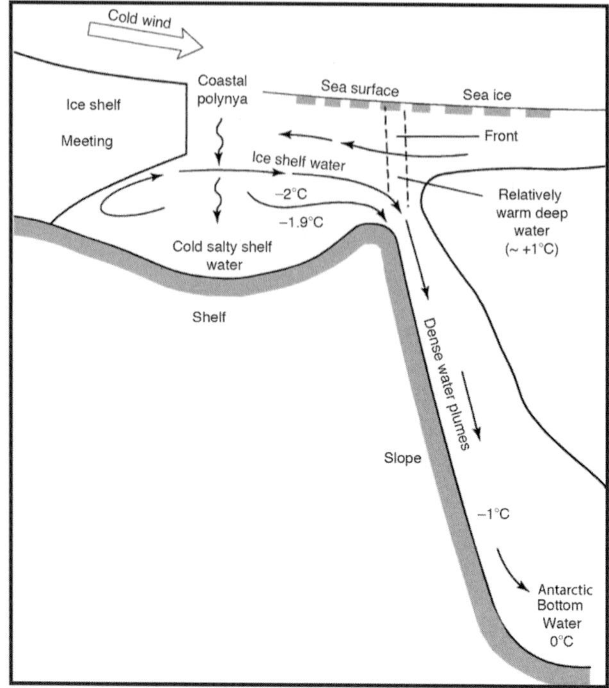

Figure 1.6 Schematic formation of AABW by convection in polynyas and further cooling under ice sheets before cascading down the slope. (Adapted from Gordon, 2013)

measurements have been limited to hydrography, which indicates that, after sinking during surface convection, shelf water is further cooled by the undersides of floating ice sheets. With temperatures as low as –2°C, the dense water flows over the shelf edge and sinks, reaching great depths when salinity exceeds 34.61.

Sinking is enhanced by cabbeling and thermobaricity (Foster, 1972; Klocker and McDougall, 2010), which result from nonlinearities in density as a function of temperature, salinity, and pressure (Sections 2.6 and 2.9). Descending the continental slope, the sinking water forms density currents entraining warmer ambient water from offshore. Entrainment approximately doubles transport to $\approx 29 \times 10^6$ m^3 s^{-1} (Talley, 2013) and can warm it to 0°C.

1.3.2 Formation of North Atlantic Deep Water (NADW)

The Indian and Pacific oceans have shallow overturning cells, with water formed in the north ultimately sinking to about 1 km, but only the North Atlantic produces water that sinks into the abyss. More accessible for observations than the Antarctic shelf, the Nordic Seas include two basins, Greenland and Norwegian,

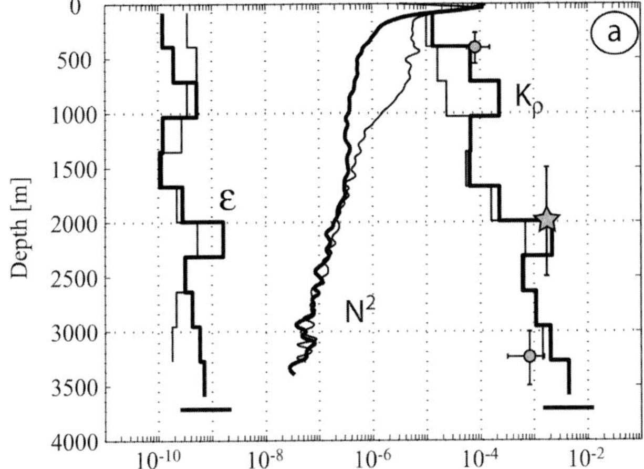

Figure 1.7 Dissipation, stratification, and diapycnal diffusivity in Greenland Basin (thick) and the Norwegian Sea (thin). Circles are K_ρ from SF_6 tracers (Watson et al., 1999), and the star is from CFC spreading in Greenland Basin (Visbeck and Rhein, 2000). (From Naveira Garabato et al., 2004)

and several ridges. Both basins are weakly stratified, particularly the western, in which stratification characteristic of the abyss comes within a few hundred meters of the surface (Figure 1.7). Owing to the weak stratification, low dissipation rates produce large diapycnal diffusivities below the shallow pycnocline, even during summer. These K_ρ estimates agreed well with inferences from spreading rates of chlorfluorocarbon (CFC) (Visbeck and Rhein, 2000) and sulfur hexafluoride (SF_6) (Watson et al., 1999).

Using springtime observations of SF_6 released in Greenland Basin the previous summer, Watson et al. (1999) inferred that convective plumes during winter penetrated to 1,200–1,400 m in the center of the basin's gyre. They argue that deep convection occurs episodically after 50 km-diameter regions have been 'pre-conditioned' for deep convection by rotation of the Greenland Basin gyre. Diapycnal diffusivities in convective regions were an order of magnitude above those during summer, but diffusivities outside plumes were unaffected. In view of the limited area of deep convection and the modest K_ρ increase, Watson et al. suggest that high year-round K_ρ rather than convection may dominate mixing in the basin.

About half of the outflow from the Nordic Seas passes through Denmark Strait, between Iceland and Greenland. Moorings have shown the outflow to be highly variable over two to five days but more steady over longer periods (Smith, 1976; Ross, 1984). Entrainment during the initial descent increases 3 Sverdrups (3×10^6 m^3 s^{-1}) of outflow to 5 Sv, which doubles to 10 Sv when joined by the Faroe Bank outflow and eventually becomes 13 Sv after further entrainment (Dickson and Brown, 1994).

1.3.3 Bottom Water Spread and Transformation into Deep Water

Atlantic Ocean. The 5×10^6 m^3 s^{-1} (Talley, 2013) of AABW flowing northward into the Atlantic follows a convoluted path, blocked by ridges and shunted to deep basins through narrow gaps. AABW accumulates in the Argentine Basin before entering Vema Channel at 31°N (Figure 1.5). Bottom water piles up again near the equator in Brazil Basin before flowing through either Romanche or Chain Fracture Zone into Sierra Leon Basin. Finally, and to a lesser degree, water accumulates in the North Atlantic before flowing into Discovery Channel. Microstructure observations (Figure 1.8) confirm intense mixing inferred from abrupt temperature decreases across the gaps.

Pacific Ocean. Bottom water flows relatively unobstructed into the South Pacific, but one-third to one-half of AABW entering the North Pacific comes through Samoan Passage. Moorings found 6×10^6 m^3 s^{-1} for the transport (Rudnick, 1997), and, based on changes in water properties, Roemmich et al. (1996) inferred diffusivity to be $K_\rho \approx 10^{-2}$ m^2 s^{-1}. Accelerating over the two principal sills, flow exceeded 0.5 m s^{-1} over the second, where it was hydraulically controlled (Alford et al., 2013). That is, the stratification Froude number exceeded one, $Fr_N \equiv u/g'h > 1$, with h being the thickness of the flow and $g' \equiv g\Delta\rho/\rho$ as reduced gravity or specific buoyancy force. Where mixing was most intense, dissipation rates exceeded 10^{-6} W kg^{-1}, overturns were hundreds of meters high, and $K_\rho \approx 10^{-1}$ m^2 s^{-1}, diluting the density anomaly fourfold during the transit.

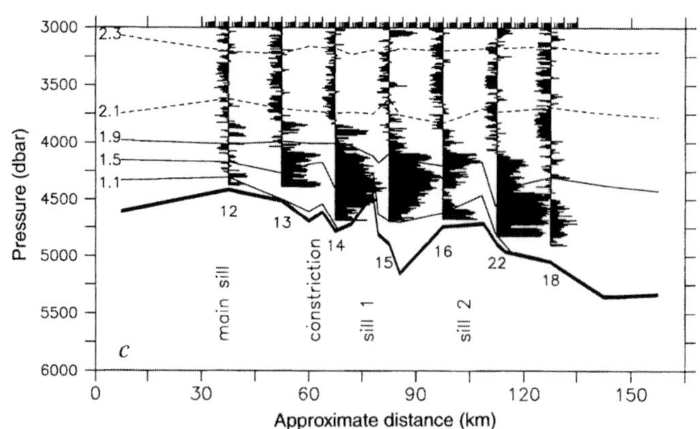

Figure 1.8 Dissipation rates in Romanche Fracture Zone from the High Resolution Profiler (HRP). The logarithmic scale is shaded about 10^{-10} W kg^{-1}. Maximum values exceed 10^{-6}. (From Polzin et al., 1996)

Indian Ocean. Of the (10–12) $\times$ 10^6 $m^3\ s^{-1}$ of AABW entering the Indian Ocean, a third flows through the Southwest Indian Ridge via the Atlantis II Fracture Zone (MacKinnon et al., 2008). The 1,000 m thick northward flow mixes deep and bottom water during a 13-day transit that is not hydraulically controlled. Changes in the deep salinity maximum imply an average diapycnal diffusivity of $K_\rho > 10^{-2}\ m^2\ s^{-1}$. Based on shear and strain finestructure, K_ρ increases from 10^{-4} at 2,000 m to 10^{-2} at 5,000 m, but the average is less than inferred from the salinity change. One possibility is that internal recirculation in the passage increases the residence time, lowering the inferred diffusivity.

In sum, mixing in gaps affects 10%–30% of the AABW. This is significant, but if these measurements are representative, most of the conversion and upwelling of AABW to deep water must be accomplished by mixing elsewhere. In the Atlantic, diapycnal mixing with AABW modifies NADW flowing south. In the Indian and Pacific Oceans, deep diapycnal mixing forms Indian Deep Water (IDW) and Pacific Deep Water (PDW), as there are no northern sources.

1.3.4 Upwelling of the Deep Water

Noting the similarity between north–south temperature–salinity (TS) relations on the surface of the eastern North Atlantic and vertical profiles at mid-latitudes, Iselin (1939) inferred that the surface water was being subducted into the interior as it flowed south, leaving a minor role for vertical mixing in forming the TS relation. Consistent with that view, Welander (1959) constructed an analytic model of an advective thermocline, commenting:

> It should be noted that the importance of diffusion processes in large-scale ocean dynamics has not yet been proved. It cannot be doubted, however, that density advection plays a fundamental role so that it seems more natural to start out from a purely advective model,

in which all diffusion effects are neglected. It appears that the model can explain the main features of the ocean density field below a boundary layer of thickness 100–200 m.

Abyssal Recipes

Welander (1959) was able to reproduce the exponential shape of the thermocline without mixing, but his model did not consider the fate of bottom water, which would continue to thicken without mixing, eventually filling ocean basins from the bottom up. Considering densities between 1 and 4 km (Figure 1.9), below those traced back to surface formation, Munk (1966) in *Abyssal Recipes* assumed that AABW slowly upwells over broad areas as it flows north through the Pacific. Noting that abyssal profiles appear in steady state, he posited a vertical heat

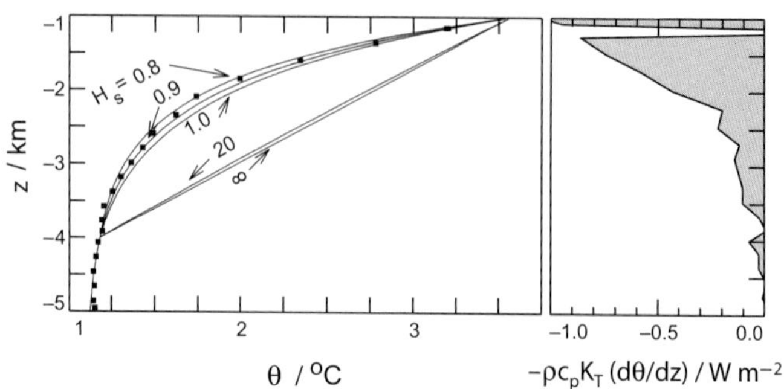

Figure 1.9 (Left) Observed potential temperature (dots) at 9.6°N, 126.9°E compared to (1.8) for $H_S = 0.8, 0.9$, and 1 km. Scale heights ≥20 km mix so strongly that temperature is linear. (Right) Vertical heat flux for the observed temperature gradients and $K_T = 10^{-4}$ W m^{-2}. (Adapted from Munk, 1966)

balance between diffusivity, K_T, produced by small-scale turbulence, and the rising cold water,

$$w_\rho \frac{\partial \overline{\theta}}{\partial z} = \frac{\partial}{\partial z}\left(K_T \frac{\partial \overline{\theta}}{\partial z}\right) \quad [\mathrm{K\ s^{-1}}]. \tag{1.7}$$

The resulting exponential solution,

$$\overline{\theta}(z) = \overline{\theta_0} - (\overline{\theta_0} - \overline{\theta_1})\left(\frac{1 - \exp(z/H_S)}{1 - \exp(H/H_S)}\right), \tag{1.8}$$

fits observations for a narrow range of scale heights, $H_S \equiv K_T/w_\rho \lesssim 1$ km. Using $w_\rho = 1.4 \times 10^{-7}$ m s^{-1} (1.2 cm day^{-1}) from the rate of bottom water formation yields $K_T = w_\rho H_S \sim 10^{-4}$ m^2 s^{-1}, often cited as the 'canonical' value for turbulent diffusivity. Much larger diffusivities, e.g. for $H_S \gtrsim 20$, would linearize the profile, and much smaller ones would result in a nearly uniform column of cold water capped by a warm surface layer.

Diapycnal and Isopycnal Coordinates

Abyssal Recipes was formulated with a vertical heat balance, ignoring the vertical components produced by quasi-horizontal motions along inclined density surfaces. Realistic estimates of these contributions equal the vertical diffusivity inferred by Munk (Solomon, 1971). Defining heat fluxes along (isopycnal) and normal (diapycnal) to density surfaces,

$$J_{\mathrm{iso}} = -\rho c_p K_{\mathrm{iso}}(\partial\theta/\partial x_{\mathrm{iso}}) \quad \text{and} \quad J_{\mathrm{dia}} = -\rho c_p K_{\mathrm{dia}}(\partial\theta/\partial x_{\mathrm{dia}}) \quad [\mathrm{W\ m^{-2}}], \tag{1.9}$$

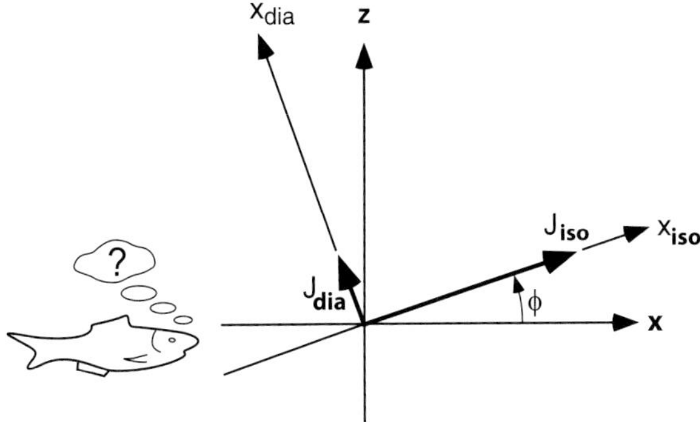

Figure 1.10 Isopycnal and diapycnal axes inclined by a small angle, ϕ, relative to the horizontal and vertical axes.

and assuming a small inclination, ϕ, from horizontal (Figure 1.10), the corresponding horizontal and vertical fluxes are

$$\begin{aligned} J_x/\rho cp &\approx \underbrace{K_{\text{iso}}\frac{\partial\theta}{\partial x}}_{10^1\ 10^{-3}=10^{-2}} - \underbrace{\phi K_{\text{iso}}\frac{\partial\theta}{\partial z}}_{10^{-3}\ 10^1\ 10^{-2}=10^{-4}} \\ J_z/\rho c_p &\approx \underbrace{\phi K_{\text{iso}}\frac{\partial\theta}{\partial x}}_{10^{-3}\ 10^1\ 10^{-3}=10^{-5}} - \underbrace{(\phi^2 K_{\text{iso}} + K_{\text{dia}})\frac{\partial\theta}{\partial z}}_{(10^{-6}\ 10^1+10^{-4})\ 10^{-1}=1.1\times10^{-5}} . \end{aligned} \tag{1.10}$$

Recasting Munk's result as diapycnal diffusivity gives $K_{\text{dia}} = 10^{-4}$ m^2 s^{-1}. A realistic choice for isopycnal diffusivity is $K_{\text{iso}} = 10$ m^2 s^{-1}, and typical inclination and gradients are $\phi = 10^{-3}$, $\partial\theta/\partial z = 10^{-1}$ K m^{-1}, and $\partial\theta/\partial x = 10^{-3}$ K m^{-1}. With these values, only 1% of the horizontal flux, J_x, is contributed by horizontal diffusivity operating on the vertical temperature gradient, but the vertical flux, J_z, receives nearly equal contributions from K_{iso} and K_{dia}, which is more commonly represented as K_ρ.

Abyssal Recipes II

In response to microstructure and tracer measurements finding K_T only one-tenth the canonical value, in *Abyssal Recipes II* Munk and Wunsch (1998) retained the earlier assumption that upwelling of bottom and deep water is driven from bottom to top solely by diapycnal mixing. Mixing, however, was attributed to 1) widespread diapycnal diffusivity of $K_\rho = 10^{-5}$ m^2 s^{-1} from background internal waves and 2) intense mixing over a small number of abyssal regions where fast flows cross rough bottoms.

To estimate the work required to lift 30×10^6 m^3 s^{-1} of bottom water each year, Munk and Wunsch kept the isopycnal model and vertical heat balance of Munk (1966) and assumed a linear equation of state, i.e. $\rho = a\theta + bS$, where a and b are constants. Allowing for vertically distributed water sources, the vertical diffusion equation becomes

$$\frac{d}{dz}\left(w(z)\rho(z) - K_\rho \frac{d\rho}{dz}\right) = A^{-1} q(z)\rho(z), \tag{1.11}$$

where A is the average area of the ocean, and $q(z) = A dw/dz$ is the volume source function. Using global data to evaluate a stratification function, $I(z) \equiv \int (1/\rho')\, d\rho'$, a simplified solution is

$$D_{\text{diss}}^{\text{abyssal}} = g\, \Delta\sigma_4\, \Gamma_{\text{mix}}^{-1}\, Q\, H_s = 2.1 \text{ [TW]}, \tag{1.12}$$

where $H_s = 1{,}300$ m was picked to match the depth-dependent calculation. As before, $\Delta\sigma_4 = 1$ kg m^{-3} is the contrast in potential density referenced to 4 km, and $\Gamma_{\text{mix}} = 0.2$.

How much of the 2.1 TW (terrawatts) comes from background mixing with $K_\rho = 10^{-5}$? The corresponding dissipation rate is $\epsilon = K_\rho\, N^2\, \Gamma_{\text{mix}}^{-1}$, and the total global dissipation rate is

$$D_{\text{diss}} = A \int_{4\text{ km}}^{1\text{ km}} \rho\epsilon^{\text{GM}}\, dz \approx g\, A\, K_{\text{vert}}^{\text{GM}}\, \Gamma_{\text{mix}}^{-1}\, \Delta\sigma_4 = 0.2 \text{ [TW]}, \tag{1.13}$$

with $A = 3.6 \times 10^{14}$ m^2 as the area of the ocean. From this, Munk and Wunsch estimated that background mixing, $K_\rho \sim 10^{-5}$ m^2 s^{-1}, supplies only 10% of the work needed to lift bottom water through the full depth of the ocean. They hypothesized that the remainder is concentrated in a few regions of rough bottoms. Klocker and McDougall (2010) suggested $D_{\text{diss}} = 0.1$ TW as a better value, noting that Munk and Wunsch did not consider cabbeling and thermobaricity (Section 2.9.3) and the $\Delta\sigma_4$ they used is inconsistent with their plot.

The Upper Meridional Overturning Cell

Talley (2013) argued that deep waters in the Indian and Pacific oceans require diapycnal mixing to bring them into the upper thermocline as it flows south. Further upwelling, however, appears to result from Ekman suction in the Southern Ocean, i.e. the horizontal divergence in surface flow driven by winds and deflected to the left by the Coriolis force. This uplift and the return surface flow forms an upper cell to the meridional circulation (Figure 1.11). Eastward winds are most intense at the latitude of Drake Passage, 63–56°S, producing a strong divergence immediately south. At the latitude of the strait, no land or subsurface feature shallower than 2,000 m blocks the eastward wind drift, preventing formation of surface slopes

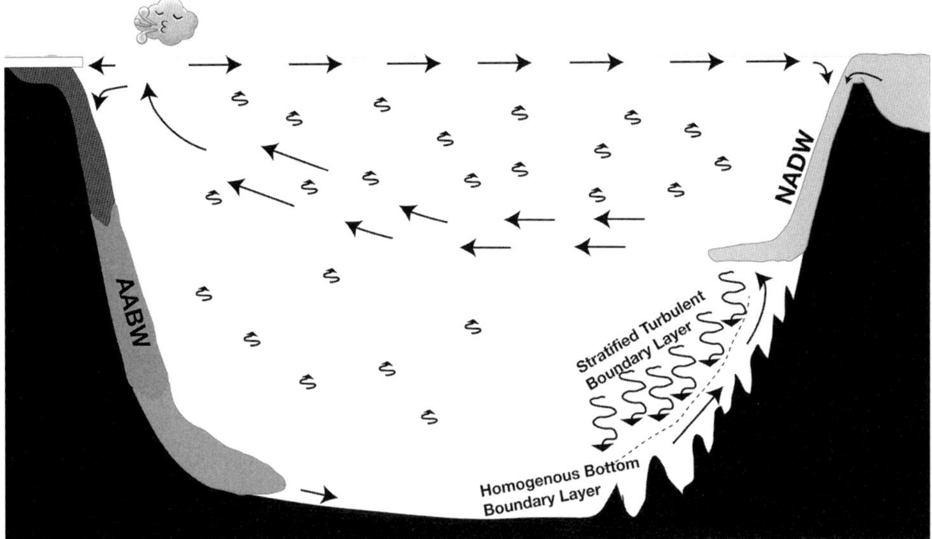

Figure 1.11 Schematic meridional overturning in the Atlantic. AABW upwells primarily in homogenous bottom boundary layers (HBBL) over rough topography on continental slopes beneath water downwelling through stratified turbulent boundary layers. To lift the AABW and the downwelling water, transport in HBBLs must be several times the rate of bottom water production. North Atlantic Deep Water (NADW) and the upwelled bottom water are brought to the surface in the upper overturning cell by Ekman suction under the Roaring Forties, supplemented by weak mixing, $K_\rho \sim 10^{-5}$ m^2 s^{-1}, in the upper ocean and over smooth abyssal bottoms.

opposing the northward Coriolis deflection. Consequently, Ekman suction appears to extend to the deep water at the base of the upper meridional cell.

Using simulations by Toggweiler and Samuels (1998) and Döös and Coward (1997), Webb and Suginohara (2001) argue that $(9-12) \times 10^6$ m^3 s^{-1} of NADW is brought to the surface in the Southern Ocean by Ekman suction (Figure 1.11). Reviewing the evidence, Marshall and Speer (2012) agree that the return path through wind-driven upwelling in the Southern Ocean is a major component of the overturning circulation. Owing to contrasts in density of the deep waters, upwelling in the Indian and Pacific sectors occurs at different distances from Antarctica.

The Lower Meridional Overturning Cell

Diapycnal mixing is needed in the lower overturning cell to lift bottom water to the upper cell. In terms of diapycnal velocity produced by mixing, upwelling requires the diapycnal buoyancy flux to become more negative upward (Section 2.7.6). This is the case in the upper pycnocline, but full-depth microstructure observations show

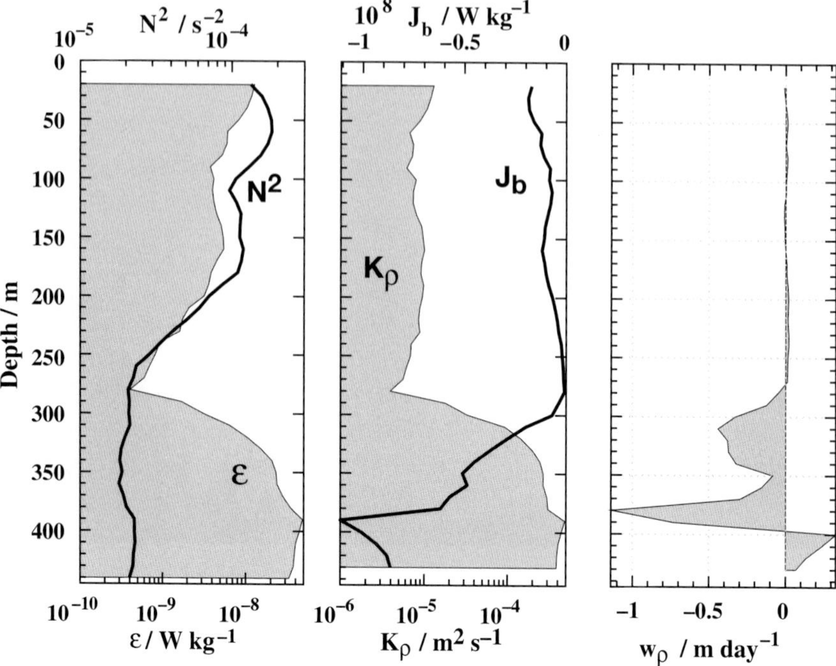

Figure 1.12 Averages at Station 7 in the Florida Strait (Figure 7.28). Above 280 m, the buoyancy flux, J_b, became more negative upward, inducing an average diapycnal upwelling of $w_\rho = +4$ mm day^{-1}. Deeper, the buoyancy flux became more strongly negative toward the bottom in a stratified turbulent boundary layer, inducing diapycnal downwelling peaking at $w_\rho < -1$ m day^{-1} where $K_\rho \sim 5 \times 10^{-4}$ m^2 s^{-1}. Forming the average profile smoothed the thin homogenous bottom boundary below 400 m, reducing estimated upwelling in it. (Data are from Winkel et al., 2002)

buoyancy fluxes nearly constant with depth in the abyss over smooth bottoms, and fluxes strongly increasing approaching rough bathymetry (Figure 8.12). The latter imply strong diapycnal downwelling several hundred meters above rough bottoms and negligible diapycnal velocity over large areas with smooth bottoms. As an example, over the steep eastern side of Florida Strait a stratified turbulent boundary layer 200 m thick induces downwelling, with peak diapycnal velocities of –1 m s^{-1} (Figure 1.12). Higher dissipation rates and weaker stratification over rough abyssal bottoms are likely to produce much larger diapycnal velocities.

Ferrari et al. (2016) argued that the upwelling occurs in thin homogenous bottom boundary layers, where the buoyancy flux decreases to meet the no-flux condition at the bottom. Because downwelling regions are much thicker than upwelling regions and upwelling must include water downwelled through the stratified turbulent layer as well as the bottom water, upwelling transport must be large; for 25 × 10^6 m^3 s^{-1}

of bottom water production, McDougall and Ferrari (2017) estimated upwelling as 100×10^6 m^3 s^{-1}. They also inferred $K_\rho \sim 5 \times 10^{-3}$ m^2 s^{-1} (ten times the peak in Figure 1.12) at the top of the bottom boundary layer to produce the required upwelling. Much needs to be done to verify this scenario, but there is little doubt that mixing over rough bottoms is key to the lower overturning cell.

1.4 Kinetic Energy and Scalar Variance Budgets

Winds and tides energize the ocean, supplemented by much smaller inputs from heating/cooling and evaporation/precipitation. Since the state of the ocean is changing very slowly, the rate of energy input should nearly equal the average viscous dissipation rate, ϵ. Similarly, rates at which large-scale temperature and salinity variability are created should nearly balance the rates at which molecular diffusion smoothes scalar variability. Important in their own right, these three parameters, ϵ, χ_T, and χ_S, are also source terms for the rate of entropy production by molecular processes (Section 2.3.2).

Global averages of χ_S and χ_T have been estimated in terms of surface parameters in three ways: Stern (1968) related χ_S to the maximum evaporation rate at the surface and the average vertical salinity gradient; Joyce (1980) estimated χ_T from the correlation between the surface temperature and heat flux; and Schanze and Schmitt (2013) combined the two approaches to calculate the net flux of density through the surface, taking account of variation of the coefficient of thermal expansion with temperature. All three approaches aim at order-of-magnitude estimates of rates of scalar dissipation that must eventually match observations in the same way that viscous dissipation must balance energy input from winds and tides.

1.4.1 Kinetic Energy

Beginning with energetics, Munk and Wunsch (1998) estimated that 2.6 TW of the 3.5 TW input from lunar and solar tides is dissipated in shallow seas, leaving 0.9 TW to maintain abyssal stratification (Figure 1.13). Only 0.2 TW of this comes from the background internal wave field, which mixes at the rate of $K_\rho \sim 10^{-5}$ W kg^{-1} (Gregg and Sanford, 1988). The remaining 0.7 TW are assigned to 'localized turbulent patches', which subsequent authors consider part of the mixing produced by near-bottom flows over rough topography. An additional 1.2 TW of wind forcing was assumed to complete the 2.1 TW needed to maintain the abyssal stratification.

Wunsch and Ferrari (2004) estimated that 1.5 TW is needed for mixing to maintain abyssal stratification, with the 0.2 TW from background internal waves supplemented by 1.3 TW of boundary turbulence and 0.2 TW of drag in deep passages,

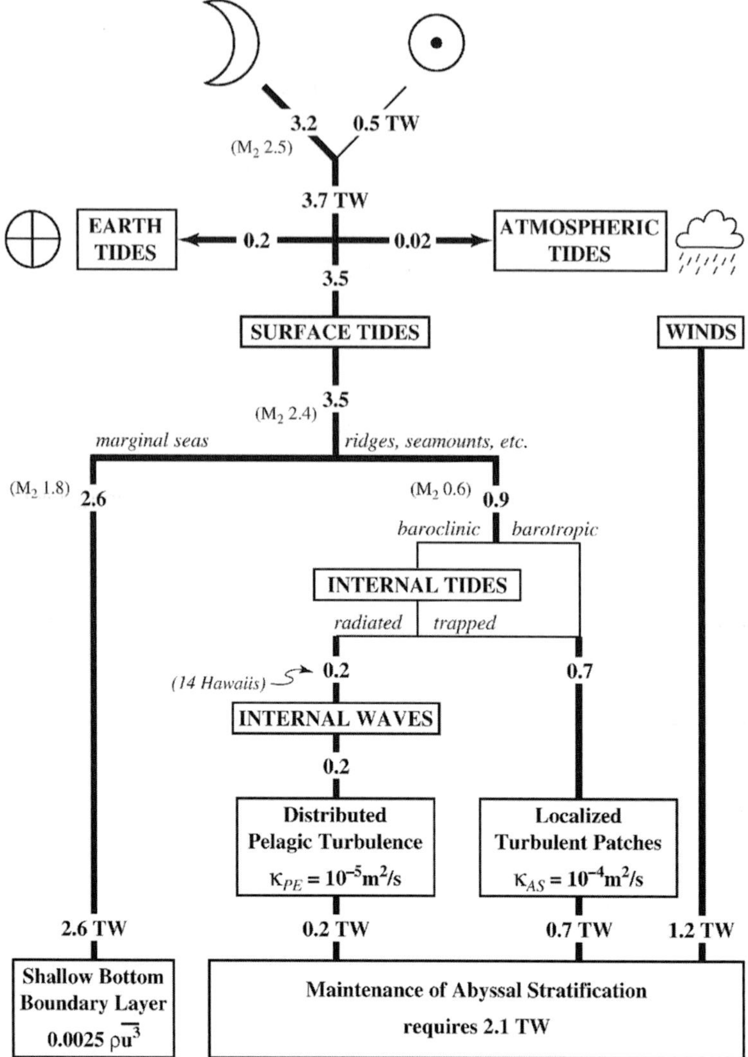

Figure 1.13 Rough energy budget for the 3.5 TW (3.5×10^{12} W) entering the ocean from lunar and solar tides, which is the only firm value in the budget. (From Munk and Wunsch, 1998)

offset by 0.2 TW returned to the general circulation from mixing. As discussed by Waterhouse et al. (2014), many flux estimates vary at least twofold. For instance, Wang and Huang (2004) estimated wind power input as 60 TW versus the 20.6 TW of Wunsch and Ferrari. Also, Klocker and McDougall (2010) calculated that only 0.6–0.7 TW is needed to maintain abyssal stratification.

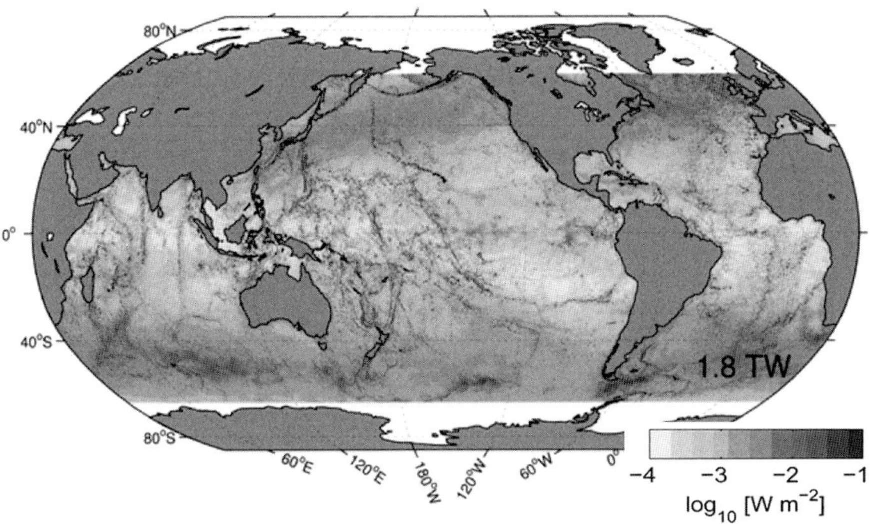

Figure 1.14 Power input to internal waves by internal tides and winds at near-inertial frequencies. Wind forcing varies smoothly over broad areas and is most intense at high latitudes. Internal tide forcing is strongest over ridges and seamounts and appears as lines of dots that are most intense at mid- and low latitudes in the western Pacific. (From Waterhouse et al., 2014. © American Meteorological Society. Used with permission) (A black and white version of this figure will appear in some formats. For the color version, please refer to the plate section)

Assuming internal waves to be in steady state, the Wunsch and Ferrari (2004) fluxes require approximately eight months to replenish an energy content of 1.4×10^{19} J. More recently, Waterhouse et al. (2014) estimated energy input to internal waves as 1.5 TW from internal tides and 0.3 TW from winds, via near-inertial internal waves. The corresponding replenishment time scale would be only three months, but dominance of the input by internal tides rather than winds points to very different spatial and temporal patterns for internal waves and the mixing they produce. Winds produce spatially smooth fields with broad maxima at high latitudes, whereas internal tide inputs are discrete peaks over seamount chains and ridges (Figure 1.14).

Dissipating the total energy input, 21.6 TW, over the entire ocean volume of 1.33×10^{18} m^3 yields an average dissipation rate of $\epsilon = 1.6 \times 10^{-8}$ W kg^{-1}. Most of this would be very shallow and result from breaking surface waves. Distributing the 0.2 TW dissipated by internal waves over the same volume would yield $\epsilon = 1.5 \times 10^{-10}$ W kg^{-1}.

In sum, understanding the energy budget is at an early stage, but finding enough to maintain the overturning circulation no longer seems a major issue.

1.4.2 χ_S from the Surface Salinity Flux

To estimate χ_S (1.3) in terms of surface evaporation and precipitation, Stern (1968) began with a steady-state salt balance multiplied by s (salinity in concentration units),

$$\rho s\, \boldsymbol{u} \cdot \nabla s = -s\, \nabla \cdot \boldsymbol{J}_S, \tag{1.14}$$

where $\boldsymbol{J}_S$ is the vector salt flux. This leads to

$$(\rho/2)\nabla \cdot \left(s^2 \boldsymbol{u}\right) = -\nabla \cdot (s \boldsymbol{J}_S) + \nabla s \cdot \boldsymbol{J}_S. \tag{1.15}$$

Integrating over the volume of the ocean and applying Stokes' theorem,

$$\frac{\rho_0}{2} \iint s_0^2 w_0\, dA_0 = -\iint s_0 J_S^0\, dA_0 - \frac{\rho}{2} \iiint \chi_S\, dV, \tag{1.16}$$

where the pressure-driven flux is neglected as insignificant compared to the turbulent flux. Zero sub- and superscripts indicate surface values.

The effective mean velocity of the surface water is

$$w_0 = E/\rho_0(1 - s_0), \tag{1.17}$$

with E positive for evaporation and negative for precipitation. The corresponding salt flux is

$$J_S^0 = -Es_0/(1 - s_0) \qquad \left[\mathrm{kg_{ss}\ m^2\ s^{-1}}\right], \tag{1.18}$$

and these expressions reduce (1.16) to

$$\iint \frac{Es_0^2}{1 - s_0}\, dA_0 = \frac{1}{2} \iiint \chi_S\, dV. \tag{1.19}$$

Applying the Reynolds decomposition to s_0 reduces the integrand to Es_0'. Because s_0' is largely produced by variability in E, the two are strongly correlated, allowing Stern to approximate the surface integral as

$$\iint \frac{Es_0^2}{1 - s_0}\, dA_0 \approx 2\bar{s}_0 \iint Es_0'\, dA_0 \approx 2\frac{E_{\max} s'_{0,\max}}{2} A_0, \tag{1.20}$$

where A_0 is the surface area. The net change in salinity is approximately the same in the horizontal as in the vertical, suggesting that $s'_{0,\max} \approx 2H\partial\bar{s}/\partial z$, where H is the scale thickness of the upper ocean.

Approximating the integral of salt dissipation by $\iiint \chi_S\, dV = \overline{\chi_S}\ H\ A_0$ and evaluating with $E_{\max}/\rho = 1$ m/year, $\bar{s} = 0.036$, and $\partial\bar{s}/\partial z = 10^{-6}$ (1 psu/km) gives

$$\overline{\chi_S} = 2\,\bar{s}\,\frac{E_{\max}}{\rho}\frac{\partial\bar{s}}{\partial x_3} \approx 2 \times 10^{-15}\ \mathrm{s}^{-1}. \tag{1.21}$$

Using the definition of χ_S (1.3), the root-mean-square salinity gradient is 10^{-3} m^{-1}, 1,000 times the mean gradient.

1.4.3 χ_T from the Surface Heat Flux

Following an approach similar to Stern's, Joyce (1980) obtained

$$\frac{2}{\rho c_p} \iint \overline{T_0 J_Q{}^0}\, dA_0 = \iiint \chi_T \, dV. \tag{1.22}$$

Average surface temperature, $\overline{T_0}$, and heat flux, $J_Q{}^0$, are highly correlated and were evaluated as zonal averages over 10° latitudinal bands to obtain $\overline{T_0 J_Q^0} \approx$ 130 W K m^{-2}. Assuming that χ_T is concentrated in a thermocline of thickness $H = 600$ m, the balance was approximated as

$$2\,\overline{T_0 J_Q^0} \;/\; \rho c_p = \chi_T \; H, \tag{1.23}$$

resulting in $\chi_T \approx 1 \times 10^{-7}$ K^2 s^{-1}, which Joyce showed is similar to values observed in the thermocline.

1.4.4 Surface Density Flux

Schanze and Schmitt (2013) represented the surface density flux as

$$J_\rho^0 = -\frac{\alpha}{c_p} J_{Qnet}^0 + \rho \beta S (E - P - R) \quad \left[\text{kg m}^{-2}\,\text{s}^{-1}\right], \tag{1.24}$$

where β is the coefficient of haline contraction (Section 2.6), c_p is the specific heat at constant pressure, and E, P, and R are rates of evaporation, precipitation, and river flow into the ocean.

Taken separately, surface integrals of the fluxes on the right are approximately zero, but together they produce a significant negative density flux across the global ocean (Figure 1.15). That is, $\iint J_{Qnet}^0 \, dA \approx 0$ and $\iint (E - P - R)\, dA_0 \approx 0$, but $\iint J_\rho^0 \, dA_0 = -7.6 \times 10^7$ kg s^{-1}, indicating a net flux of low-density water into the ocean. This results from the thermal expansion coefficient, α, being about six times larger in the tropics, where the ocean gains heat, than at high latitudes, where it loses heat. This coherence is evident from comparing panels (b) and (c). The net density flux is the integral under the curve in panel (d).

Defining z positive downward, Schanze and Schmitt represent the net flux of low-density water into the ocean as a negative density flux. In the terms of this book, however, z is positive upward at the surface, making the net flux, $\iint J_\rho^0 \, dA_0$, positive. The corresponding net buoyancy flux, $J_B^0 \equiv -(g/\rho) J_\rho^0$, is negative, representing a stabilizing flux. Recast in these terms, Schanze and Schmitt (2013) argued

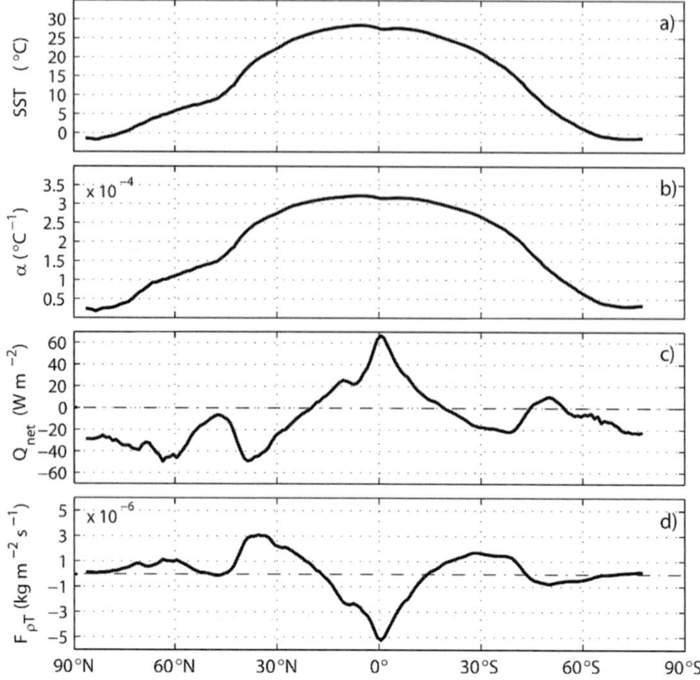

Figure 1.15 Zonal averages at the sea surface of: (a) temperature, (b) thermal expansion coefficient, (c) net heat flux, and (d) thermal component of the density flux. The sign convention for panels (c) and (d) is opposite to that used in this book. Therefore, the peak heat flux should be negative at the equator, corresponding to heat entering the ocean. Accordingly, the peak density flux should be positive at the equator, reflecting production of low-density water at the sea surface. (From Schanze and Schmitt, 2013)

that to produce a steady-state ocean, the stabilizing buoyancy flux at the surface must be balanced by a destabilizing buoyancy flux from mixing in the volume of the ocean. Following Joyce (1980) in assuming a balance between surface production of thermal variance and internal dissipation gives $\overline{\chi_T} = 8 \times 10^{-8}\ \mathrm{K^2\ s^{-1}}$, nearly the same as Joyce if the thermal variance is also dissipated over the upper 600 m.

2

Thermodynamics and Seawater Properties

2.1 Overview

Theoretical formulations of ocean mixing rest on thermodynamics, the branch of science treating heat and its relation to other forms of energy. Developing accurate expressions for seawater is particularly challenging, owing to its complexity. Here, we deal only with seawater's physical properties (Section 2.2), and those are simplified by treating it as a binary mixture of pure water and a composite of dissolved salts. Because the ocean is relatively well mixed, composite sea salt was long treated as having the same percentage of constituents everywhere in the open ocean, an approximation known as the Law of Constant Proportions. This assumption provided a means of proceeding with practical work, but it is not sufficiently accurate for some current needs, such as modeling climate change. In response, the equation of state and derivative expressions are now parameterized to include regional differences.

The First Law of Thermodynamics (Section 2.3.1), expressing the energy content of seawater, can be formulated in terms of specific enthalpy and conservative temperature, a new variable designed to replace potential temperature as a measure of heat content. The Second Law of Thermodynamics (Section 2.3.2) describes how isolated systems evolve to conserve entropy. The three components of the rate of entropy production, proportional to ϵ, χ_T, and χ_S, are important turbulence parameters quantifying the rates at which molecular viscosity and diffusivity dissipate turbulent fluctuations.

In general, molecular fluxes are functions of all thermodynamic forces with the same tensorial character (Section 2.4). For example, the molecular salt flux depends on gradients of temperature and pressure in addition to the salinity gradient. The pressure component appears dominant in most situations with laminar flow, and the thermal component, known as Soret diffusion, may be important in some places. The concentration component of temperature, however, is not significant in the ocean. Molecular diffusivities are functions of temperature, salinity, and pressure, independent of whether the flow is laminar or turbulent. The magnitudes of these coefficients, absolutely and relative to each other, strongly affect the rate and nature of ocean mixing. Flux expressions and entropy conservation lead to conditions for well mixed and equilibrium profiles (Section 2.5), revealing that the ocean is much closer to well mixed than to equilibrium.

An important part of ocean dynamics is driven by density contrasts, requiring careful examination of the equation of state, i.e. density as a function of salinity, temperature, and pressure (Section 2.6). Density varies by 5% in the ocean, mostly due to compressibility. Correlations of density fluctuations and vertical velocity produce vertical buoyancy fluxes that affect dynamics by the resultant changes in potential energy (Section 2.7). These effects are often described as diapycnal velocities proportional to vertical divergences of the buoyancy fluxes.

A measure of relative stratification is also evident from temperature plotted versus salinity. Known as θS, or TS, diagrams, plots of temperature versus salinity characterize water masses and show earlier turbulent mixing as sections with straight lines (Section 2.8). Sustained interest in quantifying lateral variability along isopycnals on θS plots, particularly where different water types are interleaved, led to several functions termed 'spice' or 'spiciness', which are approximations owing to the nonlinearity of the equation of state. Second partial derivatives of density also affect dynamics and are most familiar from the curvature of isopycnals on θS diagrams. Owing to these second-order effects, neutral surfaces, defined as the paths water follows without work being done, are not easily described (Section 2.9). They are, however, the basis for the water mass transformation equation in the latest international algorithms (Section 2.10). In the equation, mixing is specified by dianeutral and isoneutral diffusivities multiplied by different combinations of first- and second-order density derivatives.

2.2 Parameterizing Seawater

Estimation of physical processes in the ocean begins with the equation of state, i.e. density, as a function of temperature, salinity, and pressure. At present, density must be computed from in-situ measurements of temperature, electrical conductivity, and pressure. Vibrating tube densitometers measure fluid density accurately in laboratories, but useful measurements in the ocean have not been reported. Rather, density in the ocean is estimated in two stages. First, simultaneous measurements of electrical conductivity, temperature, and pressure are used to calculate salinity. Second, density is calculated from salinity, in-situ temperature, and pressure.

Reliance on in-situ measurements of electrical conductivity was enabled by the development of solid-state electronics in the 1950s. By the mid-1960s, conductivity-temperature-depth recorders (CTDs) were collecting profiles at sea. Most data were archived on magnetic tapes, although the primary product was often ink plots of salinity estimated by analog circuits during the cast. By the 1970s, parallel improvements of CTDs and computers allowed serious post-processing of raw data. To standardize procedures, UNESCO (1980) released algorithms known as the International Equation of State of Seawater–1980 (EOS–80). Much of the data discussed here was obtained under this regime (Section 2.2.1).

After several decades, better salinity algorithms were needed to improve estimates of thermal winds in climate models and to keep up with increasingly accurate measurements of temperature and electrical conductivity. These and other improvements are incorporated in the International Thermodynamic Equation of State–2010 (TEOS–10) adopted by the Intergovernmental Oceanographic Commission (Section 2.2.2).

2.2.1 The 1980 International Equation of State of Seawater

EOS–80 assumes that sea salt has the same relative proportions of constituents throughout the ocean and that the concentration of ionic species represents all solutes. For example, the Law of Constant Proportions is based on the assumption that the ratio of silica to sodium chloride is the same in the North Pacific as in the Indian Ocean. These assumptions were known not to be true in detail, but they offered a practical way of proceeding using available technology. Hence, recommended usage refers to outputs of the algorithms as practical salinity, S_P, in practical salinity units (psu), numerically equivalent to parts per thousand. In the open ocean, S_P varies over a narrow range: 32–39 psu.

The practical salinity algorithm, $S_P = S_P(c, T, p)$, is applied throughout the world ocean, where c is electrical conductivity. As instruments have become more sensitive and accurate, the assumption that species occur in the same proportions everywhere is no longer adequate. For instance, McDougall et al. (2013) note that density differences produced by variations in the composition of seawater are up to 10 times larger than the 0.003 g/kg (1:100,000) accuracy of S_P measurements. Significant errors in large-scale analyses and models result from global differences in composition and from non-ionic constituents affecting density but not electrical conductivity. The errors are particularly significant when using potential temperature, $\theta(S_P, T_{in\text{-}situ}, p, p_{ref})$, to represent heat content, where p_{ref} is an arbitrary reference pressure, often chosen at the surface.

Potential Temperature under EOS–80

Potential temperature corrects for temperature changes when water parcels change pressure without exchanging heat or salt with their surroundings,

$$\theta(S_P, T, p, p_{ref}) = T - \int_p^{p_{ref}} \Gamma_{\text{adiabatic}}\, dp \quad [^{\circ}\text{C}], \tag{2.1}$$

where

$$\Gamma_{\text{adiabatic}} = \Gamma_{\text{ad}}(S_P, T, p) = \left.\frac{\partial T}{\partial p}\right|_{S_A, p} = \frac{\alpha_T T}{\rho c_p} \quad [\text{K Pa}^{-1}] \tag{2.2}$$

is the adiabatic lapse rate, pressure is in Pa (Pascals), and K is degrees Kelvin. c_p is the specific heat at constant pressure, and $\alpha_T \equiv -(1/\rho)\partial\rho/\partial T$ is the thermal expansion coefficient as a function of in-situ temperature, T. As an example of the usefulness of potential temperature, most internal wave displacements of isopycnals involve little or no mixing. Consequently, temperature changes observed for hours to a few days on density surfaces usually do not represent variations in the heat content of the water but result from adiabatic depth cycling (Alford, 2010, figure 8). Even though θ removes temperature changes due to adiabatic and isohaline displacements, it is not conservative (McDougall, 2003). Conservative temperature, the replacement for potential temperature, is discussed in Section 2.3.1.

2.2.2 The 2010 International Equation of State of Seawater

The 2010 standard includes an algorithm for mass-based or 'absolute salinity', S_A, based on density detected by a vibrating tube densitometer. S_A is expressed as kilograms of sea salt (kg-ss) per kilogram of seawater (kg-sw) in equations, but software implementing the 2010 equation of state and data plots continue to express salinity in grams of sea salt per kilogram of seawater (IOC et al., 2010, section 2.5). Because in-situ technology has not changed, the procedure uses practical salinity in an intermediate step, and it should be retained in databases. The S_A algorithm applies a 48-term polynomial to correct S_P for geographical differences in composition and for nonionic constituents, $S_A = S_A(S_P, T, p, latitude, longitude)$. Principally involving silicate, geographical corrections differ between oceans, generally increasing with the average 'age' of the water from less than 0.01 g/kg in the North Atlantic to a maximum of 0.027 g/kg between 10 and 40 MPa in the northwest Pacific (figure 6.1 of McDougall et al., 2013). Within an ocean, the correction varies linearly with latitude north of 30°S.

Analogous to potential energy in a gravitational field, thermodynamic and chemical potentials indicate how a system can evolve without outside forcing. As noted by Fofonoff (1962), all thermodynamic properties of seawater can be computed from any of its thermodynamic potentials. Consistent, accurate determination of these properties became a reality when Feistel (2008) published the Gibbs function for seawater. Having the Gibbs function allows consistent calculation of internal energy, entropy, enthalpy, and potential enthalpy, as well as chemical potentials, the

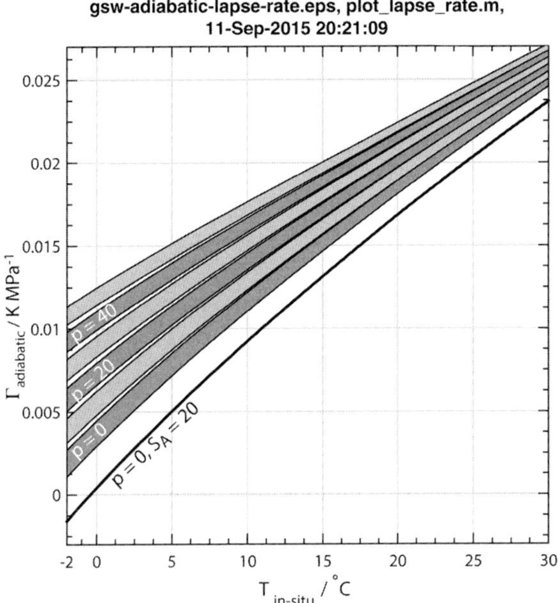

Figure 2.1 Adiabatic lapse rate for seawater using the Gibbs Seawater Toolbox (Pawlowicz et al., 2012). Absolute salinity, S_A, is 32 at the bottom and 40 at the top of each shaded range for pressures at 10 MPa intervals. The line for $p = 0$ and $S_A = 20$ includes the sign change when brackish water is cold, e.g. the surface of the Black Sea in late winter.

freezing temperature, and the latent heats of melting and evaporation. The Gibbs function and many useful seawater algorithms in the 'Gibbs Seawater Toolbox' are described by McDougall et al. (2013) and appendix A of IOC et al. (2010), which also discusses thermodynamics underlying the routines.

The Gibbs routines are the basis for much of this section, e.g. the adiabatic lapse rate in Figure 2.1.[1] For the oceanic range, the lapse rate is 0.001–0.027 K MPa^{-1}, or 0.01–0.3 K km^{-1}. The lapse rate is zero, $\Gamma_{\text{adiabatic}} = 0$, at the temperature of maximum density and is negative for lower temperatures, i.e. temperature decreases with increasing pressure. Proportional to the thermal expansion coefficient, the adiabatic lapse rate is the change in temperature required to keep specific entropy (and θ) constant during adiabatic and isohaline pressure changes (McDougall and Feistel, 2003).

[1] The international system of units (SI) expresses pressure in Pascals (Pa), but IOC et al. (2010) continue oceanographic practice with pressure in decibars (1 db = 10^4 Pa). Also, other fields would use concentration units for salinity ($s = \text{kg}_{\text{sea salt}}/\text{kg}_{\text{seawater}}$), but the Gibbs routines continue oceanographic practice with db and $S = \text{g}_{\text{sea salt}}/\text{kg}_{\text{seawater}}$). I prefer the international units and, using an intermediate oceanographic recommendation to follow international practice, keep my data in them. Here, I try to show which set is being used, but suggest checking units carefully when calculating parameters. The Gibbs routines are consistent and well documented.

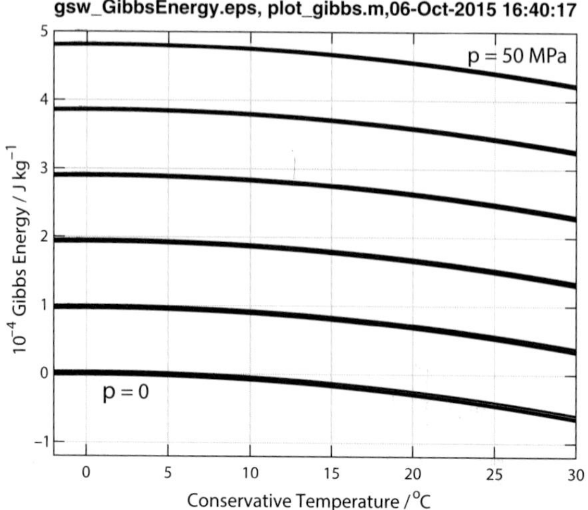

Figure 2.2 Gibbs free energy versus conservative temperature for pressure at 10 MPa intervals calculated using the Gibbs Seawater Toolbox (Pawlowicz et al., 2012). Each pressure is plotted for absolute salinity of 32, 35, and 40 g/kg, but the differences are nearly indistinguishable.

Thermodynamic Potentials

The Gibbs function is the maximum amount of nonexpansion work that can be extracted from a thermodynamically closed system, i.e. a system that can exchange heat and work with its surroundings, but not mass. The specific Gibbs function, $\tilde{g} = \tilde{g}(S_A, T, p)$, varies mostly with pressure, modestly with temperature, and very little with salinity (Figure 2.2). Specified for the full range of properties in the ocean and adjacent seas, $\tilde{g}$ is the sum of two parts, one for pure water and the other for solutes. Both parts have two arbitrary, and unknowable, constants. The freshwater constants make entropy and internal energy zero at the triple point, where pure water exists as a liquid, solid, and gas. The two saline constants set specific enthalpy, h, and specific entropy, η, to zero at standard ocean properties: 35.16504 g_{ss} kg_{sw}^{-1}, 0°C, and 0 Pa.[2]

Of particular importance to mixing studies, specific enthalpy, h, and specific entropy, η, are related to the Gibbs function by

$$\tilde{g} = \tilde{g}(S_A, T_{abs}, cp) = h - T_{abs}\eta = h - (T_0 + t)\eta \quad [\mathrm{J\,kg^{-1}}], \qquad (2.3)$$

where T_{abs} is absolute temperature.[3]

[2] Practical salinity of the reference composition of seawater is 35.165 04 psu because earlier salinities were estimated by evaporating and then weighing; some volatiles evaporated, causing the actual mass fraction to be larger than was realized (Millero et al., 1980).

[3] IOC et al. (2010) use $T_0 = 273.15$ K as the absolute temperature corresponding to 0°C and t for centigrade temperature.

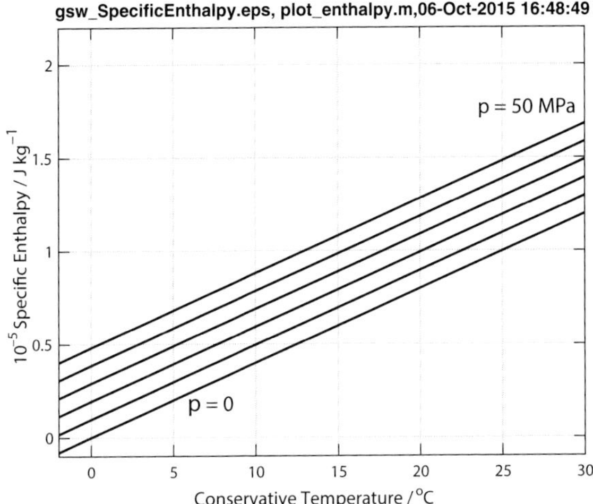

Figure 2.3 Specific enthalpy, h, versus conservative temperature, Θ, for 10 MPa pressure intervals. Curves are plotted for the oceanic range of salinity at each pressure, but they cannot be distinguished. The curves were calculated using the Gibbs Seawater Toolbox (Pawlowicz et al., 2012).

Enthalpy is a measure of the amount of heat released at constant pressure. Specific enthalpy, h, is now the basis for estimating the heat content of seawater (Section 2.3), and is obtained from $\tilde{g}$ as

$$h = \tilde{g} - T \frac{\partial \tilde{g}}{\partial T}\bigg|_{S_A,p} \quad [\mathrm{J\ kg^{-1}}]. \tag{2.4}$$

Increasing nearly linearly with temperature and moderately with pressure, h is only weakly affected by salinity (Figure 2.3). In Section 2.3, the First Law of Thermodynamics is developed in terms of specific enthalpy.

Specific entropy, η, obtained from the Gibbs function by

$$\eta = -\frac{\partial \tilde{g}}{\partial T}\bigg|_{S_A,p} \quad [\mathrm{J\ kg^{-1}\ K^{-1}}], \tag{2.5}$$

varies strongly with temperature and only weakly with pressure and salinity (Figure 2.4). A measure of the unavailability of thermal energy for mechanical work, entropy increases with the degree of disorder or randomness. In Section 2.3 the entropy equation relates the three fundamental turbulent parameters – ϵ, χ_T, and χ_S – to the thermodynamic state of a system.

The partial specific enthalpy for sea salt, $h_{S_A} \equiv (1/h)\partial h/\partial S_A\ |_{T,p}$, is a strong function of pressure, and a weak one of temperature and salinity (Figure 2.5). Also shown in the figure, h_Θ, the partial specific enthalpy for conservative temperature, varies less than 1% over the oceanic range.

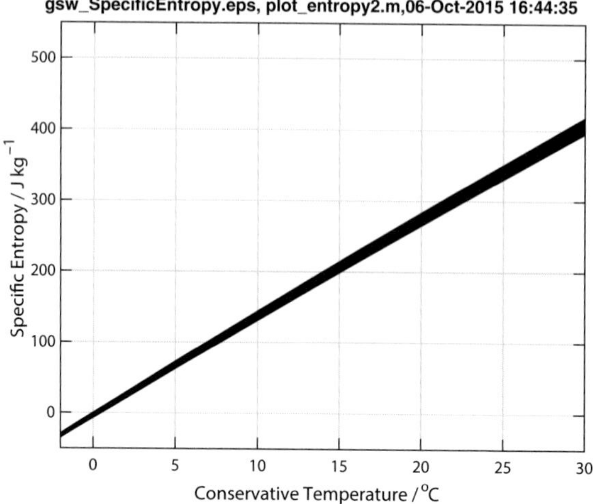

Figure 2.4 Specific entropy, η, versus Θ for absolute salinity, S_A, between 32 and 40 g/kg and pressure of 0 to 50 MPa. The curves were calculated using the Gibbs Seawater Toolbox (Pawlowicz et al., 2012).

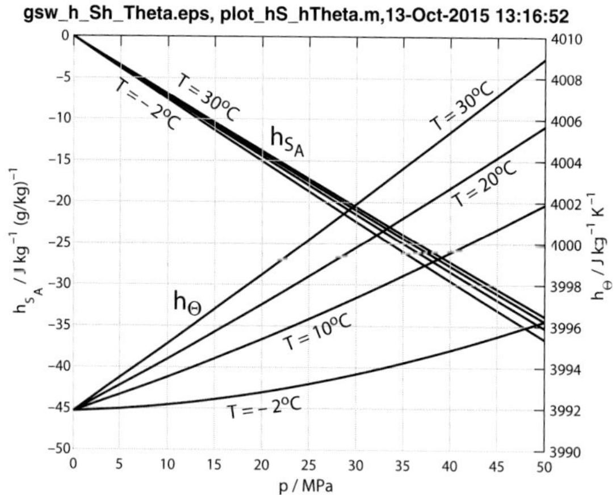

Figure 2.5 Partial specific enthalpy for sea salt, h_{S_A}, for $S_A = 35$ and $\Theta = -2$, 10, 20, and 40°C, and partial specific enthalpy for conservative temperature, h_Θ, over the same ranges. These curves were calculated using the Gibbs Seawater Toolbox (Pawlowicz et al., 2012).

Internal Energy

As the energy residing in thermal motion, internal energy is the total energy minus potential and kinetic energies. Specific internal energy, e, in terms of thermodynamic potentials is

$$
\begin{aligned}
e(S_A,t,p) &= \tilde{g} + (T_0 + t)\eta - (p + p_0)\nu \\
&= \tilde{g} - (T_0 + t)\left.\frac{\partial \tilde{g}}{\partial T}\right|_{S_A,p} - (p + p_0)\left.\frac{\partial \tilde{g}}{\partial p}\right|_{S_A,p} \quad [\mathrm{J\ kg^{-1}}].
\end{aligned} \tag{2.6}
$$

e is a nearly linear function of temperature.

2.3 First and Second Laws of Thermodynamics

For seawater, the First Law of Thermodynamics expresses the balance of energy using variables that can be measured, or calculated from measurements, with standard algorithms, ultimately leading to a heat equation applicable to the ocean (Section 2.3.1). The heat equation is presented first in terms of h (2.7) and then is reformulated in terms of specific enthalpy and conservative temperature (2.12).

Describing how isolated systems evolve, the Second Law of Thermodynamics (Section 2.3.2) is of interest primarily because the rate of entropy production, σ, contains three parameters – ϵ, χ_T, and χ_S – representing the rates at which variances are dissipated by viscosity, heat diffusion, and salt diffusion. Nonequilibrium thermodynamics uses σ to predict how systems evolve, and that approach may eventually yield unique results applicable to the ocean, but to date the potential has not been realized. The interest here is demonstrating that the three turbulence parameters are manifestations of the same thermodynamic quantity.

2.3.1 The First Law

The First Law is sometimes stated as the evolution equation for total energy, internal plus kinetic and potential. An alternative view considers the First Law as the evolution equation of 'heat' (Landau and Lifshitz, 1959; McDougall, 2003). That is the approach of TEOS–10, where it is developed as the difference between conservation of total energy and the evolution of kinetic and potential energies. That is, 'heat' is the residual.

Because enthalpy is conserved when fluid parcels mix at constant pressure, using material derivatives, IOC et al. (2010) express the First Law in terms of specific enthalpy, h,

$$
\rho\left(\frac{dh}{dt} - \frac{1}{\rho}\frac{dp}{dt}\right) = -\nabla\cdot\boldsymbol{J}_Q - \nabla\cdot\boldsymbol{J}_R + \rho\epsilon + \frac{\partial h}{\partial S_A}\rho S_{\mathrm{source}}^{\mathrm{bgc}} \quad [\mathrm{W\ m^{-3}}], \tag{2.7}
$$

where $\boldsymbol{J}_Q$ is the total vector heat flux, $\boldsymbol{J}_R$ is the vector sum of boundary and radiative fluxes, ϵ is the rate of viscous dissipation of kinetic energy per unit mass, and $S_{\mathrm{source}}^{\mathrm{bgc}}$ is the source rate of absolute salinity by remineralization. Dropping the remineralization term as negligible,

$$\frac{\partial(\rho h)}{\partial t} + \nabla \cdot (\rho \boldsymbol{u} h) - \frac{dp}{dt} = -\nabla \cdot \boldsymbol{J}_Q - \nabla \cdot \boldsymbol{J}_R + \rho\epsilon \quad [\mathrm{W\,m^{-3}}], \tag{2.8}$$

where $\boldsymbol{u}$ is the velocity vector.

Because enthalpy changes during adiabatic and isohaline processes, McDougall (2003) argues that, with negligible error, a new parameter, potential enthalpy, h°, represents the heat content of seawater, defining it as

$$h^\circ(S_A, \theta, p) = h(S_A, T, p) - \int_{p_{ref}}^{p} \alpha(S_A, \theta, p')\, dp' \quad \left[\mathrm{J\,kg^{-1}}\right], \tag{2.9}$$

with α as the specific volume of seawater. The heat equation is then

$$\frac{\partial \rho h^\circ}{\partial t} + \nabla \cdot (\rho \mathbf{u} h^\circ) = -\nabla \cdot \boldsymbol{J}_Q + \rho\epsilon \quad \left[\mathrm{W\,m^{-3}}\right]. \tag{2.10}$$

This can be cast in the form of a traditional heat equation by defining conservative temperature, Θ, to be proportional to potential enthalpy,

$$\Theta \equiv h^\circ / c_p^\circ \quad [\mathrm{K}], \tag{2.11}$$

with $c_p^\circ = 3991.868\ \mathrm{J\,kg^{-1}\,K^{-1}}$.

Conservative Temperature

Using potential temperature to represent heat content typically produces errors up to $\pm 0.1^\circ$C, but in the Amazon plume errors reach 1.4°C. Moreover, there are also systematic errors in meridional heat flux of 0.4%. Conservative temperature, Θ, reduces these errors by two orders of magnitude, and in June 2010 it was adopted as the standard for future work as part of TEOS–10 by the Intergovernmental Oceanographic Commission.

In terms of conservative temperature, the first law becomes

$$\rho\left(h_\Theta \frac{d\Theta}{dt} + h_{S_A}\frac{dS_A}{dt}\right) = -\nabla \cdot \boldsymbol{J}_Q \quad - \nabla \cdot \boldsymbol{J}_R + \rho\epsilon \quad \left[\mathrm{Wm^{-3}}\right], \tag{2.12}$$

with $h_\Theta \equiv \partial h / \partial \Theta \mid_{S_A, p}$ as the partial specific enthalpy with respect to Θ (Figure 2.5). The heat flux, $\boldsymbol{J}_Q$, is considered in Section 2.4.

As discussed in Section 2.5, conservative temperature is uniform in well-mixed layers. One of the thickest homogenous layers is mixed by geothermal heating at the bottom of the Canada Basin in the Arctic (Figure 2.6). In-situ temperature increases downward in the bottom 600 m with a slope within 10% of the adiabatic gradient.[4] Conservative temperature, Θ, is remarkably uniform, varying about

[4] Not shown in Figure 2.6, salinity from the CTD shows a slight increase with depth attributed to drift of the conductivity probe within its calibration stability.

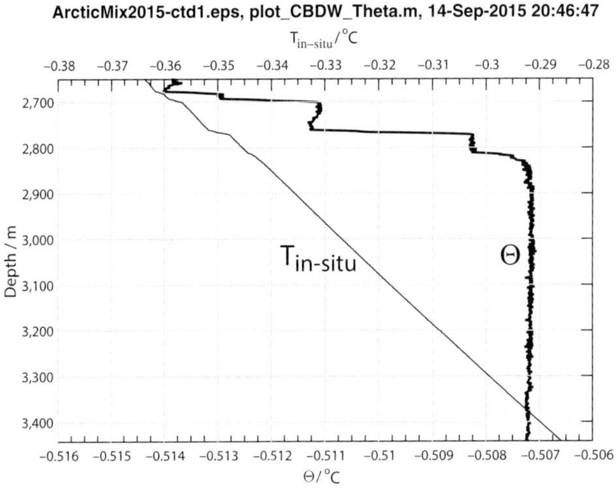

ArcticMix2015-ctd1.eps, plot_CBDW_Theta.m, 14-Sep-2015 20:46:47

Figure 2.6 Temperature in the Canada Basin (72.585 N, 144.999 W) on September 1, 2015, kindly supplied by Jen MacKinnon. Below 2,800 m, $T_{\text{in-situ}}$ increases downward at 0.92 mK m^{-1}, close to the 0.84 mK m^{-1} adiabatic gradient. Conservative temperature, Θ, is nearly constant with depth, consistent with the layer being well mixed by the geothermal heat flux. Note that the the top axis spans 0.1°C and the bottom only 0.01°C. Timmermans et al. (2003) and Carmack et al. (2012) attribute the steps capping the homogenous layer to the diffusive regime of double diffusion.

0.1 mK, reflecting sensor noise rather than temperature fluctuations. Carmack et al. (2012) use $\boldsymbol{J}_Q = 50$ mW m^{-2} as the average geothermal heat flux, corresponding to a buoyancy flux of $J_b = (g\alpha^{\Theta}/\rho c_p)\boldsymbol{J}_Q = 1.3 \times 10^{-8}$ W kg^{-1}. Convective similarity scaling for dissipation is ϵ/J_B, so J_b is a reasonable estimate of ϵ in the layer.

Heating from turbulent dissipation is negligible under most circumstances but may not be in abyssal boundary layers formed by sustained strong turbulence, e.g. $\epsilon \geq 10^{-6}$ W kg^{-1}, 100 times the estimate for the Canada Basin, was measured in the Romanche Fracture Zone (Figure 1.8) and warms at $d\Theta/dt \approx 1\mu$K hr^{-1} $\approx$ 9 mK yr^{-1}.

The corresponding equation for absolute salinity is

$$\rho \frac{dS_A}{dt} = \frac{\partial \rho S_A}{\partial t} + \nabla \cdot (\rho \mathbf{u} S_A) = -\nabla \cdot \boldsymbol{J}_S \quad [\text{g}_{\text{SS}}\, \text{m}^{-2}\, \text{s}^{-1}]. \tag{2.13}$$

An additional source term for remineralizing particulates has been omitted, as it is unlikely to be significant over short time scales.

2.3.2 The Second Law

As an entropy equation for fluids, the Second Law of Thermodynamics is

$$\rho \frac{d\eta}{dt} = -\nabla \cdot \boldsymbol{J}_\eta + \sigma_\eta \quad [\mathrm{W\,m^{-3}\,K^{-1}}], \tag{2.14}$$

where η is specific entropy (i.e. entropy per unit mass), $\boldsymbol{J}_\eta$ its flux, and σ_η the rate of entropy is produced by molecular viscosity and diffusion. The Second Law requires σ_η to be positive for irreversible processes and zero for reversible ones,

$$\sigma_\eta \geq 0. \tag{2.15}$$

Unforced systems evolve to increase their entropy as molecular processes smooth gradients.

Expanding the entropy equation with heat, salt, and momentum fluxes in terms of standard thermodynamic variables (Gregg, 1984) leads to

$$\sigma_\eta = \frac{\rho c_\mathrm{p} \kappa_T (\nabla T)^2}{T^2} + \frac{\rho \kappa_S}{(1-s)T} \left.\frac{\partial \mu_S}{\partial s}\right|_{p,T} (\nabla s - \Gamma_S \nabla p)^2 + \frac{\rho \epsilon}{T}, \tag{2.15a}$$

where κ_T and κ_S are the molecular diffusivities of heat and salt, μ_S is the chemical potential of sea salt, and $\Gamma_S = -(\alpha_s - \alpha)/(\partial \mu_S/\partial s)_{T,p}$ is the equilibrium salinity gradient. Here, salinity, s, is in concentration units, kg_{ss}/kg_{sw}.

Major contributions to σ_η occur when mixing increases $(\nabla T)^2$, $(\nabla s)^2$, and ϵ above their nonturbulent levels, often by decades. Since ∇p is constant, its contribution is negligible during mixing, allowing σ_η to be expressed more conveniently in terms of the rate at which molecular heat diffusion smoothes temperature fluctuations,

$$\chi_T \equiv 2\kappa_T \overline{(\nabla T)^2} \quad [\mathrm{K^2\,s^{-1}}], \tag{2.16}$$

the rate at which molecular salt diffusion smoothes salinity gradients,

$$\chi_S \equiv 2\kappa_S \overline{(\nabla s)^2} \quad [\mathrm{s^{-1}}], \tag{2.17}$$

and the rate at which molecular viscosity smoothes velocity gradients,

$$\epsilon = \frac{\nu}{2} \sum_{i,j=1}^{3} \left(\frac{\partial u_i}{\partial x_j} + \frac{\partial u_j}{\partial x_i} \right)^2 \quad [\mathrm{W\,m^{-2}}], \tag{2.18}$$

where ν the the kinematic viscosity of seawater. Here and elsewhere when dealing with matrices, indices i, j, and k represent x, y, z directions.

In terms of molecular dissipation rates,

$$\sigma_\eta = \frac{\rho c_\mathrm{p}}{2T^2} \chi_T + \frac{\rho\,(\partial \mu_S/\partial s)_{p,T}}{2(1-s)T} \chi_S + \frac{\rho}{T} \epsilon \quad \left[\frac{\mathrm{W}}{\mathrm{m^3 K}}\right]. \tag{2.19}$$

Relative contributions vary with circumstances (Gregg, 1984). In turbulent overturns, χ_T dominates in the thermocline and χ_S in strong haloclines, whereas in weakly stratified profiles ϵ is largest.

The rate of entropy generation is used in some studies of nonequilibrium systems to forecast evolution, in much the same way that thermodynamic and chemical potentials indicate how systems evolve when near equilibrium. This may offer a useful way to examine evolution where the dominant contribution to σ_η changes with evolution. For example, strong shear instabilities can rapidly homogenize water, leading χ_T, and presumably χ_S, to decay while ϵ remains high (Seim and Gregg, 1994). In such cases, σ_η offers a unifying framework not available from temperature or velocity only.

2.4 Molecular Fluxes and Diffusivities

As a general law of nature, thermodynamic fluxes depend on all thermodynamic forces of the same tensorial character (de Groot and Mazur, 1969). Therefore, heat and salt fluxes both depend on the same gradients, often termed Onsager forces. Using the notation of IOC et al. (2010),

$$\boldsymbol{J}_S = A\nabla(-\mu/T) + B\nabla(1/T), \tag{2.20}$$

$$\boldsymbol{J}_Q = B\nabla(-\mu/T) + C\nabla(1/T), \tag{2.21}$$

with A, B, and C as independent Onsager coefficients. The relative chemical potential, μ, is the difference between the partial chemical potentials of salt and water,

$$\mu \equiv \mu_S - \mu_W = (\partial g/\partial S_A)_{T,p} \quad [\mathrm{J\,g^{-1}}]. \tag{2.22}$$

To develop practical flux expressions, $\nabla(-\mu/T)$ is expanded as

$$\boldsymbol{J}_S = -(A(\mu/T)_T + B/T^2)\nabla T - A(\mu_{S_A}/T)\nabla S_A - A(\mu_p/T)\nabla p, \tag{2.23}$$

$$\boldsymbol{J}_Q = -(B(\mu/T)_T + C/T^2)\nabla T - B(\mu_{S_A}/T)\nabla S_A - B(\mu_p/T)\nabla p, \tag{2.24}$$

where subscripts denote differentiation, e.g. $\mu_{S_A} \equiv (\partial\mu/\partial S_A)_{T,p}$, and both fluxes depend on all three gradients. Following IOC et al. (2010), the terms are usually grouped as

$$\boldsymbol{J}_S = -\rho\kappa_S\left(\nabla S_A + \frac{\mu_p}{\mu_{S_A}}\nabla p\right) - \left(\frac{\rho\kappa_S T}{\mu_{S_A}}\left(\frac{\mu}{T}\right)_T + \frac{B}{T^2}\right)\nabla T, \tag{2.25}$$

$$\boldsymbol{J}_Q = -\frac{1}{T^2}\left(C - \frac{B^2}{A}\right)\nabla T + \frac{B\mu_{S_A}}{\rho\kappa_S T}\boldsymbol{J}_S \quad [\mathrm{W\,m^{-2}}], \tag{2.26}$$

where $\boldsymbol{J}_S$ has units of $\mathrm{g_{ss}\ m^{-2}\ s^{-1}}$. The gradient flux assumption, Fick's Law for salt diffusion, implies $A = \rho\kappa_S T/\mu_{S_A}$, which replaced A in the upper equation.

Equation (2.26) is the total heat flux. For many purposes, it is more convenient to subtract the heat of transfer, i.e. the enthalpy carried by diffusing salt, and use the reduced heat flux,

$$\boldsymbol{J}'_Q \equiv \boldsymbol{J}_Q - (\mu - T\mu_T)\boldsymbol{J}_S = -\rho c_p \kappa_T \nabla T + \frac{B'\mu_{S_A}}{\rho\kappa_S T}\boldsymbol{J}_S, \tag{2.27}$$

where $B' \equiv B + (\rho\kappa_S T^3/\mu_{S_A})(\mu/T)_T$ is a revised cross-diffusion coefficient. Known as the Dufour Effect, the second term in (2.27) is the heat flux driven by salt and pressure gradients. It is negligible, allowing the reduced heat flux to be expressed by the Fourier diffusion law,

$$\boldsymbol{J}'_Q = -\rho c_P \kappa_T \nabla T \quad [\mathrm{W\ m^{-2}}]. \tag{2.28}$$

Using B', the salt flux becomes

$$\boldsymbol{J}_S = -\rho\kappa_S\left(\nabla S_A + \frac{\mu_p}{\mu_{S_A}}\nabla p\right) - \frac{B'}{T^2}\nabla T. \tag{2.29}$$

2.4.1 Soret Diffusion

The second term, Soret diffusion, a.k.a. thermophoresis, represents the salt flux produced by temperature gradients. The mechanism producing these fluxes is now understood to result from the thermoelectric field induced by ions diffusing along the temperature gradient (Würger, 2010). Using the Soret coefficient, S_T, measured by Caldwell (1973, 1974a), $B' = \rho\kappa_S S_T S_A(1 - S_A/1000)T^2$, results in

$$\boldsymbol{J}_S = -\rho\kappa_S\left(\nabla S_A + \frac{\mu_p}{\mu_{S_A}}\nabla p\right) - \rho\kappa_S S_T S_A(1 - S_A/1000)\nabla T\left[\frac{\mathrm{g_{ss}}}{\mathrm{m^2\ s^1}}\right]. \tag{2.30}$$

Varying with temperature and pressure (Figure 2.7), the Soret coefficient, S_T, can have either sign in the ocean. In much of the thermocline, S_T is close to zero, restricting its importance to warm water at shallow depths; Figure 2.8 shows one example. It may, however, be a factor in thermohaline convection (Caldwell, 1974b); Gregg and Cox (1972) estimated Soret diffusion to be half of concentration-driven diffusion in the thermohaline intrusions they observed. No data have been reported about the concentration dependence of S_T or κ_S, but, in view of the small range of salinity in the ocean, variability is likely to be small. IOC et al. (2010) dismiss Soret diffusion as unimportant in a turbulent ocean. That, however, is an untested assumption and seems contrary to the nature of turbulent diffusion. In the thermocline, turbulence is intermittent, and the magnitude of temperature and salinity fluctuations in turbulent patches scales with the height of the largest overturn and average vertical gradients. Hence, acceleration of diffusion by turbulence stretching and sharpening gradients should remain proportional to

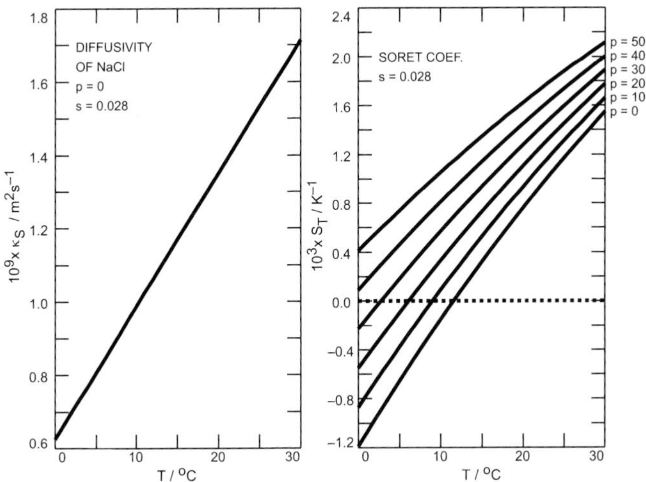

Figure 2.7 (Left) The molecular diffusivity of sodium chloride varies nearly threefold with temperature. (Right) Dependence of the Soret coefficient on temperature and pressure. Pressures, upper right, are in MPa. (From Caldwell, 1973, 1974a)

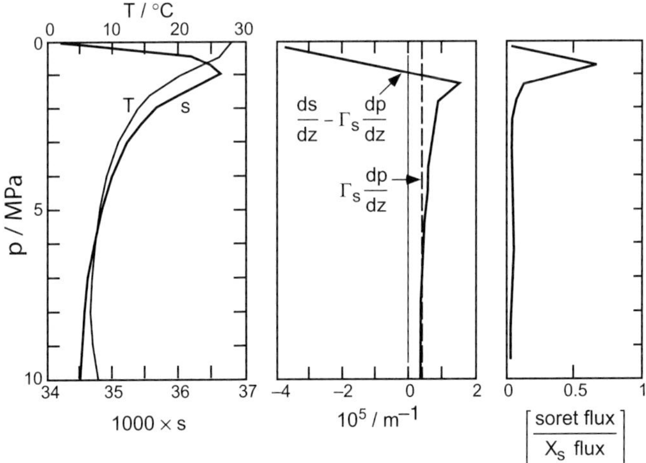

Figure 2.8 In the strong halocline of the tropical Atlantic (11° 27′N, 50° 34′ W), the Soret flux is significant along the mean gradients only for $p < 1$ MPa (right panel). In the middle panel, salt flux driven by the mean gradient is significant only for $p < 5$ MPa. (From Gregg, 1984. ©American Meteorological Society. Used with permission)

the mean gradients. Where Soret diffusion is important for molecular diffusion in laminar flow, it should also be important during turbulent events. It would have no net effect if the mixing homogenized the profile, but that rarely happens outside surface and bottom boundary layers.

2.4.2 Salt Diffusion

Requiring migration of relatively large ions, salt diffuses much more slowly than heat diffuses by molecular collisions. Using a sodium chloride solution with $s = 0.0285$, Caldwell (1973, 1974a) reported a small diffusivity, κ_S, varying strongly with temperature (Figure 2.7). Later measurements showed no dependence on pressure (Caldwell, 1973, 1974a).

In addition to ∇s, the salt flux (2.30) also is driven by the pressure gradient. The equilibrium gradient, $\Gamma_S = \mu_p/\mu_{S_A} = (2.7 - 4.2) \times 10^{-10}$ Pa^{-1}, corresponds to practical salinity gradients of 0.27–0.42 psu per 100 m. In most of the ocean, the pressure gradient term dominates diffusion driven by average salinity gradients. For example, in the tropical Atlantic the salinity gradient exceeds the equilibrium salinity gradient only in the shallow halocline (Figure 2.8); below 6 MPa, the pressure effect is dominant. During turbulence, large increases in small-scale salinity gradients overwhelm the pressure term, which is important only in the absence of turbulence. Since the pressure term has a very small divergence, salt conservation is

$$\frac{dS_A}{dt} = -\nabla \cdot \left[-\kappa_S \left(\nabla S_A - \Gamma_S \nabla p\right)\right] \doteq \kappa_S \nabla^2 S_A. \tag{2.31}$$

2.4.3 Thermal Diffusivity

Thermal diffusivity depends mostly on temperature, but varies less than ±10% about $\kappa_T = 1.45 \times 10^{-7}$ [m^2 s^{-1}] (Figure 2.9, left). Changes with salinity and pressure are even less. Pressure and salinity are larger factors for specific heat at constant pressure, c_p, but the entire span is only 5% (Figure 2.9, right).

2.4.4 Viscous Diffusion

The viscous contribution to entropy production is evaluated using the viscous stress tensor,

$$\Pi_{ij} = -\rho\nu \left(\frac{\partial u_i}{\partial x_j} + \frac{\partial u_j}{\partial x_i}\right) \quad \text{[Pa]}, \tag{2.32}$$

where ν is the kinematic viscosity, which varies more than twofold in the ocean (Figure 2.10). The product of the driving force and flux for kinetic energy,

$$\Pi : \nabla\nabla \boldsymbol{u} = -\rho\epsilon \quad [\text{W m}^{-2}], \tag{2.33}$$

is taken as the definition of the viscous dissipation rate, ϵ.

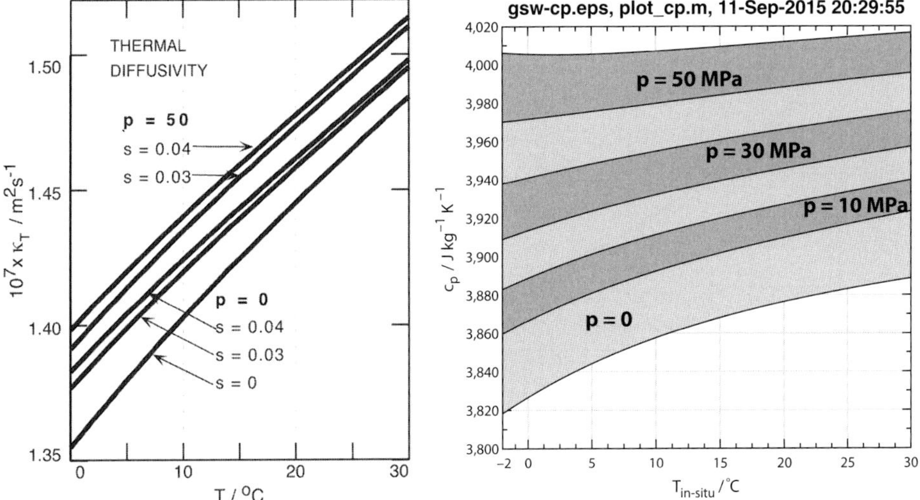

Figure 2.9 Thermal diffusivity (left) and specific heat of seawater at constant pressure (right), computed with the Gibbs Seawater Toolbox (Pawlowicz et al., 2012). Absolute salinity, S_A, varies from 32 to 40 within each shaded band at constant pressure.

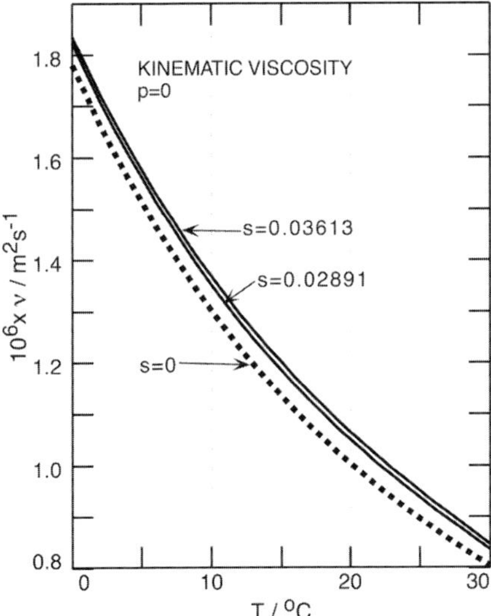

Figure 2.10 Kinematic viscosity, ν, of seawater, computed from the equation of state and the dynamic viscosity of Miyake and Koizumi (1948). It varies more than twofold with temperature, precluding using a constant value when calculating dissipation rates.

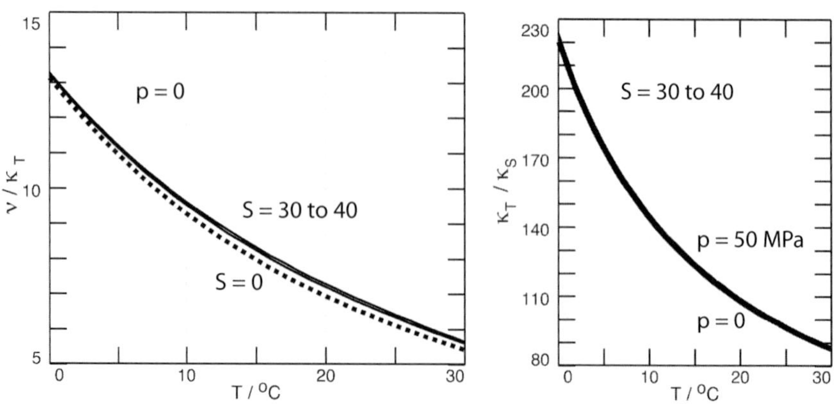

Figure 2.11 (Left) Prandtl number of seawater, $Pr \equiv \nu/\kappa_T$. (Right) The ratio of thermal and haline diffusivities in seawater, κ_T/κ_S.

2.4.5 Consequences of Diffusivity Contrasts

Contrary to views many years ago that the molecular properties of seawater are of little significance to ocean physics, differences between ν, κ_T, and κ_S strongly affect mixing in the ocean and are characterized by the dimensionless Prandtl and Schmidt numbers,

$$Pr \equiv \nu/\kappa_T, \quad Sc \equiv \nu/\kappa_S. \tag{2.34}$$

The Prandtl number, varying between 5.5 and 13, is largest at low temperatures (Figure 2.11, left). Reflecting the weak diffusivity of salt compared to heat (Figure 2.11, right), the Schmidt number ranges from 500 to 3,000 and is also largest in cold water. Not surprisingly, ratios of the smallest turbulent scales of velocity, heat, and salt can be expressed in terms of Prandtl and Schmidt numbers.

The contrast in molecular properties is also shown using dimensional analysis to compare molecular length and time scales for viscosity and diffusion,

$$l_{\mathrm{mol}} \equiv (D\tau)^{1/2} \quad [\mathrm{m}], \quad \tau_{\mathrm{mol}} \equiv l^2/D \quad [\mathrm{s}], \tag{2.35}$$

where D is the molecular viscosity or diffusivity. For example, for $l = 1$ cm, τ_{mol} is 100 seconds for viscous diffusion, 12 minutes for thermal diffusion, and 28 hours for salt diffusion (Table 2.1). These large contrasts affect mixing in several ways:

Dissipation length scales: In well-developed turbulence, energy cascades from the scales where it is created, say by shear instability, toward smaller scales until gradient variances are so large that ϵ, χ_T, and χ_S match fluxes coming from larger scales. Velocity fluctuations with dissipation rate ϵ are smoothed at length scales of $(\nu^3/\epsilon)^{1/4}$ (Section 3.6.1). Corresponding temperature fluctuations are smoothed at

Table 2.1 *Molecular diffusive length, $l_{\text{mol}} \equiv (D\tau)^{1/2}$, and time, $\tau_{\text{mol}} \equiv (l^2/D)^{1/2}$, scales for velocity, heat, and salt in seawater. D is molecular viscosity or diffusivity, as appropriate.*

	l_{mol}			τ_{mol}			
Time	Velocity	Heat	Salt	Length	Velocity	Heat	Salt
Second	1 mm	0.4 mm	32 μm	1 mm	1 sec	7.1 sec	17 min
Minute	9 mm	3 mm	0.2 mm	1 cm	100 sec	12 min	28 hr
Hour	6 cm	2 cm	2 mm	0.1 m	2.8 hr	19.8 hr	4.8 day
Day	0.3 m	0.1 m	9 mm	1 m	12 day	83 day	31.8 yr

scales smaller by $Pr^{-1/2} \sim 1/3$ times the velocity scale. Salinity fluctuations must be driven to scales of $(\kappa_S/\kappa_T)^{1/2} \sim 1/30$ of the temperature dissipation scale.

Double diffusion: Owing to the large contrast in diffusivities, profiles with opposing contributions to stratification from temperature and salinity are diffusively unstable and can generate vertical convection if the net density contrast is small (Chapter 4). Where warm, saline water overlies cooler, fresher water, as in the middle of ocean gyres, diffusive instability can generate alternating, centimeter-wide rising and sinking columns known as salt fingers. Where warm, saline water lies below cooler, fresher water, convection can form sequences of well-mixed layers, most prominently at high latitudes.

Differential diffusion: Mixing in the pycnocline occurs episodically, producing overturns that usually do not mix to completion; few homogenous or nearly homogenous layers are found. Partial mixing should leave finely granulated salinity fluctuations that gradually settle out. Contradicting the practice of representing heat and salt fluxes with the same diapycnal eddy diffusivity, this differential diffusion (Gargett, 2003) is discussed in more detail in Section 3.9.1.

2.5 Equilibrium and Well-Mixed States

If isolated from external forcing, thermodynamic flows and forces would gradually decay to zero, leading to equilibrium gradients

$$\nabla T_{\text{in-situ}}\big|_{\text{EQ}} = 0, \quad \nabla S\big|_{\text{EQ}} = \Gamma_S \nabla p, \quad \nabla \mathbf{u}\big|_{\text{EQ}} = 0. \tag{2.36}$$

If equilibrium were not isothermal, temperature gradients would drive heat conduction that could perform work, violating Kelvin's formulation of the Second Law. Accordingly, conservative temperature, Θ, increases downward at equilibrium. Salinity has a strong equilibrium gradient, $\mathrm{d}S_P/\mathrm{d}z\ |_{\text{EQ}} \approx -3.5$ ppt km^{-1}, increasing with depth due to gravitational settling. Pytkowicz (1963) estimates that relaxation of the salt field would require millions of years. While this was occurring, the ions composing sea salt would not maintain constant proportions, resulting in separate equilibrium profiles for each species.

At the other extreme, prolonged, vigorous mixing would homogenize the ocean so well that fluid parcels could be interchanged without altering macroscopic distributions. To be well mixed, the ocean must be uniform in entropy, salinity, and velocity,

$$\nabla \eta \big|_{\mathrm{WM}} = 0, \quad \nabla S \big|_{WM} = 0, \quad \nabla \boldsymbol{u} \mid_{WM} = 0, \tag{2.37}$$

leading to

$$\nabla T \mid_{WM} = -\rho \Gamma_Q \mathbf{g} \quad \text{and} \quad \nabla \Theta \mid_{WM} = 0. \tag{2.38}$$

A well-mixed ocean has constant absolute salinity, S_A, and conservative temperature, Θ; in-situ temperature increases downward at the adiabatic lapse rate, e.g. in the Canada Basin (Figure 2.6). Figure 2.12 contrasts the two reference states, emphasizing that well mixed states are produced and maintained by vigorous stirring. The average salinity profile is much closer to well mixed than to equilibrium.

To investigate homogenizing the ocean, Gille (2004) integrated $\partial\theta/\partial t = K\partial^2\theta/\partial z^2$, the vertical diffusion equation, with zero-gradient boundary conditions at the surface and at 3,000 m. In addition to being isolated from vertical heat and salt fluxes, the domain had horizontal advection shut off. Using a vertical eddy coefficient of $K = 10^{-4}$ m^2 s^{-1}, integration had to proceed for 200 years to

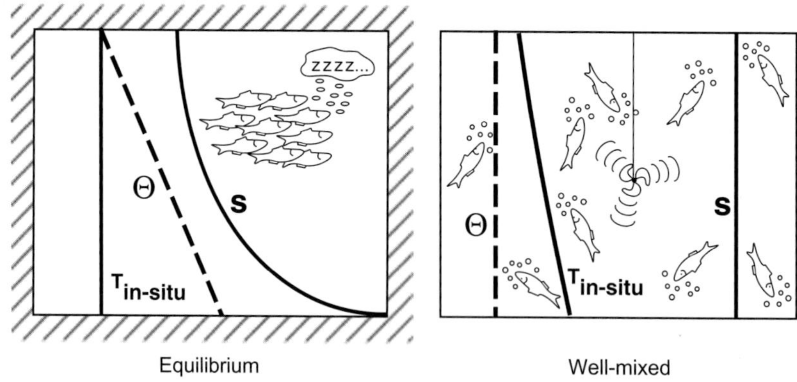

Figure 2.12 Schematic equilibrium and well-mixed states of the ocean. All thermodynamic flows and forces cease at equilibrium, leading to a profile that is isothermal, $T_{\text{in-situ}}$ = cst., and strongly stratified by salinity, while conservative temperature, Θ, increases downward. In contrast, well-mixed states must be maintained by active stirring to keep the profile isohaline and at uniform Θ, while $T_{\text{in-situ}}$ increases downward at the adiabatic gradient.

produce a well-mixed ocean, much shorter than the time for an ocean isolated from external forcing to reach equilibrium.

2.5.1 Potential Density under TEOS–10

By definition, well-mixed profiles have no static stability; in-situ density changes only due to compressibility. Consequently, compressibility must be removed from observations to determine static stability. As an approximation, potential density, ρ_Θ, is defined as the density a water parcel would have if displaced adiabatically, i.e. without changing heat or composition, from its in-situ pressure to a reference pressure. Under TEOS–10,

$$\rho_\Theta \equiv \rho\left(S_A, \Theta\left(S_A, \Theta, p, p_{\mathrm{ref}}\right), p_{\mathrm{ref}}\right) \quad [\mathrm{kg\ m^{-3}}]. \tag{2.39}$$

For convenience, potential density is written as $\sigma_\Theta \equiv \rho_\Theta - 1000$.

As evident from comparing potential and in-situ density, compressibility dominates in-situ density (Figure 2.13), and σ_θ is almost homogenous by comparison. In well-mixed profiles, potential density would be constant. But, as discussed in the next section, owing to density nonlinearities, this is only approximate and significant errors can occur when p_{ref} differs greatly from p. For instance, the near-bottom inversion of σ_θ in the insert is spurious and results from integrating nonlinearities

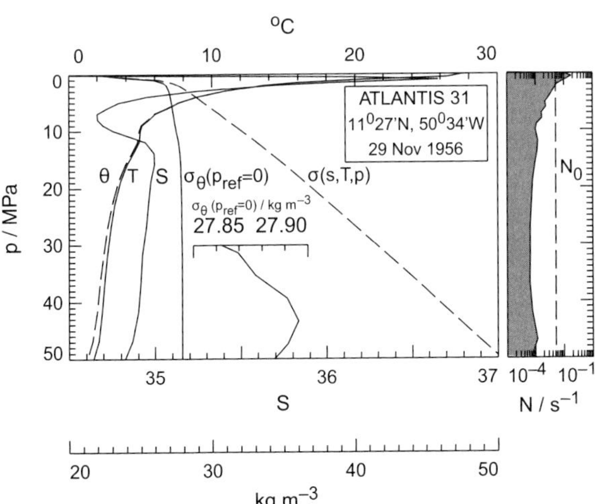

Figure 2.13 Hydrographic profile from the tropical Atlantic versus practical variables with $\sigma(S, T, p)$ as in-situ density and σ_θ referred to the surface. Expanding σ_θ near the bottom (insert) reveals an apparent density inversion that is an artifact of nonlinearities in the equation of state accentuated using a distant reference pressure. The profile was in fact stable, as seen by the buoyancy frequency, N. T and S from this profile are also in Figure 2.8.

of the equation of state across 45 MPa and 25°C. Noting this effect, Lynn and Reid (1968) suggested using a deep reference pressure to track abyssal water.

2.6 Equation of State and Its Derivatives

Known as the equation of state, density is a function of absolute salinity, conservative temperature, and pressure. It varies ±3% in the pelagic ocean (Figure 2.14), mostly due to compressibility. The 32° temperature range produces about twice the change in density as the 8 ppt salinity span.

The temperature of maximum density and the freezing temperature both decrease with increasing salinity, crossing at $S_A = 24.7$ ppt. Consequently, where surface salinity exceeds 24.7, water freezes before reaching the temperature of maximum density. Elsewhere, cold surface water sinks until it equilibrates. Owing to the fresh water released by melting ice, at the end of summer both situations can be found in the Arctic, short distances apart.

Because the equation of state is nonlinear, a second-order expansion is needed to express incremental changes,

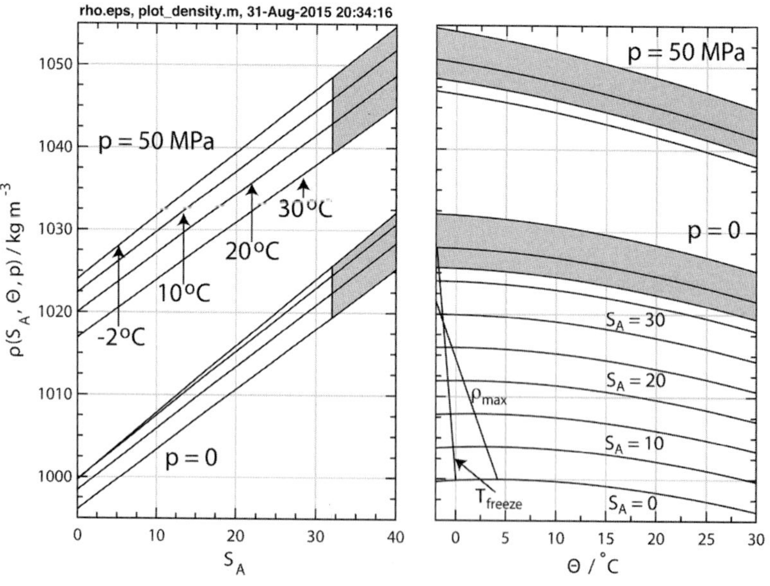

Figure 2.14 In-situ density versus absolute salinity and conservative temperature. The density range in the pelagic ocean is shaded. Temperatures of maximum density and of freezing cross at $S_A = 24.7$. Densities were computed with the Gibbs Seawater Toolbox (Pawlowicz et al., 2012).

$$
\begin{aligned}
\frac{1}{\rho}\delta\rho = & \underbrace{\gamma^{\Theta}\delta p + \frac{1}{2}\frac{\partial\gamma^{\Theta}}{\partial p}\delta p^2}_{Compressibility} + \underbrace{-\alpha^{\Theta}\delta\Theta + \beta^{\Theta}\delta S_A}_{First\text{-}order\ \Theta\, S_A} \\
& \underbrace{-\frac{1}{2}\frac{\partial\alpha^{\Theta}}{\partial\Theta}\delta\Theta^2 - \frac{\partial\alpha^{\Theta}}{\partial S_A}\delta\Theta\delta S_A + \frac{1}{2}\frac{\partial\beta^{\Theta}}{\partial S_A}\delta S_A^2}_{Second\text{-}order\ \Theta\, S_A} \\
& \underbrace{-\frac{\partial\alpha^{\Theta}}{\partial p}\delta\Theta\delta p + \frac{\partial\beta^{\Theta}}{\partial p}\delta S_A\delta p}_{Second\text{-}order\ \Theta\, p\, S_A}\,. \qquad (2.40)
\end{aligned}
$$

Isothermal compressibility (Figure 2.15a),

$$
\gamma^{\Theta} \equiv \frac{1}{\rho}\frac{\partial\rho}{\partial p}\bigg|_{\Theta,\,p} = \frac{1}{\rho c^2} \quad \left[\mathrm{Pa}^{-1}\right], \qquad (2.41)
$$

is inversely proportional to the product of in-situ density and the speed of sound, c. Compressibility varies $\pm 10\%$ and is largest in cold surface water.

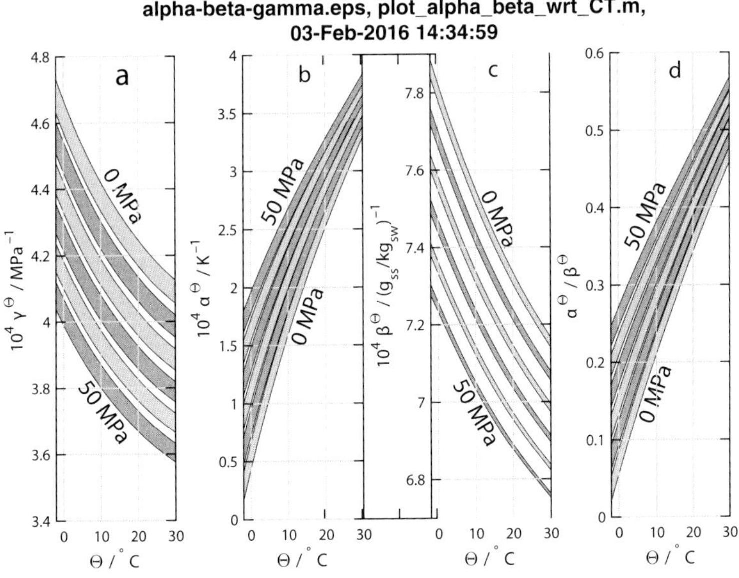

Figure 2.15 Coefficients of (a) compressibility, (b) thermal expansion, and (c) haline contraction, and (d) the ratio $\alpha^{\Theta}/\beta^{\Theta}$. Shaded bands are for pressure at intervals of 10 MPa and S_A = 32–40, with the higher salinity at the bottom of the bands.

The coefficient of thermal expansion (Figure 2.15b),

$$\alpha^{\Theta} \equiv -\frac{1}{\rho}\frac{\partial \rho}{\partial \Theta}\bigg|_{S_A,\,p} \quad [\mathrm{K}^{-1}], \tag{2.42}$$

is defined with a negative sign, a relic from publishing coefficients in tables. It varies more than tenfold, principally with temperature, significantly with pressure, and very little with salinity. Decreasing moderately with increasing pressure, α^{Θ} is largest in warm water and smallest at shallow depths at high latitude.

The coefficient of haline contraction (Figure 2.15c),

$$\beta^{\Theta} \equiv \frac{1}{\rho}\frac{\partial \rho}{\partial S_A}\bigg|_{\Theta,\,p}, \tag{2.43}$$

is the fractional change in density per gram of sea salt divided by kilograms of seawater. Largest in cold water at low pressure, β^{Θ} varies less than $\pm 10\%$.

The ratio of thermal expansion to haline contraction, $\alpha^{\Theta}/\beta^{\Theta}$, varies greatly (Figure 2.15d). In shallow, cold water, temperature changes have a minor effect on density; to equal the density change of $\Delta S_A = 1$ ppt requires a 20°C temperature change near 0°C, but only 2°C in warm water.

As discussed in more detail in Section 2.9, second-order $\Theta\ S_A$ changes are dominated by the first term, which produces the cabbeling instability, an increase in density during mixing. Second-order $\Theta\ p\ S_A$ terms are also dominated by their first term, known as thermobaricity, owing to density variations induced by temperature changing with varying pressure.

2.6.1 Buoyancy and Stratification

The stratification ratio for profiles is expressed as the density ratio

$$R_\rho^{\Theta} \equiv \frac{\text{Thermal stratification}}{\text{Haline stratification}} = \frac{\alpha^{\Theta}\partial\Theta/\partial z}{\beta^{\Theta}\partial S_A/\partial z}. \tag{2.44}$$

(The equivalent ratio using practical salinity is expressed as R_ρ.) Among other applications, the density ratio identifies where double diffusion is likely to occur.[5] For example, east of Barbados (Figure 2.16) $R_\rho > 2$ between 200 and 300 m, as it is below 600 m, except in the temperature-salinity (TS) inversion, demonstrating that thermal stratification is more than twice as strong as the destabilizing salinity

[5] Ranges of R_ρ for stable stratification vary with how profiles are stratified. Profiles with hot, salty water over cold, fresh have $R_\rho > 1$. Cold, fresh over hot, salty has $0 < R_\rho < 1$. Some authors invert the definition (sometimes without comment) in these cases. In this book, R_ρ^{-1} is used instead. Finally, profiles with hot, fresh water over cold, salty have $-\infty < R_\rho < 0$.

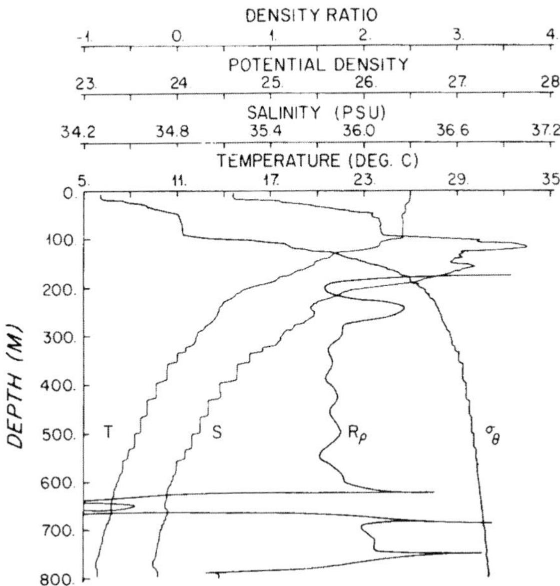

Figure 2.16 Stratification in the salt fingering staircase east of Barbados. The staircase occurs where $R_\rho \lesssim 1.6$. (From Schmitt et al., 1987)

gradient. A prominent thermohaline staircase produced by salt fingering occurs between 300 and 600 m, where R_ρ drops below 2. Similar staircases are rare and have been found only where $R_\rho \lesssim 1.6$ (Chapter 4).

In stable profiles, buoyancy forces restore displaced water parcels to their initial vertical positions. Defining density anomalies relative to their surroundings as $\rho' \equiv \rho - \rho_0$, where ρ_0 is a reference density, often taken as a local average, the specific buoyancy force is

$$b \equiv -(g/\rho_0)\rho' \quad [\mathrm{N\,kg^{-1}}], \tag{2.45}$$

where g is gravitational acceleration. Thus, b is positive upward when $\rho' < 0$. For a small vertical displacement, ζ, Hesselberg and Sverdrup (1914) approximated the equation of motion as

$$\frac{d^2\zeta}{dt^2} = b = -g\frac{\rho'}{\rho_0} \quad [\mathrm{N\,kg^{-1}}]. \tag{2.46}$$

Because displacements are nearly isentropic and isohaline, the density of a displaced parcel, ρ', varies only due to compressibility,

$$\rho' \approx \zeta\left[\left.\frac{\partial\rho}{\partial z}\right|_{\text{in-situ}} - \left.\frac{\partial\rho}{\partial p}\right|_{\Theta,S_A}\frac{dp}{dz}\right] = \zeta\left[\left.\frac{\partial\rho}{\partial z}\right|_{\text{in-situ}} - \rho^2 g\gamma^{\Theta}\right], \tag{2.47}$$

and (2.46) can be written as

$$\frac{d^2\zeta}{dt^2} = -N^2\zeta, \tag{2.48}$$

with

$$N^2 \equiv -\frac{g}{\rho}\left[\left.\frac{\partial\rho}{\partial z}\right|_{\text{in-situ}} - g\rho^2\gamma^{\Theta}\right] \quad \left[\text{s}^{-2}\right]. \tag{2.49}$$

Known as the Brunt–Vaisala frequency, or the buoyancy frequency, N is the natural frequency for oscillations, small and not so small, in stratified profiles, and therefore the maximum frequency of internal waves. Also, N^{-1} is the time scale for overturns in stratified water. Although these equations illustrate the significance of the buoyancy frequency, (2.49) is an approximation. Under TEOS–10, an exact form is

$$N^2 = g\left(\alpha^{\Theta}\frac{\partial\Theta}{\partial z} - \beta^{\Theta}\frac{\partial S_A}{\partial z}\right) \quad \left[\text{s}^{-2}\right]. \tag{2.50}$$

Discussed by McDougall and Barker (2012), alternative forms are

$$N^2 = g\alpha^{\Theta}\left(\frac{R_\rho{}^{\Theta} - 1}{R_\rho{}^{\Theta}}\right)\frac{\partial\Theta}{\partial z} = g\beta^{\Theta}(R_\rho{}^{\Theta} - 1)\frac{\partial S_A}{\partial z}. \tag{2.51}$$

2.7 Buoyancy Flux and Potential Energy

2.7.1 Buoyancy Flux

Density fluctuations correlated with vertical motions, w', produce vertical fluxes of density and of specific buoyancy force,

$$J_\rho \equiv \overline{\rho' w'} \quad \left[\text{kg m}^{-2}\,\text{s}^{-1}\right], \tag{2.52}$$

$$J_b \equiv -(g/\rho_0)J_\rho = -(g/\rho_0)\overline{\rho' w'} \quad \left[(\text{N/kg})(\text{m/s}) = \text{W kg}^{-1}\right]. \tag{2.53}$$

With z positive upward, positive buoyancy fluxes raise light and depress heavy water, e.g. during convection at the sea surface and double diffusion (Figure 2.17). Produced by turbulence in stratified profiles, negative buoyancy fluxes raise dense and suppress light water. Therefore, as a source term in the turbulent kinetic equation (6.44), J_b produces turbulence when positive and suppresses it when negative.

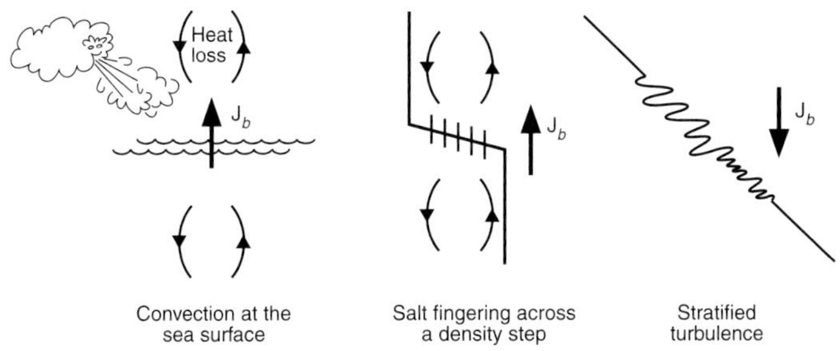

Figure 2.17 Schematic buoyancy fluxes produced by convection at the sea surface, by salt fingering, and by turbulence in stratified profiles.

2.7.2 Potential Energy of an Incompressible Fluid

If seawater is considered incompressible, e.g. under the Boussinesq approximation, its potential energy, pe, is limited to its gravitational potential energy, gpe. Relative to a vertical reference level, z_0, a volume has

$$gpe(z, z_0) \equiv g \int_{z_0}^{z} \rho(x, y, \xi)\, \xi \, dV \quad [\mathrm{J}]. \tag{2.54}$$

Using $z = 0$ and $z_0 = -3,750$ m, Oort et al. (1994) calculated the gpe of the world ocean as 2×10^{25} J. Concerned with potential energy over small domains, most mixing studies calculate gpe per horizontal area. Thus, for linear stratification, $\rho(z) = \rho_0(1 - zN^2/g)$, the potential energy relative to the center of a layer h meters thick is

$$gpe = g \int_{-h/2}^{h/2} z\rho(z)\, dz = -\frac{\rho_0 N^2 h^3}{12} \quad \left[\mathrm{J\ m^{-2}}\right]. \tag{2.55}$$

Homogenizing a layer increases its gpe to zero relative to its center by

$$\Delta gpe = (1/12)\rho_0 N^2 h^3 \quad \left[\mathrm{J\ m^{-2}}\right]. \tag{2.56}$$

The increase can also be calculated as the work required to lift the entire mass to the center of the layer, $z = 0$, from the initial center of gravity at $z_{cg} = -N^2h^2/12g = -23$ µm for $N = N_0$ and $h = 10$ m. The work is $\Delta gpe = 2.1$ J m^{-2}.

2.7.3 Available Potential Energy of an Incompressible Fluid

Most potential energy is inert and cannot be converted into kinetic energy. Known as available potential energy, ape, the part that can be converted is defined as the difference between the gpe of the existing profile and the gpe after the density field is rearranged adiabatically to increase monotonically with depth. The potential

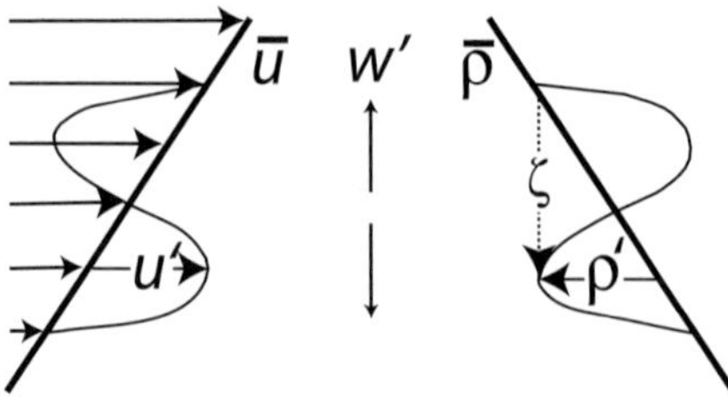

Figure 2.18 Schematic overturn of linear density and velocity profiles. u', w', and ρ' are energy-containing scales, as is ζ, the vertical displacement from the reference, or equilibrium, position, shown here for a parcel originating at the top of the overturn.

energy of the rearranged or reference profile, gpe_r, is the minimum possible for that density field (Lorenz, 1955). For a volume of incompressible fluid,

$$ape \equiv gpe - gpe_r = g \int \rho(z)(z - z_r)\, dV \quad [\mathrm{J}], \tag{2.57}$$

where z_r is the height of $\rho(z)$ in the reference profile. gpe_r is not constant unless the system is at thermodynamic equilibrium. Otherwise, it increases due to molecular diffusion, albeit slowly. As an example of applying (2.57), Moum et al. (2007b) reported that a nonlinear internal wave crossing a continental shelf had a peak *ape* density exceeding 0.2 J kg^{-1}, about four times its kinetic energy density.

Available potential energy produced by density overturns is of particular interest in mixing studies. A density parcel displaced adiabatically by ζ from its equilibrium depth (Figure 2.18) differs from density at its new depth by

$$\rho'(z) = -\zeta \frac{\partial \overline{\rho}}{\partial z} = \frac{\rho_0 N^2 \zeta}{g} \quad \left[\mathrm{kg\,m^{-3}}\right]. \tag{2.58}$$

Averaged over an overturn, *ape* (Section 2.7.5) is

$$ape = \frac{1}{2} N^2 \overline{l_{th}^2} = \frac{1}{2N^2}\left(\frac{g}{\rho_0}\right)^2 \overline{(\rho')^2} \quad \left[\mathrm{J\,kg^{-1} = m^2\,s^{-2}}\right], \tag{2.59}$$

where l_{th} is known as the ‘Thorpe scale’ in recognition of the sorting technique introduced in Thorpe (1977).

2.7.4 Potential Energy of a Compressible Fluid

In addition to *gpe*, the potential energy of compressible fluids also includes a pressure-volume term representing work done by surface pressure, p_s. In addition, some authors (Curry and Webster, 1999; Tailleux, 2009) include internal energy as a third component of potential energy. As expressed by Tailleux (2013) for a compressible fluid with a nonlinear dependence of density on temperature and pressure, potential energy is

$$pe = \underbrace{g \int_0^z \rho(x,y,\xi)\,\xi\,dV}_{gpe} + \underbrace{\int e(\eta,\upsilon)\,dV}_{ie} + p_s V \quad \text{[J]}, \tag{2.60}$$

with $\upsilon = 1/\rho$ as specific volume. In a closed system, gpe varies in response to the buoyancy flux and mixing,

$$\frac{dgpe}{dt} = \underbrace{g \int \rho w\,dV}_{BuoyFlux} + \left.\frac{dgpe_r}{dt}\right|_{\text{mixing}} \quad \text{[W]}. \tag{2.61}$$

Unlike the buoyancy flux, which can have either sign, the mixing contribution always increases potential energy. The derivative of the second term has three components (2.62). The first, viscous dissipation of kinetic energy, dke, always increases internal energy, but the work of expansion or contraction, $Wexp$, can have either sign,

$$\frac{die}{dt} = \underbrace{\int \rho\epsilon\,dV}_{dke} - \underbrace{\int \rho p \frac{D\upsilon}{Dt}\,dV}_{Wexp} - \underbrace{p_s \int \rho \frac{D\upsilon}{Dt}\,dV}_{Wexp_s} \quad \text{[W]}. \tag{2.62}$$

$Wexp_s$ is the work of expansion or contraction by surface pressure. Effects of dke and $Wexp$ on internal energy are equal and opposite to their effects on kinetic energy,

$$\frac{dke}{dt} = \underbrace{-\int \rho\epsilon\,dV}_{-dke} + \underbrace{\int \rho p \frac{D\upsilon}{Dt}\,dV}_{WExp} \underbrace{-g \int \rho w\,dV}_{-BuoyFlux} \quad \text{[W]}, \tag{2.63}$$

where

$$ke \equiv \int (\rho/2)\boldsymbol{u}^2\,dV \quad \text{[J]}. \tag{2.64}$$

Central to the dynamics of stratified turbulence, exchanges between potential, internal, and kinetic energy are pursued further in Section 3.3.

2.7.5 Available Potential Energy of a Compressible Fluid

Arguing that equation (2.57) is not sufficiently accurate to express the energetics of the general circulation over global space and the time scales of a compressible ocean over an irregular bottom, Huang (2005) proposed a formalism including available internal energy, aie,

$$ape = \underbrace{g\int \rho(z - z_r)\,dV}_{agpe} + \underbrace{\int \rho(e - e_r)\,dV}_{aie} + \underbrace{p_s(V - V_r)}_{Wexp} \quad [\mathrm{J}], \qquad (2.65)$$

where V is the total volume of the ocean, and $de = -pdv$ for the reversible adiabatic and isohaline exchanges required to compute the reference state. Owing to the large effect of nonlinearities in the equation of state, calculation of the reference state must be done iteratively, working up from great depths and sorting all densities at the same pressure after the previous density sort. This expression is the upper limit of the energy that can be converted into kinetic energy if forcing is stopped.

As seen in Figure 2.19, the Southern Ocean contributes most of the ape; maintained by strong wind stress, this is the largest volume of dense water that would slump to form bottom water if forcing stopped. Deep water formation in the North Atlantic makes a weaker positive contribution. Water at low and mid-latitudes has negative $agpe$ and would rise, pushed up by the denser water. The averaged ape density is 1.5 J kg^{-1}, compared to -0.85 J kg^{-1} for aie, which results from cold seawater being more compressible than warm water. Thus, the net ape is -0.62 J kg^{-1} (Huang, 2005).

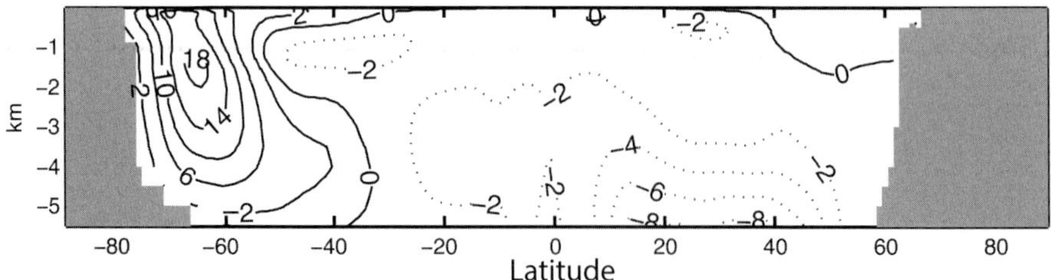

Figure 2.19 ape for the Atlantic in units of 10^{13} J m^{-2}. Positive ape at high latitudes is associated with formation of Antarctic Bottom Water (AABW) and North Atlantic Deep Water (NADW). They form the bottom water of the reference state, pushing up the negative ape water throughout the interior at mid- and low latitudes. (From Huang, 2005)

2.7.6 Diapycnal Velocity

The diapycnal velocity of a water parcel with Eulerian velocity $\boldsymbol{u}$ is its velocity across a density, a.k.a. isopycnal, surface,

$$\boldsymbol{u}_\rho \equiv (\boldsymbol{u} \cdot \hat{\boldsymbol{n}}_\rho)\hat{\boldsymbol{n}}_\rho - \tilde{\boldsymbol{u}}_\rho \quad [\mathrm{m\,s^{-1}}], \tag{2.66}$$

where $\hat{\boldsymbol{n}}_\rho$ is a unit vector normal to the density surface, directed toward increasing density, and

$$\tilde{\boldsymbol{u}}_\rho = -\frac{1}{|\nabla \rho|}\frac{\partial \overline{\rho}}{\partial t}\hat{\boldsymbol{n}}_\rho \tag{2.67}$$

is the velocity of the surface normal to itself (Ferrari et al., 2016). Continuity (3.10) allows diapycnal velocity to be expressed in terms of the divergence of density and buoyancy fluxes,

$$\boldsymbol{u}_\rho = -\frac{\nabla \cdot \boldsymbol{J}_\rho}{|\nabla \rho|}\hat{\boldsymbol{n}}_\rho = \frac{\rho_0}{g}\frac{\nabla \cdot \boldsymbol{J}_b}{|\nabla \rho|}\hat{\boldsymbol{n}}_\rho. \tag{2.68}$$

Diapycnal velocity is positive when it goes from dense to less dense water. Marshall et al. (1999) gives the full three-dimensional expression, and Ferrari et al. (2016) note that diapycnal velocity is nearly vertical and thus proportional to the vertical divergence of density and buoyancy fluxes,

$$w_\rho \simeq \frac{1}{|\partial \rho/\partial z|}\frac{\partial \boldsymbol{J}_\rho^z}{\partial z} = -\frac{1}{N^2}\frac{\partial J_b^z}{\partial z}. \tag{2.69}$$

Upwelling across isopycnals is induced when the density flux increases upward, causing dense water to become less dense. Equivalently, upwelling occurs when the buoyancy flux becomes more negative upward, as is usually the case for turbulence in the pycnocline. Alternatively, downwelling across isopycnals accompanies the density flux decreasing downward, as less dense water becomes more dense. This corresponds to the buoyancy flux becoming more negative downward, as found when the intensity of stratified turbulence increases toward the seafloor. Figure 1.12 illustrates both cases.

2.8 Temperature–Salinity Diagrams and Spiciness

Since Helland-Hansen introduced temperature–salinity (θS) diagrams during a 1916 meeting in Oslo, oceanographers have gained insight into the evolution of water properties by plotting profiles as temperature versus salinity. Temperature–salinity pairs, points in θS space, are known as water types. Layers hundreds of meters thick formed by localized processes are designated as water masses and form characteristic θS curves. Talley et al. (2011) discuss the major water masses and their role in the global circulation. One of the most prominent, Antarctic Bottom Water (AABW), spreads northward throughout much of the ocean.

When water having masses m_1 and m_2, temperatures Θ^1 and Θ^2, and salinities S_A^1 and S_A^2 mix, the mixture lies on the straight line,

$$\overline{\Theta} = (m_1\Theta^1 + m_2\Theta^2)/(m_1 + m_2)$$
$$\overline{S_A} = (m_1 S_A^1 + m_2 S_A^2)/(m_1 + m_2), \tag{2.70}$$

connecting $\Theta^1 S_A^1$ and $\Theta^2 S_A^2$.

As an example of the utility of θS diagrams, two well-defined water masses mix strongly during brief transits through the Bosphorus, the narrow strait connecting the Black Sea to the Sea of Marmara. In late summer, warm, brackish ($S \approx 17$) water from the Black Sea enters on the surface from the north, while cool, salty water from Marmara flows northward along the bottom (Figure 2.20, colour version on p. XXX). Mixing lines are evident in volumetric θS plots of water masses at opposite ends of the strait. During its 16-hour transit, the core of the salty water

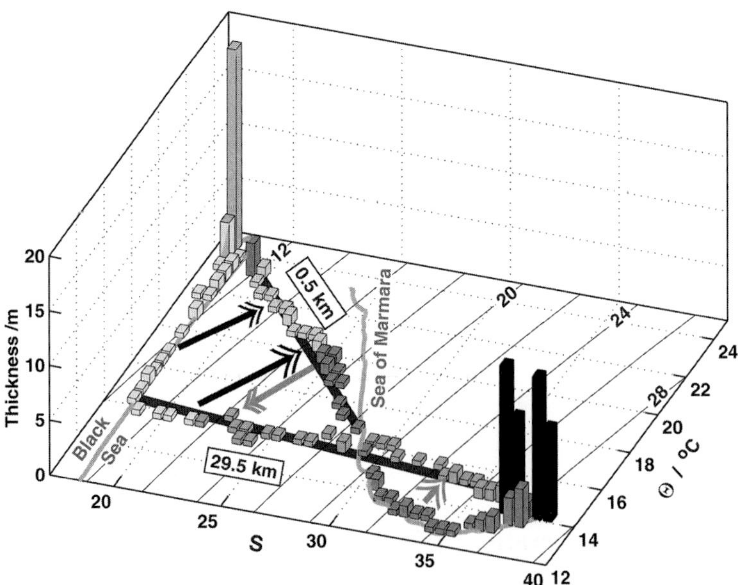

Figure 2.20 Volumetric θS diagram for the two-layer exchange flow through the Bosphorus. Numbered gray contours are σ_θ levels, and the heights of the columns are proportional to the thickness of water binned with $\Delta\theta = 0.5°C$ and $\Delta S = 0.5$ psu. (From Gregg and Özsoy, 2002) (A black and white version of this figure will appear in some formats. For the color version, please refer to the plate section)

freshens ($\nabla S = -1.4$ psu) but retains its thickness, shown by the two dark red columns being transformed to the two dark brown columns. Entering as a surface mixed layer 18 m thick, the core Black Sea water (tall green column) exits as a more saline ($\nabla S = +3.6$ psu) surface layer less than 5 m thick (dark blue column). The horizontal black line parallel to the x-axis is a mixing line at the north end of the strait connecting the outflowing salty water to the shallow thermocline in the Black Sea, and the diagonal black line is another mixing line connecting the outflowing low-salinity water to the interface above the entering Marmara water. Interfaces at both ends of the Bosphorus contain recycled water.

2.8.1 Spice

Since Stommel (1962), there have been attempts to quantify along-isopycnal variability on θS diagrams, motivated by the desire to examine 'passive' variations among water samples having the same density. To construct curves orthogonal to σ_θ isolines to measure the independent dimension in θS space, Veronis (1972) constructed a τ_{spice} function orthogonal to isopycnals,

$$d\tau_{\text{spice}} = -\alpha^{\Theta} d\Theta + \beta^{\Theta} dS_A. \tag{2.71}$$

After Munk (1981) referred to similar lines as having 'isospiciness', 'spice' and 'spiciness' have been terms for isopycnal variability.

Jackett and McDougall (1985) argue that no definition of orthogonality yields a dynamically passive variable. Equivalently, $d\tau_{\text{spice}}$ cannot be a total derivative because the cross-derivatives are not equal. Noting that $\beta^{\Theta}\delta S_A = -\alpha^{\Theta}\delta\Theta$ along each isopycnal, McDougall and Krzysik (2015) relax orthogonality in favor of requiring variations of spiciness along isopycnals to be proportional to the isopycnal integral of β^{Θ}. This function measures the strength, in density units, of isopycnal water-mass contrasts driving double-diffusive interleaving. That is,

$$\int_{\sigma_\Theta} d\tau_{\text{spice}} = 2\int_{\sigma_\Theta} \beta^{\Theta}\, dS_A. \tag{2.72}$$

Using this approach, spiciness functions are included in the Gibbs Seawater Toolbox for pressures of 0, 10, and 20 MPa, shown as contours in Figure 2.21. Along 20°C isolines, τ_{spice} is almost orthogonal to σ_Θ. At each pressure, $\tau_{\text{spice}} = 0$ at $S_A = 35.16504$ and $\Theta = 0$°C. Because τ_{spice} would change as a result of isohaline and adiabatic heaving, McDougall and Krzysik (2015) strongly recommend against interpolating τ_{spice} to intermediate pressures.

Although widely mentioned, little quantitative use has been made of spice. One exception was part of a study of lateral variability in which Ferrari and Rudnick (2000) examined wavelets of spice. Another used spice curvature as an indicator of frontal interleaving (Shcherbina et al., 2009). Nonetheless, McDougall (personal

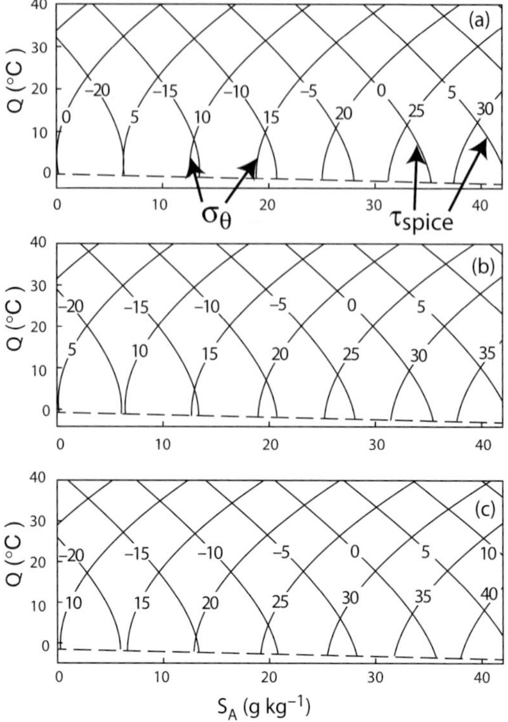

Figure 2.21 ΘS_A diagrams with contours of potential density, σ_θ, and spiciness, τ_{spice}, for pressures of 0 MPa (a), 10 MPa (b), and 20 MPa (c), with $\tau_{\text{spice}} = 0$ at $S_A = 35.16504\ g_{ss}/kg_{sw}$ and $\Theta = 0°C$. Dashed lines show the freezing temperature of seawater. (Adapted from McDougall and Krzysik, 2015)

communication, 2015) recommends using the anomaly of absolute salinity along a neutral density surface instead of spice.

2.9 Neutral Surfaces, Thermobaricity, and Cabbeling

Contours of salinity, temperature, and dissolved chemicals show that lateral motions often occur with weak mixing and little expenditure of work over large distances. Consequently, beginning with Wust (1933) and Montgomery (1938), potential density has been used for tracing water masses, e.g. Figure 1.5. Although the errors were well known, e.g. the spurious deep density inversion in Figure 2.13, they were ignored for many applications before McDougall (1987b). Building on Pingree (1972) and Ivers (1975), McDougall pointed out the size of the errors between properties mapped on potential density and on true isentropic surfaces, which McDougall terms 'neutral surfaces'. Shown schematically in Figure 2.22,

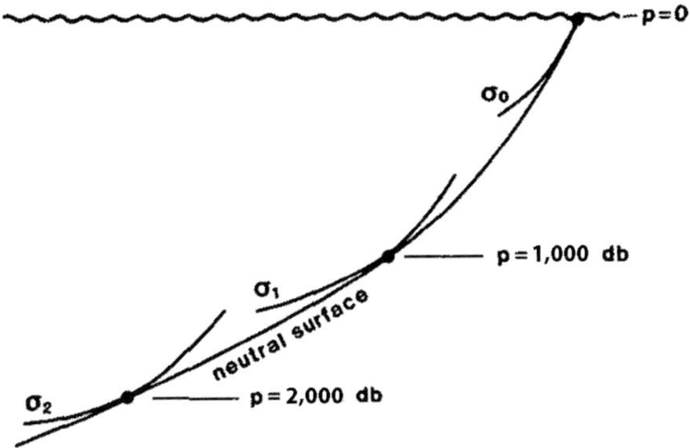

Figure 2.22 Schematic depth versus distance section with a neutral surface and potential density surfaces tangent to it at the surface (σ_0), 1,000 db (σ_1), and 2,000 db (σ_2). (McDougall, 1987a)

neutral surfaces are paths followed by water parcels moving laterally without mixing, i.e. they are adiabatic and isohaline. A potential density surface is a good local approximation, but, owing to the nonlinearities of the equation of state, the reference pressure used to compute potential density must be updated frequently.

The neutral gradient operator is defined by

$$\nabla_n \equiv \left.\frac{\partial}{\partial x}\right|_n \hat{\mathbf{i}} + \left.\frac{\partial}{\partial y}\right|_n \hat{\mathbf{j}}, \tag{2.73}$$

where $\hat{\mathbf{i}}$ and $\hat{\mathbf{j}}$ are unit vectors in the x and y directions. For reference, the conventional gradient operator is $\nabla = \nabla_n + \partial/\partial z|_{x,y}\ \hat{\mathbf{k}}$, where $\partial z|_{x,y}$ is the conventional Cartesian derivative, and $\hat{\mathbf{k}}$ is the unit vector normal to the neutral surface. With this operator, neutral surfaces are defined by

$$\frac{1}{\rho}\nabla_n \rho = \gamma^{\Theta}\nabla_n p. \tag{2.74}$$

Density change along neutral surfaces results only from compressibility. Using $\rho(S_A, \Theta, p)$, this definition is equivalent to

$$\alpha^{\Theta}\nabla_n \Theta = \beta^{\Theta}\nabla_n S_A. \tag{2.75}$$

Because no work is required to move water along neutral surfaces, they are natural paths for lateral mixing. Following Jackett and McDougall (1997), a neutral surface is found in a set of hydrographic profiles by choosing potential density in one profile and searching the others to find the same potential density when both potential densities are referenced to the same average pressure. Intersections of potential density surfaces are lines of constant Θ and S_A in three-dimensional space.

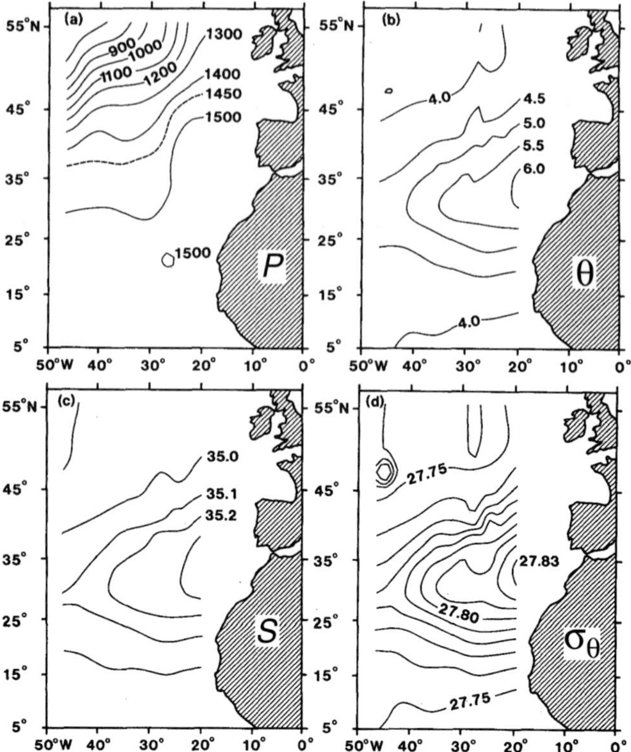

Figure 2.23 Properties on a neutral surface in the North Atlantic. Pressure on the surface (a) increases from 700 db off Nova Scotia to over 1,500 db off Gibraltar, while potential temperature (b) increases from less than 4°C in the northwest to over 6°C. Practical salinity (c) increases a few ppt and σ_θ (d) increases about 0.1 kg m^{-3}. Panels (b)–(d) show the tongue of warm saline water exiting the Mediterranean. (Adapted from McDougall, 1987a)

At a particular location, neutral surfaces can differ in depth by hundreds of meters from any of the σ_θ surfaces intersecting them elsewhere, and properties mapped on them often differ greatly from those along any of the tangent potential density surfaces (Figure 2.23). Panel (a) shows the pressure of a neutral surface in the North Atlantic that has $\sigma_\theta = 27.75$ in the southwest corner. Relatively flat in the southern half of the region, the surface rises from 1,500 db off Gibraltar to less than 700 db off Nova Scotia. Potential temperature (b), practical salinity (c), and potential density (d) decrease from Gibraltar, suggesting a source spreading out from the strait, as is the case. Panel (a) demonstrates that the Gibraltar tongue remains near the same depth for hundreds of kilometers. Dye injected on the neutral surface at 700 db off Nova Scotia and advected along it would arrive off Gibraltar near 1,500 db (Figure 2.24). Using potential density at the injection depth

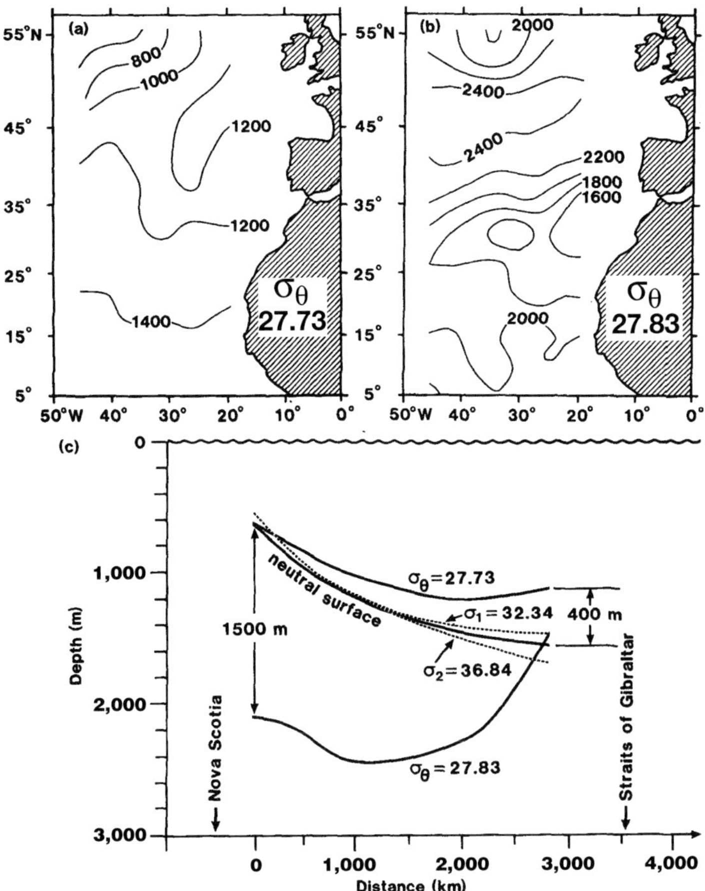

Figure 2.24 Pressure (db) on (a) $\sigma_\theta = 27.73$ and (b) $\sigma_\theta = 27.83$ and how these surfaces vary in depth (c). Dashed lines are σ_1 referenced to 1,000 db and σ_2 referenced to 2,000 db. Both closely approximate the neutral surface in the western part of the section. (From McDougall, 1987a)

($\sigma_\theta = 27.73$) would predict 1,200 m off Gibraltar, where it would be on a potential density surface of $\sigma_\theta = 27.83$.

While examining the implications of lateral motions along neutral surfaces, McDougall and Jackett (1988) traced water around a gyre on a neutral surface and found it returning about 10 m deeper after one circuit. This vertical migration does not by itself involve mixing, but vertical flux balances, e.g. Munk (1966), would treat it as a turbulent diffusive flux. In addition to vertical migration of particles moving around gyres on neutral surfaces, the second-order density derivatives also cause parcels to migrate vertically off the surfaces. These effects are known as thermobaricity and cabbeling.

2.9.1 Thermobaricity

By a factor of ten, $-(\partial\alpha^{\Theta}/\partial p)\delta\Theta\delta p$ dominates the second-order Θ p S_A terms in (2.40). Known as 'thermobaricity' (McDougall, 1987b), this second derivative of density expresses the pressure dependence of the thermal expansion coefficient, α^{Θ}, or equivalently, the temperature dependence of compressibility, γ^{Θ}. As a consequence of thermobaricity, lateral motions, e.g. by mesoscale eddies, shift water off neutral surfaces. The schematic in Figure 2.25 shows water sinking below a sloping neutral surface, but the displacements can be upward or downward depending on the sign of $\delta\Theta\delta p$.

Thermobaric displacements do not require mixing, but they produce a weak dianeutral velocity that McDougall (1987b) gives as

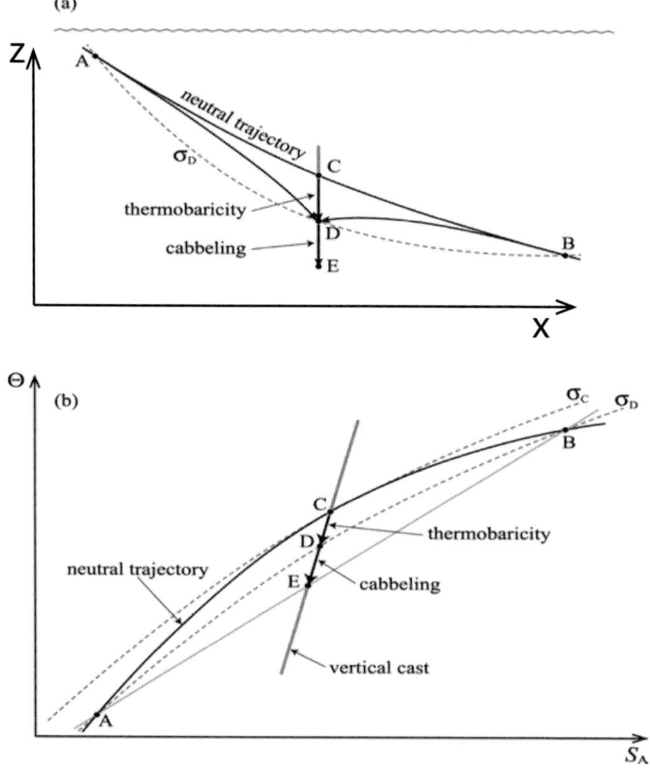

Figure 2.25 (a) Schematic paths showing thermobaricity sinking parcels to D when brought adiabatically from A and B toward C, followed by cabbeling when the water is mixed after arriving at D. The example equally well could have supposed rising due to thermobaricity, but in both cases cabbeling would produce sinking. (b) The same sequence versus absolute salinity and conservative temperature. (Adapted from IOC et al., 2010)

$$e^{Tb} = -\frac{g}{N^2} K_{\text{iso}} \nabla_n \cdot \nabla_n p \left(\frac{\partial \alpha^{\Theta}}{\partial p} - \frac{\alpha^{\Theta}}{\beta^{\Theta}} \frac{\partial \beta^{\Theta}}{\partial p} \right) \quad \left[\text{m s}^{-1} \text{ K}^{-1} \right], \qquad (2.76)$$

where K_{iso} is diffusivity along the neutral surface, a.k.a. epineutral diffusivity. Dianeutral velocities are of the order 10^{-7} m s^{-1}, comparable to the average upwelling velocity linked to $K_z \approx 10^{-4}$ m^2 s^{-1} by Munk (1966).

Themobaricity appears to play an important role in forming AABW (Section 1.3.1) by adding to the density contrast propelling plumes down the continental slope of the Weddell Sea (Killworth, 1977). At the top of the slope, the plume is colder than its surroundings, and thus more dense and more compressible. The primary thermobaric term becomes positive as the plume sinks below the surrounding water, increasing the density contrast during descent.

In terms of absolute salinity and conservative temperature, IOC et al. (2010) define the thermobaric coefficient as

$$T_b^{\Theta}(S_A, \Theta, p) = \left. \frac{\partial \alpha^{\Theta}}{\partial p} \right|_{S_A, \Theta} - \left. \frac{\alpha^{\Theta}}{\beta^{\Theta}} \frac{\partial \beta^{\Theta}}{\partial p} \right|_{S_A, \Theta} \quad [\text{K}^{-1} \text{ Pa}^{-1}]. \qquad (2.77)$$

A term in the water mass transformation expression (Eq. 2.83), T_b^{Θ} is mainly a function of temperature and secondarily of pressure (Figure 2.26).

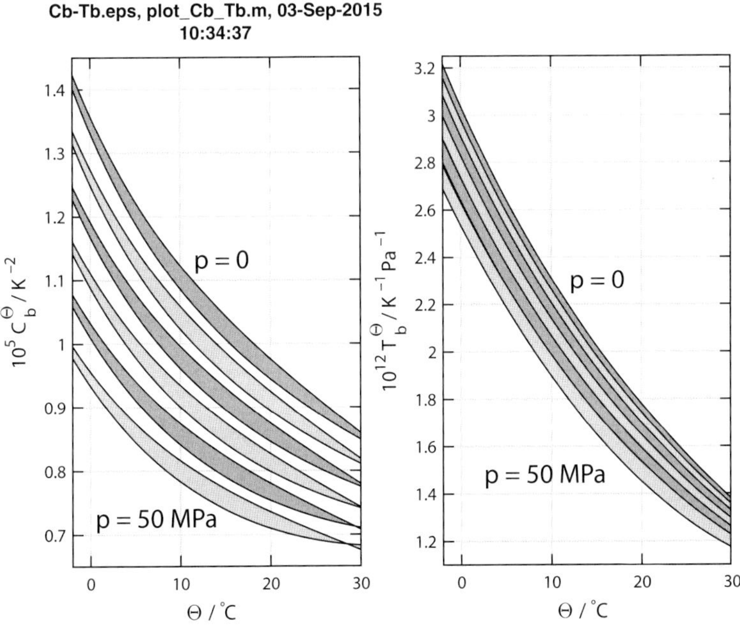

Figure 2.26 Cabbeling (left) and thermobaric (right) coefficients in the Gibbs Seawater Toolbox (IOC et al., 2010). Shaded bands at 10 MPa pressure intervals span salinity ranges of 32–40, with higher salinities at the bottoms of the bands.

2.9.2 Cabbeling

Second-order $\Theta\, S_A$ terms in (2.40) are dominated by $-(1/2)(\partial\alpha^\Theta/\partial\Theta)\delta\Theta^2$, the temperature dependence of α^Θ. At the surface, α^Θ changes tenfold due to temperature variations. Changes are less in the abyss, owing to the narrower temperature range and the decrease of $\partial\alpha^\Theta/\partial\Theta$ in cold water. Because $\partial\alpha^\Theta/\partial\Theta$ and $\delta\Theta^2$ are positive, this term is negative and increases density when waters mix, an effect sometimes referred to as 'the densification of mixing.' Cabbeling, a.k.a. contraction on mixing (Fofonoff, 2001), always causes water to sink below its original neutral surface. Because mixing is conservative in Θ and S_A, water types mix along straight lines connecting them on ΘS_A diagrams. Owing to the curvature of isopycnals, reflecting the nonlinearity of density, when water parcels originating on the same isopycnal mix, their product will be more dense. Densification accompanying cabbeling produces a negative density flux (heavy water sinking), and hence a positive buoyancy flux, $\boldsymbol{J}_B = -(g/\rho)\boldsymbol{J}_\rho$. This flux is opposite to the negative buoyancy flux resulting from turbulent mixing in stratified profiles and has large-scale consequences.

In terms of absolute salinity and conservative temperature, IOC et al. (2010) define the cabbeling coefficient as

$$C_b^\Theta \equiv \left.\frac{\partial\alpha^\Theta}{\partial\Theta}\right|_{S_A,p} + 2\frac{\alpha^\Theta}{\beta^\Theta}\left.\frac{\partial\alpha^\Theta}{\partial S_A}\right|_{\Theta,p} - \left(\frac{\alpha^\Theta}{\beta^\Theta}\right)^2 \left.\frac{\partial\beta^\Theta}{\partial S_A}\right|_{\Theta,p} \quad [\mathrm{K}^{-2}]. \tag{2.78}$$

C_b^Θ is largest at low temperatures and salinities, where isopycnal curvature is greatest (Figure 2.26). McDougall (1987b) demonstrates that dianeutral velocities produced by cabbeling tend to be about three times larger than those resulting from thermobaricity. Effects attributed to cabbeling include:

Formation of AABW. Using a finite-amplitude simulation, Foster (1972) concluded that cabbeling is likely involved in the formation of AABW. The scope, however, seems considerably larger, as a numerical simulation by Klocker and McDougall (2010) finds widespread cabbeling and thermobaricity along the Antarctic Circumpolar Front, forming 6×10^6 m^3 s^{-1} of dense water.

Formation of North Pacific Intermediate Water (NPIW). Cabbeling accounts for at least half of the density difference between the Oyashio winter mixed layer, where the NPIW originates, and what is observed at the salinity minimum in the North Pacific (Talley and Yun, 2001).

Frontal circulation. Theoretical analysis of a temperature–salinity front that had no lateral density contrast estimated the sinking velocity due to cabbeling as $w_\rho = -K_{\mathrm{iso}}\,(\partial\alpha/\partial T)(\partial T/\partial y)^2/(\partial\rho/\partial z)$, where y is the cross-frontal direction (Garrett and Horne, 1978). For a typical front, they estimated $\approx$ 1 m/day, which scale analysis indicates could be supported by horizontal convergence, given a large temperature contrast across the front.

Balancing the negative buoyancy flux at the sea surface. Coherence of high thermal expansion coefficients and heat fluxes in the tropics contrasted with smaller thermal expansion coefficients and heat loss at high latitudes produces a net negative buoyancy flux (positive density flux upward) at the sea surface (Section 1.4.4). Since the ocean is not becoming more buoyant at a corresponding rate, internal processes must be producing an equivalent positive buoyancy flux. This can only be due to cabbeling, which produces a positive buoyancy flux in contrast to turbulent mixing in stratified profiles. The decrease in specific volume accompanying the density increase must be considered for accurate predictions of sea level rise.

Limiting the magnitude of vertical temperature gradients in the thermocline. Fofonoff (1998, 2001) demonstrated that cabbeling establishes an upper limit to thermal stratification that can occur without being actively forced; vertical displacement of gradients larger than

$$\left.\frac{dT}{dz}\right|_{\text{diff}} = \frac{\frac{\rho g}{p}\alpha^{\Theta} + \rho g(-\alpha^{\Theta}\gamma^{\Theta} + \frac{\partial \alpha^{\Theta}}{\partial p}) - (-\alpha^{\Theta}\beta^{\Theta} + \frac{\partial \alpha^{\Theta}}{\partial S})\frac{dS}{dz}}{(\alpha^{\Theta})^2 + \partial\alpha^{\Theta}/\partial T} \quad \left[\frac{\text{K}}{\text{m}}\right] \tag{2.79}$$

will release more gpe_r by cabbeling than is produced by molecular diffusion. Comparisons with observed profiles demonstrate depth ranges throughout the ocean where stratification matches this limit. Examples from main thermoclines in the subtropical North Atlantic and North Pacific show measured gradients fluctuating about diffusive limits (Figure 2.27).

Fofonoff argued that large-scale forcing acts to increase vertical gradients, e.g. downwelling surface and upwelling deep water, until they reach the critical gradient, where molecular diffusion triggers cabbeling, which in turn releases *ape*, decreasing the stratification and increasing gpe_r. Stratification stable to cabbeling instability must have a positive stability parameter,

$$E_{\text{thermal}} \equiv -\frac{d}{dz}\left(\frac{p}{\rho}\alpha^{\Theta}\right)\frac{dT}{dz}. \tag{2.80}$$

Sections of neutral stability are constant in depth plots of $(p/\rho)(\partial\alpha^{\Theta}/\partial T)$.

2.9.3 Dianeutral Velocity

Vertical motions induced by mixing combined with density nonlinearities produce diapycnal and dianeutral velocities through density and neutral surfaces, respectively. Velocities of 10 μm s^{-1} are too small to measure, but they are a natural way of quantifying vertical motions of the global overturning circulation. As a thought experiment, in Figure 2.25 equal volumes of water encased in insulating

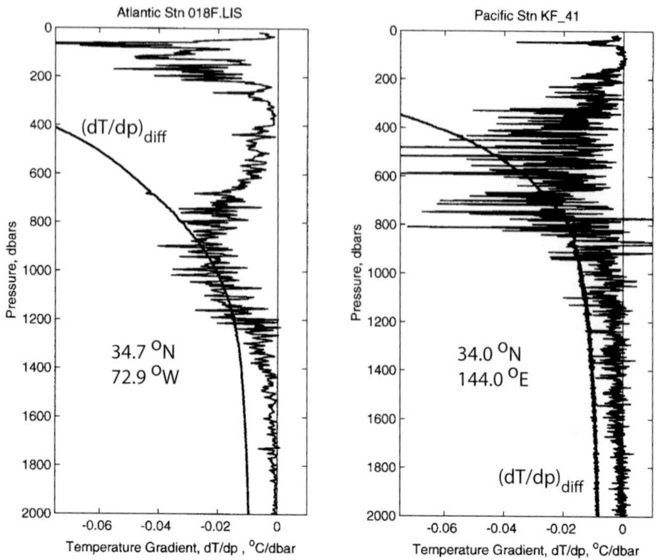

Figure 2.27 Temperature gradients in the subtropical Atlantic and Pacific versus the diffusive limit set by cabbeling. (Adapted from Fofonoff, 2001. ©American Meteorological Society. Used with permission)

plastic bags at A and B begin on same potential density surface, σ_D, and on the same neutral surface (ACB) that slopes across isopycnals. If the bags are brought toward C, they will retain their original conservative temperature, Θ, and absolute salinity, S_A, but due to thermobaricity they will sink below the neutral surface to arrive at D. Because the parcels have not mixed when brought to the same geographical position, D, they remain at A and B on the ΘS_A diagram (b) and still lie on σ_D. If the bags are then mixed, the mixture must fall on the straight line connecting them on the ΘS_A diagram, in this case at the midpoint, assuming equal volumes are mixed. Owing to the ΘS_A curvature, increased density of the mixture causes it to sink to E. The additional sinking is termed 'cabbeling'.

McDougall (1987b) and IOC et al. (2010) express the time-averaged dianeutral velocity, $\overline{w_{\rm n}}$, as

$$\overline{w_{\rm n}} N^2/g = -K_{\rm iso}\left(C_b^{\Theta}\nabla_n\hat{\Theta}\cdot\nabla_n\hat{\Theta} + T_b^{\Theta}\nabla_n\hat{\Theta}\cdot\nabla_n P\right)$$
$$+\,\alpha^{\Theta}\left(K_\rho\hat{\Theta}_z\right)_z - \beta^{\Theta}\left(K_\rho\hat{S}_{A_z}\right)_z - \beta^{\Theta}\hat{S}^{S_A}_{\rm source}. \qquad (2.81)$$

Hats denote weighting by thickness, and $K_{\rm iso}$ is the isoneutral turbulent diffusivity, i.e. along neutral tangent planes. Orthogonal to the planes, K_ρ is the dianeutral turbulent diffusivity. Proportional to K_{iso}, the first term is the dianeutral velocity induced by lateral turbulence moving water through dianeutral gradients of temperature (cabbeling) and of temperature and pressure (thermobaricity). Because $\nabla_n\hat{\Theta} \cdot \nabla_n\hat{\Theta}$ is always positive, cabbeling always produces sinking. Proportional to $\nabla_n\hat{\Theta} \cdot \nabla_n P$, thermobaricity can take either sign.

Divergences of dianeutral fluxes of heat and salt drive the second and third terms, which are also proportional to the coefficients of thermal expansion and haline contraction, respectively. Depending on the dianeutral curvature of temperature and salinity, these terms can take either sign. Also, note that α^{Θ} changes sign in cold, fresh water. The last term expresses effects of particulate remineralization and can often be neglected.

Dianeutral velocity can also be recast as a density conservation equation (IOC et al., 2010),

$$\overline{w_{\text{n}}} N^2 = -\, g K_{\text{iso}} \left(C_b^{\Theta} \nabla_n \hat{\Theta} \cdot \nabla_n \hat{\Theta} + T_b^{\Theta} \nabla_n \hat{\Theta} \cdot \nabla_n P \right) - g \beta^{\Theta} \hat{S}^{S_A}_{\text{source}} \tag{2.82}$$

$$+ \frac{1}{A}\left(A K_\rho N^2\right)_z - K_\rho N^2 \frac{R_\rho}{(R_\rho - 1)} \left[\frac{\alpha_z^{\Theta}}{\alpha^{\Theta}} - \frac{\beta_z^{\Theta}}{\beta^{\Theta}} \frac{1}{R_\rho} \right],$$

where A is the depth-dependent area of the density surface. A positive term, cabbeling always generates a negative dianeutral velocity, but thermobaricity can cause rising or sinking. IOC et al. (2010) stresses that this is the complete equation relating upwelling, $\overline{w_{\text{n}}}$, to the mixing problem addressed by Munk (1966) and Munk and Wunsch (1998). In attributing all of the upwelling to K_ρ, they neglected the effect of K_{iso} on dianeutral velocity and were not aware of the effects depending on the density ratio in the last term.

2.10 Water Mass Transformation

Produced by individual, identifiable processes, water masses have tight ΘS relationships (Talley et al., 2011). Building on McDougall (1984) and Graham and McDougall (2013), IOC et al. (2010) expresses water mass transformation using diffusivities along, $K_{\text{isoneutral}}$, and perpendicular to, $K_{\text{dianeutral}}$, neutral surfaces. Using the TEOS–10 formulation of the water mass transformation equation, the evolution of Θ along a neutral tangent plane is given by (2.83), where all variables are weighted by thickness and density between two neutral surfaces close together in the vertical. For example, density-weighted conservative temperature is $\overline{\rho\Theta(\rho)}/\overline{\rho}$. C_b^{Θ} is the cabbeling coefficient (2.78), and T_b^{Θ} is the thermobaricity coefficient (2.77).

$$\left.\frac{\partial\hat{\Theta}}{\partial t}\right|_n + \hat{\mathbf{v}}\cdot\nabla_n\hat{\Theta} = \underbrace{\gamma_z\nabla_n\cdot(\gamma_z^{-1}K_{\text{iso}}\nabla_n\hat{\Theta})}_{\text{I}}$$
$$+ K_{\text{isoneutral}}\, gN^{-2}\hat{\Theta}_z\left(\underbrace{C_b^{\Theta}\nabla_n\hat{\Theta}\cdot\nabla_n\hat{\Theta}}_{\text{II}} + \underbrace{T_b^{\Theta}\nabla_n\hat{\Theta}\cdot\nabla_n P}_{\text{III}}\right)$$
$$+ \underbrace{K_{\text{dianeutral}}\,\beta^{\Theta} gN^{-2}\hat{\Theta}_z^3\frac{d^2\hat{S}_A}{d\hat{\Theta}^2}}_{\text{IV}} + \underbrace{\frac{\beta^{\Theta}}{\alpha^{\Theta}}\frac{R_\rho}{(R_\rho-1)}\hat{S}^{\text{bgc}}_{\text{source}}}_{\text{V}} \quad (2.83)$$

The terms represent:

I. Lateral turbulent diffusion, $K_{\text{isoneutral}}$, along the neutral surface.
II. The combined effect of $K_{\text{isoneutral}}$ and cabbeling, C_b^{Θ}, in moving water off or onto the surface. The effect is nonlinear and proportional to the square of the gradient of conservative temperature on the neutral surface.
III. The combined effect of $K_{\text{isoneutral}}$ and thermobaricity in moving water off or onto the surface. Also nonlinear, the effect is proportional to along-surface components of temperature and pressure gradients.
IV. Dianeutral turbulent diffusion perpendicular to the neutral surface. This is the total effect of diapycnal, or dianeutral, mixing on water mass transformation. Another way of expressing the effect of dianeutral velocity moving heat off the neutral surface, this term shows that three-dimensional turbulence affects Θ evolution only to the degree that the ΘS_A relation is curved. Otherwise, water mixes without changing ΘS_A characteristics. $K_{\text{dianeutral}}$ can be approximated by $K_\rho = 0.2\epsilon N^{-2}$.
V. A heat source (or sink) as a biogeochemical source $S^{\text{bgc}}_{\text{source}}$. If important, heating by turbulent dissipation would appear as $S^{\epsilon}_{\text{source}} = \epsilon/c_p$.

The corresponding conservation equation for absolute salinity is

$$\left.\frac{\partial\hat{S}_A}{\partial t}\right|_n + \hat{\mathbf{v}}\cdot\nabla_n\hat{S}_A = \gamma_z\nabla_n\cdot(\gamma_z^{-1}K\nabla_n\hat{S}_A)$$
$$+ K_{\text{iso}}gN^{-2}(\hat{S}_A)_z\left(C_b^{\Theta}\nabla_n\hat{\Theta}\cdot\nabla_n\hat{\Theta} + T_b^{\Theta}\nabla_n\hat{\Theta}\cdot\nabla_n P\right)$$
$$+ K_\rho\alpha^{\Theta}gN^{-2}\hat{\Theta}_z^3\frac{d^2\hat{S}_A}{d\Theta^2} + \frac{R_\rho}{(R_\rho-1)}S_{S_A}^{\text{source}} \quad (2.84)$$

The terms have the same interpretation as those in (2.83).

Water mass transformation rates at steady state are depicted schematically in Figure 2.28. Because partial derivatives with time are assumed to be zero, net transformation rates, given by the advective terms on the left sides of (2.83) and

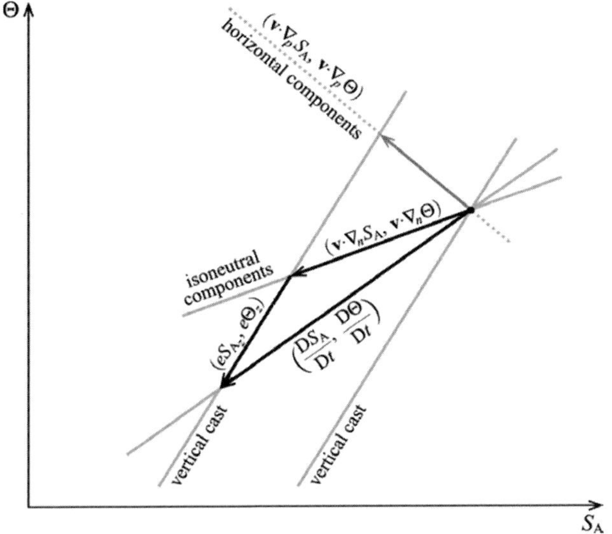

Figure 2.28 Schematic water mass transformation rates at steady state showing material derivatives, DS_A/Dt and $D\Theta/Dt$, as vector sums of advection along a neutral surface, $\mathbf{v} \cdot \nabla_n S_A$ and $\mathbf{v} \cdot \nabla_n \Theta$, and dianeutral advection by cabbeling and thermobaricity. At steady state, the material derivatives are projections of horizontal advection, $\mathbf{v} \cdot \nabla_p S_A$ and $\mathbf{v} \cdot \nabla_p \Theta$, represented by the gray arrow to their right. (From IOC et al., 2010)

(2.84), are shown as gray arrows perpendicular to the vertical casts. They are vector sums of lateral advection along the neutral surface and dianeutral advection by cabbeling and thermobaricity. The former can be estimated from observations, in contrast to the latter, which must be estimated and not with confidence.

As an example of the use of equation 2.83, McDougall (1984) reevaluated inferences of K_ρ in the North Pacific thermocline by White and Bernstein (1981), who compared two large-scale hydrographic surveys a year apart. The surveys were used to infer temporal changes and advective transport, i.e. the left-hand side of the water mass conversion equation. Diffusivities were inferred by assuming that the material derivative of temperature was balanced solely by the flux from small-scale turbulence, here termed 'dianeutral turbulent flux'. Evaluation of the White and Bernstein data for the $\sigma_\Theta = 27.2$ surface by McDougall found the factor multiplying K_ρ in term IV to equal 0.12. Consequently, the K_ρ found by White and Bernstein is too small by a factor of $1/0.12 = 8.3$ and should be 1.7×10^{-4} m^2 s^{-1}. Because the profile is diffusively stable in this depth range, enhanced turbulence is the only possibility to balance observed advection and temporal change. At the salinity minimum, $\sigma_\Theta = 27.0$, the term multiplying K_ρ was 0.98, and no correction was needed. On the $\sigma_\Theta = 26.6$ surface, above the salinity minimum, the factor multiplying K_ρ was –0.25, requiring $K_\rho = -0.8 \times 10^{-4}$ m^2 s^{-1}, which is rather implausible. Instead, McDougall notes that $R_\rho = 3.6$, demonstrating that salt fingering is possible.

Term V for salt fingering is

$$S^{\mathrm{sf}}_{\mathrm{source}} = \frac{\beta^\Theta}{\alpha^\Theta} J^{\mathrm{sf}}_S \frac{R_\rho - R^{\mathrm{sf}}_{\mathrm{flux}}}{R_\rho - 1}, \tag{2.85}$$

where J^{sf}_S is the salt flux produced by fingers, and $R^{\mathrm{sf}}_{\mathrm{flux}}$ is the salt fingering flux ratio (Chapter 4). McDougall concluded that the salt flux source also had the wrong sign. Hence, isopycnal mixing cannot be neglected on this surface.

3
Turbulence

3.1 Overview

Notoriously difficult to define, turbulence is often described by its major characteristics: random, dissipative, diffusive, high Reynolds number, with three-dimensional vorticity (Tennekes and Lumley, 1972). These are treated in standard references: Pope (2000), Kundu and Cohen (2004), Hinze (1975), and Monin and Yaglom (1971, 1975). Assuming some of that background, this chapter focuses on aspects of turbulence central to understanding and measuring viscous and scalar dissipation rates in the ocean and their accompanying turbulent diffusivities.

Owing to the complexity and nonlinearity of turbulence in stratified fluids, dimensional analysis is essential for identifying dominant terms in equations, estimating magnitudes, and developing useful nondimensional parameters (Section 3.2). Applying the Reynolds decomposition to the equations of motion (Section 3.3) and direct numerical simulations show how the evolution of turbulence varies with intensity and its effects on mixing. Because many turbulent parameters cannot now be measured in the ocean, the turbulent kinetic equation and scalar equivalents for Θ'^2 and S'^2, buoyancy flux and potential energy (Section 3.4), are used to infer what cannot be measured from what can be.

Section 3.5 considers the production of turbulence by surface stress and convection because some results carry over to mixing in the pycnocline where double diffusion and breaking internal waves are the major sources, as explored in later chapters. Once generated, strong turbulence acquires universal characteristics owing to the role of inertial interactions in transferring energy to smaller scales (Section 3.6). These interactions in turn produce similar cascades of temperature and salinity variance (Section 3.7) that are dissipated by molecular diffusivity at rates χ_T and χ_S.

Strong stratification characteristic of the pycnocline decouples vertical and horizontal flows, producing a horizontal energy cascade that forms vortical motions, a.k.a. pancake eddies, as turbulence decays (Section 3.8). Mixing rarely goes to completion, and large differences between molecular diffusivities of heat and salt appear to form granulations of salt parcels slowly restratifying after temperature fluctuations have been smoothed by diffusion (Section 3.9). Also, regions that have been homogenized or partially homogenized relax laterally, reducing the increase in potential energy produced by the mixing.

Intermittence of turbulence in homogenous fluids is exacerbated in the pycnocline, where turbulence occurs episodically in patches (Section 3.10). Probability distributions are usually close to lognormal and those statistics are good guides for estimating sampling requirements. Average dissipation rates can be estimated directly from microstructure measurements, but diapycnal diffusivities require assumptions simplifying the turbulence and applying the turbulent kinetic energy equation and the corresponding scalar variance equations (Section 3.11). In addition, estimates from viscous dissipation rates require multiplication by a measure of mixing efficiency (Section 3.12), which is the most uncertain part of the process.

3.2 Dimensional and Scale Analysis

Owing to the complexity and nonlinearity of turbulence, dimensional and scale analysis have long guided interpretations of observations. To determine the number of independent dimensionless parameters needed for turbulence in stratified water, Mater and Venayagamoorthy (2014) characterized turbulence by its kinetic energy, $tke \sim u'^2$, and by its dissipation rate,

$$\epsilon = \frac{tke}{\tau} = \frac{u'^2}{l/u'} = \frac{u'^3}{l} \quad \left[\frac{\mathrm{m}^2}{\mathrm{s}^3} = \frac{\mathrm{W}}{\mathrm{kg}}\right]. \tag{3.1}$$

Surprisingly accurate, this estimate, known as Taylor scaling (Taylor, 1935), is widely used. Characterizing water by its three molecular properties and the environment by its buoyancy frequency and shear gives seven dimensional parameters containing two fundamental units, length and time:

$$\begin{gathered}\text{Turbulence: } tke\ [\mathrm{m^2\,s^{-2}}],\ \epsilon\ [\mathrm{m^2\,s^{-3}}]\\ \text{Water: } \nu, \kappa_T, \kappa_S \quad [\mathrm{m^2\,s^{-1}}]\\ \text{Environment: } N, Shear \quad [\mathrm{s^{-1}}]\,.\end{gathered}$$

By the Buckingham π theorem (Kundu and Cohen, 2004, p. 168), seven dimensional parameters and two fundamental units can support $7 - 2 = 5$ dimensionless parameters. The most fundamental, the Reynolds number, relates energy-containing scales u' and l by

$$Re \equiv \frac{\text{inertial force}}{\text{viscous force}} = \frac{u'^2/l}{\nu u'/l^2} = \frac{u'l}{\nu}. \quad (3.2)$$

Having ϵ but not u', Ivey and Imberger (1991) used Taylor scaling, $u' = (\epsilon l)^{1/3}$, to express the Reynolds number as $Re_\epsilon = \epsilon^{1/3} l^{4/3}/\nu$. This and similar derivative expressions of fundamental dimensionless variables are based on additional assumptions and thus should not be considered equivalent to the original dimensionless variable, hence the subscript.

When buoyancy limits energy-containing scales, the turbulent velocity scale becomes $u_b \equiv Nl$, termed the 'buoyancy velocity', and Taylor scaling is $\epsilon = u_b^3/l$. With these, and rewriting $u_b l$ as $(u_b^3/l)/(u_b/l)^2$, the buoyancy Reynolds number is

$$Re_b \equiv \frac{u_b l}{\nu} = \frac{u_b^3/l}{\nu (u_b/l)^2} = \frac{\epsilon}{\nu N^2}. \quad (3.3)$$

Because ϵ is often measured more easily than u', Re_b is the most common reference for turbulent intensity in the pycnocline. When shear dominates, the more appropriate measure may be the shear Reynolds number,

$$Re_S \equiv \frac{\epsilon}{\nu \; Shear^2}. \quad (3.4)$$

With $u_b = Nl$, (3.3) gives the vertical length scale where buoyancy limits overturns as

$$l_O \equiv (\epsilon/N^3)^{1/2}. \quad (3.5)$$

Referred to as the Ozmidov scale (Ozmidov, 1965), l_O estimates root-mean-square overturning scales in the pycnocline (Dillon, 1982).

In addition to the key dimensionless variables Re, $Pr \equiv \nu/\kappa_T$, and $Sc \equiv \nu/\kappa_S$, the gradient Richardson number,

$$Ri_g \equiv \frac{N^2}{(\partial u/\partial z)^2} \sim \frac{N^2}{(u/l)^2} \sim \frac{\text{buoyancy force}}{\text{inertial force}} \sim \left(\frac{\tau_{shear}}{\tau_{buoyancy}}\right)^2 = \frac{1}{Fr_N^2}, \quad (3.6)$$

is needed to describe the stability of stratified profiles to shear. It is usually computed directly from simultaneous profiles of stratification and shear. The square root of its inverse, the stratification Froude number, $Fr_N \equiv u/Nl$, is estimated most directly from streamwise velocity, stratification, and length scale. When shear, S, is large, consider instead the shear Froude number,

$$Fr_S \equiv u/Sl. \quad (3.7)$$

When the energetics of stratified turbulence are considered, a key parameter is the flux Richardson number,

$$Ri_{\text{flux}} \equiv \frac{J_b}{\overline{u'w'}(\partial \overline{u}/\partial z)} = \frac{\text{rate of work against buoyancy}}{\text{rate of turbulence production}}. \tag{3.8}$$

Values are usually between 0 and 1, but Ri_{flux} can be negative when restratification overwhelms production.

3.3 Energetics

In stratified profiles, turbulence usually begins with a buoyancy flux, J_b, extracting kinetic energy from the background flow. By lifting dense and depressing light water, the buoyancy flux creates available potential energy, ape, that destabilizes stratification. Subsequent collapse of the instability generates turbulent kinetic energy, tke, which is ultimately dissipated into heat by viscosity at rate ϵ and into increased potential energy by diffusion at rate χ_{pe}. The coupled equations for these energy exchanges are developed by applying Reynolds decomposition to the equations of motion.

3.3.1 Equations of Motion

Most equations of motion for stratified flows (Riley et al., 1981; Kundu and Cohen, 2004) include density fluctuations, $\rho' \equiv \rho - \rho_0$, only in the buoyancy term of the momentum equation,

$$\frac{\partial \boldsymbol{u}}{\partial t} + \boldsymbol{u} \cdot \nabla \boldsymbol{u} = -\frac{\nabla p}{\rho_0} + \boldsymbol{g}\frac{\rho'}{\rho_0} + \nu \nabla^2 \boldsymbol{u} \tag{3.9}$$

$$\nabla \cdot \boldsymbol{u} = 0 \tag{3.10}$$

$$\frac{\partial \rho}{\partial t} + \boldsymbol{u}_h \cdot \nabla_h \rho + w\frac{\partial \overline{\rho}}{\partial z} = \kappa_\rho \nabla^2 \rho, \tag{3.11}$$

where ρ_0 is a reference density, often the average, $\overline{\rho}$. Known as the Boussinesq approximation, this simplifies analysis, but Tailleux (2009) argues that it is accurate only in some parts of the ocean, e.g. in warm surface water. Elsewhere, particularly in cold water, density nonlinearities can produce significant errors in potential energy changes due to mixing. With this caveat, the Boussinesq approximation will be used for turbulence.

When the buoyancy force is negligible, scaling the momentum equation with u' and l leads to

$$\frac{\partial \hat{\boldsymbol{u}}}{\partial \hat{t}} + \hat{\boldsymbol{u}} \cdot \nabla \hat{\boldsymbol{u}} = -\nabla \hat{p} + \frac{1}{Re}\nabla^2 \hat{\boldsymbol{u}}, \tag{3.12}$$

where 'hats' indicate non-dimensional variables. For Reynolds numbers larger than ≈ 100, the viscous term, second on the right, is negligible. If the stratification Froude number, Fr_N, is small, flows are also inviscid and hydrostatic.

3.3.2 The Reynolds Decomposition

In the simplest situation, turbulence develops from infinitesimal perturbations to steady laminar flow. If the perturbations grow to finite size, instantaneous velocity can be represented as the sum of steady, $\overline{u}$, and fluctuating, u', components,

$$u = \overline{u} + u' \quad \text{with} \quad \overline{\overline{u}u'} = 0 \quad \text{and} \quad \overline{u'u'} \neq 0. \tag{3.13}$$

Averages of fluctuations times the mean, $\overline{u'\overline{u}}$, are zero, but averages of products of fluctuations, e.g. $\overline{u'w'}$, are usually finite (Monin and Yaglom, 1971, section 3.1). Known as Reynolds decomposition (Reynolds, 1895), this procedure is based on the assumption, often implicit, that space and time scales of average, or low-frequency, flow are clearly separated by spectral gaps from the turbulence. For instance, acoustic anemometers on bows of research ships average momentum fluxes in the atmospheric boundary layer for 10–15 minutes, long relative to the ≈ 60 s lifetime of turbulent eddies, and shorter than the hourly to daily changes in large-scale winds producing the turbulence.

In stratified profiles, the spatial scales of turbulence overlap those of internal waves, with changes in spectral slope rather than spectral gaps marking shifts in dominant dynamics. Doppler shifting smears internal wave and turbulent frequencies in Eulerian measurements, but in Lagrangian frames they naturally separate at the buoyancy frequency, N, (D'Asaro and Lien, 2000a,b), with internal waves at lower and turbulence at higher frequency (Figure 6.7). Owing to these difficulties, application of the Reynolds decomposition to stratified profiles is more often a pedagogical tool rather than a procedure justified by data. For instance, the Reynolds decomposition applied by Osborn and Cox (1972) includes everything not turbulent in the mean, including internal waves and mesoscale eddies.

3.3.3 Turbulent Kinetic Energy

Applying the Reynolds decomposition to velocity and density in the momentum equation, averaging, and subtracting the average from the full equation yields a relation for fluctuations of specific momentum, e.g. $\partial u'/\partial t$. Multiplying by u' and averaging gives the equation for specific turbulent kinetic energy, $tke \equiv \frac{1}{2}\overline{u'_j u'_j}$, which in component form is

$$\frac{\partial \frac{1}{2}\overline{u'_j u'_j}}{\partial t} + \frac{\partial}{\partial x_i}\left[\overline{u'_i\left(\frac{p'}{\rho_0} + \frac{1}{2}u'_j u'_j\right)} + \overline{\nu\left(\frac{\partial u'_i}{\partial x_j} + \frac{\partial u'_j}{\partial x_i}\right)u'_j}\right]$$

$$= \underbrace{-\overline{u'_i u'_j}\,\frac{\partial \overline{u_i}}{\partial x_j}}_{\text{Shear Prod.}} \underbrace{-\overline{\rho' u'_3}\,\frac{g}{\rho_0}}_{J_b} \underbrace{- \nu\overline{\left(\frac{\partial u'_i}{\partial x_j} + \frac{\partial u'_j}{\partial x_i}\right)\frac{\partial u'_i}{\partial x_j}}}_{\epsilon} \quad \left[\frac{\mathrm{W}}{\mathrm{kg}}\right]. \tag{3.14}$$

Turbulence is produced by the Reynolds stress, $\overline{u'_i u'_j}$, working against mean shear, $\partial \overline{u_i}/\partial x_j$. In stratified profiles the buoyancy flux, J_b, is negative, extracting tke, but during convection it is positive, increasing tke. Always positive, the dissipation rate, ϵ, is a sink. Usually negligible, divergence terms express the rate of work by the total dynamic pressure and viscous stresses.

For a simple divergenceless shear flow (Figure 2.18),

$$\frac{\partial\, tke}{\partial t} = -\overline{u'w'}\,\frac{\partial \overline{u}}{\partial z} + J_b - \epsilon \quad \left[\frac{\mathrm{W}}{\mathrm{kg}}\right]. \tag{3.15}$$

Work by turbulent stress against mean shear increases tke by extracting kinetic energy from the mean flow. For example, upward vertical velocity in a positive mean shear produces a deficit in horizontal velocity, making $-\overline{u'w'}\ \partial\overline{u}/\partial z > 0$. Downward motion generates positive u', again making production positive. For negative mean shear, changes in signs combine, demonstrating that Reynolds stress again increases tke. Consequently, on average, shear production is positive.

Separating (3.15) into orthogonal components reveals that tke production occurs only in the streamwise component (Stewart, 1959),

$$\text{streamwise} \quad \frac{\partial \overline{\frac{1}{2}(u')^2}}{\partial t} = \overline{\frac{p'}{\rho_0}\frac{\partial u'}{\partial x}} - \epsilon_x - \overline{u'w'}\,\frac{\partial \overline{u}}{\partial z} \tag{3.16}$$

$$\text{spanwise} \quad \frac{\partial \overline{\frac{1}{2}(v')^2}}{\partial t} = \overline{\frac{p'}{\rho_0}\frac{\partial v'}{\partial y}} - \epsilon_y \tag{3.17}$$

$$\text{vertical} \quad \frac{\partial \overline{\frac{1}{2}(w')^2}}{\partial t} = \overline{\frac{p'}{\rho_0}\frac{\partial w'}{\partial z}} - \epsilon_z + J_b. \tag{3.18}$$

Initially, u' is much larger than the other components, but the pressure-shear correlation, $\overline{(p'/\rho_0)\partial u'/\partial z}$, is usually negative, transferring tke to the other components. This comes from convergences in streamwise fluctuations, $\partial u'/\partial z < 0$, increasing pressure fluctuations, i.e. $p' > 0$, and divergences decreasing p'. Applying continuity,

$$\overline{(p'/\rho_0)\partial u'/\partial z} = -\overline{(p'/\rho_0)\partial v'/\partial z} - \overline{(p'/\rho_0)\partial w'/\partial z}. \tag{3.19}$$

The transfer acts to equalize energy among components, i.e. to make the turbulence isotropic; should fluctuations make $\overline{(v')^2}$ or $\overline{(w')^2}$ dominant, the same mechanism will transfer energy from them. Dissipation occurs in all components, but vertical turbulence is suppressed most strongly, owing to the additional negative term, J_b.

Direct numerical simulations of a Kelvin–Helmholtz instability demonstrate the tendency toward isotropy (Figure 3.1), as also found in grid turbulence (Liu, 1995). Streamwise and vertical kinetic energies grow rapidly in the initial overturning billows. The spanwise component is energized only when the other components are near their maxima and the billows are close to collapse. As collapse begins,

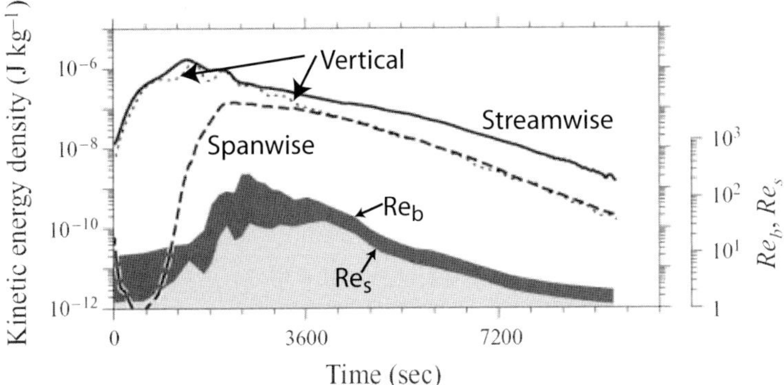

Figure 3.1 Evolution of *tke* components in a direct numerical simulation of a Kelvin–Helmholtz instability with $Ri_0 = 0.08$ and $Re_0 = 1965$. During growth, the streamwise component slightly exceeded the vertical component, and during decay, it remained larger than streamwise and vertical components. (Adapted from Smyth, 1999)

all three components are nearly equal, but soon vertical and spanwise energy lag the streamwise component, owing to continued Reynolds stress production during decay. Dissipation in the streamwise component, ϵ_x, decreases production by draining energy from $(u')^2$ (Stewart, 1959).

3.3.4 Viscous Dissipation Rate

The full form of the dissipation rate contains 12 terms (3.20). Three (1, 5, 9) are shear variances in the direction of the fluctuation, six (2, 3, 4, 6, 7, 8) are variances of cross-shears, and the remaining three are products of cross-shears (Browne et al., 1987). For a mean flow in the x direction, u', v', and w' are streamwise, spanwise, and vertical components. For a horizontal tow

$$\epsilon = \nu \Bigg[\underbrace{2\overline{\left(\frac{\partial u'}{\partial x}\right)^2}}_{1} + \underbrace{\overline{\left(\frac{\partial v'}{\partial x}\right)^2}}_{2} + \underbrace{\overline{\left(\frac{\partial w'}{\partial x}\right)^2}}_{3} + \underbrace{\overline{\left(\frac{\partial u'}{\partial y}\right)^2}}_{4} + \underbrace{2\overline{\left(\frac{\partial v'}{\partial y}\right)^2}}_{5}$$
$$+ \underbrace{\overline{\left(\frac{\partial w'}{\partial y}\right)^2}}_{6} + \underbrace{\overline{\left(\frac{\partial u'}{\partial z}\right)^2}}_{7} + \underbrace{\overline{\left(\frac{\partial v'}{\partial z}\right)^2}}_{8} + \underbrace{2\overline{\left(\frac{\partial w'}{\partial z}\right)^2}}_{9}$$
$$+ \underbrace{2\overline{\left(\frac{\partial u'}{\partial y}\right)\left(\frac{\partial v'}{\partial x}\right)}}_{10} + \underbrace{2\overline{\left(\frac{\partial u'}{\partial z}\right)\left(\frac{\partial w'}{\partial x}\right)}}_{11} + \underbrace{2\overline{\left(\frac{\partial v'}{\partial z}\right)\left(\frac{\partial w'}{\partial y}\right)}}_{12} \Bigg] \quad (3.20)$$

aligned with the mean flow, u' is axial, while v' and w' are normal. For a profile, w' is axial, and u' and v' are normal. Hot-films and pitot tubes measure axial flow, and airfoils detect normal components (Chapter 5).

When dissipation scales are isotropic, terms 1–9 in (3.20) are equal; 10–12 equal each other, are negative, and are half of the other terms (Libby, 1996, section 4.7). This yields the dissipation rate from one component as

$$\epsilon_{\text{isotropic}} = \underbrace{15\nu\,\overline{(\partial u'/\partial x)^2}}_{\text{axial}} = \underbrace{(15/2)\nu\,\overline{(\partial u'/\partial z)^2}}_{\text{normal}} \quad \left[\mathrm{W\,kg^{-1}}\right]. \tag{3.21}$$

Grant et al. (1962) applied the axial form to hot-film measurements on a towed body; Osborn (1980) used the normal expression with airfoils on profilers. Working on a submersible, Gargett et al. (1984) tested isotropy by comparing normal spectra from airfoils and axial spectra from a hot-film.

3.4 Scalar Variances

Turbulence in stratified water produces fluctuations in temperature, salinity, and density that may affect dynamics by their buoyancy fluxes. Equations for scalar variances are developed in the same way as those for turbulent kinetic energy. Applying the Reynolds decomposition, $\Theta = \overline{\Theta} + \Theta'$, to conservative temperature in the heat equation, averaging, and applying continuity gives the mean heat equation, divided by ρc_p, as

$$\frac{\partial \overline{\Theta}}{\partial t} + \overline{\boldsymbol{u}} \cdot \nabla\overline{\Theta} = -\nabla \cdot \overline{\mathbf{u}'\Theta'} + \nabla \cdot \left(\kappa_T \nabla\overline{\Theta}\right) \quad \left[\mathrm{K\,s^{-1}}\right]. \tag{3.22}$$

The terms on the right are divergences of turbulent and molecular fluxes. Subtracting (3.22) from the full Reynolds Θ equation, multiplying by Θ', averaging, and rearranging yields

$$\frac{\partial \overline{\Theta'^2}}{\partial t} + \nabla \cdot \left[\overline{\mathbf{u}'\Theta'^2} - \kappa_T \nabla\overline{\Theta'^2}\right] = -2\overline{\mathbf{u}'\Theta'} \cdot \nabla\overline{\Theta} - \chi_{\mathrm{T}} \quad \left[\mathrm{K^2\,s^{-1}}\right], \tag{3.23}$$

which is the turbulent heat equation divided by ρc_p. As with the *tke* equation, the divergence term does not include transport by the mean flow.

Divergence terms in (3.23) express transport of thermal variance by turbulence and molecular diffusion. The first term on the right is $\overline{\Theta'^2}$ production, analogous to the Reynolds stress in (3.14). Temperature fluctuations result from water parcels being displaced along mean gradients, usually vertically, $-\overline{w'\Theta'}\,\partial\overline{\Theta}/\partial z$. Production tends to be positive, increasing $\partial\overline{(\Theta')^2}/\partial t$. For example, with $\partial\overline{\Theta}/\partial z > 0$, $w' > 0$ produces $\Theta' < 0$, making the term, including the negative sign, positive, increasing $\overline{\Theta'^2}$.

The last term in (3.23) is the rate of diffusive smoothing of temperature fluctuations, a.k.a. the thermal dissipation rate,

$$\chi_{\mathrm{T}} \equiv 2\kappa_T \left(\overline{(\partial\Theta'/\partial x)^2} + \overline{(\partial\Theta'/\partial y)^2} + \overline{(\partial\Theta'/\partial z)^2} \right) \quad [\mathrm{K}^2\ \mathrm{s}^{-1}], \tag{3.24}$$

The three components are equal when diffusive-scale temperature gradients are isotropic. In practice, only the component along the probe direction can be measured. It is multiplied by 3, assuming isotropy.

Applying the same procedure to the salt balance,

$$\frac{\partial \overline{S_A{}'^2}}{\partial t} + \nabla \cdot \left[\overline{\mathbf{u}}\overline{S_A{}'^2} + \overline{\mathbf{u}' S_A{}'^2} - \kappa_S \nabla \overline{S_A{}'^2} \right] = -2\overline{\mathbf{u}' S_A{}'} \cdot \nabla \overline{S_A} - \chi_{\mathrm{S}}\,, \tag{3.25}$$

leads to terms analogous to those in the thermal variance equation.

Representing density as a linear sum of the two constituents,

$$\rho' = -\alpha^{\Theta}\Theta' + \beta^{\Theta} S_A', \tag{3.26}$$

leads to

$$\begin{aligned} &\frac{\partial \overline{\rho'^2}}{\partial t} + \nabla \cdot \Big[\overline{\boldsymbol{u}}\overline{\rho'^2} + \overline{\boldsymbol{u}'\rho'^2} - \rho_0^2(\beta^{\Theta})^2 \kappa_S \nabla \overline{(S_A')^2} - \rho_0^2(\alpha^{\Theta})^2 \kappa_T \nabla \overline{(\Theta')^2} \\ &\quad + 2\rho_0^2 \alpha^{\Theta} \beta^{\Theta} \left(\kappa_T \overline{s' \nabla \Theta'} + \kappa_S \overline{\Theta' \nabla s'} \right) \Big] \\ &= -2\overline{\rho' \boldsymbol{u}'} \cdot \nabla \overline{\rho} + 2\rho_0^2 \alpha^{\Theta} \beta^{\Theta} \left(\kappa_T + \kappa_S \right) \overline{\nabla S_A' \cdot \nabla \theta'} - \chi_{\rho}. \end{aligned} \tag{3.27}$$

The terms are equivalent to those in the temperature and salinity variance equations, except for cross-terms involving molecular diffusion, which may be insignificant because major contributions to $\nabla S_A'$ should occur at smaller scales than those dominating $\nabla\Theta'$. No measurements, however, are available for testing. The density dissipation rate is

$$\chi_{\rho} \equiv \rho_0^2 (\alpha^{\Theta})^2 \chi_{\mathrm{T}} + \rho_0^2 (\beta^{\Theta})^2 \chi_{\mathrm{S}} \quad \left[(\mathrm{kg/m^2})^2\ \mathrm{s}^{-1} \right]. \tag{3.28}$$

Because $\alpha^{\Theta}/\beta^{\Theta} \approx 0.02$ at polar temperatures (Fig 2.15), thermal dissipation may make a negligible contribution to χ_{ρ} at shallow depths at high latitudes.

3.4.1 Potential Energy

Potential energy can be expressed as the sum of reference, a.k.a. background, and available, *ape*, components (Section 2.7). Neglecting compressibility, molecular diffusion, and transport by the mean flow, (3.27) leads to

$$\frac{\partial ape}{\partial t} + \nabla \cdot \overline{\boldsymbol{u}'\, ape} = -J_b - \chi_{ape} \quad \left[\mathrm{W\ kg^{-1}} \right]. \tag{3.29}$$

Negative buoyancy fluxes produced by stratified turbulence increase *ape*, while positive buoyancy fluxes during convection decrease it. Molecular diffusion, parameterized by

$$\chi_{ape} \equiv (g/\rho_0 N)^2 \chi_\rho \quad [\mathrm{W\,kg^{-1}}], \tag{3.30}$$

always drains *ape*.

3.4.2 Buoyancy Flux

An equation for the buoyancy flux, $J_b \equiv -(g/\rho_0)\overline{\rho' w'}$, is formed by cross-multiplying ρ' and w' equations and averaging,

$$\underbrace{\frac{\partial}{\partial t}\overline{\rho' w'}}_{(1)} + \underbrace{\overline{w'^2}\frac{d\bar{\rho}}{dz}}_{(2)} + \underbrace{\frac{d}{dz}\overline{\rho' w'^2}}_{(3)} + \underbrace{\overline{\rho' w'}\frac{d\bar{u}}{dz}}_{(4)} = -\underbrace{\frac{1}{\rho_0}\overline{\rho'\frac{\partial p'}{\partial z}}}_{(5)} - \underbrace{\frac{g}{\rho_0}\overline{\rho'^2}}_{(6)}. \tag{3.31}$$

Jones and Musonge (1988, eq. 3) give the three-dimensional form of this equation without buoyancy forces, which omits the last term. Rewriting in terms of the specific buoyancy force (2.45) and of *ape* (2.57),

$$\underbrace{\frac{\partial J_b}{\partial t}}_{(1)} + \underbrace{\overline{w'^2}N^2}_{(2)} + \underbrace{\frac{d}{dz}\overline{b' w'^2}}_{(3)} + \underbrace{J_b\frac{d}{dz}\bar{u}}_{(4)} = -\underbrace{\frac{1}{\rho_0}\overline{b'\frac{\partial p'}{\partial z}}}_{(5)} + \underbrace{2N^2 APE}_{(6)} \quad \left[\frac{\mathrm{W\,kg^{-1}}}{\mathrm{s}}\right]. \tag{3.32}$$

Term 2 expresses production, driving the buoyancy flux negative in stratified profiles and positive in unstable situations. The vertical divergence of the buoyancy flux, term 3, is generally considered of first order and decreases J_b when positive. It is sometimes represented in models by $\overline{b' w'^2} = -K_\rho d(\overline{b' w'})/dz$. Positive shear (term 4) drives buoyancy fluxes toward zero during stratified turbulence and convection. Sometimes referred to as 'pressure scrambling' (Jones and Musonge, 1988), term 5 expresses the effect of pressure fluctuations in redistributing buoyancy. Finally, restratification, term 6, always decreases the magnitude of the buoyancy flux during stratified turbulence. As noted by Stewart (1959), buoyancy flux and *ape* are out of phase and induce oscillations as stratified turbulence decays: from (3.29) a negative buoyancy flux increases *ape*, and from (3.32) increased *ape* drives the negative buoyancy toward zero. These oscillations have been observed in wakes of stratified grid flows, e.g. Stillinger et al. (1983).

3.5 Production

Shear instability, convective instability, surface stress, and free convection are the primary mechanisms producing turbulence in the ocean. Details vary, e.g. whether internal waves or density currents drive shear instabilities, but common aspects often form the basis for numerical simulations. How much and when details matter is an issue. In the open-ocean pycnocline, dissipation rates produced by breaking internal waves can be predicted from wave energetics (Section 7.7), removing the need to understand details. Those details, however, matter in other situations or when translating dissipation rates to diapycnal diffusivities, particularly for mixing at low Reynolds numbers. Shear production was discussed briefly in Section 3.3.3 and is considered in more detail in Section 7.7. Here, two simple cases from boundary layers are discussed because the approaches have been carried over into stratified situations.

3.5.1 Surface Stress

Steady flow over flat surfaces can homogenize boundary layers, similar to flows over rough plates (Kundu and Cohen, 2004, section 13.11). On the seafloor, stress is transmitted by pressure drag on roughness ranging from fine sediment to boulders. At the sea surface, ripples and waves supply the roughness. Above the roughness, a constant-stress layer forms with velocity as a function of height, z, and the friction velocity

$$u_* \equiv \sqrt{\tau_0/\rho} = \left(\overline{-u'w'}\right)^{1/2} \quad [\mathrm{m\,s^{-1}}], \tag{3.33}$$

with τ_0 as the surface stress. In laterally homogenous constant-stress layers, the *tke* equation reduces to shear stress balancing viscous dissipation,

$$0 = \overline{u'w'}\,\frac{\partial \overline{u}}{\partial z} + \epsilon. \tag{3.34}$$

From dimensional analysis, the mean shear is $\partial \overline{u}/\partial z = u_*/k_v z$, where von Karman's constant, $k_v \sim 0.4$, is included for historical continuity, and z is height above the surface. The inverse dependence on height leads to a logarithmic mean velocity profile, in which assuming that diameters of turbulent eddies scale with height returns Taylor scaling,

$$\overline{u} = (u_*/k_v)\ln z + \mathrm{cst}, \quad \epsilon = u_*^3/k_v z \quad [\mathrm{W\,kg^{-1}}]. \tag{3.35}$$

Repeated profiling in a turbulent tidal channel with airfoils mounted on a vorticity meter confirmed the balance between local production and dissipation in the bottom half of the channel (Figure 3.2).

3.5.2 Free Convection

Convection driven by geothermal heat fluxes through the seafloor and surface buoyancy losses produce the thickest mixed layers in the ocean, sometimes extending

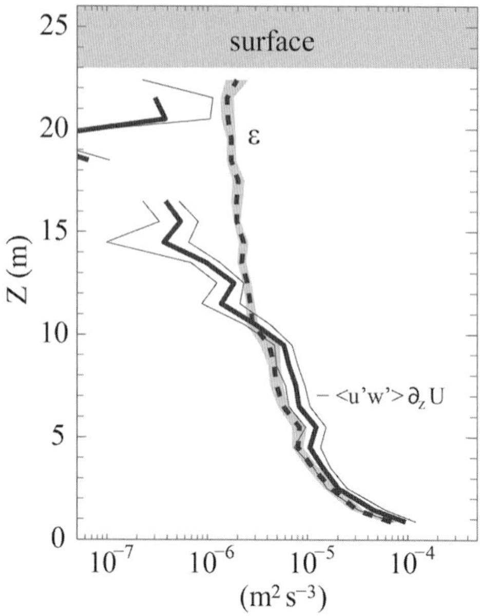

Figure 3.2 Dissipation and shear production rates in a turbulent tidal channel. Dissipation was measured with airfoils. Shear production was estimated from the mean shear and the mean momentum flux measured with an electromagnetic vorticity meter (Section 5.3.1). Dissipation balanced production (3.34) within 10 m of the bottom. Shading and thin lines are 95% confidence limits. There were two log layers: from 4.5 to 12 m, with $u_* = 43 \pm 4$ mm s^{-1}, and below 3 m, with $u_* = 24 \pm 6$ mm s^{-1}. (From Sanford and Lien, 1999)

hundreds or even thousands of meters from the boundary. For pure, steady convection, the *tke* equation (3.14) reduces to

$$0 = J_b - \epsilon. \tag{3.36}$$

Positive during convection and often $\sim 10^{-8}$ to $\sim 10^{-7}$ W kg^{-1} at the sea surface during nights at mid-latitude, the surface buoyancy flux, J_b^0, establishes a nearly constant dissipation rate throughout convecting layers (Figure 3.3); the length scale increases with distance from the boundary and the velocity scale varies as $u' \sim (J_b{}^0 z)^{1/3}$, so that Taylor scaling is $\epsilon \sim u'^3/z \sim J_b{}^0$.

Free convection regimes often exist below/above constant-stress regimes in surface/bottom mixed layers. Because dissipation rates are additive, total dissipation can be summed as

$$\epsilon(z) = 0.58 J_b^0 + 1.76 u_*^3/(k_v z) \quad \left[\text{W kg}^{-1}\right] \tag{3.37}$$

(Lombardo and Gregg, 1989). Breaking surface waves produce additional dissipation at the top of the layer (Gregg, 1987; Annis and Moum, 1995).

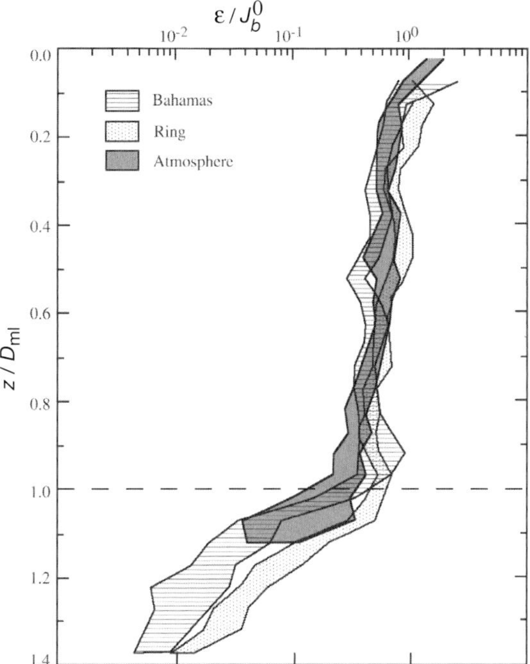

Figure 3.3 Dissipation rates, ϵ, normalized by the surface buoyancy flux, J_b^0, versus normalized mixed layer depth, z/D_{ml}. Free convection was the dominant process, except close to the surface, where stress production produced the semi-logarithmic decrease in dissipation with depth. Ocean data were from the Bahamas and a warm-core ring. Atmospheric observations were over land in Minnesota and the United Kingdom. (Adapted from Shay and Gregg, 1986. © American Meteorological Society. Used with permission)

3.6 The Kolmogorov Energy Cascade

Kolmogorov (1941) hypothesized that steady turbulence has universal statistical properties at lengths much smaller than generation scales. Inherently nonlinear, inertial transfer, $\boldsymbol{u}' \cdot \nabla \boldsymbol{u}'$, produces a net energy flux toward smaller scales, each stage progressively decoupling from larger scales. Kolmogorov proposed that after several decades of scale transfer, the turbulence is in local statistical equilibrium, with its properties dominated by inertial transfer, giving the turbulence a universal character. At steady state, the rate of energy transfer through the spectrum equals the rate at which it is generated and dissipated, making the dissipation rate, ϵ, the primary parameter throughout the equilibrium range. When shear variance, $(\nabla \boldsymbol{u}'(l))^2$, at small length scale l is large, kinematic viscosity, ν, becomes a second parameter governing turbulence in what is termed the 'viscous subrange'. This section explores the extension and verification of Kolmogorov's ideas to

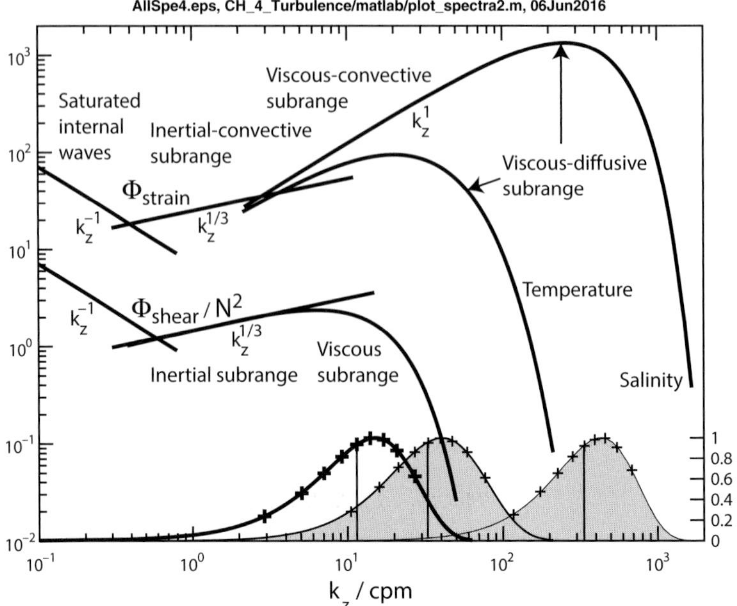

Figure 3.4 One-dimensional spectra of shear and scalar strain for $\epsilon = 1 \times 10^{-8}$ W kg^{-1} and $N = 0.052$ s^{-1}, corresponding to $K_\rho = 7.3 \times 10^{-5}$ m^2 s^{-1} and $Re_b = 364$. The Panchev and Kesich (1969) shear spectrum is normalized by N^2, and the Kraichnan (1968) scalar gradient spectra (with $q_k = 7.3$) by dT/dz^2 and dS/dz^2 plus an arbitrary offset. At the bottom, normalized variance-preserving spectra are marked at intervals of 10% with vertical lines at 50%.

understand the sequence of domains within the equilibrium range and its equivalent for scalars, both illustrated as a function of wavenumber in Figure 3.4, with variance-preserving forms at the bottom of the plot.[1]

3.6.1 The Universal Equilibrium Range

Kolmogorov (1941) formalized his insights with two hypotheses that were updated when Kolmogorov (1962) added another. Paraphrasing:

1. For length scales $l \ll l_e$, where l_e is the energy-containing scale, probability distributions of normalized velocity and time differences over separations l depend only on Re and are the same in all turbulent flows.
2. When $Re \gg 1$, these distributions are independent of Re.

[1] To have areas on a plot versus a logarithmic x-axis show equal contributions to the variance, the variance-preserving spectrum of $\Phi(k)$ is constructed by plotting $k\Phi(k)$ on a linear y-axis versus $\log k$ on the x-axis, reflecting $\int k_z \Phi(k_z)\, d\ln k_z = \int \Phi(k_z)\, dk_z$.

3. Probability distributions of velocity differences tend to be lognormally distributed because they are products of velocity differences over progressively larger scales.

Assuming $l \ll l_e$ insures that inertial transfer has operated through a sequence of progressively smaller eddies to statistically decouple equilibrium scales l from production. Owing to the universal character of inertial interactions, the statistical character of turbulence is the same, whether generated in a convecting surface layer or in a large overturn in the pycnocline. In practice, spectra close to universal forms are observed at lengths within a factor of two or so of energy-containing scales. Consistent with being decoupled from production scales, the equilibrium range tends to be isotropic due to the tendency of the pressure-shear correlation to move energy from more to less energetic components (Section 3.3.3).

The Inertial Subrange

Because inertial transfer dominates the mid-range of equilibrium length scales, the spectrum of three-dimensional kinetic energy, $E_{ke}(k)$, depends only on ϵ and the three-dimensional scalar wavenumber, $k \equiv (k_1^2 + k_2^2 + k_3^2)^{1/2}$. Dimensional analysis of $E_{ke} \propto (\mathrm{m/s})^2/\mathrm{m}^{-1} \propto \epsilon^c k^d$ yields

$$E_{ke}(k) = a\epsilon^{2/3}k^{-5/3} \quad \left[\mathrm{m^2\ s^{-2}/m^{-1} = J\ kg^{-1}/m^{-1}}\right], \tag{3.38}$$

in radians per meter, m^{-1}, with universal constant a. When energy is distributed isotropically, contours of equal energy are spheres about the origin in three-dimensional wavenumber space. When turbulence occurs in stratified fluids, it is also useful to distinguish between vertical, k_z, and horizontal, k_h, wavenumbers, where $k = (k_2^z + k_h^2)^{1/2}$.

One-dimensional versions, needed to compare with observations, are obtained by assuming that velocity is isotropic and solenoidal, i.e. with $\nabla \cdot \boldsymbol{u} = 0$. For a probe aligned with the x_1-axis, one-dimensional spectra are

$$\Phi_{ke}^{\mathrm{axial}}(k_1) = \int_{k_1}^{\infty} \left(1 - \frac{k_1^2}{k^2}\right) \frac{E_{\mathrm{ke}}(k)}{k}\, dk = \frac{18}{55} a\epsilon^{2/3} k_1^{-5/3} \tag{3.39}$$

$$\Phi_{ke}^{\mathrm{normal}}(k_1) = \frac{1}{2}\int_{k_1}^{\infty} \left(1 + \frac{k_1^2}{k^2}\right) \frac{E_{\mathrm{ke}}(k)}{k}\, dk = \frac{4}{3}\Phi_{ke}^{\mathrm{axial}}(k_1) \left[\frac{(\mathrm{m/s})^2}{\mathrm{m}^{-1}}\right] \tag{3.40}$$

(Monin and Yaglom, 1975, section 12.3). Grant et al. (1962) verified (3.39) with a hot-film on a streamlined body towed through a 100 m-deep tidal channel. The current, estimated at 1 $\mathrm{m\ s^{-1}}$ near the end of a strong tide, was presumed to have previously homogenized the water column. The large Reynolds number, $Re \sim 1\ \mathrm{m\ s^{-1}} \times 10^2\ \mathrm{m}/10^{-6}\ \mathrm{m^2\ s^{-1}} = 10^8$, provided universal conditions to test

Kolmogorov's predictions. Subsequent laboratory and geophysical observations give $a = 1.53$ (Sreenivasan, 1995).

Observations are often presented as one-dimensional shear spectra,

$$\Phi_{shear}(k_1) = \mathcal{F}\left(\frac{dv}{dx_1}\right)\left(\mathcal{F}^*\frac{dv}{dx_1}\right) = (ik_1)\mathcal{F}(v)(-ik_1)\mathcal{F}^*(v) = k^2\Phi_{vel}(k_1), \tag{3.41}$$

where $\mathcal{F}$ is a Fourier transform, $*$ indicates complex conjugate, i is the imaginary unit, and the transform of a one-dimensional derivative is ik_1 times the transform of the variable. In the inertial subrange, one-dimensional Kolmogorov shear spectra are

$$\Phi^{\text{axial}}_{shear}(k_1) = k_1^2\Phi^{\text{axial}}_{vel}(k_1) = (18/55)a\epsilon^{2/3}k_1^{1/3} \tag{3.42}$$

$$\Phi^{\text{normal}}_{shear}(k_1) = k_1^2\Phi^{\text{normal}}_{vel}(k_1) = (4/3)\Phi^{\text{axial}}_{shear}(k_1) \quad [\text{s}^{-2}/\text{m}^{-1}]. \tag{3.43}$$

To convert shear spectra from radians per meter (m^{-1}) to cycles per meter (cpm), $\Phi^{\text{cpm}}_{shear}(k_1) = (2\pi k_1)^2\Phi^{\text{rad}}_{vel}(k_1)$. Strong turbulence in stratified profiles can approach universal behavior when Reynolds numbers are large. In Figure 3.5 the inertial subrange extends over at least a decade and a half.

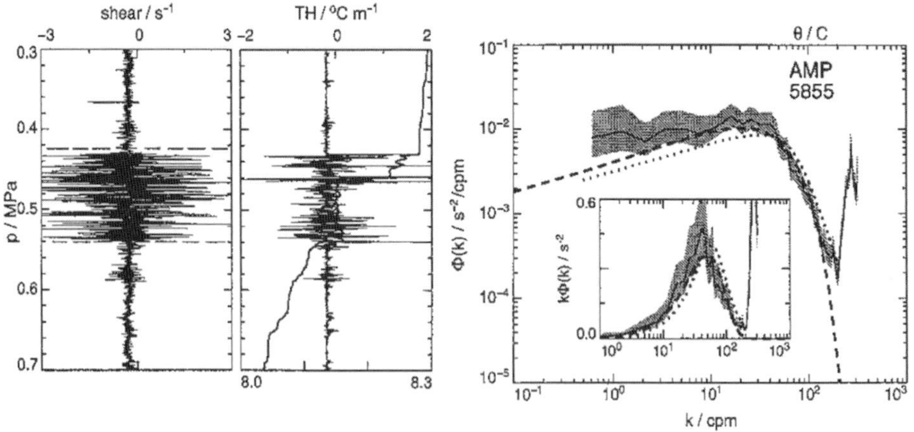

Figure 3.5 Airfoil (left) and thermistor (middle) profiles for the overturning billows in Figures 1.2 and 1.4. The Reynolds number was 3×10^6. Dashed lines show the depth range of the spectrum. (Right) Shear spectrum with 95% confidence limits and universal shear spectra for $\epsilon = 7.7 \times 10^{-6}$ W kg^{-1} by Panchev and Kesich (1969) (dashed) and Nasmyth (1970) (dotted). The sharp rise at 200 cpm resulted from amplification of noise by filters correcting airfoil attenuation. The insert has the spectra in variance-preserving form. (Adapted from Seim and Gregg, 1994)

Lagrangian Form

Dissipation rates and inertial subrange spectra have also been obtained from Lagrangian floats (Section 5.4.4), as spectra of float acceleration and by attaching acoustic Doppler velocimeters (ADVs) (Lien and D'Asaro, 2006). Dimensional analysis by Corrsin (1963) and Tennekes and Lumley (1972) was used to predict the inertial subrange spectrum of the float as

$$\Phi_{\text{accel}}(\omega) = \beta_{\text{accel}}\,\epsilon \quad \left[(\text{m/s}^2)^2/\text{s}^{-1}\right]. \tag{3.44}$$

The low end is defined by ω_0, the large-eddy frequency, and the high end by the size of the float. Spectra of acceleration in a turbulent tidal channel, calculated by twice-differentiating pressure, confirm the prediction, with a flat inertial subrange for the higher dissipation rates followed by a high-frequency rolloff produced by the response being an average over the float's length (Figure 3.6, left). As dissipation rates decrease, only the rolloff remains. Observations indicate $\omega_0 \approx N/2$ and $\beta_{\text{accel}} = 1.9 \pm 0.1$ for $Re_\lambda > 100$ (Lien and D'Asaro, 2002). To obtain independent ϵ estimates, Lien and D'Asaro (2006) invoked Taylor's hypothesis to translate time series of vector velocity in the small volume measured by the ADV to scalar

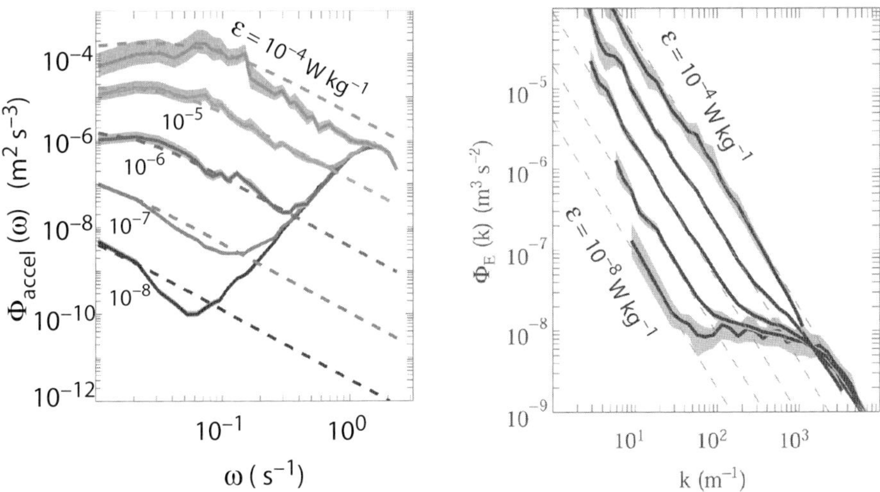

Figure 3.6 (Left) Acceleration spectra of a Lagrangian float in Puget Sound and Tacoma Narrows. Set by the float length, the rolloff from the flat inertial subrange shifted to lower frequency as spectral amplitude decreased. The steep rise at high frequency was due to noise. (Right) Spectra of total kinetic energy from the ADV versus three-dimensional wavenumber along progressive vector velocity displacements. Dashed lines are $k^{-5/3}$ inertial subranges for different ϵ. (Adapted from Lien and D'Asaro, 2006. © American Meteorological Society. Used with permission)

wavenumber along progressive displacement vectors. The spectra exhibit $k^{-5/3}$ slopes characteristic of the inertial subrange (Figure 3.6, right).

The Viscous Subrange

In the inertial subrange, shear variance rises rapidly with increasing wavenumber (Figure 3.5, insert) until viscosity smoothes the fluctuations. The resulting exponential decrease in spectral amplitude reduces variance contributions until they are negligible at five times the peak wavenumber. Dimensional analysis using ϵ and ν gives Kolmogorov length, velocity, and time scales as

$$l_\eta = (\nu^3/\epsilon)^{1/4} \ [\mathrm{m}], \ v_\eta = (\nu\epsilon)^{1/4} \left[\mathrm{m\,s^{-1}}\right], \ \tau_\eta = (\nu/\epsilon)^{1/4} \left[\mathrm{s^{-1}}\right]. \qquad (3.45)$$

In radians, the Kolmogorov wavenumber is the inverse, $k_\eta = (\epsilon/\nu^3)^{1/4}$.

Using Taylor scaling (3.1) in l_η gives the ratio of energy-containing to viscous length scales as $l/l_\eta = Re^{3/4}$. Ratios vary from 177 for $Re = 10^3$ in moderately strong pycnocline patches to 10^6 for $Re = 10^8$ in the tidal channel sampled by Grant et al. (1962). When the wavenumber bandwidth is large, multiple sensors are needed to detect all scales (Chapter 5).

Because the shape of the spectrum in the viscous subrange is not included in Kolmogorov's theory, the spectrum measured by Grant et al. (1962) at high Reynolds number was normalized by Kolmogorov parameters and used as a reference for many years (Nasmyth, 1970; Oakey, 1982). Subsequently, Wesson and Gregg (1994) showed that the theoretical form by Panchev and Kesich (1969) is equivalent to the Nasmyth spectrum for moderate-to-strong turbulence (Figs. 3.49 and 3.7a). Twenty percent of the cumulative variance is reached at $k/k_\eta = 0.2$, and nearly all is included by $k/k_\eta = 1$ (Figure 3.7a).

Spectra from turbulent patches appear as high-wavenumber extensions of spectra dominated by internal waves (Figure 7.23). The transition occurs at the scales of the largest overturns (Figure 7.50). For the weak turbulence produced by internal waves at the Garrett and Munk (1972) background, conditions are far from Kolmogorov's assumptions, and average spectra show only inflections at turbulent scales (Figure 7.48). Similarly, Sanchez et al. (2011) found excellent agreement with the universal form in only 69 of 1,000 spectra of turbulence in a lake.

Owing to the very large range of dissipation rates in the ocean, the viscous subrange shifts over three decades of wavenumber (Figure 3.8), with significant variance between $\sim$0.2 cpm and $\sim$6,000 cpm. The lowest wavenumbers are often contaminated by vehicle motion. Because present probes cannot resolve fluctuations much smaller than 10 mm (100 cpm), Wesson and Gregg (1994) used the Panchev and Kesich (1969) spectrum to extrapolate spectra to higher wavenumbers, arguing that uncertainties are less than would result from ignoring very high dissipation rates. The corrections can be large, e.g. 19.6 when $\epsilon = 10^{-1}\ \mathrm{W\,kg^{-1}}$.

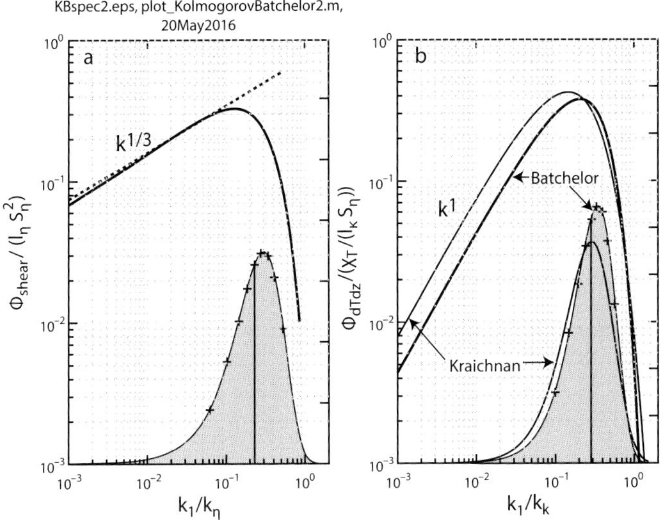

Figure 3.7 Non-dimensional spectra of (a) normal shear (Panchev and Kesich, 1969) and (b) temperature gradient (Batchelor, 1959; Kraichnan, 1968) versus one-dimensional wavenumber, k_1, normalized by Kolmogorov, k_η, and Batchelor, k_κ, wavenumbers (3.49). Pluses, '+' , on variance-preserving spectra are at intervals of 0.1 in fractional cumulative variance, with vertical lines at 0.5. Scalar spectra were evaluated with $q_B = 4.4$ and $q_K = 7.9$ (Sanchez et al., 2011). $S_\eta \equiv (\epsilon/\nu)^{1/2}$ is the Kolmogorov shear scale.

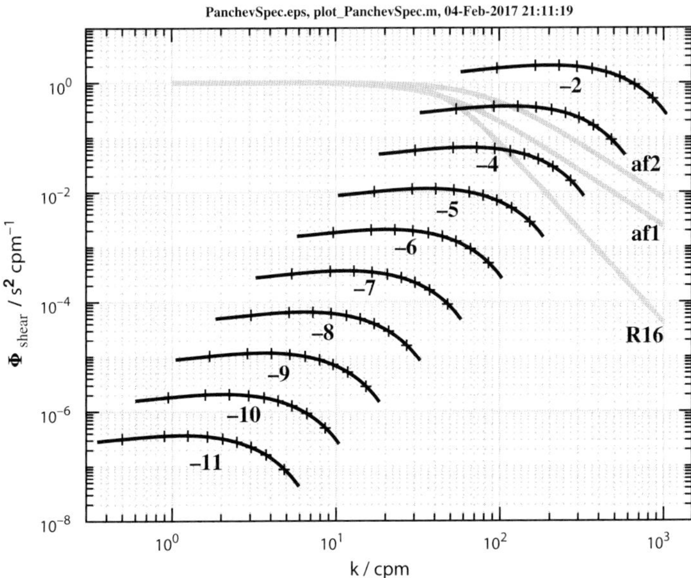

Figure 3.8 Panchev and Kesich (1969) turbulent shear spectra for most of the ϵ range in the ocean, as base-10 logarithms with units of W kg^{-1}. Each spectrum spans 5–95% of the variance contributing to ϵ, with ticks at multiples of 10%. Gray curves are spatial response functions for two airfoils (Macoun and Lueck, 2004) and the R16 Russian probe (Lai et al., 2000).

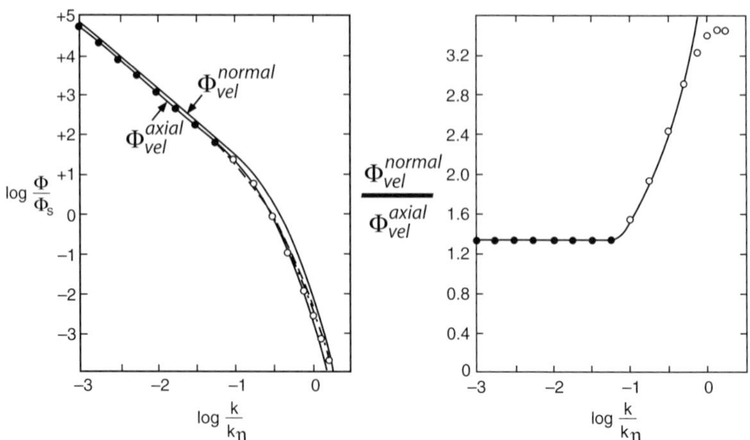

Figure 3.9 Universal velocity spectra for turbulence with large Re_b (lines) compared with observations, dots, from Knight Inlet. In the inertial subrange, the normal component is 4/3 times the axial component, as expected for Kolmogorov spectra. (Adapted from Gargett et al., 1984)

3.6.2 Small-Scale Isotropy

As observations revealed low mixing rates in much of the upper pycnocline, it was recognized that assuming isotropy was producing overestimates of ϵ for at least some data. Because diapycnal diffusivities were much lower than predicted by Munk (1966), this was not a problem until greater accuracy was desired and Oakey (1982) proposed using ratios of ϵ and χ_T to estimate Γ_{mix}. Gargett ct al. (1984) began the quantification of isotropy by operating a submersible near the Knight Inlet sill, with a hot-film and two airfoils to measure streamwise, spanwise, and vertical fluctuations. When $Reb \geq 4.3 \times 10^4$, spectra were consistent with isotropy in inertial and viscous subranges (Figure 3.9). Departures from dissipation-scale isotropy were found when $Re_b \lesssim 200$. Using airfoils sensing vertical and spanwise velocity on a research submarine off California, Yamazaki and Osborn (1990) applied the theory of axisymmetric turbulence (Batchelor, 1946), in this case about a vertical axis, to estimate departures from isotropy at low Reynolds numbers. They found ϵ estimates from the two components diverging as Re_b decreased below 100, with the ratio rising to $\sim$1.5 at $Re_b = 10$.

Using direct numerical simulations to examine isotropy in Kelvin–Helmholtz instabilities, Smyth and Moum (2000) confirmed isotropy for all shear components when $Re_b \gtrsim 10^2$ (Figure 3.10). Some components, however, are accurate to much lower buoyancy Reynolds numbers. For example, vertical profiles of shear in the spanwise direction, v'_z, have very small errors for Re_b as low as one. Other components yield slight underestimates at $Re_b = 10^2$, e.g. streamwise sampling of streamwise shear of vertical velocity, w'_x.

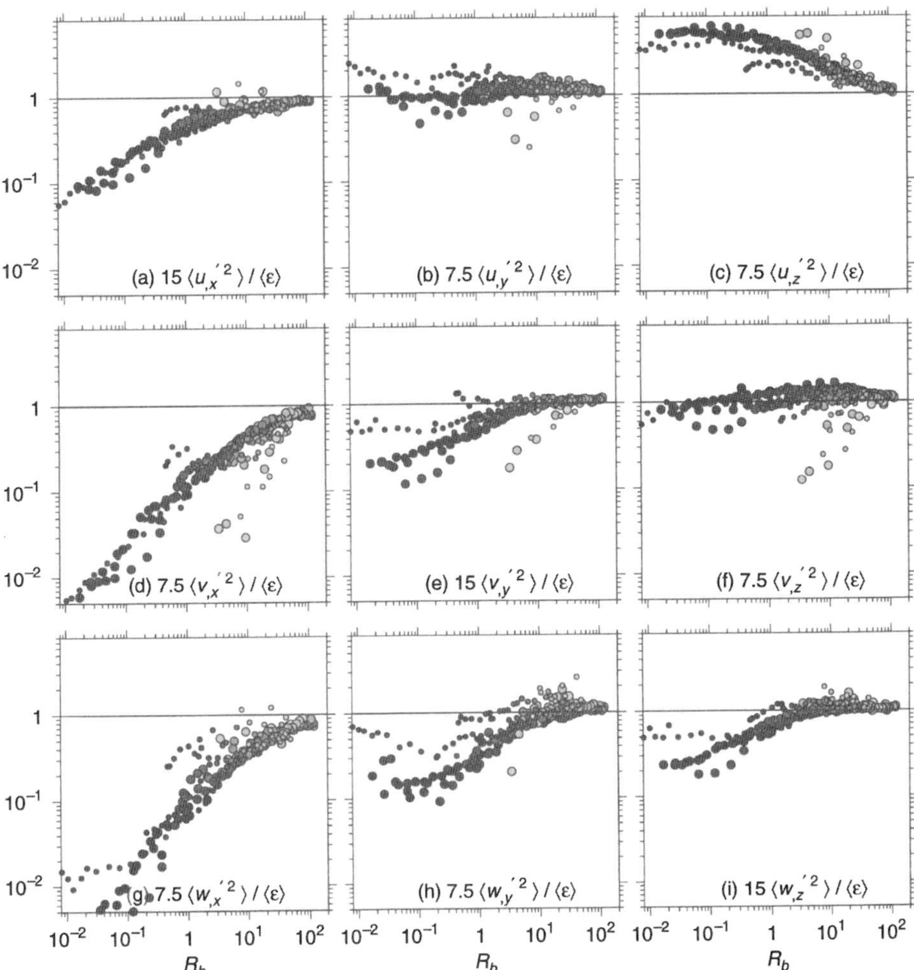

Figure 3.10 Ratios of true ϵ to estimates from one shear component scaled assuming isotropy. The comparisons are from direct numerical simulation of a Kelvin–Helmholtz instability with x and y the streamwise and spanwise directions. Comparisons are arranged by flow (rows) and measurement (columns) directions. For example, airfoils on profilers measure streamwise shear, u'_z (c), and spanwise shear , v'_z (f), and a hot-film or pitot tube measures vertical shear, w'_z (i). Relying on u'_z overestimates true ϵ for $Re_b \lesssim 10^2$; the v'_z estimate is accurate to very low Re_b; while w'_z estimates are accurate for $Re_b \gtrsim 3$. (Adapted from Smyth and Moum, 2000)

3.7 The Advective Scalar Cascade

When buoyancy forces are not significant in the inertial subrange, temperature and salinity fluctuations are passively advected without significantly affecting the turbulent strain creating them. The equilibrium range for scalars has three subranges. The largest scales are termed the ‘inertial-advective’, or ‘inertial-convective’, subrange because they are produced by inertial subrange strain distorting background scalar profiles. In this subrange, scalar spectra have the same form as the inertial subrange, $k^{-5/3}$ (Figure 3.4). As viscosity cuts off the strain spectrum to form the viscous subrange, the scalar spectrum develops the viscous-advective subrange, which overlaps the viscous subrange at low wavenumbers and extends to higher wavenumbers until cutoff in the diffusive subrange.

3.7.1 Inertial-Advective Subrange

Analogous to inertial transfer, in the inertial-advective subrange, scalar variance is transferred to smaller scales by the advective term, $\boldsymbol{u} \cdot \nabla T$ (Batchelor, 1959). At steady state, scalar dissipation rates χ_T and χ_S define the transfer rates of temperature and salinity variance through the spectrum, similar to the role of ϵ as the transfer rate of kinetic energy. Consequently, the temperature spectrum, $E_T(k) \propto$ K^2 s^{-1}, is proportional to χ_T, ϵ, and k, where K is degrees Kelvin. Dimensional analysis gives the three-dimensional temperature spectrum as

$$E_\text{T}(k) = c_T \chi_\text{T} \epsilon^{-1/3} k^{-5/3} \quad \left[\text{K}^2/\text{m}^{-1}\right], \tag{3.46}$$

where c_T is a constant and k is the three-dimensional wavenumber. Not surprising in view of the similarity in dynamics, the wavenumber dependence is the same as that of the inertial subrange. Assuming spectral isotropy and applying spectral relations from Monin and Yaglom (1975, section 12.1) leads to the one-dimensional version,

$$\Phi_T(k_1) = \int_{k_1}^{\infty} \frac{E_T(k)}{k}\, dk = (3/5) c_T \chi_\text{T} \epsilon^{-1/3} k_1^{-5/3} \quad [\text{K}^2\ \text{m}^{-1}]. \tag{3.47}$$

Known as the Obukhov-Corrsin constant, $c_T = (5/3)0.4$ (Sreenivasan, 1996).

Lagrangian Form

Dimensional analysis gives the Lagrangian frequency spectrum of the time derivative of potential density, σ_Θ, in the inertial-advective subrange as

$$\Phi_{D\sigma_\Theta/Dt}(\omega) = \beta_\sigma \chi_\sigma \quad \left[(\text{kg/m}^3)^2\text{s}^{-1}\right], \tag{3.48}$$

where β_σ is a universal constant (D'Asaro and Lien, 2007). The first measurements, obtained in strongly turbulent tidal channels, show the inertial-advective subrange as the flat portion of the spectrum (Figure 3.11). Rising spectral levels at frequencies greater than ω_L were produced by density features advected past the float. The observations yield $\beta_\sigma = 0.6$ to within a factor of two. Other scalars would be treated in a similar manner.

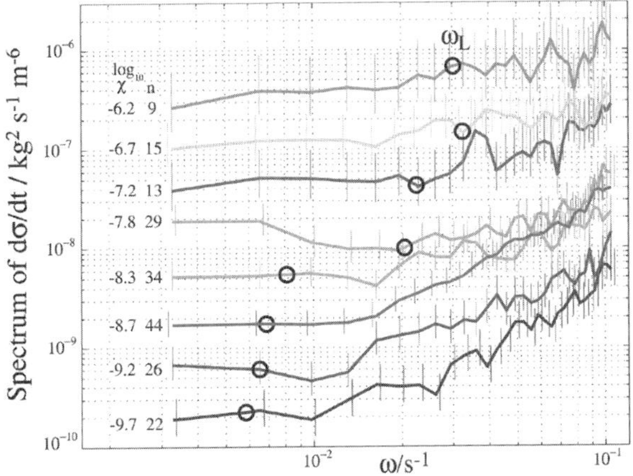

Figure 3.11 Average spectra of the time derivative of potential density at a Lagrangian float for different χ_σ. Circles indicate $\omega_L(\omega)$, where the float response begins to attenuate the inertial subrange. (Adapted from D'Asaro and Lien, 2007. © American Meteorological Society. Used with permission)

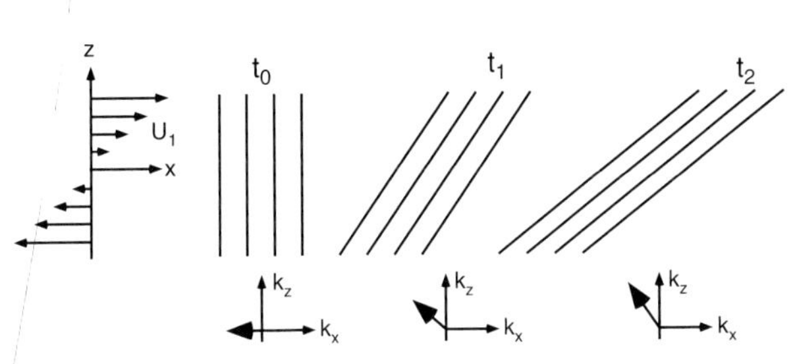

Figure 3.12 As vertical shear rotates and compresses isotherms, their wavenumber vector (large arrowheads, bottom row) grows and rotates toward vertical, ultimately aligning with k_3, the direction of greatest compression, a.k.a. the axis of the least principal rate of strain.

3.7.2 Viscous-Advective and Viscous-Diffusive Subranges

With molecular diffusivity, κ, smaller than viscosity, little turbulent scalar diffusion occurs in seawater for fluctuations larger than the Kolmogorov scale, l_η. At smaller scales, strain within the decaying velocity field increases scalar gradients, shifting variance to higher wavenumbers as isoscalars are compressed and rotated (Figure 3.12). For example, normalizing the principal axes of strain, α, β, γ, by the Kolmogorov strain rate, $S_\eta = (\epsilon/\nu)^{1/2}$, yields $(\hat{\alpha}, \hat{\beta}, \hat{\gamma}) = (0.5, 0, -0.5)$ for the situation in Figure 3.12. Scalar surfaces are extended laterally in the direction of the greatest principal rate of strain, $\hat{\alpha}$, and they are compressed along the least principal rate of strain, $\hat{\gamma}$.

Assuming that the diffusive length scale, l_κ, is proportional to the strain rate, $S_\eta = (\epsilon/\nu)^{1/2}$, and to the molecular diffusivity, κ, dimensional analysis yields the diffusive, or Batchelor, length scale,

$$l_\kappa = (\kappa/S_\eta)^{1/2} = l_\eta(\kappa/\nu)^{1/2} \quad [\mathrm{m}] \tag{3.49}$$

(Batchelor, 1959). For temperature and salinity, $l_{\kappa_T} \approx l_\eta/3$ and $l_{\kappa_S} \approx l_\eta/30$.

The temperature spectrum in the viscous-convective subrange is proportional to S_η, k, and χ_T, leading by dimensional analysis to

$$E_T^B(k) \propto \chi_\mathrm{T} S_\eta^{-1} k^{-1} \quad [\mathrm{K}^2\,\mathrm{m}^{-1}]. \tag{3.50}$$

The k^{-1} slope is the signature of uniform straining (Batchelor, 1959). A linearized heat equation extends the spectrum to include the diffusive rolloff,

$$E_\mathrm{T}(k) = \frac{(q_b/S_\eta)\chi_\mathrm{T}\exp(-q\kappa_\mathrm{T}(\nu/\epsilon)^{1/2}k^2)}{k} \quad \left[\frac{\mathrm{K}^2}{\mathrm{m}^{-1}}\right]. \tag{3.51}$$

The Batchelor spectrum has one undetermined parameter, q_b, the time scale for turbulence to sharpen scalar gradients. By assuming steady parallel shear within fluid elements, Batchelor (1959) estimated $q_b = -1/\hat{\gamma} = 2$.

In contrast to slowly changing strain modeled by Batchelor (1959), Kraichnan (1968) assumed uncorrelated, rapidly changing strain. For temperature gradients, Sanchez et al. (2011) write the one-dimensional forms as

$$\Phi_{dTdz}(k1) = \sqrt{q_t/2}\,\chi_\mathrm{T}\,\kappa_T^{-1}\,k_{\kappa_T}^{-1}\,f_{dTdz}(y) \quad [(\mathrm{K/m})^2/\mathrm{m}^{-1}], \tag{3.52}$$

with non-dimensional functions

$$f_{dTdz}^{\mathrm{Batchelor}}(y) = y\left(\exp(-y^2/2) - y\sqrt{\pi/2}\left[1 - \mathrm{erf}(y/\sqrt{2})\right]\right), \tag{3.53}$$

$$f_{dTdz}^{\mathrm{Kraichnan}}(y) = y\exp(-\sqrt{3}\,y), \tag{3.54}$$

where $y \equiv k_1/(k_{\kappa_T}/\sqrt{2q})$. The Kraichnan gradient spectrum retains the k_1^1 slope in the viscous-convective subrange, but its magnitude is slightly higher, and it rolls off less rapidly (Figure 3.7b). Half of the gradient variance occurs for $k_1 \leq 0.29k_\kappa$ (Batchelor) and $k_1 \leq 0.22k_\kappa$ (Kraichnan).

The viscous-convective subrange was first observed in a laboratory channel (Gibson and Schwarz, 1963), and Grant et al. (1968b) measured the full high-wavenumber temperature spectrum by towing a cold-film in a tidal channel. Comparing with simultaneous ϵ measurements gave $q_\mathrm{B} = 3.9 \pm 1.5$, twice Batchelor's estimate. Elliott and Oakey (1976) reported spectra from the thermocline rolling off less rapidly than Batchelor's form, and numerical simulations (Bogucki et al., 1997; Smyth, 1999) and measurements (Sanchez et al., 2011) found Kraichnan's spectrum a better fit to observations. Defining $q_e \equiv -1/<\hat{\gamma}_e>$, Smyth (1999)

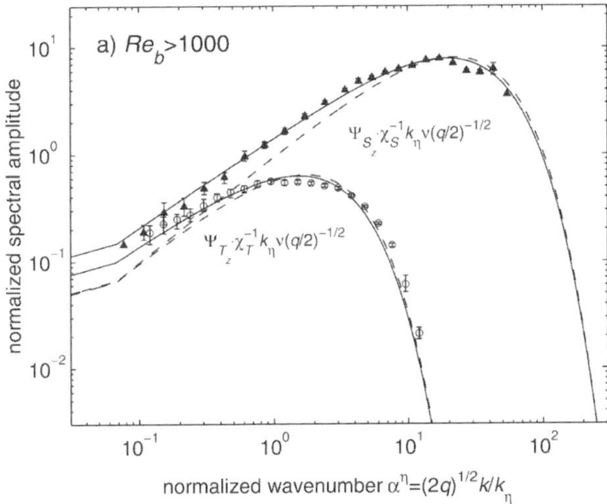

Figure 3.13 Spectra of temperature and salinity gradients from > 150 turbulent patches with $Re_b > 1{,}000$. Solid lines show Kraichnan spectra modified to have $k^{0.85}$ slopes in the viscous-convective subrange; dashed lines use the original k^1 slope. Both were evaluated with $q_k = 7.5$. (From Nash and Moum, 2002. © American Meteorological Society. Used with permission)

recommends $q_e = 7.3 \pm 0.4$ when $Re_b \gtrsim 100$ and $Re_S \gtrsim 10$. As Re_b decreased below 10, Smyth found q_e rising steeply with large scatter. From a large ensemble of simulations, Sanchez et al. (2011) also found Kraichnan's spectrum the better fit and recommend $q_k = 7.9 \pm 2.5$ and $q_b = 4.4 \pm 0.8$.

Dillon and Caldwell (1980) estimated ϵ by comparing the rolloff of temperature gradient spectra with Batchelor's spectrum. Gregg and Sanford (1980, 1981) and Gargett (1985) expressed reservations, arguing that, owing to the uncertainites about thermistor response corrections and the rolloff shape of theoretical spectra, these estimates should not be given the same weight as ϵ from shear measurements.

Nash and Moum (2002) obtained the first salinity gradient spectra resolving the viscous-convective subrange and extending into the diffusive rolloff (Figure 3.13). Spiking common in salinity records was avoided by computing spectra and cospectra of electrical conductivity from a tiny four-electrode 'Head' probe (Section 5.2.3) and temperature from a Fastip thermistor (Section 5.2.2). These were combined using coefficients from a linearized dependence of conductivity on temperature and salinity. Unexpectedly, spectra of temperature and salinity had slightly flatter slopes in the viscous-convective subrange than either Kraichnan or Batchelor spectra. Nash and Moum argued that this may be a signature of differential diffusion (Section 3.9.1) generating a reverse cascade of variance for fluctuations greater than $\approx$3 mm.

3.8 The Horizontal Cascade in Strong Stratification

Strong stratification decouples vertical and horizontal flows into thin vortical motions cascading energy toward smaller scales and ultimate dissipation (Section 6.7). Inertial transfer dominates this cascade, producing horizontal spectra with $k_h^{-5/3}$ slopes similar to spectral slopes in the three-dimensional Kolmogorov cascade. Numerical simulations and laboratory experiments find these dynamics when

$$Fr_h \equiv \frac{u}{Nl_h} \ll 1 \quad \text{and} \quad Re_h \equiv \frac{ul_h}{\nu} \gg 1 \quad \text{with} \quad Fr_h^2 Re_h \gg 1, \tag{3.55}$$

where Fr_h and Re_h are horizontal Froude and Reynolds numbers (Billant and Chomaz, 2001). Horizontal spectra consistent with a horizontal cascade have been found in the ocean, but the extent and importance of vortical mode interactions relative to the Kolmogorov cascade remain to be determined. The same or closely related phenomena approached from the perspective of internal waves are discussed in Chapters 6 and 7.

3.8.1 Instabilities and Pancake Eddies

Dyed wakes of spheres towed through stratified profiles reveal thin vortex pairs propagating outward from opposite sides of late wakes (Figure 6.27). Although the vortices resemble a von Karmen vortex street, numerical simulations and dimensional analysis demonstrate that these vortical motions, a.k.a. pancake eddies, are natural features of strongly stratified turbulence (Riley et al., 1981). For example, they have been observed as 'zigzag' instabilities slicing vertically uniform vortex pairs into rotating disks (Figure 3.14). Submesoscale geostrophic vortices have also been invoked to explain rapid lateral dispersion of tracers over of 1–10 km in the mid-ocean thermocline (Ledwell et al., 1993; Sundermeyer and Price, 1998;

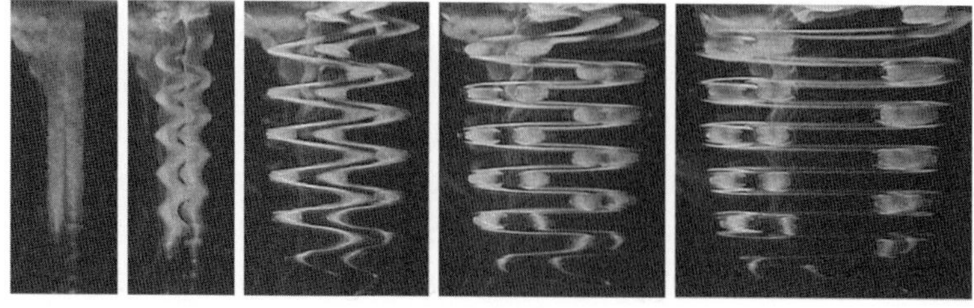

Figure 3.14 Frontal views of the breakdown of two counter-rotating dyed vortices by the zigzag instability when $Fr_h = 0.19$. The instability began with sinusoidal deformations displacing the columns with minimal effect on their internal structure. Eventually the columns were sliced into separate dipole pairs moving nearly opposite to the original direction. (Adapted from Billant and Chomaz, 2000a)

Polzin and Ferrari, 2004). Numerical simulations of rapid spreading in a coastal thermocline are consistent with formation of vortices by collapse of thin mixed layers (Lelong and Sundermeyer, 2005; Sundermeyer and Lelong, 2005).

Riley et al. (1981) developed equations for strongly stratified flows that were restated in non-dimensional form by Brethouwer et al. (2007) as

$$\frac{\partial \hat{\boldsymbol{u}}_h}{\partial \hat{t}} + \hat{\boldsymbol{u}}_h \cdot \nabla_h \hat{\boldsymbol{u}}_h + \frac{Fr_h^2}{(l_v/l_h)^2} \hat{w} \frac{\partial \hat{\boldsymbol{u}}_h}{\partial \hat{z}} = -\nabla_h \hat{p} + \frac{1}{Re_h} \frac{1}{(l_v/l_h)^2} \frac{\partial^2 \hat{\boldsymbol{u}}_h}{\partial \hat{z}^2} \tag{3.56}$$

$$0 = -\frac{\partial \hat{p}}{\partial \hat{z}} - \hat{\rho} \tag{3.57}$$

$$\nabla_h \cdot \hat{\boldsymbol{u}}_h + \frac{Fr_h^2}{(l_v/l_h)^2} \frac{\partial \hat{w}}{\partial \hat{z}} = 0 \tag{3.58}$$

$$\frac{\partial \hat{\rho}}{\partial \hat{t}} + \hat{\boldsymbol{u}}_h \cdot \nabla_h \hat{\rho} + \frac{Fr_h^2}{(l_v/l_h)^2} \hat{w} \frac{\partial \hat{\rho}}{\partial \hat{z}} = \hat{w} + \frac{1}{Re_h Sc} \frac{1}{(l_v/l_h)^2} \frac{\partial^2 \hat{\rho}}{\partial \hat{z}^2}, \tag{3.59}$$

where hats indicate non-dimensional variables. Although these equations assume $Fr_h \ll 1$ and $Re_h \gg 1$, viscous and diffusive terms are retained to show the influence of l_v/l_h (Billant and Chomaz, 2000b).

Because the relative magnitudes of vertical advection and viscous terms in the horizontal momentum equation depend on Fr_h^2 and Re_h, $Fr_h^2\, Re_h$ is the fundamental parameter for strongly stratified turbulence, with $Fr_h \ll 1$ and $Re_h \gg 1$. Note that if $\epsilon = u^3/l_h$, i.e. horizontal Taylor scaling, holds, as assumed by the arguments for strongly stratified turbulence, then

$$Fr_h^2\, Re_h = \left(\frac{u}{Nl_l}\right)^2 \frac{ul_h}{\nu} = \frac{u^3/l_h}{\nu N^2} = \frac{\epsilon}{\nu N^2} = Re_b. \tag{3.60}$$

The non-dimensional equations (3.56–3.59) are self-similar with respect to $\hat{z}N/U$, leading Billant and Chomaz (2000b) and Brethouwer et al. (2007) to conclude that when no vertical length scale is imposed, $l_v \sim U/N$ (equivalent to taking the buoyancy velocity as $u_b = Nl$, in Section 3.2). This implies that $l_v/l_h = (U/N)/l_h = Fr_h$ and that vertical advection terms are dynamically important.

3.8.2 Forward Energy Cascade and Inertial Subrange Spectra

Adopting $l_v \sim U/N$ scaling (Section 3.2), implying $Fr_v \sim 1$, Lindborg (2006) hypothesized a forward energy cascade, from large to small scales, when $Fr_h \lesssim 0.02$. Three-dimensional but highly anisotropic, this regime is driven by inertial transfer characterized by the horizontal version of Taylor scaling of velocity and the kinetic energy spectrum,

$$u \sim (l_h \epsilon)^{1/3} \quad \text{and} \quad \Phi_{ke} \sim (l_h \epsilon)^{2/3}. \tag{3.61}$$

In the limit of $Fr_h \ll 1$, the forward cascade causes buoyancy forces and l_v to adjust as horizontal inertial interactions move energy toward smaller scales. The hypothesis is considered valid for vertical scales constrained by $l_0 Fr_h^{-1/2} \lesssim l_v \lesssim l_0$, where l_0 is the Ozmidov scale.

Driven by the same terms in the momentum equation, horizontal inertial interactions have the same dimensional scaling as three-dimensional interactions in Kolmogorov turbulence. Consequently, spectra of kinetic and potential energy can be expressed as

$$\Phi_{KE}(k_h) = c_1\, \epsilon^{2/3} k_h^{-5/3} \quad [\mathrm{J\,kg^{-1}}\, m^{-1}] \tag{3.62}$$

$$\Phi_{PE}(k_h) = c_2\, \epsilon_{pe} \epsilon^{-1/3} k_h^{-5/3} \quad [\mathrm{J\,kg^{-1}}\, m^{-1}] \tag{3.63}$$

Numerical simulations for strongly stratified non-rotating flow are consistent with these forms and yield coefficients of $c_1 \simeq c_2 = 0.51 \pm 0.02$ (Lindborg, 2006). Scaling vertical spectra of strongly stratified turbulence as $\Phi_{ke}(k_z) \propto u^2 l_z$ yields

$$\Phi_{ke}(k_z) \sim N^2 k_z^{-3}. \tag{3.64}$$

Spectra with this slope and approximate amplitude are observed in the ocean at wavelengths less than those of Kolmogorov turbulence. There are, however, two other models for this range, based on a buoyancy subrange of turbulence and saturated internal waves (Section 7.8).

Evidence for a distinct regime of strongly stratified turbulence comes from dimensional analysis, laboratory experiments, and numerical simulations. The strongest indirect evidence for vortical motions is the Riley and Lindborg (2008) reinterpretation of the $k_h^{1/3}$ slopes of horizontal shear spectra of Klymak and Moum (2007). Collected near the Hawaiian Ridge, the spectra (Figure 3.15) were

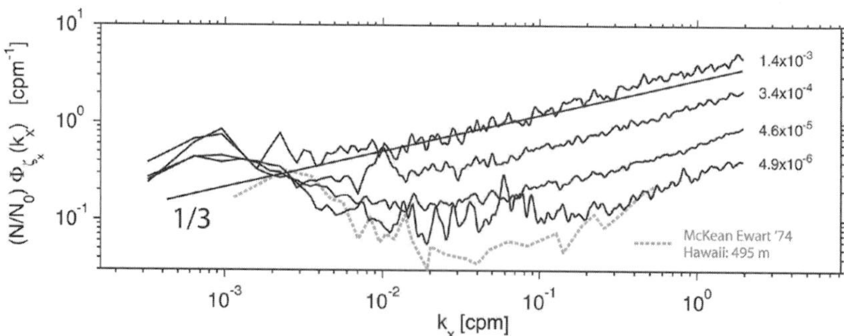

Figure 3.15 Spectra of isopycnal slope normalized by N/N_0 versus horizontal wavenumber. Binned by $K_\rho = 0.2\epsilon/N^2$, the spectra were taken of horizontal tows near Kaena Ridge by Klymak and Moum (2007) and reinterpreted by Riley and Lindborg (2008). (From Riley and Lindborg, 2008. © American Meteorological Society. Used with permission)

interpreted as signatures of very energetic Kolmogorov turbulence by Klymak and Moum. Riley and Lindborg (2008), however, argued that horizontal and three-dimensional subranges blend seamlessly at Ozmidov wavenumbers 0.83 and 14 cpm for the upper and lower spectra.

3.9 Evolution and Decay

Turbulence in surface mixed layers is often statistically steady, changing over hours as synoptic weather systems evolve. Owing to the shallow depths, these changes can be followed by rapid ship-based profiling. Evolution of mixing in the pycnocline, however, is more elusive, as it occurs episodically in thin patches, lasts several buoyancy periods, and extends laterally for meters to a few kilometers. Owing to the absence of repeated observations, understanding of the mixing evolution in the pycnocline relies heavily on laboratory and numerical studies of what are believed to be the same mechanisms.

Laboratory mixing studies have focused on turbulence generated by flows through grids and by simple mechanisms, such as Kelvin–Helmholtz shear instabilities. The turbulence, however, is often too intense for direct ϵ measurements, limiting estimation of turbulence to indirect measures, such as increases in potential energy.

Numerical analyses have examined shear and convective instabilities. In particular, Kelvin–Helmholtz and Holmboe shear instabilities have been the focus of direct numerical simulation (DNS) that seeks to resolve all relevant length scales. If an event occupies $1/M$ of a digital domain of N points, the maximum wavenumber bandwidth is $N/2M$, and for low-order numerical methods often closer to $N/4M$ (Gregg et al., 2018). Thus, with $N = 1{,}024$ the maximum bandwidth is $1{,}024/(2 \times 2) = 256$, and the largest Reynolds number is $Re = (l/l_\eta)^{4/3} = 256^{4/3} = 1{,}626$. This suffices for most pycnocline events produced by background internal waves but is much less than that found in energetic patches, e.g. Figure 1.2. To examine more energetic turbulence, many simulations have been run with a Prandtl number of 1, precluding realistic examination of scalar diffusion.

Shear instabilities grow vertically by vortex pairing until stratification limits further growth at the Ozmidov scale, l_O. Details vary with the mechanism, but, in general, secondary instabilities and gravity combine to generate turbulence as overturns collapse, after which universal turbulent dynamics are presumed to govern, modulated by background shear. The rarity of homogenized layers in the pycnocline demonstrates that most mixing ends before going to completion. When mixing is not sustained by continuing strong shear, the decrease of turbulent intensity during decay displays universal properties as functions of Re_b. Issues of concern for mixing include differing diffusion rates for heat and salt, and loss

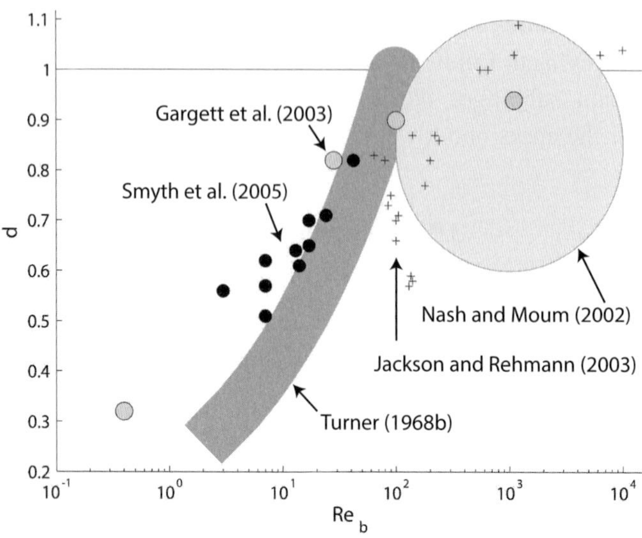

Figure 3.16 Ratio of the cumulative diapycnal diffusivities of heat and salt, $d \equiv \lim K_s^c(t)/K_t^c(t)$, versus buoyancy Reynolds number. Values from Turner (1968b), Gargett et al. (2003), and Jackson and Rehmann (2003) were obtained in tanks. Those of Nash and Moum (2002) came from the ocean. Smyth et al. (2005) estimates were from DNS of Kelvin–Helmholtz instabilities with $Pr = 7$ and $Sc = 50$. (Adapted from Smyth et al., 2005. © American Meteorological Society. Used with permission)

of the increased potential energy due to mixing as homogenized regions collapse and spread.

3.9.1 Differential Diffusion

The large contrast in molecular diffusivities between heat and salt was long considered to have no effect on turbulent mixing. Recent laboratory experiments and numerical simulations, however, have shown this not to be the case for low-intensity turbulence common in the pycnocline (Gargett, 2003). For instance, salt mixes slower than heat in tanks stirred with oscillating grids, but the ratio increases to one as turbulence becomes more intense (Figure 3.16). Direct numerical simulations show similar results, with the turbulent diffusivity ratio rising to $\sim$1 when $Re_b \gtrsim 100$. Much remains to be learned, but for weak turbulence, differential diffusion should be considered in addition to anisotropy when estimating diapycnal diffusivities.

3.9.2 Mixed Layer Collapse

Estimates of mixing usually neglect the loss of potential energy as homogenized regions collapse after mixing ceases. When a profile with initial stratification N_0

is partially mixed to stratification N_p over area a and subsequently collapses to area A, the change in potential energy is

$$\Delta gpe_r = \frac{\rho_0 h^3 a (N_0^2 - N_p^2)}{24}\left[1 + \frac{a}{A} + \frac{N_p^2}{N_0^2}\left(1 - \frac{a}{A}\right)\right] \quad [\mathrm{J}] \tag{3.65}$$

(Hebert, 1988a). For complete homogenization, i.e. $N_p = 0$, and no spreading, $a/A = 1$, the increase in potential energy is $\Delta gpe_r = (1/12)\rho_0 h^3 a N_0^2$. If spreading proceeds as $a/A \to 0$, the increase in potential energy is reduced by half, $\Delta gpe_r \to (1/24)\rho_0 h^3 a N_0^2$.

Simulating collapse of a mixed region with a half-width L, Lelong and Sundermeyer (2005) found that, after initial radial spreading, further expansion and circulation are strongest when $R/L \approx 1$, where R is the deformation radius. During the first half-inertial period, over 70% of the initial potential energy was converted to kinetic energy of the vortex before damped oscillations began generating internal waves. As the oscillations died, neglecting mixing from vortex shear and radiating internal waves,

$$pe_0 = \underbrace{pe_{vortex}}_{0.45\, pe_0} + \underbrace{ke_{vortex}}_{0.20\, pe_0} + \underbrace{pe_{iw}}_{0.10\, pe_0} + \underbrace{ke_{iw}}_{0.25\, pe_0}, \tag{3.66}$$

with the ratio pe_{vortex}/pe_0 close to the limit of (3.65) for $a/A = 0$.

3.10 Intermittence and Statistics

Turbulence in the pycnocline occurs episodically in thin patches. Some sequential intervals of 0.5 m to several meters used to compute ϵ and χ_T are filled with turbulence, while others have varying amounts. Probability densities of these ensembles are highly skewed and roughly lognormal. In most cases nonparametric methods, e.g. the bootstrap, are used to estimate confidence limits. Trying to understand the statistics begins with the variability of statistically steady homogenous turbulence (internal intermittence) and expands to the statistics of processes forcing turbulence (external intermittence). In the following, internal intermittence within patches of statistically homogenous turbulence is considered, followed by the external intermittence of breaking internal waves (Section 7.6).

3.10.1 Lognormality of Dissipation in Fully Developed Homogenous Turbulence

Turbulent cascades are similar to breakage processes, such as sequence of collisions – boulders→rocks→pebbles – that determine distributions of sand grains on beaches. Proportional to products of independent random variables, breakage variables have lognormal probability densities, in contrast to sums of independent random variables that are described by normal probability densities (Montroll and Shlesinger, 1982).

As formulated by Gurvich and Yaglom (1993) and reviewed by Yamazaki and Lueck (1990), Kolmogorov's third hypothesis leads to lognormal probability distributions of the turbulent dissipation rate, ϵ_r, averaged over length scale r with $l_\eta \ll r \ll L$, where L is the length of the domain over which ϵ_r is statistically homogenous. Within these constraints, the average dissipation rate, ϵ_N, can be written as a product of ratios,

$$\epsilon_N = \langle\epsilon\rangle \left(\frac{\epsilon_1}{\langle\epsilon\rangle}\right)\left(\frac{\epsilon_2}{\epsilon_1}\right)\cdots\left(\frac{\epsilon_N}{\epsilon_{N-1}}\right). \tag{3.67}$$

Consequently, the natural logarithm of ϵ_N is a sum,

$$\ln\epsilon_N = \ln\langle\epsilon\rangle + \ln\left(\frac{\epsilon_1}{\langle\epsilon\rangle}\right) + \ln\left(\frac{\epsilon_2}{\epsilon_1}\right) + \cdots + \ln\left(\frac{\epsilon_N}{\epsilon_{N-1}}\right). \tag{3.68}$$

If these ratios are independent and identically distributed, by the Central Limit Theorem, $\ln\epsilon_N$ has a normal distribution independent of the $\ln(\epsilon_{i+1}/\epsilon_i)$ distributions (Rice, 1988). Using scaled microscale shear meeting the Gurvich and Yaglom criteria, Yamazaki and Lueck (1990) demonstrated statistically significant lognormality in an 8 m patch of strong turbulence in the pycnocline. Larger ensembles are usually approximately lognormal.

Defining μ and σ as expected value and standard deviation of the logarithm of positive random variable x,

$$\mu \equiv E\left[\ln x\right], \quad \sigma \equiv E\left[(\ln x - \mu)^2\right], \tag{3.69}$$

the lognormal probability density function of x is

$$\ln\mathcal{N}(x;\mu,\sigma) = \frac{1}{x\sigma(2\pi)^{1/2}}\exp\left(-\frac{(\ln x - \mu)^2}{2\sigma^2}\right), \quad x > 0. \tag{3.70}$$

Low-order moments (Table 3.1) show that lognormal distributions are skewed, with modes significantly smaller than means, which are controlled by extended high-magnitude tails. Confidence limits for lognormal variables are multiplicative, e.g. if L_n and U_n are lower and upper 95% confidence limits for n samples, then the 95% confidence interval is $[L_n\,x_{\text{mle}}, U_n\,x_{\text{mle}}]$, with x_{mle} as the maximum likelihood estimate of the mean. For typical σ, tens to a hundred or more samples are needed to estimate means within a factor of 2 (Figure 3.17). Much tighter confidence limits require more samples than are likely to be obtained from a homogenous, steady population using present sampling capabilities.

Table 3.1 *Low-order moments and 95% confidence limits of lognormal probability distributions for n samples.*

Median	$\exp(\mu)$
Mode	$\exp\left(\mu - \sigma^2\right)$
Mean, x_{mle}	$\exp\left(\mu + \sigma^2/2\right)$
Variance	$\exp\left(2\,\mu + \sigma^2\right)\left(\exp(\sigma^2) - 1\right)$
95% confidence limits	$x_{\text{mle}}\exp(\pm 1.96\eta),\ \eta \equiv (\sigma^2/n + \sigma^4/(2(n+1))$

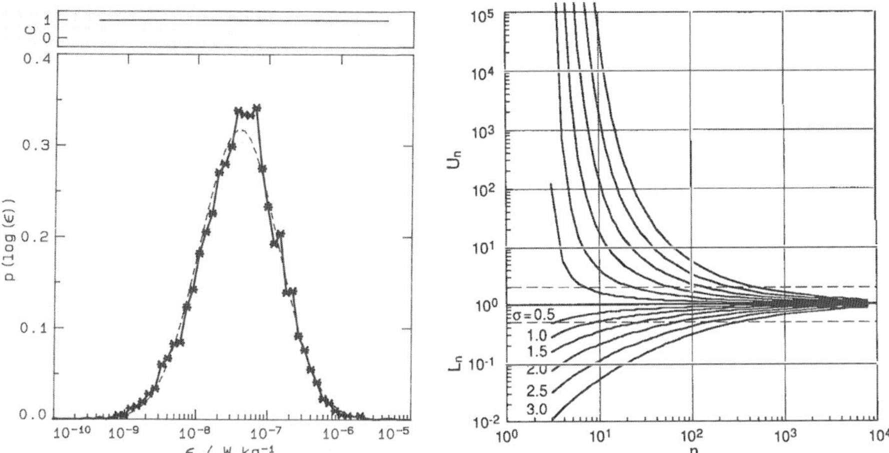

Figure 3.17 (Left) Probability density of 0.5 m estimates of $\log_{10}\epsilon$ in the free-convection region of an oceanic mixed layer (Figure 3.3). At the top, C = 1 when theoretical and observed distributions agree with 95% confidence. (Right) Multiplicative upper, $\mathrm{U_n}$, and lower, $\mathrm{L_n}$, 95% confidence limits for lognormal variable x as functions of the number of samples, n, and s, the standard deviation of the natural logarithm, $\ln x$, for $s = 0.5$ to $s = 3.0$. Horizontal dashed lines mark $\mathrm{U_n} = 2$ and $\mathrm{L_n} = 0.5$. (From Gregg et al., 1993. © American Meteorological Society. Used with permission)

3.10.2 Statistics of Bulk Dissipation Rates Driven by Internal Wave Shear

Assumptions of steady, fully developed turbulence leading to lognormality are not satisfied by ensembles of microstructure profiles from the pycnocline. Nonetheless, empirical distributions are highly skewed and approximately lognormal, with $\sigma_{\ln\epsilon} \sim 2.9$ when internal waves are at GM. Elevating wave intensity fourfold increases the spread to $\sigma_{\ln\epsilon} \sim 3.2$ (Gregg et al., 1993). The approximate lognormality can be traced to characteristics of the internal waves producing the turbulence. When internal waves are at GM, obtaining independent ϵ estimates requires averaging over 10 m, corresponding to the wavenumber where the slope of the vertical shear spectrum changes from $k_z^{\sim 0}$ to $k_z^{\sim -1}$ (Sec 7.8). This transition

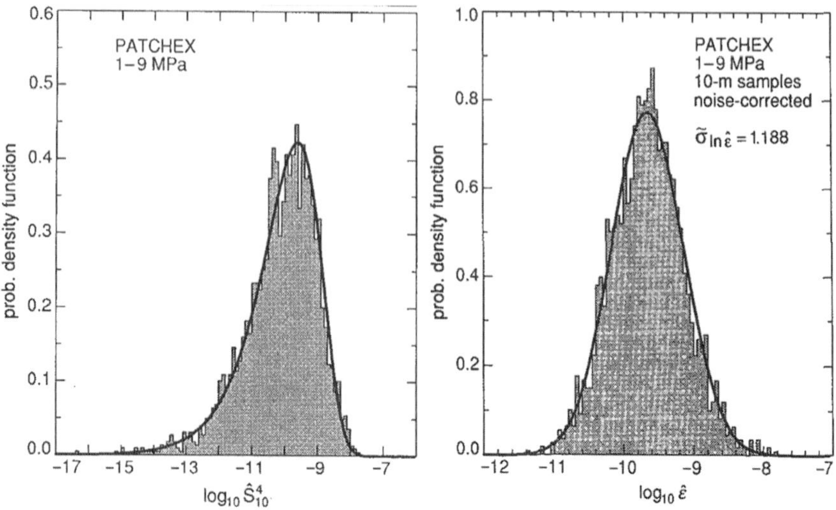

Figure 3.18 Probability densities of 10 m averages of $\hat{S}^4_{10} \equiv S^2_{10}/\langle N^2\rangle$ (left) and $\hat{\epsilon} \equiv \epsilon_{10}/\langle N^2\rangle$ (right) when internal waves were close to GM during the Patches Experiment. Corrected for noise, the $\hat{\epsilon}$ distribution was lognormal within 95% confidence. (Adapted from Gregg et al., 1993. © American Meteorological Society. Used with permission)

between internal wave and saturated regimes shifts to 15 m when internal waves are four times GM. To obtain adequate statistics where stratification varies with depth, both 10 m shear and ϵ are normalized as $\hat{S}_{10} \equiv S_{10}(N_0/\langle N^2\rangle^{1/2})$ and $\hat{\epsilon}_{10} \equiv \epsilon_{10}(N_0^2/\langle N^2\rangle)$.

Ten-meter dissipation rates do not satisfy Gurvich-Yaglom criteria for lognormality, but, nonetheless, their distributions are lognormal or nearly so (Figure 3.18). Gregg et al. (1993) hypothesized that the lognormality is a consequence of internal wave dissipation being proportional to $\hat{S}^4_{10}$ (Sec 7.6), which has a distribution close to lognormal, save for a low-magnitude tail. Absence of an equivalent tail in $\ln\epsilon$ was attributed to averaging and limited bandwidth of the measurements. Under this hypothesis, $\sigma_{\ln S^4_{10}} = 2.57$ is the upper bound for $\sigma_{\ln\hat{\epsilon}}$. In practice, confidence limits of dissipation rates are estimated by lognormal statistics or by nonparametric measures such as the bootstrap (Efron and Gong, 1983).

3.11 Estimating Diapycnal Turbulent Fluxes

Numerical models of ocean circulation use turbulent eddy diffusivities to represent fluxes produced by small-scale mixing. Inherently perpendicular to the surfaces being mixed, the fluxes are termed 'diapycnal', or more generally, 'diascalar'. For example, the diapycnal diffusivity of buoyancy is

$$K_\rho \equiv -J_b/\overline{N^2} \quad [\mathrm{m^2\ s^{-1}}], \tag{3.71}$$

where $J_b \equiv -(g/\rho)\overline{\rho' w'}$ is the buoyancy flux. Because the correlations producing turbulent fluxes, e.g. $\overline{u' w'}$, $\overline{\Theta' w'}$, have not yet been measured reliably, the following sections discuss indirect methods using parameters that can be measured, principally ϵ and χ_T.

3.11.1 The Dissipation Method for Momentum Diffusivity

For steady and spatially homogenous turbulence in a simple vertical shear flow, the turbulent kinetic energy (tke) equation (6.44) reduces to

$$0 = -\overline{u'w'}\,\frac{\partial \overline{u}}{\partial z} + J_b - \epsilon \quad \left[\mathrm{W\,kg^{-1}}\right]. \tag{3.72}$$

When this technique was developed for the atmospheric boundary layer, $\overline{u'w'}$ could not be measured directly; rather, measuring ϵ was possible, and $\partial\overline{u}/\partial z$ was inferred for a logarithmic boundary layer. Consolidating the unknowns into the flux Richardson number (3.8) reduced the tke balance to

$$(1 - R_{\text{flux}})\frac{\partial \overline{u}}{\partial z} = -\epsilon. \tag{3.73}$$

Expressing the vertical turbulent momentum flux in terms of an eddy coefficient times the mean shear,

$$\overline{u'w'} = -K_M \frac{\partial \overline{u}}{\partial z} \quad \left[\mathrm{m^2\ s^{-2}}\right], \tag{3.74}$$

leads to

$$K_M = \frac{1}{(1 - R_{\text{flux}})}\,\frac{\epsilon}{(\partial\overline{u}/\partial z)^2} \qquad \left[\mathrm{m^2\ s^{-1}}\right]. \tag{3.75}$$

This applies only where turbulence is produced by breakdown of the mean shear. It does not apply where turbulence is produced by breaking internal waves, even if a mean shear is also present. If $R_{\text{flux}} \leq 0.2$, uncertainites in it are often less important than those of ϵ. There are few, if any, places in the pycnocline where turbulence is produced by mean shear, but the approach was adapted to stratified profiles.

3.11.2 The Osborn-Cox Method for Thermal Diffusivity

Assuming that thermal microstructure in the thermocline is at a steady balance between production by vertical overturning and diffusive smoothing, Osborn and

Cox (1972) neglected horizontal divergences and simplified the temperature variance equation (3.23) to a balance between production and diffusive smoothing,

$$2\,\overline{w'\Theta'}\,\frac{\partial\overline{\Theta}}{\partial z} = \chi_{\mathrm{T}}. \tag{3.76}$$

Defining the thermal eddy coefficient as

$$\overline{w'\Theta'} = -K_T\frac{\partial\overline{\Theta}}{\partial z} \tag{3.77}$$

leads to the eddy coefficient as the ratio of the rate of diffusive smoothing to the square of the mean temperature gradient,

$$K_T = \frac{\chi_{\mathrm{T}}}{2\left(\partial\overline{\Theta}/\partial z\right)^2} \quad \left[\mathrm{m^2\,s^{-1}}\right]. \tag{3.78}$$

Recognizing Cox's contributions to measuring temperature microstructure, the Cox number is the ratio of turbulent and molecular thermal diffusivities,

$$\text{Cox number} \equiv \frac{K_T}{\kappa_{\mathrm{T}}} = \frac{\overline{\nabla\Theta'^2}}{\left(\partial\overline{\Theta}/\partial z\right)^2} = \frac{\overline{(\partial\Theta'/\partial x)^2} + \overline{(\partial\Theta'/\partial y)^2} + \overline{(\partial\Theta'/\partial z)^2}}{\left(\partial\overline{\Theta}/\partial z\right)^2}. \tag{3.79}$$

Integrating the Cox number over a full three-dimensional monotonic temperature field yields $dgpe_r/dt$, the increase in background potential energy the Osborn-Cox model seeks to estimate (Winters and D'Asaro, 1996). As applied to one-dimensional profiles, the technique is an approximation rather than an exact result.

Other sources of uncertainty in applying Osborn-Cox include:

(1) *The degree of isotropy*. The one variance component, $\overline{(\partial\Theta'/\partial z)^2}$, measured is multiplied by 3, assuming isotropy. Direct numerical simulations show isotropy when the Cox number exceeds 100, but not when it is smaller (Figure 3.19).

(2) *Spatial resolution of the gradients*. Glass-coated thermistors frequently cannot resolve the smallest signals at typical vehicle speeds of 1 m s^{-1}. Nominal probe transfer functions correct for attenuation, but often only partially.

(3) *Lateral convergence of* Θ'^2 in the presence of strong lateral gradients and horizontal diffusivity violates the assumption of local production of θ'^2 by overturning if applied across the vertical span of the intrusion (Gregg, 1975b; Alford et al., 2005), but the model can be applied to turbulence on intrusion boundaries using a triple decomposition of temperature (Joyce, 1977).

(4) *Differences in algorithms*. Some χ_T estimates are obtained by integrating spectra after correction for probe response. Others fit universal spectra to corrected spectra. Two universal forms have been used (Batchelor, 1959; Kraichnan, 1968), with different values for the empirical scaling constant, q.

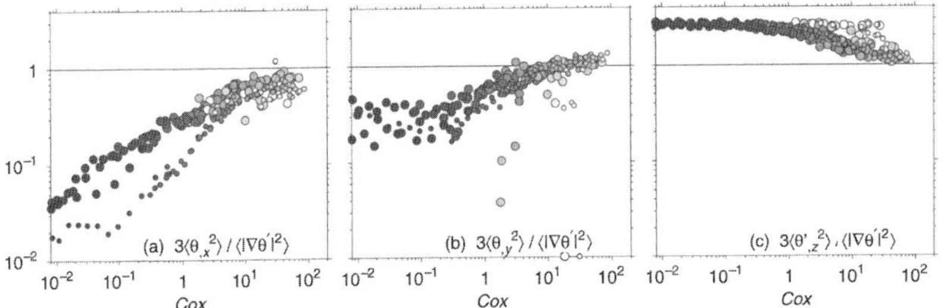

Figure 3.19 DNS ratios of Cox numbers estimated by multiplying one temperature gradient component to true values using all three. From left to right, streamwise, spanwise, and vertical components in a Kelvin–Helmholtz instability. The smallest symbols are for $Pr = 7$, the largest for $Pr = 1$. Using vertical gradients overestimates Cox numbers when Cox $\lesssim$ 20. (Adapted from Smyth and Moum, 2000)

(5) *Estimation of* $\partial\overline{\Theta}/\partial z$. The difficulty of estimating the mean gradient of overturns occurring in irregular stratification can produce the largest errors (Smyth et al., 2001).

3.11.3 The Osborn (1980) Method for Diapycnal Diffusivity

The diapycnal eddy coefficient, K_ρ, represents the buoyancy flux produced by turbulent dissipation in a constant-N profile,

$$J_b = -K_\rho N^2 \quad \left[\mathrm{W\,kg^{-1}}\right]. \tag{3.80}$$

At steady state, in terms of the buoyancy flux and the flux Richardson number, the *tke* equation (3.15) reduces to

$$\frac{J_b}{R_{\mathrm{flux}}} = J_b - \epsilon. \tag{3.81}$$

Replacing the buoyancy flux with (3.80) gives

$$K_\rho = \left(\frac{R_{\mathrm{flux}}}{1 - R_{\mathrm{flux}}}\right)\frac{\epsilon}{N^2} = \Gamma_{\mathrm{mix}}\frac{\epsilon}{N^2} \quad [\mathrm{m^2\,s^{-1}}], \tag{3.82}$$

where Γ_{mix} is the mixing coefficient for K_ρ (Section 3.12). Citing $R_{\mathrm{flux}} \approx 0.15$ from Ellison (1957), Osborn (1980) treated 0.2 as an upper bound for Γ_{mix}.

3.12 Mixing Efficiency

The mixing coefficient, Γ_{mix}, is a measure of mixing efficiency, defined as the increase in reference, or background, potential energy divided by the energy expended to produce the mixing,

$$\text{mixing efficiency} \equiv \frac{\Delta gpe_r}{\text{energy expended to produce the mixing}}, \tag{3.83}$$

where Δgpe_r is the increase in reference, or background, gravitational potential energy produced by the mixing. Changes in potential energy can be easily measured in laboratory tanks and numerical models, but not so directly in the pycnocline, owing to the distortion of profiles by shear and the difficulty of bounding mixing patches. On the other hand, dissipation can be measured relatively well in the pycnocline, but dissipation scales in tank experiments are usually too small to measure. As a result, there are many variations of the variables used to evaluate efficiency (Gregg et al., 2018). Here, we focus on the mixing coefficient widely reported from oceanic measurements.

Assuming that eddy coefficients for heat, salt, and density are equal, $K_T = K_S = K_\rho$, Oakey (1982) combined estimates from (3.78) and (3.82) to express the mixing coefficient as

$$\Gamma_{\chi\epsilon} \equiv \frac{\chi_T N^2}{2\epsilon(\partial\overline{\Theta}/\partial z)^2}. \tag{3.84}$$

Using data from 5 to 100 m in Rockall Trough, Oakey obtained $\Gamma_{\chi\epsilon} = 0.24$. Subsequent reports range from 0.0022 to 0.48 (Gregg et al., 2018), the former from a fish aggregation where the low efficiency was attributed to turbulent generation at dissipation scales, and the latter from the 15 m overturn in Figure 1.2. Most values scatter around 0.2, and many have tight error bars, but there is no consistency in the means from from one ensemble to the next. Owing to many sources of biases and errors, the scatter in $\Gamma_{\chi\epsilon}$ is not surprising for the ratio of two partially correlated lognormal variables often measured with inadequate spatial resolution.

Some variability in $\Gamma_{\chi\epsilon}$ estimates may reflect dependencies on unreported background parameters: N^2, Ri or Fr, R_ρ, and Re or Re_b. As an example, in reanalyzing data from the North Atlantic Tracer Release Experiment (NATRE), Ruddick et al. (1997) found K_ρ consistent with salt fingering when $R_\rho \lesssim 1.5$, but not when it was substantially larger (Figure 3.20). The path forward with this coefficient will require sensors that fully resolve gradients producing ϵ, χ_T, and χ_S, better understanding of isotropy, and measurement of background variables affecting efficiency. In addition, the underlying assumption $K_T = K_\rho$ needs to be re-examined. It appears to be true when turbulence is intense and homogenizes water rapidly, but equal diffusivities are unlikely during the weak mixing most commonly found in the pycnocline. Also, mixing efficiency needs to include the loss in potential energy as mixed and partially mixed regions collapse.

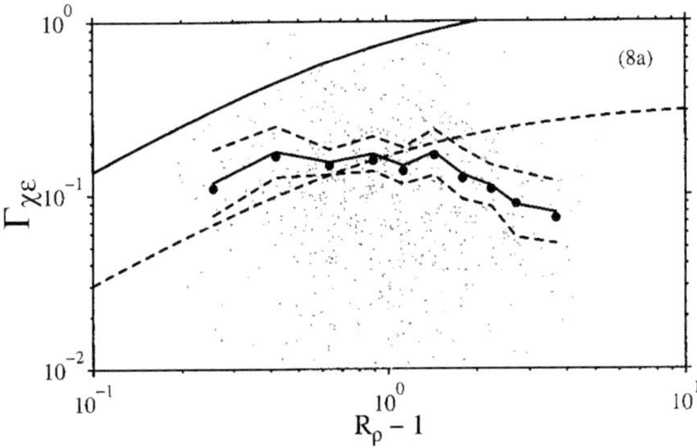

Figure 3.20 $\Gamma_{\chi\epsilon}$ with the mean (solid) and 95% confidence limits (dashed) versus $R_\rho - 1$. Curving solid and dashed lines are for salt fingering with flux ratios $R^{sf}_{flux} = 0.6$ and 0.25. (Adapted from Ruddick et al., 1997. © American Meteorological Society. Used with permission)

3.12.1 Comparisons with Tracer Diffusivities

Confidence in applying Osborn and Cox (1972) and Osborn (1980) to the pycnocline was established by comparisons with the rate of thickening of artificial tracers injected into the pycnocline. The impetus came from Davis (1994a) arguing that applying fluxes estimated by the Osborn-Cox method to the general circulation is inconsistent because the model assigns variability from internal waves, mesoscale circulation, and interannual variability to the mean field. Analyses such as Munk (1966), however, are based on averages over several years and include small-scale turbulence, internal waves, mesoscale circulation, and interannual variability in the eddy field. Davis (1994b) concluded that linking Osborn-Cox to the general circulation is ad hoc rather than an exact result. This conclusion led to NATRE, which injected an artificial tracer into the pycnocline of the northeast Atlantic. Section 5.8 discusses details of the release and its analysis.

NATRE was followed with similar releases and microstructure measurements over abyssal hills in Brazil Basin and in the Antarctic Circumpolar Current. In addition, dye was used for a short-term release over the New England continental shelf. Owing to these comparisons in very different oceanic regimes (Table 3.2), inferences of diapycnal diffusivity from microstructure measurements have been accepted as accurate, within roughly a factor of 2. The success of these comparisons is somewhat surprising, as diffusivity from the tracers is an average over the length of the experiment, 18 months for NATRE, and microstructure was sampled a few times, usually for only a few weeks each time. The comparisons agreed as well as they did because breaking internal waves dominated the mixing, and the wave field

Table 3.2 *Tracer releases with simultaneous microstructure sampling: 1) In the subtropical gyre of the North Atlantic, with simultaneous temperature microstructure from Sherman and Davis (1995). Nearly the same value was measured with airfoils just before the release by Toole et al. (1994). 2) Over abyssal hills in the Brazil Basin. 3) Over the New England continental shelf, with microstructure from Oakey and Greenan (2004). 4) In the Antarctic Circumpolar Current, west of the Drake Passage. The tracers were* SF_6, *except for #3 where dye was used.*

Location	Depth m	$10^5 K_{tracer}$ $m^2\ s^{-1}$	$10^5 K_{micro}$ $m^2\ s^{-1}$	Citation
25.7°N, 28.3°W[1]	300	1.1	1.1 ± 0.2	Ledwell et al. (1998)
23°S, 20°W[2]	4,000	5	5	Ledwell et al. (2000)
40.5°N, 60.5°W[3]	12–18	0.18 ± 0.04	0.15 ± 0.05	Ledwell et al. (2004)
58°S, 107°W[4]	1,500	1.3 ± 0.02	0.75 ± 0.07	Ledwell et al. (2011)

was nearly steady for months, and statistically uniform over horizontal distances of tens to hundreds of kilometers.

3.13 Perspectives

Length scales of weak and moderate turbulence are routinely resolved in the pycnocline. Favorable comparisons of average spectra of these measurements with theoretical forms confirm the universal character of the turbulence at inertial and dissipation scales. If more accurate estimates of pycnocline diffusivity are needed, we must progress from measuring mostly uncorrelated profiles to three-dimensional observations of patch generation and evolution to understand mechanisms of breaking, isotropy, mixing efficiency, and restratification after mixing ends.

4

Double Diffusion

4.1 Overview

Before 1960 the 100-fold difference between the molecular diffusivities of heat and salt was considered of no consequence in a strongly turbulent ocean. Stommel et al. (1956) inadvertently began changing this view by proposing to use copper pipes as perpetual salt fountains in profiles where both temperature and salinity decrease with depth and nutrient concentrations increase. After priming with water from its lower end, heat conducted through through the pipe walls, J_Q, would rapidly bring water inside the pipe to the temperature outside, resulting in buoyant upward flow of nutrient-rich water driven by low salinity in the pipe compared to outside (Figure 4.1). Groves (1959) challenged flow estimates, arguing that dumping 290 kg of Peruvian guano per year would enrich nutrients as much as a salt fountain. Events, however, took a happier turn. During a few exciting hours in 1960, related by Arons (1981) and Gregg (1991), Stommel and several colleagues at Woods Hole discovered double diffusion in salt water, showing that natural perturbations can trigger enhanced convection driven by the large contrast between heat and salt diffusivities.

Temperature and salinity can stably stratify water in three ways (Figure 4.2). 1) Because heat and salt both stratify the interface when hot, fresh water lies over cold, salty water, the rapid diffusion of heat relative to salt smoothes the edges of the interface while the core of the interface barely changes. Producing no motion, this configuration is diffusively stable. 2) When hot, salty water overlies cold, fresh water, rapid thermal diffusion cools the bottom of the upper layer while warming the top of the lower layer. Both generate density instabilities across the interface; cooling the upper layer allows its salt to sink into the less saline and less dense lower layer; warming the lower layer allows its fresher water to rise into the upper layer. These motions form tall, thin columns, a.k.a. salt fingers, of alternately sinking and rising water, reviewed by Kunze (2003). 3) When hot, salty water is on the bottom,

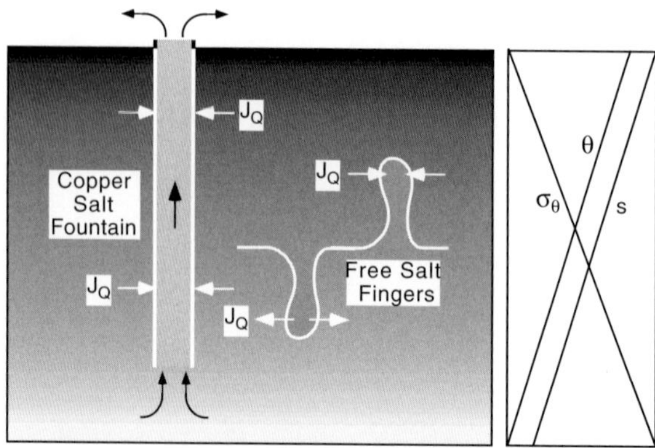

Figure 4.1 Schematic of a perpetual salt fountain and natural salt fingering in typical mid-gyre profiles. Once primed, upward flow in the fountain is driven by heat diffusion through its copper walls. Natural perturbations can also initiate fingering.

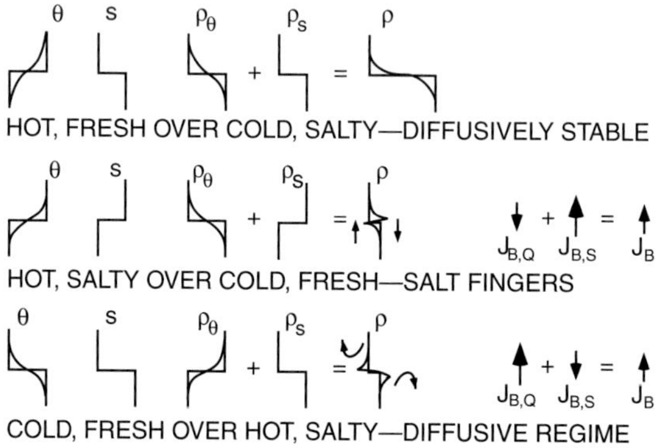

Figure 4.2 By smoothing the edges of the interface, thermal diffusion generates density inversions across the interface when hot, salty overlies cold, fresh. When cold, fresh is on top, both layers are destabilized but not the interface. Net buoyancy fluxes, J_b, are positive in both convective cases.

heat diffusion destabilizes both layers internally, generating convection in both, but water does not cross the interface. These modes are referred to as 'diffusive layers' or the 'diffusive regime', reviewed by Kelley et al. (2003).

Strong fingers in continuously stratified profiles lengthen until they become unstable, limiting further growth. Their buoyancy flux, however, destabilizes water above and below, forming layers bounding the zone of initial fingering. Additional fingering sites then develop at the top and bottom of the layers to produce a growing staircase. A parallel sequence occurs in profiles with temperature and salinity increasing downward (Figure 4.3). Because sharply defined staircases are

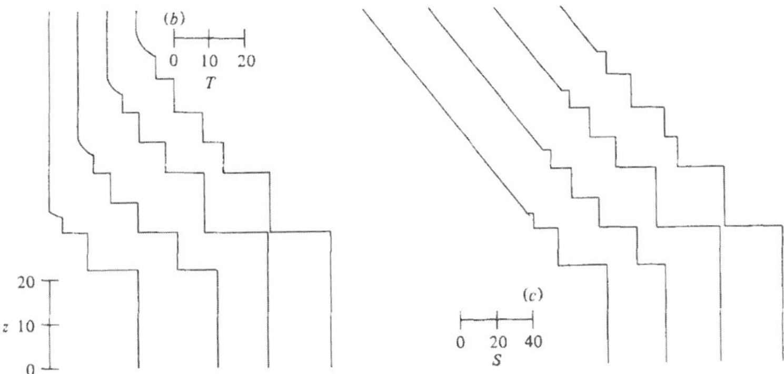

Figure 4.3 Evolution of temperature (left) and salinity (right) in a tank heated at the bottom after being filled with water having uniform temperature and salinity linearly increasing downward. As convection thickens the bottom layer, a linear temperature gradient grows above it, eventually overturning to form a new layer. Subsequently, lower layers merge as the staircase grows upward. (From Huppert and Linden, 1979)

rare and occur only where R_ρ is less than 2, their discovery in diffusively unstable profiles convinced skeptics that double diffusion occurs naturally, at least in some places.

After considering common aspects of double diffusion in Section 4.2, salt fingering is considered in detail in Section 4.3 and its staircases in Section 4.4. The diffusive regime is explored in Section 4.5 and its staircases in Section 4.6. Finally, the role of both regimes on upper and lower boundaries of thermohaline intrusions is discussed in Section 4.7. Fingering where staircases are lacking is examined in Section 8.6 as one aspect of a survey of mixing in open-ocean pycnoclines.

4.2 Double-Diffusive Convection

Adapting studies of thermal convection, theoretical work began with linear stability analysis and fluxes expressed in terms of laboratory 4/3 laws (Table 4.1) on the assumption that fluxes are independent of layer thickness. Fingering fluxes are in terms of $\beta \Delta S$ because the salinity component dominates the buoyancy flux. Likewise, fluxes through diffusive layers are in terms of $\alpha \Delta T$. As discussed in more detail in Sections 4.3 and 4.5, the 4/3 flux laws are consistent with laboratory experiments, but they overestimate oceanic fluxes in most comparisons with microstructure.

The intensity of convection is expressed by the Nusselt number,

$$Nu \equiv \frac{\text{Convective heat flux}}{\text{Solid plane heat flux}} = \frac{J_Q}{\rho c_p \kappa_T (\Delta T / H)}, \tag{4.1}$$

Table 4.1 *Buoyancy fluxes in* W kg^{-1} *in terms of the laboratory 4/3 law, flux ratios* $R^{\mathrm{sf}}_{\mathrm{flux}}$ *and* $R^{\mathrm{dr}}_{\mathrm{flux}}$, *and coefficients* c^{sf} *and* c^{dr}, *all functions of* R_ρ.

Salt Fingering	Diffusive Regime
$J^{\mathrm{sf}}_{b,S} = c^{\mathrm{sf}}\kappa_{\mathrm{T}}{}^{1/3}(g\beta\Delta S)^{4/3}$	$J^{\mathrm{dr}}_{b,Q} = c^{\mathrm{dr}}(\kappa_{\mathrm{T}}{}^2/\nu)^{1/3}(g\alpha\Delta\theta)^{4/3}$
$J^{\mathrm{sf}}_{b,Q} = -R^{\mathrm{sf}}_{\mathrm{flux}} J_{\mathrm{b,S}}$	$J^{\mathrm{dr}}_{b,S} = -R^{\mathrm{dr}}_{\mathrm{flux}} J^{\mathrm{dr}}_{b,Q}$
$J^{\mathrm{sf}}_{b} = J^{\mathrm{sf}}_{\mathrm{b,S}}(1 - R^{\mathrm{sf}}_{\mathrm{flux}})$	$J^{\mathrm{dr}}_{b} = J^{\mathrm{dr}}_{b,Q}(1 - R^{\mathrm{dr}}_{\mathrm{flux}})$

where H is layer thickness. With solid plane fluxes proportional to the 4/3 power of the driving gradient, the fluxes vary as $Nu^{1/3}$.

Double diffusive convection varies with the thermal Rayleigh number,

$$Ra_T \equiv \frac{\text{Buoyancy force}}{\text{Viscous force}}\left(\frac{\nu}{\kappa_T}\right) = \frac{g\alpha\Delta T H^3}{\nu\kappa_T}, \tag{4.2}$$

and the corresponding saline Rayleigh number, $Ra_S \equiv g\beta\Delta S H^3/(\nu\kappa_S)$.[1] During strong convection, temperature and salinity differences across the layer are concentrated in thin boundary layers. Rayleigh numbers estimated for diffusive staircases in the Arctic and Lake Kivu, Africa span $10^4 \leq Ra_T \leq 10^{10}$ (Hieronymus and Carpenter, 2016), a wider range than achieved in laboratory and numerical studies. This discrepancy is cited as the reason for differences between laboratory fluxes and independent estimates of fluxes in some oceanic staircases. Owing to the strong temperature dependence of the thermal expansion coefficient, α, the buoyancy flux in diffusive layers also varies strongly with temperature, as well as with density and flux ratios. This has large effects where diffusive layering is most prominent, in cold Arctic and Antarctic waters and in hot geothermal basins.

4.2.1 Linear Stability Analysis

To explore double diffusion, Stern (1960), Veronis (1965, 1968), and Baines and Gill (1969) added a linear salinity gradient to extend linear stability analysis of thermal convection to two components (Fig. 4.4). Assuming vertical advection is opposed by lateral viscosity and diffusion, the equations for momentum, heat, and salt conservation are:

$$\frac{\partial u}{\partial t} = -\frac{1}{\rho_0}\frac{\partial p}{\partial x} + \nu\nabla^2 u \tag{4.3}$$

[1] As with R_ρ, β and ΔS must both be expressed in terms of ppt or concentration units.

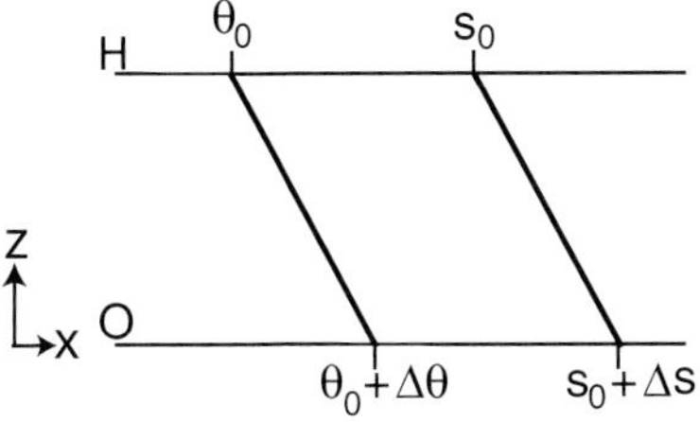

Figure 4.4 Schematic of temperature and salinity decreasing linearly upward in water between two parallel plates H apart.

$$\frac{\partial w}{\partial t} = -\frac{1}{\rho_0}\frac{\partial p}{\partial z} + g(\alpha\theta - \beta S) + \nu\nabla^2 w \tag{4.4}$$

$$\frac{\partial \theta}{\partial t} - w\frac{\Delta\theta}{H} = \kappa_{\mathrm{T}}\nabla^2\theta \tag{4.5}$$

$$\frac{\partial S}{\partial t} - w\frac{\Delta S}{H} = \kappa_{\mathrm{S}}\nabla^2 S. \tag{4.6}$$

Assuming normal mode solutions,

$$\psi = \psi_0 \, \exp(\sigma t) \, \sin(\pi k_x x)\sin(\pi k_z z) \tag{4.7}$$

$$(\theta, s) = (\theta_0, s_0) \, \exp(\sigma t) \, \cos(\pi k_x x) \, \sin(\pi k_z z), \tag{4.8}$$

with a stream function, e.g. $u = \partial\psi/\partial z$, and removing pressure by cross-differentiation, these equations are combined into a characteristic cubic for σ, the growth rate of infinitesimal disturbances. One root is always real, representing growing convection. Neutral boundaries, $\sigma_{\mathrm{imag}} = 0$, identify steady flow and yield

$$Ra_T = \frac{27\pi^4}{4} + Ra_S, \tag{4.9}$$

which reduces to the critical Rayleigh number for thermal convection. For small Ra_T and Ra_S, this boundary is ZXY in Figure 4.5. Other roots are real or complex conjugates.

Since positive Ra_T and Ra_S correspond to temperature and salinity decreasing upwards, static stability occurs in the shaded region below $Ra_T = Ra_S$, along which $R_\rho = 1$. Left of ZXY there is always one solution with a positive real root representing exponential growth. Below $Ra_T = Ra_S$, i.e. in the third quadrant, this growth occurs in stable stratification and results from salt fingering. Salt fingering is possible everywhere in the lightly shaded portion of the third quadrant, which is nearly all of the quadrant owing to the exaggerated slope of ZXY. More quantitatively, Rayleigh numbers in the sea usually far exceed the critical value.

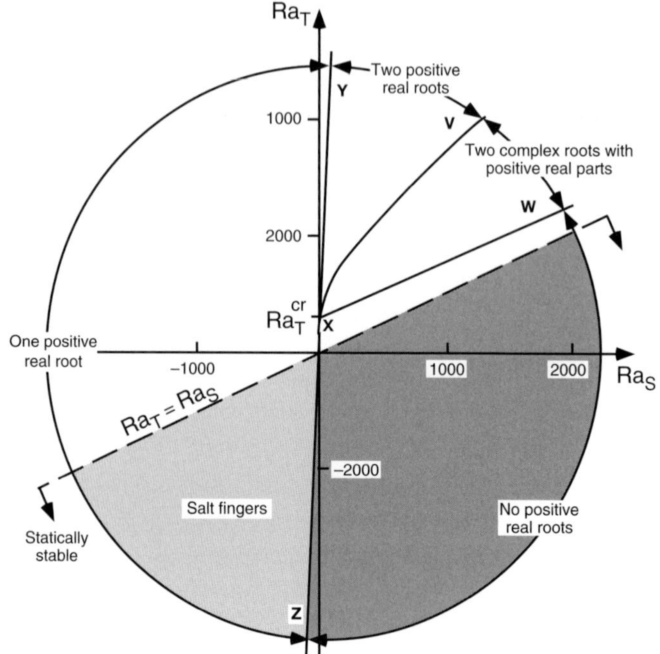

Figure 4.5 Stability diagram for double diffusion at low Rayleigh numbers and fixed wavenumber. Arrows on the circle mark the domains of roots of the characteristic equation. The slope of ZXY is exagerated to separate it from the vertical axis. (Based on analysis by Baines and Gill, 1969)

Consequently, in practice (4.9) reduces to $Ra_T = Ra_S$, and salt fingering is possible when

$$1 < R_\rho < \kappa_\mathrm{T}/\kappa_\mathrm{S}. \tag{4.10}$$

Since $90 < \kappa_\mathrm{T}/\kappa_\mathrm{S} < 220$, this condition is satisfied nearly everywhere hot, salty water overlies more dense, cold, less saline water.

Roots above $Ra_T = Ra_S$ represent static instabilities modified by heat and salt diffusion. Only direct modes, positive real roots, exist above $Ra_T = Ra_S$ and to the left of ZXY. Growing oscillations, termed ‘overstable modes’, are found by solving for $\sigma^2_{\mathrm{imag}} \neq 0$, e.g. XW in Figure 4.5. Solutions in YXV produce two direct modes and a growing oscillating instability. XW is the boundary of the overstable oscillations, a prediction confirmed observationally by Shirtcliffe (1967). At larger Ra_S than shown in the figure, XW crosses $Ra_T = Ra_S$. The instability then occurs in stable profiles as the diffusive regime. In ZXW, all solutions have negative real roots and decay exponentially, indicating both static and diffusive instability. Except for the tiny wedge between ZX and the negative Ra_T axis, this stable region results from hot, fresh above cold, salty water. Subsequent direct numerical simulations find boundary layer growth affecting the instability so the oscillatory phase is insignificant compared to convective breakdown (Carpenter et al., 2012a). Baines and Gill (1969) also considered the instability of convection with stabilizing solute and destabilizing heat gradients. Convection always began with oscillations in separate cells when stability was marginal, transporting solute more rapidly than heat.

4.2.2 The Turner Angle

To show where double diffusion is likely without the ambiguity of R_ρ, Ruddick (1983) proposed the Turner angle, which the Gibbs Seawater Toolbox implements as

$$Tu = \text{atan2d}\left(\underbrace{\alpha^{\Theta}\partial\Theta/\partial z + \beta^{\Theta}\partial S_A/\partial z}_{Y},\ \underbrace{\alpha^{\Theta}\partial\Theta/\partial z - \beta^{\Theta}\partial S_A/\partial z}_{X}\right), \qquad (4.11)$$

where atan2d is the four-quadrant arctangent function in degrees. As shown in Figure 4.6 (upper), statically stable profiles occur when $-90^\circ < Tu < +90^\circ$. Fingering is possible between $+45^\circ$, where it is unlikely, and $+90^\circ$, where it is very likely. Diffusive layers are possible from -45° to -90° and most likely near -90°. The lower plot shows rapid changes in regime with depth through multiple thermohaline intrusions.

4.2.3 Global Patterns

You (2002) finds 30% of the ocean with temperature and salinity profiles suitable at some depth for salt fingering. These conditions, Turner angles of $+60^\circ$ to $+80^\circ$ in Figure 4.7, are found in the upper kilometer of the major ocean basins at mid-latitudes. North Atlantic profiles are suitable for fingering at most depths, but in the North and South Pacific potential fingering is limited to the upper pycnocline. Deep fingering is possible, however, beneath mid-depth salinity maxima in the South Atlantic and South Indian oceans. These profiles result from 1) excess evaporation, principally in subtropical gyres, 2) interleaving of North Atlantic Deep Water (NADW) and Circumpolar Water (CW), and 3) overflows from saline marginal seas, e.g. outflow from the Mediterranean Sea into the Atlantic and from the Red Sea into the Indian Ocean.

The possibility of diffusive layering other than on intrusion boundaries is more restricted, occurring at some depths in 14% of profiles as a consequence of 1) surface cooling and ice melting at high latitudes, 2) excess precipitation, and 3) intrusions of cool, fresh water from high latitudes into the ocean interior. These

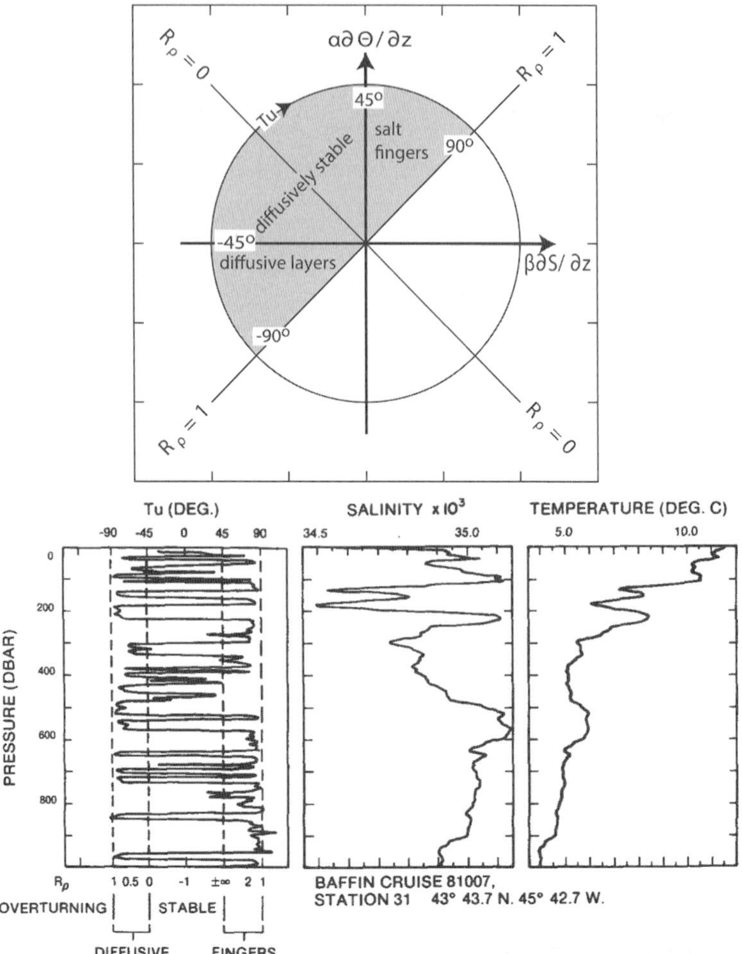

Figure 4.6 (Upper) Turner angle, Tu, on a gradient ΘS diagram with statically stable regions shaded. Fingering is most likely close to $+90^\circ$ and diffusive layering close to -90°. (Lower) The Turner angle through multiple thermohaline intrusions. (From Ruddick, 1983)

occur at all longitudes in the Southern Ocean around Antarctica and throughout the Arctic (Kelley et al., 2003).

4.3 Salt Fingering

Early understanding of fingering came from laboratory experiments with fingers extending across a thin interface between homogenous convecting layers (Turner, 1973), for which $(\Delta S)^{4/3}$ flux laws apply. Viewed at high resolution, fingers are complex (Figure 4.8), with bulbous noses produced by water velocity in the

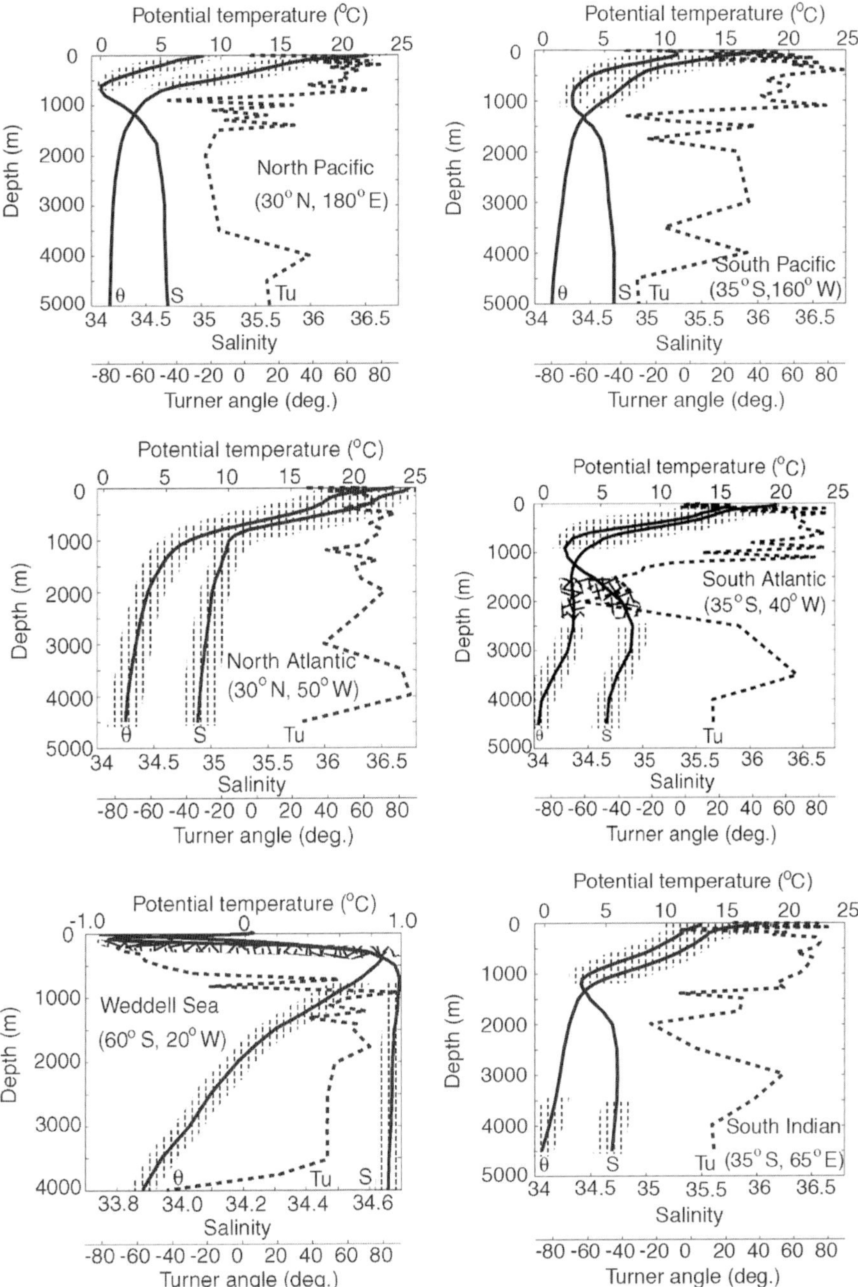

Figure 4.7 Potential temperature, salinity, and Turner angles for representative profiles. Vertical lines show where fingering is possible and crosses where double-diffusive layering can occur. (Adapted from You, 2002)

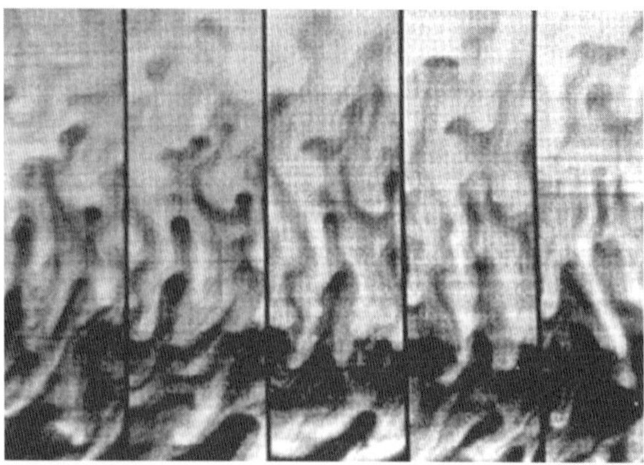

Figure 4.8 Images 19 mm wide and at 5 s intervals of fluorescent dye in a fingering interface illuminated by a vertical sheet of light. The sequence began 320 s after the interface was formed with $R_\rho = 2.46$. (From Taylor and Bucens, 1989, using images kindly supplied by John Taylor)

fingers exceeding the rate at which the fingers lengthen. Later experiments included diffusively unstable linear profiles. To avoid sidewall heat losses and to slow the rate of interactions, some experiments replaced temperature with sugar, which has a diffusivity one-third that of salt. Analytic theory addressed major features of the experiments – finger diameter, length, velocity, and fluxes – as well as the evolution of linear profiles into staircases with fingering interfaces separated by convecting layers. Comparing predicted characteristics with oceanic measurements tests the degree to which double diffusion in laboratories can be extrapolated to the ocean and brackish lakes.

4.3.1 Analytic Modeling

The equations of motion (Stern, 1960) are formulated with horizontal viscosity opposing buoyancy forcing of vertical momentum, and with molecular diffusion balancing vertical advection to change temperature and salinity,

$$\frac{\partial w}{\partial t} - \nu \nabla_h^2 w = b = g(\alpha \delta T - \beta \delta S) \tag{4.12}$$

$$\frac{\partial \delta T}{\partial t} - \kappa_T \nabla_h^2 \delta T + w \frac{\partial T}{\partial z} = 0 \tag{4.13}$$

$$\frac{\partial \delta S}{\partial t} - \kappa_S \nabla_h^2 \delta S + w \frac{\partial S}{\partial z} = 0. \tag{4.14}$$

Steady solutions, with time derivatives set to zero, yield properties of equilibrium fingers. To study growing fingers, following Stern (1975), Kunze (1987) applied

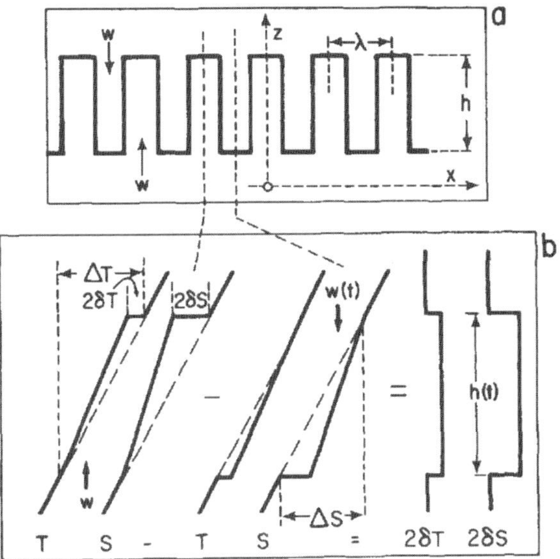

Figure 4.9 Schematic of fingers with horizontal wavelength λ and height $h(t)$. Thick lines in vertical cross-section (a) mark strong gradients at finger ends and between adjacent fingers. Profiles in (b) show temperature and salinity in rising and sinking fingers with more fractional change in salinity, $2\delta S/\Delta S$, than in temperature, $2\delta T/\Delta T$. Contrasts between adjacent fingers are on the right. (From Kunze, 1987)

the full equations to the geometry in Figure 4.9. With σ for the growth rate and h for finger height, continuity and geometry give

$$w = \frac{\sigma h}{2}, \ \frac{\partial T}{\partial z} = \left\langle \frac{\overline{\partial T}}{\partial z} \right\rangle (1 - \delta T), \ \frac{\partial S}{\partial z} = \left\langle \frac{\overline{\partial S}}{\partial z} \right\rangle (1 - \delta S) \tag{4.15}$$

for velocity and scalar gradients in the fingers, with δT and δS as nondimensional temperature and salinity anomalies. With these constraints, the three equations of motion contain five unknowns: δ_T, δ_S, σ, h, and $k_h = 2\pi/\lambda_h$ for the horizontal wavenumber of the fingers. Assuming sinusoidal structure and nondimensionalizing leaves three unknowns in terms of k_h because finger height, h, cancels out. Only one root of the cubic is real, corresponding to growth for $1 < R_\rho < \kappa_t/\kappa_S$. For representative density ratios, the fastest growth rate,

$$\sigma_{\text{fg}} = \frac{1}{2}\sqrt{\frac{(\kappa_T - R_\rho \kappa_S) g \beta \langle \overline{\partial S/\partial z} \rangle}{\nu}} \left(\sqrt{R_\rho} - \sqrt{R_\rho - 2} \right), \tag{4.16}$$

corresponds to a narrow peak versus horizontal wavenumber, k_h (Figure 4.10).

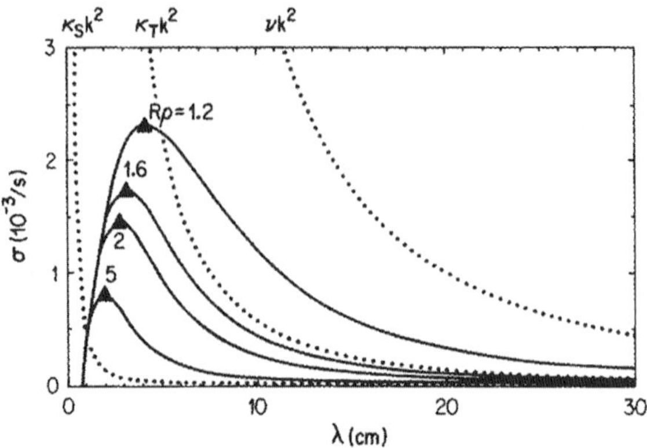

Figure 4.10 Growth rate, σ, versus horizontal wavelength, λ_h, for four values of R_ρ, including 1.6 for the fingering staircase east of Barbados. Triangles mark the fastest growth rates, and dotted lines are molecular decay scales, as marked. (From Kunze, 1987)

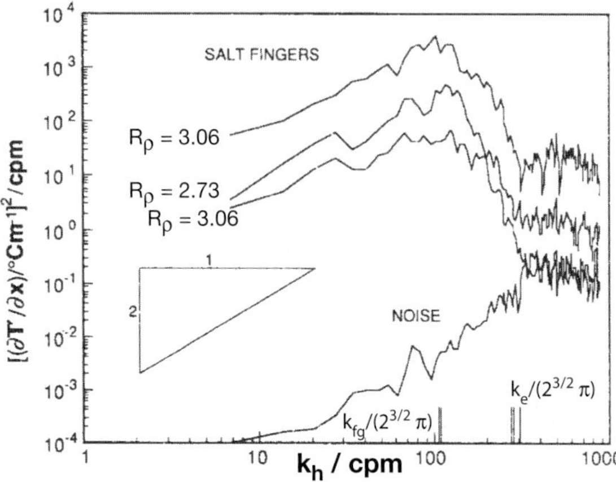

Figure 4.11 Spectra of horizontal temperature gradients through the fingers in Figure 4.8 are more peaked than turbulent spectra. k_{fg} and k_e are wavenumbers of fastest-growing and equilibrium fingers. Upper spectra are progressively offset by one decade. (From Taylor and Bucens, 1989)

Finger Width

The fingers growing most rapidly are thicker than equilibrium fingers,

$$k_{fg} = \left(\frac{g\alpha \overline{T}_z}{\nu\kappa_T}\right)^{1/4}\left(1 - \frac{1}{R_\rho}\right)^{1/4}, \; k_e = \left(\frac{g\alpha \overline{T}_z}{\nu\kappa_T}\right)^{1/4}\left(\frac{\kappa_T/\kappa_S}{R_\rho} - 1\right)^{1/4}. \quad (4.17)$$

Horizontal temperature gradient spectra of laboratory fingers show k_{fg} corresponding to the peak and k_e to the effective upper bound of the spectra (Figure 4.11).

For $R_\rho = 1.6$ in the Barbados staircase, $\lambda_h = 30\text{–}40$ mm. Because fingers become thinner with increasing density ratio and mean gradients, the large gradients needed for laboratory tanks produce fingers considerably thinner than expected in typical ocean gradients.

Finger Length

Fingers are observed to reach a maximum length, but linear models lack the physics to constrain growth. In response, Stern (1969) hypothesized collective instability when the finger buoyancy Reynolds number is of order 1, i.e. $Re_b \sim bw/\nu N^2 \sim \mathcal{O}(1)$. Buoyancy released at finger ends was assumed to generate internal waves disrupting the fingers. As an alternative, Kunze (1987) limited growth by assuming breakdown at a critical value of the gradient Richardson number between fingers,

$$Ri_{\text{finger}} \equiv N^2/(\nabla_h w)^2 \sim N^2/(k_h w)^2 = 1/4. \tag{4.18}$$

This criterion is identical to Stern's, provided $\partial w/\partial t$ is negligible. The corresponding maximum length is

$$h_{\max} = \frac{2^{3/2}(C_w Ri_{\text{finger}})^{-1/2}\nu^{3/4}}{\left(\kappa_T g\beta \langle\partial S/\partial z\rangle\right)^{1/4}} \left(\sqrt{R_\rho} + \sqrt{R_\rho - 1}\right)^{3/2}, \tag{4.19}$$

with $C_w = 1/2$ for horizontal velocity represented by a single sinusoid and 1/4 for a double sinusoid. Evaluated for the Barbados staircase, (4.19) gives lengths of a few tens of centimeters for fingers extending fully across an interface and slightly less than 1 m within a 2 m interface (Figure 4.12). Most interfaces in the staircase are considerably thicker (Section 4.4).

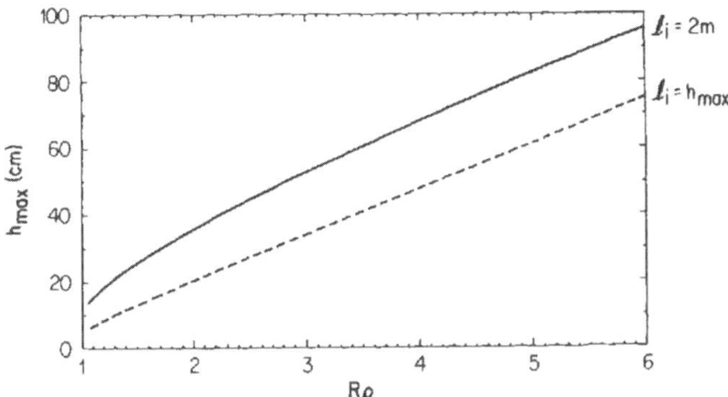

Figure 4.12 Maximum finger length versus R_ρ in an interface with $\Delta S = 0.1$ ppt, for a 2 m-thick interface (solid) and for fingers extending fully across interfaces (dashed). (From Kunze, 1987)

The time required for fingers to grow from initial perturbation to maximum length increases from one buoyancy period at $R_\rho = 1.2$ to five at $R_\rho = 2$. The maximum velocity during growth,

$$w_{\max} = (2C_w Ri)^{-1/2} \left(\nu \kappa_T g \beta \langle \overline{\partial S/\partial z} \rangle \right)^{1/4} \left(\sqrt{R_\rho} + \sqrt{R_\rho - 1}\right)^{1/2}, \qquad (4.20)$$

gives $\sim$0.1 mm s^{-1} for the Barbados staircase (Kunze, 1987). Using a tracer, Linden (1973) noted that velocities in fingers greatly exceed the rate of finger lengthening, i.e. $w > \partial h/\partial t$, explaining the bulbous ends on fingers.

4.3.2 Fluxes

Because direct flux measurements, e.g. $\overline{S'w'}$, are beyond present technology, fluxes across fingering interfaces in tanks have been estimated from rates of change in temperature and salinity in layers bounding the interfaces. These laboratory fluxes, however, are much larger than those inferred by measuring ϵ in oceanic staircases and assuming that the turbulent kinetic energy equation reduces to $J_b \approx \epsilon$ when shear production is negligible. Although fingering is involved in both settings, the environments are markedly different. Laboratory experiments begin with fingers extending across thin interfaces, which may subsequently thicken until the fingers become self-limiting. The $(\Delta S)^{4/3}$ laws thus describe fluxes across interfaces no more than tens of centimeters thick filled with fingers extending completely across them. Fingering in the ocean occurs in continuous gradients, requiring flux laws in terms of gradients rather than ΔT and ΔS. This is the case even in staircases like the one east of Barbados, where most interfaces are 2–5 m thick. Why interfaces are thicker than the maximum finger length is not understood. Also, unlike the initial laboratory experiments, fingering in the ocean occurs in the presence of vertical shear, which changes them into salt sheets parallel to the direction of the shear.

Laboratory Measurements Fitted to the 4/3 Law

In spite of scatter and offsets between experiments, laboratory estimates of finger flux ratios show a consistent trend, rising from $\sim$0.5 at large R_ρ toward 1 as $R_\rho \to 1$ (Figure 4.13). Owing to the speed of the mixing as R_ρ decreases, flux ratios are difficult to measure accurately. The fingering coefficient, c^{sf}, also rises steeply as $R_\rho \to 1$ and is equally difficult to measure (Figure 4.14).

Fluxes in Continuous, Unsheared Gradients

Assuming that fingering is dominated by the fastest-growing fingers, Kunze (1987) modeled fluxes in continuous gradients as

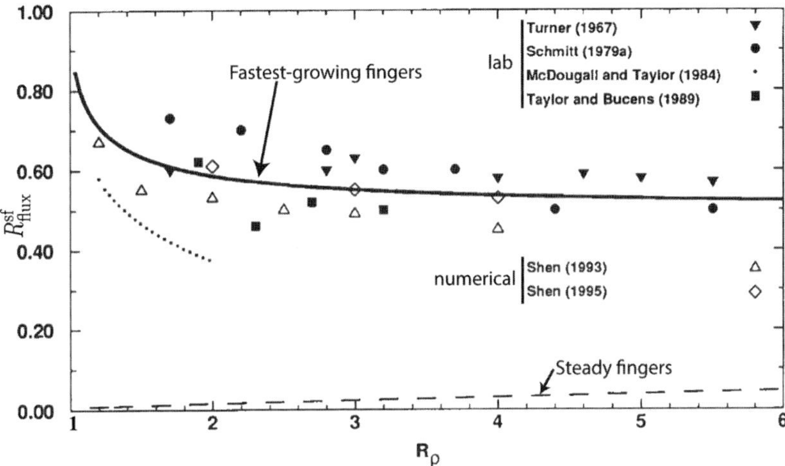

Figure 4.13 Laboratory, numerical, and theoretical salt finger flux ratios. (Adapted from Kunze, 2003)

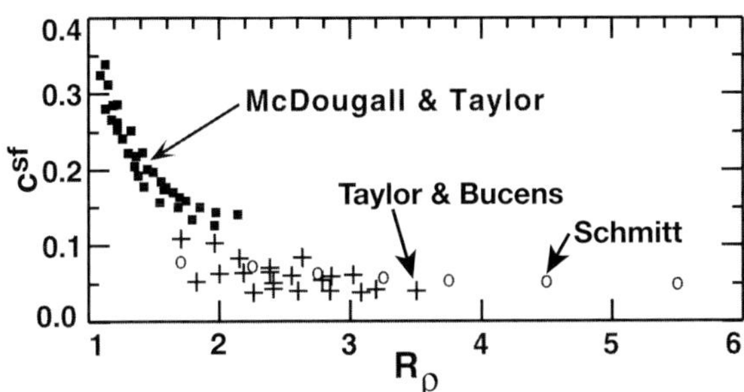

Figure 4.14 Laboratory measurements of the salt finger flux coefficient by Schmitt (1979), McDougall and Taylor (1984), and Taylor and Bucens (1989).

$$J^{sf}_{b,S}\Big|_{max} = \frac{C_S}{2RiC_w}\nu g\beta\left\langle\overline{\partial S/\partial z}\right\rangle\left[\left(\sqrt{R_\rho}+\sqrt{R_\rho-1}\right)^2\right] \tag{4.21}$$

$$J^{sf}_{b,Q}\Big|_{max} = \frac{C_T}{2RiC_w}\nu\beta\left\langle\overline{\partial S/\partial z}\right\rangle\left[\sqrt{R_\rho}\left(\sqrt{R_\rho}+\sqrt{R_\rho-1}\right)\right], \tag{4.22}$$

with $C_T = C_S = C_w = 1/2$ for sinusoidal fingers and 1/4 for double sinusoids. Dividing (4.21) by $g\beta\left\langle\overline{\partial S/\partial z}\right\rangle$ gives the diffusivity, K_S (Figure 4.15), and including R_ρ in the denominator when dividing (4.22) yields K_T.

Eddy Coefficients

K_S formed from (4.21) rises from $< 10^{-6}$ m^2 s^{-1} at $R_\rho = 1.2$ to slightly less than 10^{-5} m^2 s^{-1} at high density ratios (Figure 4.15). More in keeping with laboratory flux laws, numerical simulations show the opposite, with K_S decreasing

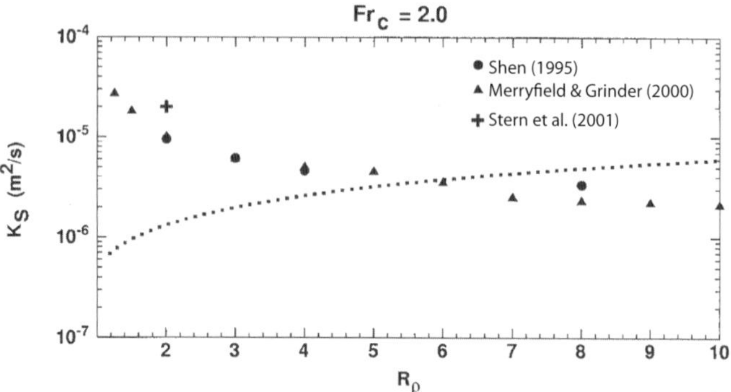

Figure 4.15 K_S for salt fingering in a constant-N profile for (4.21) as a dotted line, and in simulations. (Adapted from Kunze, 2003. © American Meteorological Society. Used with permission)

from 3×10^{-5} for R_ρ close to 1, to 2×10^{-6} at $R_\rho = 10$. Limits on domain size restricted most κ_S/κ_T simulations to two dimensions. Extrapolating three-dimensional runs with larger diffusivity ratios to more realistic values indicates that the two-dimensional runs were underestimates by factors of 2–3 (Stern et al., 2001).

To compare with diapycnal diffusivities produced by turbulence, McDougall (1988) assumed $\epsilon = J_b$ and used $N^2 = g(\alpha\overline{\Theta_z} - \beta\overline{S_z}) = g\beta\overline{S_z}(R_\rho - 1)$ to obtain eddy coefficients for salt fingering as

$$K_S^{\rm sf} \equiv -\frac{J_S^{\rm sf}}{\rho < S_z >} = \frac{R_\rho - 1}{1 - R_{\rm flux}^{\rm sf}} \frac{\epsilon}{N^2} \quad [{\rm m^2\ s^{-1}}] \tag{4.23}$$

$$K_T^{\rm sf} \equiv -\frac{J_Q^{\rm sf}}{\rho c_p < \Theta_z >} = \frac{R_{\rm flux}^{\rm sf}}{R_\rho} K_S \quad [{\rm m^2\ s^{-1}}] \tag{4.24}$$

$$K_\rho^{\rm sf} \equiv -\frac{J_b^{\rm sf}}{N^2} = -\frac{\epsilon}{N^2} \quad [{\rm m^2\ s^{-1}}]. \tag{4.25}$$

Unlike the component fluxes, the net buoyancy flux, $J_b^{\rm sf}$, is up-gradient, making profiles more stratified, not less, because the energy source for the instability is the available potential energy released by the contrast in molecular diffusivities of heat and salt. Because of this, by analogy to buoyancy-driven turbulence at boundaries, fingering has $\Gamma_{\rm mix} = 1$, larger than for shear-generated turbulence with the same dissipation rate and stratification. Inoue et al. (2008) tested these fluxes with the North Atlantic Tracer Release Experiment (NATRE) measurements, but simulations by Kimura and Smyth (2011) demonstrated that dissipation-scale velocity in sheared fingering is not isotropic, calling into question comparisons with microstructure estimates based on isotropy.

Applying these diffusivities to staircases is ad hoc, owing to the assumption that turbulent production and dissipation occur against uniform background stratification, in contrast to the actual steppy stratification in staircases. Also, ϵ and χ_T are not distributed uniformly, and there is substantial mean and fluctuating shear (Section 4.4) not considered in the derivation. Moreover, additional errors result from assuming isotropy when calculating dissipation rates in anisotropic fingers (Flanagan et al., 2014; Radko, 2014).

Cox Numbers

Closely related to fluxes, Kunze (1987) gives fingering Cox numbers as

$$< \mathrm{COX}_T > = \frac{\nu\sqrt{R_\rho - 1}\left(\sqrt{R_\rho} + \sqrt{R_\rho - 1}\right)}{2\sigma t_{\max} C_w \, Ri_{\text{finger}} \, \kappa_T \, R_\rho} \tag{4.26}$$

$$< \mathrm{COX}_S > = \frac{\nu\sqrt{R_\rho - 1}\left(\sqrt{R_\rho} + \sqrt{R_\rho - 1}\right)^3}{2\sigma t_{\max} C_w \, Ri_{\text{finger}} \, \kappa_T}, \tag{4.27}$$

using

$$\sigma t_{\max} = \ln\left[\left(\frac{8\nu}{C_w Ri_{\text{finger}} \kappa_T}\right)^{1/2} (R_\rho - 1)^{1/4} \left(\sqrt{R_\rho} + \sqrt{R_\rho - 1}\right)^{3/2}\right]. \tag{4.28}$$

The average Cox number for temperature is nearly constant at 8 for $R_\rho \geq 2$ and asymptotes to zero as $R_\rho \rightarrow 1$. By contrast, $< \mathrm{COX}_S >$ rises quadratically from zero at $R_\rho = 1$ and passes 8 at $R_\rho = 5$.

4.3.3 Effects of Turbulence and Shear

Early experiments with salt fingers revealed that fingers were easily disrupted by light stirring (Stern, 1960), leading to skepticism about their importance in oceanic pycnoclines believed to be turbulent. To explore whether the fluxes were shut off or just diminished by turbulence, Gregg (1968) inferred fluxes across a fingering interface in a tank with oscillating grids stirring both layers bounding the interface. Increasing the frequency of oscillation raised the level of turbulence at the interface, causing the flux ratio to rise from $R^{\text{sf}}_{\text{flux}} \sim 0.5$, characteristic of fingering, to the density ratio across the interface, the ratio expected for mechanically-driven mixing. The reduction in fingering flux was gradual rather than abrupt. Linden (1971) quantified the the reduction in terms of $\overline{(w')^2}/\overline{(u')^2}$, where w' is the velocity of undisturbed fingers, and u' is the rms turbulent velocity. Salt flux had a minimum when the ratio was $\sim$0.3; mechanical mixing was dominant when the ratio was $\lesssim 0.05$.

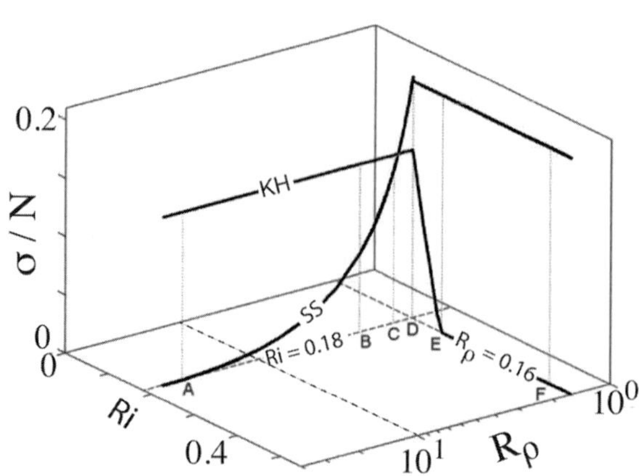

Figure 4.16 Normalized growth rates, σ/N, for salt sheets (SS) and Kelvin–Helmholtz instabilities for $Pr = 7$, $\kappa_S/\kappa_T = 0.04$, and $\lambda_S/2h = 0.1$. Growth rates are equal at $R_\rho = 2$ and $Ri = 0.18$ (C). (Adapted from Smyth and Kimura, 2011. © American Meteorological Society. Used with permission)

Mean shear stretches fingers into salt sheets aligned with the mean flow while retaining the same horizontal wavenumber as the fastest-growing fingers (Linden, 1974). Growth rates are similar to those for Kelvin–Helmholtz instabilities over common ranges of gradient Richardson numbers and density ratios (Figure 4.16). As R_ρ decreases at constant Ri, the salt sheet growth rate increases exponentially to exceed the Kelvin–Helmholtz rate when $R_\rho \lesssim 2$. Increasing the gradient Richardson number to 0.25 brings the Kelvin–Helmholtz rate to zero without affecting salt sheet growth. In addition, as interfaces thicken both rates decrease, as $h^{-1/2}$ for sheets and h^{-1} for shear instability (Smyth and Kimura, 2011).

Direct numerical simulations of thin interfaces with realistic diffusivities find two secondary instabilities as sheets grow (Kimura and Smyth, 2007). As alternate salt sheets rise and sink, compression of their tips against ambient fluid outside the interface amplifies shear more than stratification. The resulting tip instability generates quasi-periodic propagating modes, P in Figure 4.17. A separate buckling instability has vertical wavelength $\approx 1.8\lambda_{fg}$ and is slightly tilted along the mean flow. The two secondary instabilities combine to make the flow turbulent, reducing the effective salt diffusivity below fluxes for simple sheared fingering.

Exploring the effect of shear on fingering across thin interfaces with direct numerical simulations showed dependencies on the density ratio and the gradient Richardson number, even when $Ri_g > 1/4$ (Kimura and Smyth, 2011). A simulation with $R_\rho = 1.6$ and $Ri_g = 6$ found linear growth after the interface was formed until $\sigma_{\text{fg}} t \sim 2$. This was followed by exponential growth until the flow transitioned to turbulence at $\sigma_{\text{fg}} t \sim 8$. The degree of isotropy decreased when R_ρ

Table 4.2 *Characteristics of the major salt fingering staircases.*

Location	Depth m	R_ρ	Area km^2
Mediterranean outflow	1,250–1,500	1.3	6.3×10^4
East of Barbados	300–600	1.6	5×10^5
Tyrrhenian Sea	600–2,500	1.15	5×10^4

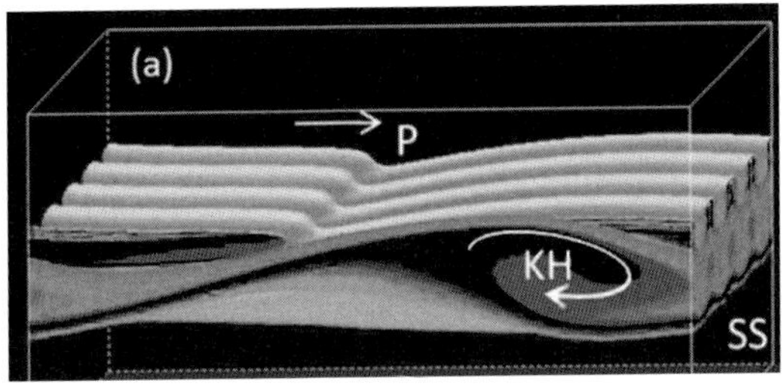

Figure 4.17 Salt buoyancy for C in Figure 4.16. P is the propagating disturbance, and SS are rising and sinking salt sheets. (Adapted from Smyth and Kimura, 2011. © American Meteorological Society. Used with permission)

was increased, with temperature and salinity variances coming increasingly from spanwise, or cross-flow, derivatives. Analyzing what a profiler would measure revealed nearly zero contribution to the true dissipation rate during linear growth and only 32% to 53% after the transition to turbulence, owing to anisotropy in the velocity dissipation range. Because scalar fields were nearly isotropic, profiler estimates of the mixing coefficient, Γ_{mix}, were twice the true value. Allowing Kelvin–Helmholtz instabilities when $Ri_g > 1/4$ showed that when growth rates of salt sheets and Kelvin–Helmholtz were equal, the sheets dominated because R_ρ of the thickening interface grew faster than Ri_g (Smyth and Kimura, 2011). The mixing coefficient, Γ_{mix}, approached 0.2 when shear dominated and exceeded 0.5 when salt sheets were dominant.

4.4 Salt Fingering Staircases

Extensive salt fingering staircases are rare, having been found in only three locations where large-scale circulation produces density ratios $R_\rho \lesssim 1.6$ over several hundred meters (Table 4.2). The first to be found is on the lower side of the tongue of warm, salty Mediterranean water in the eastern subtropical North Atlantic. The largest staircase field is immediately north of South America and east of Barbados. Shallower than the others, it is also the most intensely sampled. The staircase with the largest vertical span is in the Tyrrhenian Sea, west of central Italy.

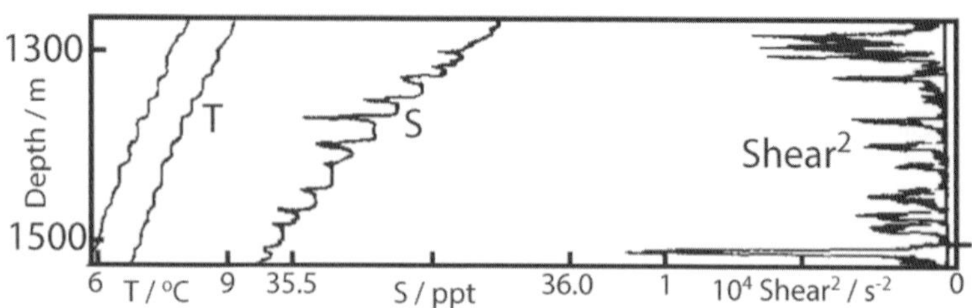

Figure 4.18 Temperature (from a CTD and a free-fall profiler), salinity, and shear through the fingering staircase beneath the Mediterranean outflow. Salinity spikes are artifacts of mismatched sensor dynamic responses. (Adapted from Simpson et al., 1979)

4.4.1 Mediterranean Outflow into the North Atlantic

Discovered in an early salinity-temperature-depth (STD) profile (Tait and Howe, 1968), the staircase beneath the Mediterranean outflow was identified as double diffusive by its uniqueness: only double diffusion is known to produce extensive stairsteps. A later free-fall profile revealed shear concentrated across the steps (Figure 4.18). The staircase has been traced laterally for tens of kilometers, most recently as continuous records in backscatter from high-gain processing of 10–100 Hz seismic pulses normally used to profile bottom sediment. One record from a towed acoustic array shows a Mediterranean eddy (meddy) disrupting the staircase (Biescas et al., 2010). Elsewhere, the staircase was coherent for up to 50 km.

4.4.2 East of Barbados

The staircase east of Barbados is between 300 and 600 m, where R_ρ has a minimum of 1.6 between warm, saline Subtropical Underwater and cool, fresh Antarctic Intermediate Water. Owing to its shallow depth, this staircase has been probed by two intensive expeditions. The first, in 1985, mapped the field with air-expendable probes (Figure 4.19). Conductivity-temperature-depth (CTD) sections traced layers 400 km laterally, several of which were also in surveys eight months later (Schmitt et al., 1987). Temperature and salinity in the layers became warmer, saltier, and denser along the mean current toward the northwest (Figure 4.20). Changes along layers were surprisingly uniform, with $R_\rho^{\text{layer}} = \alpha\Delta\theta/\beta\Delta S = 0.85$. Noting that these changes result from vertical divergences in fluxes across the interfaces, McDougall (1991) concluded that the striking uniformity of along-layer changes cannot be due solely to the flux ratio for double diffusion. Other processes must be involved, the simplest of which is modest advection of mass through the interfaces due to cabbeling.

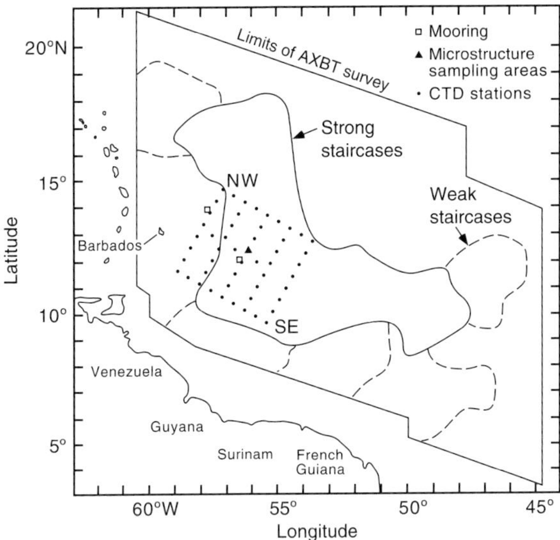

Figure 4.19 Extent of fingering staircases in fall 1985 from aircraft AXBT surveys overlaid on positions of CTD profiles and intensive sampling with a thermistor chain and towed and dropped microstructure sensors. In 2001, SF_6 was released near the northeast corner of the 1985 CTD grid. (From Schmitt et al., 1987)

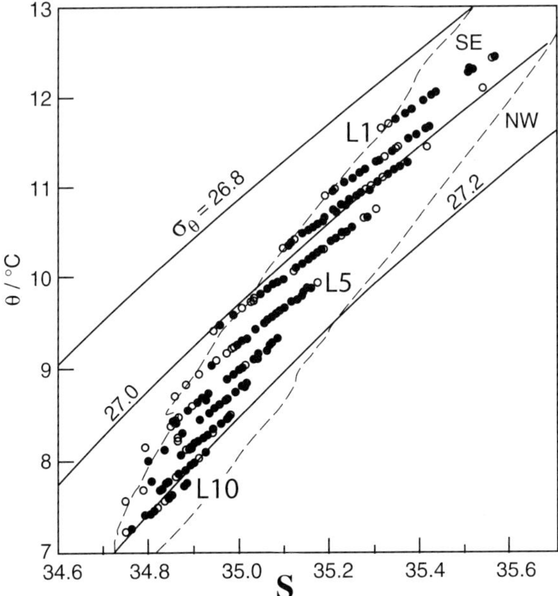

Figure 4.20 Trends in potential temperature and salinity in the Barbados staircase during spring 1985. Layers $\geq$ 10 m thick are solid, and those 5–10 m are open. Dashed lines are θS relations from southeast and northwest corners of the CTD survey. Crossing potential density surfaces, layers became warmer and saltier from southeast to northwest. L1, etc. correspond to layer numbers in Figure 4.23. (Adapted from Schmitt et al., 1987)

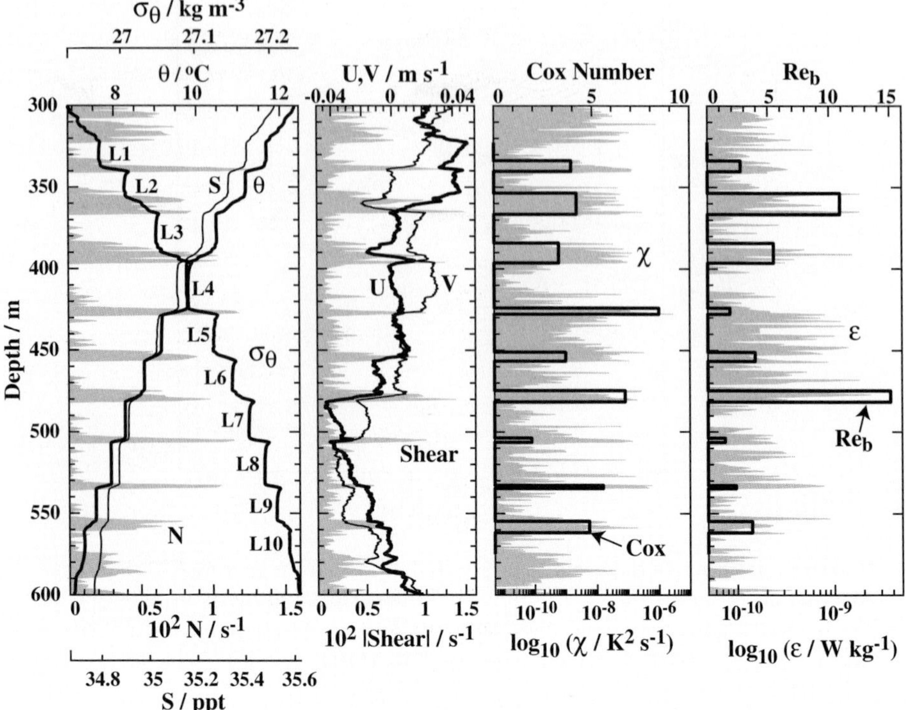

Figure 4.21 Salt fingering staircase east of Barbados in 1985. Averages over interfaces, the Cox number and Re_b are overestimates, as they use ϵ and χ computed assuming isotropy. (Replotted from Gregg and Sanford, 1987)

Layers were 11–26 m thick and contained substantial shear variance, even though most of the shear was across the steps (Figure 4.21). In addition, mean shear was significant across L7–L9. Dissipation rates were so low that they were contaminated by instrument noise; the average across the staircase, $\epsilon = 1.9 \times 10^{-10}$ W kg^{-1}, was only slightly larger than the layer average, 1.4×10^{-10} W kg^{-1}. Layers 5 and 6, however, stand out, with ϵ and χ_T several times the others. By contrast, applying laboratory $(\Delta S)^{4/3}$ flux laws to the interfaces gave $J_b^{\mathrm{sf}} = (0.8\text{–}4.1) \times 10^{-8}$ W kg^{-1}. Neglecting shear production of turbulence, similarity scaling like that applied to convecting boundary layers gives $\epsilon \sim J_b$ in the layers. The large discrepancy with laboratory flux laws was the impetus for the Kunze (1987) model of finite-length salt fingers in linear gradients.

Thermistor-chain tows revealed coherent structures in the layers, some of which were salt-stabilized temperature inversions (Figure 4.22). Marmorino (1991) identified them as intrusions, but they were consistent with convective plumes having high horizontal-to-vertical aspect ratios, e.g. plumes in atmospheric boundary layers have aspect ratios of 5–50 (Agee, 1987).

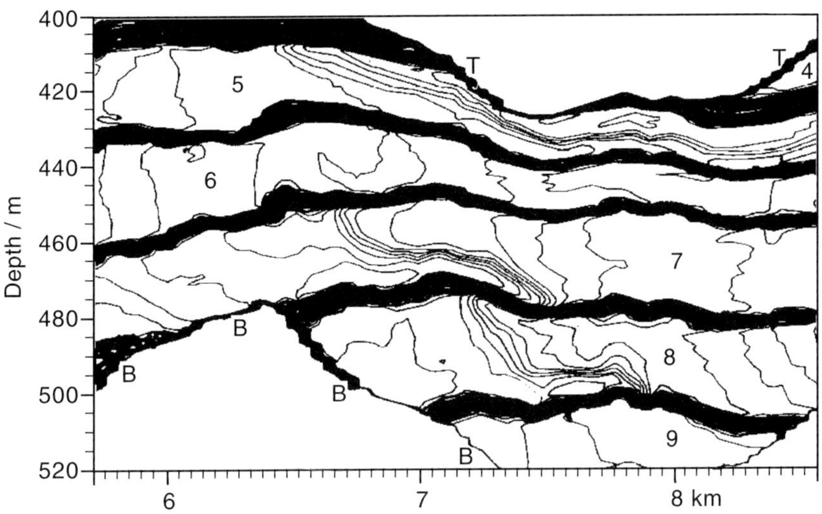

Figure 4.22 Temperature contours in the Barbados staircase from towed thermistors ~0.5 m apart. Layers 5, 7, and 8 show exchange flows, with cold tongues flowing right at the bottom and warm tongues moving left at the top. (From Marmorino, 1991)

Interfaces were 2–13 m thick, much greater than maximum finger lengths of a few tenths of a meter. Consequently, $(\Delta S)^{4/3}$ flux laws did not apply, except in a few cases where interfaces contained homogenous sublayers separated by thin interfaces (Figure 4.23). Mean and fluctuating shear peak in the interfaces, but gradient Richardson numbers were 1.3 to 5. Dissipation rates averaged 4.9×10^{-10} W kg^{-1}, higher than in the layers, but were likely overestimates owing to assuming isotropy where there were no overturns. Matching buoyancy Reynolds numbers of $1.5 \lesssim Re_b \lesssim 14$ correspond to weak turbulence consistent with fingering, as do Cox numbers of 2.5–9. Horizontal tows found dissipation rates exceeding a noise level of 5×10^{-10} W kg^{-1} in only one interface (Lueck, 1987).

Profiles found the interfaces full of roughly sinusoidal temperature fluctuations having wavelengths of $\approx$ 60 mm, corresponding to peaked temperature and conductivity gradient spectra in horizontal tows (Lueck, 1987; Marmorino, 1987a). Shadowgraph images (Figure 4.24) show horizontal banding that Kunze et al. (1987) interpreted as evidence of fingers strongly tilted by shear, consistent with the reduction in vertical fluxes.

In 2001 sulfur hexafluoride, SF_6, was released in the staircase to compare the bulk diffusivity of the tracer with microstructure estimates (Schmitt et al., 2005). Sampling 10 months later found several tracer clouds, one far to the west in the

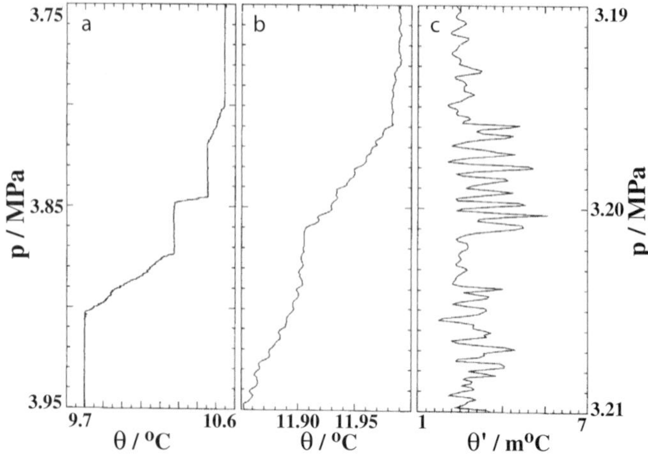

Figure 4.23 An interface between layers 3 and 4 had two homogenous sublayers separated by an interface thin enough to be spanned by fingers. Thicker interfaces had ~60 mm fluctuations from tilted fingers. (Adapted from Gregg and Sanford, 1987)

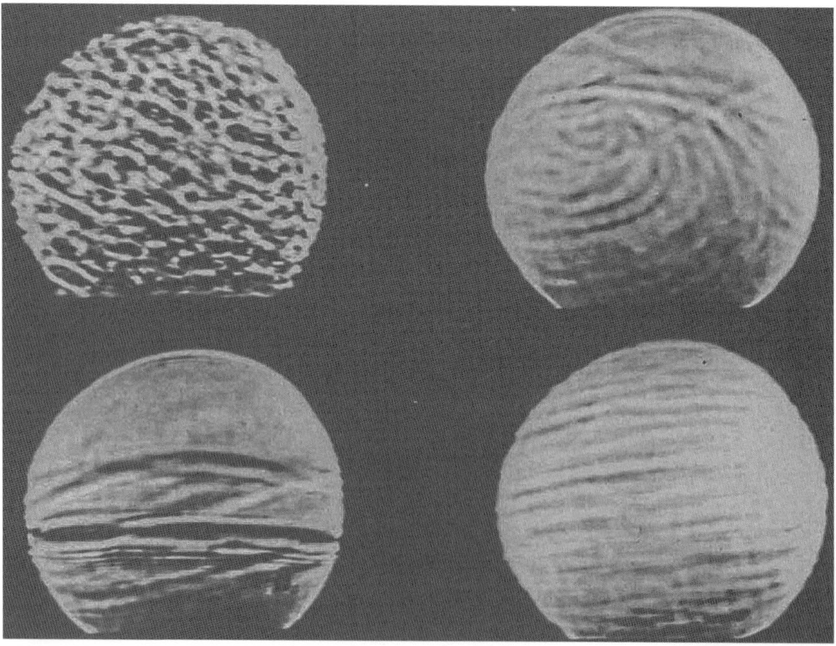

Figure 4.24 Shadowgraphs from the Barbados staircase. Each image was 0.1 m in diameter and a view through 0.6 m of water. They are interpreted as showing stratified turbulence (upper left), a shear billow (upper right), a diffusive regime interface (lower left), and tilted salt fingers (lower right). (Adapted from Kunze et al., 1987)

Caribbean and another to the east. The average tracer diffusivity using N computed by smoothing the staircase was $K_{SF_6} = (0.8\text{–}0.9) \times 10^{-4}$ m^2 s^{-1}. Because the molecular diffusivity of SF_6 is within 10% of the molecular salt diffusivity, κ_S, the bulk eddy coefficient of the tracer was equated to that of salt. The bulk diffusivity, however, does not represent the local intensity of fingering in interfaces that are more strongly stratified than the smoothed profile used to calculate tracer diffusivity. The equivalent interfacial diffusivity can be estimated from the equality of salt flux across the interface and the bulk flux,

$$K_S^{\text{interface}} \, (\partial S/\partial z)_{\text{interface}} = K_S^{\text{bulk}} \, (\partial S/\partial zs)_{\text{bulk}} \tag{4.29}$$

(E. Kunze, personal communication, 2019). The ratio of the gradients was about 10, yielding $K_S^{\text{interface}} \sim 10^{-5}$ m^2 s^{-1}, the same order as the numerical estimates at low R_ρ in Figure 4.15.

For comparison with the tracer diffusivity, ϵ measurements with (4.23) gave $K_S = 2.40 \times 10^{-4}$ m^2 s^{-1} with $(0.8\text{–}6.6) \times 10^{-4}$ for 95% confidence limits (Schmitt et al., 2005). The authors obtained better agreement with the tracer by using $K_S = (R_\rho / R_{\text{flux}}^{\text{sf}})(\chi_\theta / (2\overline{\theta_z}^2))$, inverting the McDougall (1988) expression (4.24) for K_T^{sf} and evaluating it with the Osborn and Cox (1972) expression (3.78) for fully developed turbulence. This, however, might be an overestimate because major contributions to χ_θ come from interfaces containing anisotropic microstructure.

4.4.3 Tyrrhenian Sea

In a basin between Italy, Sardinia, and Sicily, the Tyrrhenian Sea has a salt-fingering staircase below Levantine Intermediate Water and above Western Mediterranean Deep Water (Zodiatis and Gasparini, 1996). Extending from 600 to 2,500 m, the staircase has the largest vertical span known and is well defined in the center of the basin, where circulation is weak. Some layers are coherent for 150 km, but year-to-year changes are dramatic, with a few very thick layers replacing many thinner ones (Figure 4.25).

4.5 Diffusive Layering

Unlike salt fingering, possible over wide areas but difficult to identify without staircases, the diffusive regime is possible over limited domains and is inherently tied to staircases. Some staircases have a hundred or more meter-thick steps, while others have several steps tens of centimeters high (Figure 4.26). Much thinner than layers, interfaces are usually smooth when $R_\rho^{-1} > 2$, producing fluxes by molecular diffusion. At lower density ratios, occasional overturns enhance fluxes. Diffusive staircases are found in profiles with low R_ρ^{-1} below cool, low-salinity water and

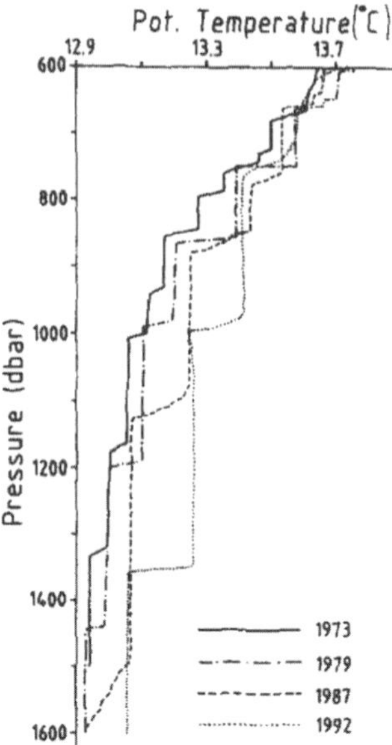

Figure 4.25 Potential temperature profiles spanning 19 years through the upper part of the salt fingering staircase in the central Tyrrhenian basin. Between 1973 and 1992 average temperature and salinity between 600 and 1,600 m increased 0.173 K and 0.045 ppt, while their differences decreased by $\Delta\theta = 0.76$ K and $\Delta S = 0.18$ ppt. Layer thickness increased as the number of layers decreased: 10 in 1973, 8 in 1984, 6 in 1987, and 4 in 1992. (Adapted from Zodiatis and Gasparini, 1996)

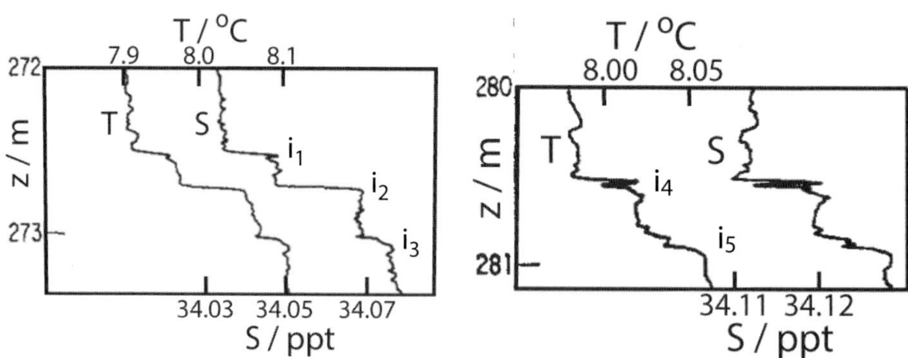

Figure 4.26 Diffusive layers on a coastal intrusion with interfaces having $R_\rho^{-1} = 1.07$ (i_5) to 1.34 (i_1). After correction for thermistor response, interface i_2 was 8 mm thick. This resolution was possible owing to the 40–80 mm s^{-1} fall rate of the profiler. Layer thicknesses are 0.16 m (i_1–i_2) to 0.26 m (i_2–i_3). Interface i_4 had a small overturn. (Adapted from Gregg and Cox, 1972)

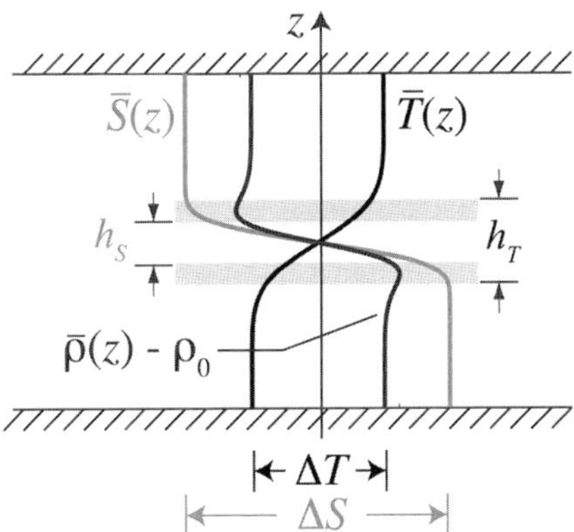

Figure 4.27 Schematic diffusive interface with $h_S < h_T$ and diffusive boundary layers (shaded) with thickness $b = (h_T + h_S)/2$. The difference produces density inversions in the boundary layers, which drive convection in layers bounding the interface. (From Carpenter et al., 2012b. © American Meteorological Society. Used by permission)

above warmer and saltier water, as well as in geothermally heated saline lakes and deep basins.

The first diffusive staircase, produced at Woods Hole by leaving a salt-stratified tank overnight on a hot-water radiator, was followed by more systematic studies (Turner and Stommel, 1964; Turner, 1968a). Heating the bottom of a salt-stratified tank destabilizes adjacent water to produce a convecting layer that eventually produces another above it. While the steady heat flux is maintained, additional layers grow upward to form staircases in spite of lower layers merging. Dominated by molecular diffusion, the interfaces have salinity steps thinner than those of temperature (Figure 4.27).

The one-dimensional dynamics of these laboratory experiments seem relevant to diffusive staircases extending across lakes and deep basins. Applicability to open-ocean staircases, however, remains uncertain, owing to possible lateral effects. Also, oceanic staircases have much weaker stratification and fluxes but higher Rayleigh numbers than laboratory staircases. No observations have documented development of oceanic staircases from smooth gradients.

4.5.1 Fluxes

Laboratory Measurements

Kelley (1990) fit laboratory fluxes with

$$c_{\mathrm{K}}^{\mathrm{dr}} = 0.0032 \exp(4.8 {R_\rho}^{0.72}), \quad R_{\mathrm{flux}}^{\mathrm{dr}} = \frac{{R_\rho}^{-1} + 1.4({R_\rho}^{-1} - 1)^{3/2}}{1 + 14({R_\rho}^{-1} - 1)^{3/2}} \tag{4.30}$$

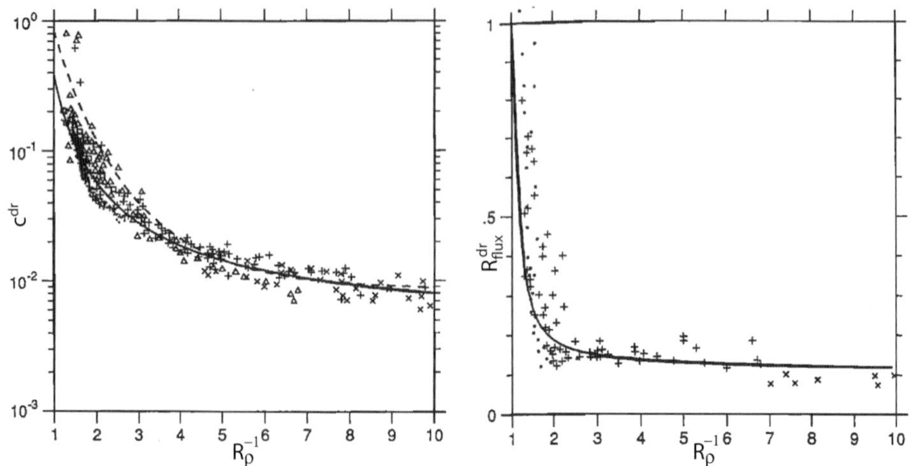

Figure 4.28 Coefficient and flux ratios from laboratory measurements of the diffusive regime with symbols from different authors. Solid lines are (4.30), and the dashed line on the left is (4.31). (From Kelley, 1990)

for $1 < R_\rho{}^{-1} < 10$ (Figure 4.28). For $R_\rho{}^{-1} < 4$, c^{dr} is bounded from above by a Marmorino and Caldwell (1976) expression, quoted by Kelley (1990) as

$$c^{\mathrm{dr}}_{\mathrm{MC}} = 0.00859 \exp\left(4.6 \exp\left(-0.54(R_\rho{}^{-1} - 1)\right)\right). \tag{4.31}$$

As $R_\rho{}^{-1} \rightarrow 1$, both the coefficient, c^{dr}, and the flux ratio, $R^{\mathrm{dr}}_{\mathrm{flux}}$, rise steeply, so that the heat flux exceeds the solid plane value, i.e. $Nu > 1$, only when $R_\rho{}^{-1} < 2$ (Turner, 1965). Details vary between experiments (Crapper, 1975) and may indicate dependence on other parameters, such as the magnitude of the heat flux (Marmorino and Caldwell, 1976). Kelley (1990) notes that the coefficient in $Nu \propto (\Delta T)^{1/3}$ may be in error, but the difference in fluxes is less than experimental uncertainty.

As $R_\rho^{-1} \rightarrow 1$ across diffusive interfaces, shadowgraphs reveal increasing turbulence, leading Fernando (1989) to develop flux parameterizations for low stability that include $\overline{w^2}$. Vertical velocity, w, is not a routine microstructure measurement, but the approach emphasizes the need for turbulent measurements to relate staircases in the laboratory to those in the ocean.

Ocean Measurements

Estimated heat fluxes in oceanic diffusive staircases are more consistent with laboratory fluxes than are estimates in fingering staircases. Molecular diffusion across thin interfaces is a major component of the fluxes, requiring vertical resolution of centimeter-scale temperature steps with slow free-fall profiling. Descending at 0.04–0.08 m s^{-1} and applying a measured dynamic response function resolved

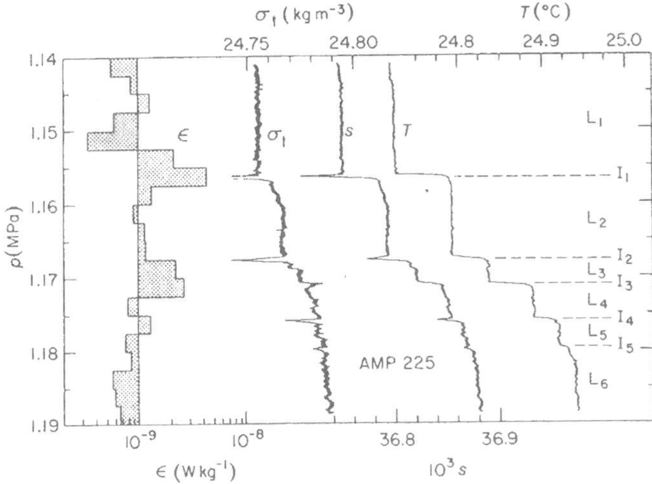

Figure 4.29 Diffusive staircase below a cold, fresh intrusion in the Bahamas. From I_1 to I_5: R_ρ^{-1} and $10^9 J_b^{dr}$ from the 4/3 flux law were 1.57/4.0, 1.54/2.3, 1.52/2.8, 1.36/1.3, and 1.17/0.5. The top layer is a temperature minimum below water unstable to fingering. Spikes in salinity and density are instrument artifacts. (From Larson and Gregg, 1983)

diffusive layers on coastal intrusions to $\sim$10 mm (Fig. 4.26). Even though the density ratio was close to 1, some interfaces were linear with molecular heat fluxes within a factor of 2 of flux expressions derived by Huppert (1971) from Turner's laboratory measurements (Gregg and Cox, 1972). Some interfaces, however, contained overturns, as expected for intermittent entrainment when $R_\rho^{-1} \sim 1$.

Adding airfoils to microstructure profilers allows fluxes to be estimated in layers using the turbulent kinetic energy equation, which reduces to $\epsilon = J_b$ when shear production is negligible. Dissipation rates through the diffusive staircase in Figure 4.29 are within a factor of 2 of laboratory fluxes, again demonstrating much better agreement than is found for fingering staircases.

The fast fall rates maximizing airfoil signals preclude resolving thin temperature interfaces in the most recent microstructure profiling, particularly when only nominal dynamic calibrations are available for the thermistors. To overcome this, Sommer et al. (2013b) repeatedly sampled steady diffusive staircases in Lake Kivu with a thermistor and an SBE7 dual-needle conductivity probe (Sections 5.2.2, 5.2.3). They estimated dynamic responses of the probes by dropping at 0.19, 0.38, and 0.89 m s^{-1} (Figure 5.9). Using the measured response functions, molecular fluxes through interfaces agreed well with Kelley (1990), particularly for $2 < R_\rho^{-1} < 4$. Most attempts to resolve interfaces in high-latitude staircases, however, have used thermistors lacking dynamic calibrations.

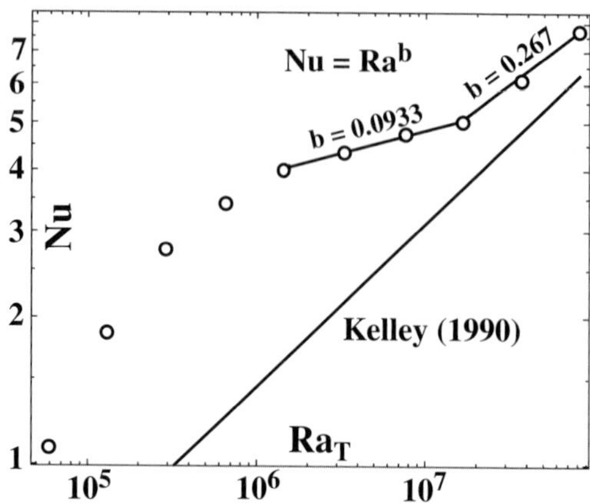

Figure 4.30 Nusselt number versus Rayleigh number from 2D (except for $Ra = 1.5 \times 10^6$) numerical simulations of a diffusive interface bounded by two convecting layers (circles), compared with the flux expression of Kelley (1990). The simulations exhibit two power-law ranges. (Adapted from Hieronymus and Carpenter, 2016. © American Meteorological Society. Used with permission)

Numerical

Direct numerical simulations (DNS) of a diffusive interface find that rigid boundaries reduce fluxes by factors of 2–3 compared to periodic boundaries, offering a possible explanation of why laboratory laws underestimate fluxes in oceanic staircases (Flanagan et al., 2013). With grid spacing less than half the salinity Batchelor scale for $\kappa_T/\kappa_S = 0.01$, other DNS find Nusselt numbers exceeding Kelley (1990) at all Ra_T (Figure 4.30). Power-law behavior appeared only over limited Ra_T ranges and with coefficients less than 1/3.

Convection in layers is sustained by quasi-steady plumes from persistently unstable boundary layers surrounding the interface (Carpenter et al., 2012a), and their shear sharpens the interface to maintain strong molecular diffusion across it. Because full, or possibly even partial, isotropy seems unlikely when small overturns on interfaces enhance molecular diffusion, estimates of anisotropy are needed from DNS to improve fluxes estimated from microstructure profiles through diffusive staircases.

Direct numerical simulations by Hieronymus and Carpenter (2016) find that molecular conduction at low Rayleigh numbers,

$$J_{b,cond}^{\mathrm{dr}} \equiv g(\alpha\kappa_T \Delta T/H + \beta\kappa_S \Delta S/H) \quad [\mathrm{W\,kg^{-1}}], \tag{4.32}$$

produces $\sim 45\%$ of the total flux when $Ra_T = 10^5$. As convection becomes more intense, the fraction drops to $\sim 17\%$ at $Ra_T = 10^8$, and the Nusselt number becomes $Nu = (J_{b,conv} + J_{b,cond})/J_{b,cond}$.

4.5.2 Layer Thickness

Kelley (1984) argued that, with κ_T and $\overline{N}$ as the appropriate dimensional parameters, thickness depends on R_ρ as

$$H^{\mathrm{dr}} = \left[2.5 \times 10^8 R_\rho^{-1.1}(\nu/\kappa_t)(R_\rho^{-1} - 1)\right]^{1/4} (\kappa_T/\overline{N})^{1/2} \quad [\mathrm{m}]. \tag{4.33}$$

A reasonable fit to many observations, the expression overestimates heights in an Arctic staircase (Figure 4.31, left). Fernando (1987) proposed that convecting layers in diffusive staircases reach a critical thickness when kinetic and potential energies of their eddies are of the same order, rendering them unable to entrain overlying fluid. Similarity scaling of convecting boundary layers gives the vertical velocity variance in terms of height and the buoyancy flux driving convection as $w^2 \sim (J_b H)^{2/3}$. Eddies are limited when $w^2 \approx \Delta b H$, where the buoyancy step across the interface is $\Delta b \approx N^2 H$. Combining gives diffusive layer thickness as

$$\begin{aligned} H^{\mathrm{dr}} &\propto (J_b^{\mathrm{dr}}/N^3)^{1/2} \\ &\approx 14(J_b^{\mathrm{dr}}/N_S^3)^{1/2}(1 - R_\rho)^{-3/4} \quad [\mathrm{m}], \end{aligned} \tag{4.34}$$

where N is an average over a scale that smoothes the staircase, N_S is the smoothed initial stratification due to salinity, and, to be consistent with Figure 4.31, $R_\rho \equiv \beta\Delta S/\alpha\Delta T$. Fernando (1989) expressed the buoyancy frequency in terms of

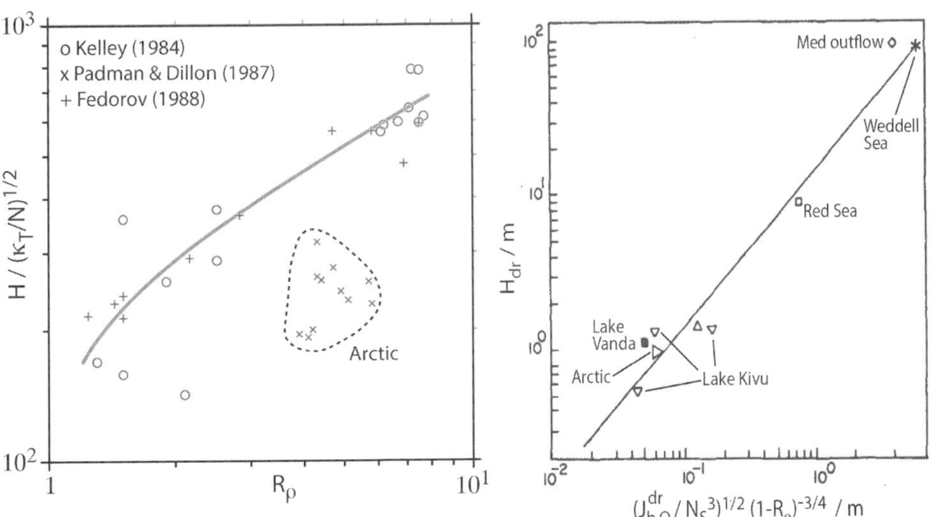

Figure 4.31 (Left) Scaled thickness of diffusive layers compared with (4.33). (Adapted from Kelley et al., 2003.) (Right) Layer thicknesses compared with (4.34). (From Fernando, 1989. These authors use $R_\rho \equiv \beta\Delta S/\alpha\Delta T$. © American Meteorological Society. Used with permission)

the initial salinity gradient with a correction for temperature. The scaling constant came from Figure 4.31 (right), which exhibits good agreement over a wide span of thicknesses. With the total kinetic energy (tke) equation reducing to $\epsilon = J_b$ when shear production is negligible, this scaling reproduces the Ozmidov scale in a different setting and should also apply to fingering staircases.

4.6 Diffusive Staircases

In contrast to short diffusive staircases on thermohaline intrusions (Section 4.7), those found at high latitudes are extensive, often spanning sections 100 m or thicker, where cold, fresh surface water lies over warm, saline water formed at low latitudes. Short staircases occur in some lakes and ocean basins, where geothermal heat warms bottom water or hot brine is released. Microstructure in these basins has yet to be reported, but measurements in upper-ocean staircases at high latitudes find dissipation rates of $\mathcal{O}(10^{-10}\ \mathrm{W\,kg^{-1}})$, too weak to be distinguished from airfoil noise. Consequently, ϵ in these layers has been estimated by fitting theoretical forms to high wavenumber spectra of temperature gradients. This seems useful when dynamic probe responses are known, but can be problematic otherwise. Many staircases have R_ρ^{-1} substantially larger than 2 with fluxes dominated by molecular conduction across smooth interfaces. Interfaces with low density ratios are more likely to have overturns enhancing fluxes, but the limited separation between energy-containing and dissipation scales makes isotropy unlikely. Where heat fluxes have been estimated, $\mathcal{O}(0.1\ \mathrm{W\,m^{-2}})$ is the most common value, except in the eastern Arctic, where some fluxes are more than ten times larger. Owing to its small size and extensive sampling over several decades, Lake Kivu has the only staircases described by full heat and salt budgets.

4.6.1 Lake Kivu

An African rift lake 500 m deep, Lake Kivu has multiple diffusive staircases (Figure 4.32) at the depths of entering groundwater (Sommer et al., 2019). Dissolved carbon dioxide adds ~40% to the salt stratification, and dissolved methane decreases it by ~ 15%. In addition to the lake being an excellent natural laboratory for diffusive layering (Figure 4.33), heat and salt balances have been studied intensively to assess the threat of a catastrophic overturn releasing deep dissolved CO_2 and CH_4. On a cheerier note, the methane is being extracted commercially, initially to power a brewery.

Profiles contain more than 300 diffusive layers, typically ~0.5 m thick and separated by interfaces 0.18 m thick with 4.7 mK temperature changes (Figure 4.33). Including effects of dissolved gas, density ratios are $R_\rho^{-1} = 2.0$–4.5 for much of

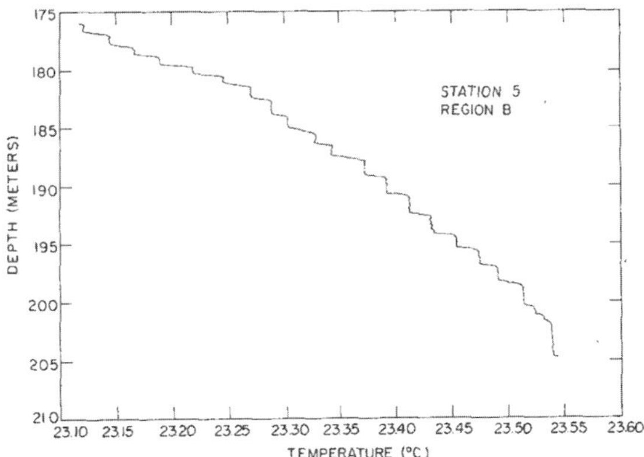

Figure 4.32 Diffusive steps just above the mid-depth homogenous layer in Lake Kivu. (From Newman, 1976. © American Meteorological Society. Used with permission)

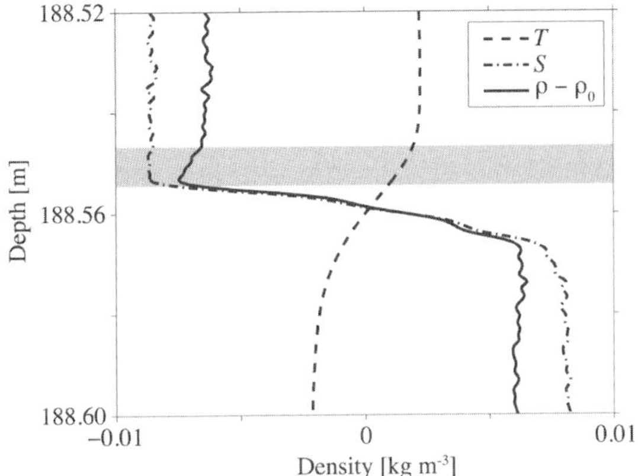

Figure 4.33 Temperature, salinity, and density through a diffusive step in Lake Kivu observed with a high-resolution profiler. Shading shows an unstable boundary layer, typical above T and S interfaces having different thicknesses. (From Carpenter et al., 2012b. © American Meteorological Society. Used with permission)

the deep water. Layers are coherent for several hundred meters, sometimes a few kilometers, giving aspect ratios of $\sim 1{,}000$. Direct numerical simulations demonstrate that for $R_\rho^{-1} > 3$ molecular fluxes through interfaces equal the total vertical fluxes (Sommer et al., 2013b).

As found in many oceanic staircases, dissipation rates were close to or lower than airfoil noise. Rather, ϵ was estimated by fitting Batchelor's theoretical form to temperature spectra from dynamically calibrated thermistors and dual-needle conductivity probes (Sommer et al., 2013b,a). 'Unshaped' spectra were omitted, and signals from interfaces were not included because they are mostly nonturbulent. Coupling dissipation rates and direct numerical simulations (Section 4.5.1) yielded heat and salt budgets demonstrating that subsurface inflows are essential; without them, the staircases would decay in a few decades.

4.6.2 Deep Basins

Many deep basins contain water heated geothermally to form homogenized bottom layers hundreds of meters thick. Some form when bottom water more saline than overlaying water is heated from below, and others are maintained by hot brine seeps. Most support diffusive staircases. The bottom layer in the Black Sea appears to be an exception, but Kelley (1989) estimated that there may be layers too thin to be resolved with existing data. More typically, the Canada Basin (Figure 2.6) contains a homogenous layer of local seawater up to 1 km thick heated by a geothermal heat flux of 40–60 mW m^{-2} (Timmermans et al., 2003). It is capped by two to three layers 10–60 m thick with temperature and salinity decreasing upward and $R_\rho^{-1} \approx 1.6$. Interfaces are 2–16 m thick and have molecular fluxes too small to transmit the geothermal heat fluxes. Nor have overturns in the interfaces been observed, leading Timmermans et al. (2003) to hypothesize that most of the flux escapes from the sides of the basin with enough passing vertically to maintain the staircase.

The Atlantis II Basin in the Red Sea (Figure 4.34) appears to be controlled by hot brine from the sea floor (Swift et al., 2012). High fluxes, $\sim$2.5 W m^{-2}, from the bottom layer warm the four diffusive layers, but the rate is decreasing, presumably in response to cooler brine entering.

4.6.3 Arctic

Most diffusive staircases in the Arctic are on the upper boundary of warm saline Atlantic water (Figure 4.35) coming through the Fram Strait, flowing eastward off Russia, and then along the Eurasian side of the Lomonosov Ridge and into the Makarov and Canada Basins (Guthrie et al., 2015). The core of the inflow forms a subsurface TS maximum descending below 300 m along its path.

Discovered by Neal et al. (1969), diffusive staircases above the core of Atlantic water have been observed at many sites, often from drifting ice stations. On the Eurasian side, staircases are sporadic. Shallower staircases in regions of high stratification have some layers separated by interfaces containing 5–6 sublayers, and others exhibit varying degrees of shear-driven mixing (Polyakov et al., 2019). Deeper layers in less stratified water are tens of meters thick and have estimated heat fluxes of 3–4 W m^{-2}, in contrast to $\mathcal{O}(0.1)$ W m^{-2} for the shallower layers.

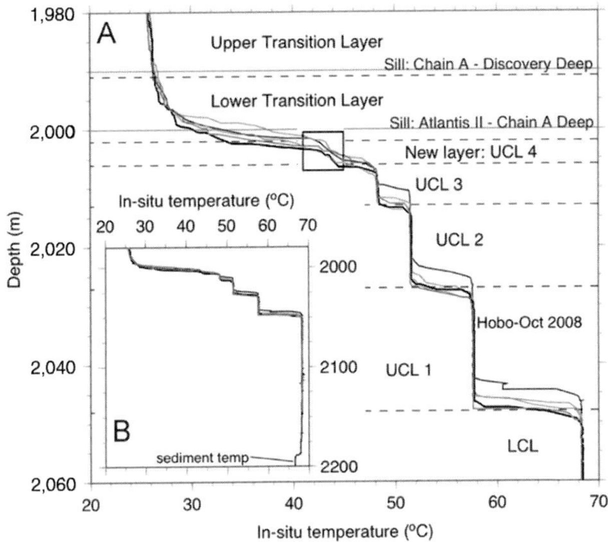

Figure 4.34 Diffusive staircase in the Atlantis II Basin of the Red Sea. Upper layers UCL1–UCL4 top the 135 m-thick lower layer, LCL. The sediment is 1.7°C cooler than the bottom water. Salinity (not shown) increases from 50 ppt at 2,000 m to 250 ppt at the bottom, far beyond the oceanic salinity range. R_ρ^{-1} increases from $\approx$ 2 at the staircase top to $\approx$ 18 at the lower interface. (Adapted from Swift et al., 2012)

Farther along the Atlantic water path in the Amundsen Basin, Guthrie et al. (2015) examined several hundred interfaces with $R_\rho^{-1} \sim 3.45$ and $Ra_T \sim 10^9$. Heat fluxes averaged 0.36 ± 0.02 W m^{-2}, corresponding to $Nu = 10 - 100$, and were close to laboratory fluxes if $Nu \propto Ra_T^{0.29}$. In the Beaufort Sea, at the western end of the path, staircases are ubiquitous, extending $\sim 1,000$ km across the basin and fading at its edges, presumably in response to higher background turbulence (Timmermans et al., 2003). Layers are 1–2 m thick, and $R_\rho^{-1} = 4 - 6$ (Padman and Dillon, 1987). Airfoils detect some signals in interfaces, but turbulence levels are so low that few useful measurements have been reported. Overall, the heat fluxes are unlikely to impact surface heat budgets and are thus not factors in melting the surface ice pack (Timmermans et al., 2008).

4.6.4 Antarctic

Although conditions for diffusive layering surround the Antarctic, observations of staircases are sparse and limited to the Weddell Sea (Figure 4.36), where density ratios are smaller than in the Arctic, and nonlinearities in the equation of state affect fluxes from double diffusion (McDougall, 1981). Shaw and Stanton (2014)

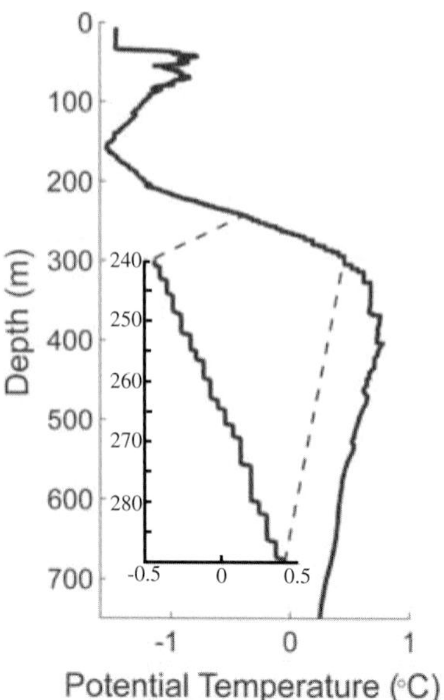

Figure 4.35 Typical upper-ocean temperature profile over the Canada Basin with a diffusive staircase above the temperature maximum at the core of the saline Atlantic Water. Layers are 1–20 m thick, and some have been tracked for hundreds of kilometers under a drifting ice island. Many have smooth interfaces consistent with Fickian diffusion. Laboratory flux laws give diapycnal heat fluxes of 0.05–0.3 W m^{-2}. (Adapted from Timmermans et al., 2008. © American Meteorological Society. Used with permission)

estimated mixing rates from microstructure profiles with an FP07 thermistor and a dual-needle conductivity probe, using a correction factor from Washburn et al. (1996) when salinity fluctuations cannot be ignored. As in the Arctic, airfoil signals were too weak to be useful. Eddy diffusivities were estimated using Osborn and Cox (1972), which the authors note is a delicate issue in staircases. As a function of density ratio, the bulk thermal diffusivity rose steeply past $K_T = 10^{-4}$ m^2 s^{-1} as R_ρ^{-1} decreased below 1.2. Double diffusion, however, was not the only mixing process, as mixing rates also depended on the Froude number defined as $Fr \equiv |d\overline{u}/dz|/N$.

4.7 Thermohaline Intrusions

Sections of early bathythermograph profiles revealed temperature inversions extending several kilometers near fronts (Spilhaus, 1939). Adding salinity disclosed profiles with all three forms of stable stratification alternating from one to another over vertical separations of meters (Stommel and Federov, 1967). As observations accumulated in the Pacific, patterns emerged showing a geography of large inversions running north–south near coasts and east–west along the subtropical

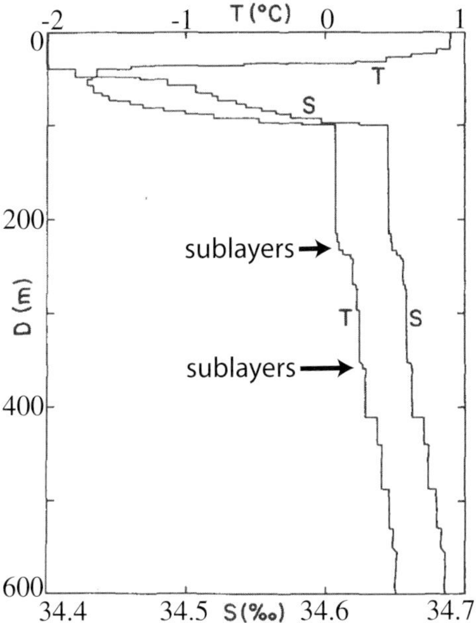

Figure 4.36 Diffusive staircase in the Weddell Sea spanning high and low mean gradients in a transition from cold winter surface water to warmer and more saline deep water. R_ρ^{-1} was 1.39 in the upper layers, 77–100 m, and 1.03 below 100 m. Some interfaces have sublayers similar to those in the eastern Arctic. (Adapted from Foster and Carmack, 1976. © American Meteorological Society. Used with permission)

front (Roden, 1964). Understanding them has evolved slowly, owing to the difficulty of making detailed observations.

4.7.1 Mapping Intrusions

Complex interleavings are often found across fronts (Figure 4.37), with some intrusions rising and others sinking across density surfaces, raising the prospect that advection may produce significant diapycnal fluxes not reflected in microstructure measurements. An extended section across the subtropical front in the North Pacific with a depth-cycling CTD shows salinity minima clustered near two branches of the front (Figure 4.38a). Mapping these minima, however, requires subjective criteria, e.g. S' over Δz, that must vary with the background stratification. Examining intrusions in terms of spiciness, τ (Section 2.8), finds that salinity minima and maxima are subsets of a wider class of distortions described by a more objective parameter, the diapycnal curvature of spiciness,

$$\tau_{\sigma\sigma} \equiv \left(\frac{d^2\tau}{d\sigma^2}\right)_{\text{profile}} \approx 2\beta\rho S_{\sigma\sigma} \approx 2\alpha\rho\theta_{\sigma\sigma} \tag{4.35}$$

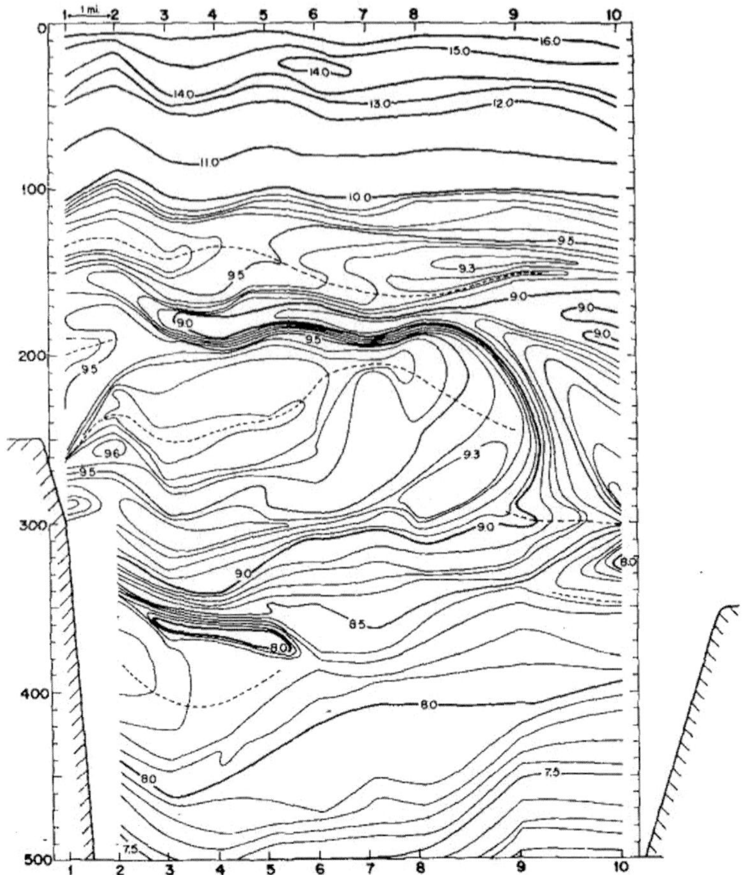

Figure 4.37 Temperature contours from a line of CTD profiles over the San Diego Trough show complex interleaving between the cool, fresh California Current offshore (right) and warm, saline equatorial water inshore. (From Gregg, 1975a. © American Meteorological Society. Used with permission)

(Shcherbina et al., 2009). Spiciness curvature is a dynamic quantity, multiplied by the diapycnal diffusivity, K_ρ, in evolution equations for salinity and temperature. Moreover, its maxima and minima are coherent over much greater distances than are salinity minima (Figure 4.38b).

Slow lateral spreading of intrusions was observed for two years using acoustic floats to track Meddy Sharon, a lens of warm saline Mediterranean water in the North Atlantic (Figure 4.39). Sequential surveys found steady erosion of the meddy as thermohaline intrusions penetrated toward its core at average speeds of 1 mm s^{-1}. By comparing temperature and salinity changes observed three times over one year before the intrusions reached the core, Hebert (1988b) inferred

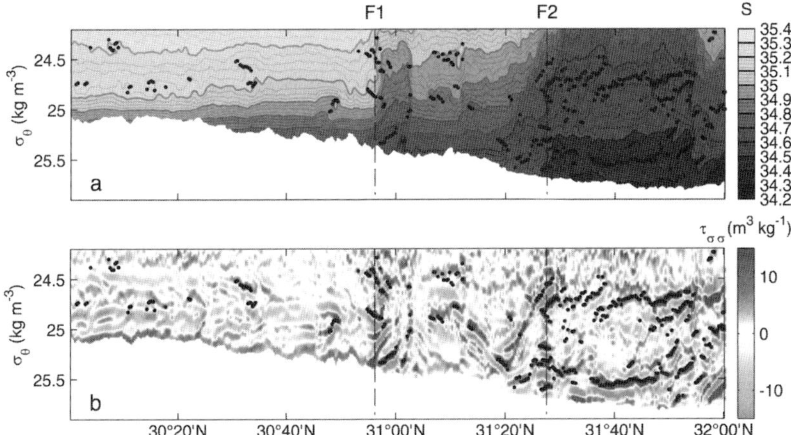

Figure 4.38 Salinity (a) and diapycnal spiciness curvature (b) along 158°W across two branches, F1 and F2, of the Subtropical Front. Salinity minima, black dots, are coherent over parts of $\tau_{\theta\theta}$ maxima and minima (shading). (Adapted from Shcherbina et al., 2009. © American Meteorological Society. Used with permission.) (A black and white version of this figure will appear in some formats. For the color version, please refer to the plate section)

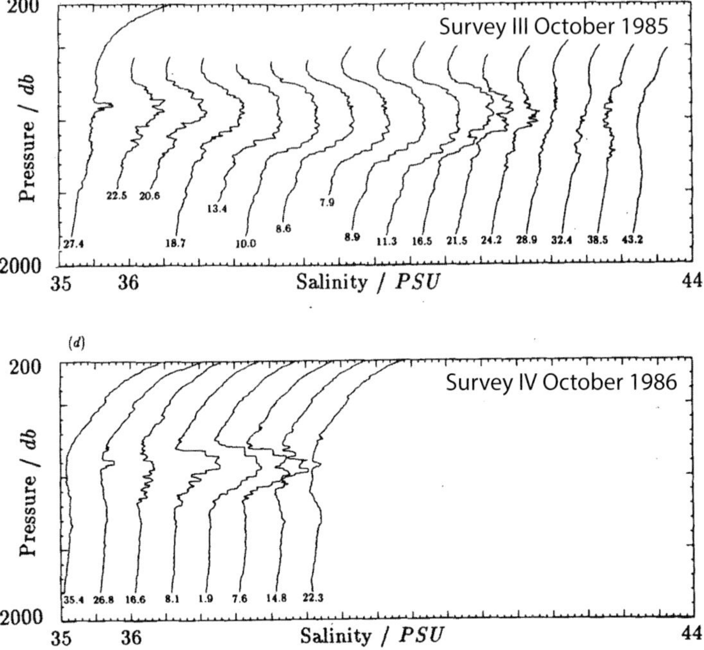

Figure 4.39 Salinity profiles offset by 0.5 ppt across Meddy Sharon, with radial distances from the center in km. Survey III found intrusions penetrating the center. A year later the meddy was half its initial size. Some profiles had short diffusive staircases above and fingering staircases below the meddy core, but most exhibited multiple maxima and minima attributed to sequences of intrusions. (Adapted from Armi et al., 1989. © American Meteorological Society. Used with permission)

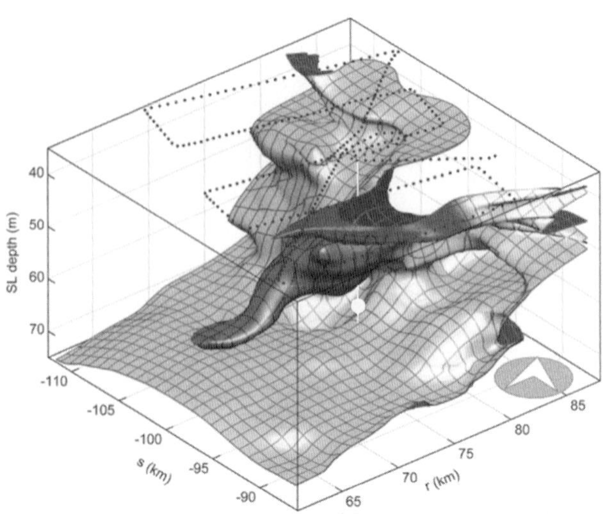

Figure 4.40 The 34.80 isohaline versus semi-Lagrangian depth of a cold, fresh intrusion with multiple tongues sinking with varying slopes across isopycnals. It extends 10–15 km across and 5 km along the front. The float is shown as a white sphere. (From Shcherbina et al., 2010. © American Meteorological Society. Used with permission)

diapycnal buoyancy fluxes of $\mathcal{O}(10^{-9})$ W kg^{-1} from salt fingering below the θS maximum of the meddy. Only fluxes predicted by Kunze (1987) for thick interfaces agreed with the inferences; laboratory flux laws and Kunze's expression for thin interfaces were 10 times too large.

Understanding how intrusions form requires repeated observations in three dimensions. The first three-dimensional mapping with a CTD mounted on a depth-cycling towed body found a cold, fresh tongue 2 km wide extending 10 km across a shallow front (Gregg, 1980b). The map could not be repeated after Long Range Aid to Navigation (LORAN) navigation faded. Years later, GPS solved the navigation problem, and neutrally buoyant, acoustically tracked floats allowed sequential observations of intrusions on the subtropical front in the North Pacific. One set of maps found a tongue-like feature (Figure 4.40). Although followed for several days, changes were smaller than the resolution of the maps. Contours, however, show that earlier the intrusion had descended across isopycnals, consistent with, but not proof of, salt fingering above the TS minimum dominating the buoyancy fluxes into the intrusion.

Some sections of the front were crinkled but lacked intrusions. Where intrusions were found, all were cold and fresh, coming from the north. Except for a few tongues, many intrusions appeared as folds (Figure 4.41). That is, their along-front dimensions were greater than their displacement across the front. This one extended

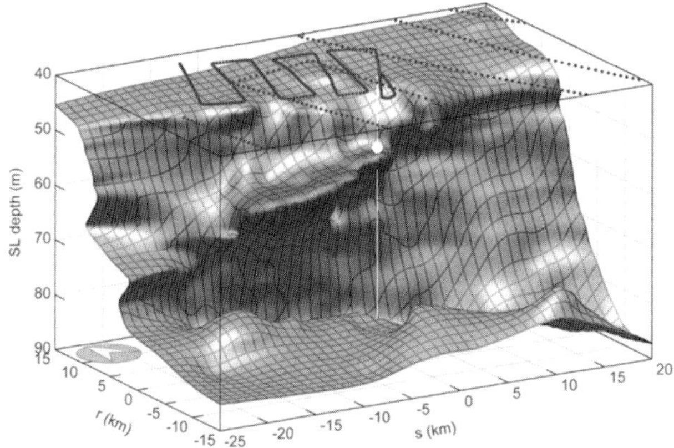

Figure 4.41 A cold, fresh intrusion as a fold of the 35.0 ppt isohaline crossing the front. The salinity minimum was tagged with a neutrally buoyant float shown as a white sphere. (From Shcherbina et al., 2010. © American Meteorological Society. Used with permission)

30 km along and 10 km across the front. Followed for a week, changes due to internally driven evolution could not be distinguished from advective distortions.

4.7.2 Mixing

Microstructure profiles show recurring patterns in profiles with multiple intrusions (Figure 4.42). Minima tend to be homogenous, particularly when bounded by a sharp interface above and several diffusive steps below, implying that the minimum is receiving fluxes from fingers and diffusive layering. The steps usually end above the next θS maximum, where there is little or no mixing. Above the minima some profiles have strong centimeter-scale fluctuations consistent with tilted salt fingers or sheets, e.g. 276–279.5 m in Figure 8.28, which shows the region just below the right side of Figure 4.42. Estimated fluxes would have 'filled in' the temperature minimum at 255 m (C) in 10 hours. Formed by lateral advection rather than by vertical overturning, temperature fluctuations in intrusions are inconsistent with the Osborn and Cox (1972) model, leading Joyce (1977) to develop a triple Reynolds decomposition with intrusion thickness as the intermediate length scale.

4.7.3 Formation Driven by Double Diffusion

To test the proposal of Stern (1967) that thermohaline intrusions could result from lateral instability to double diffusion, laboratory experiments by Turner (1978) compared diffusively stable and unstable intrusions into background profiles with the same N^2. In sharp contrast to diffusively stable intrusions, diffusively unstable

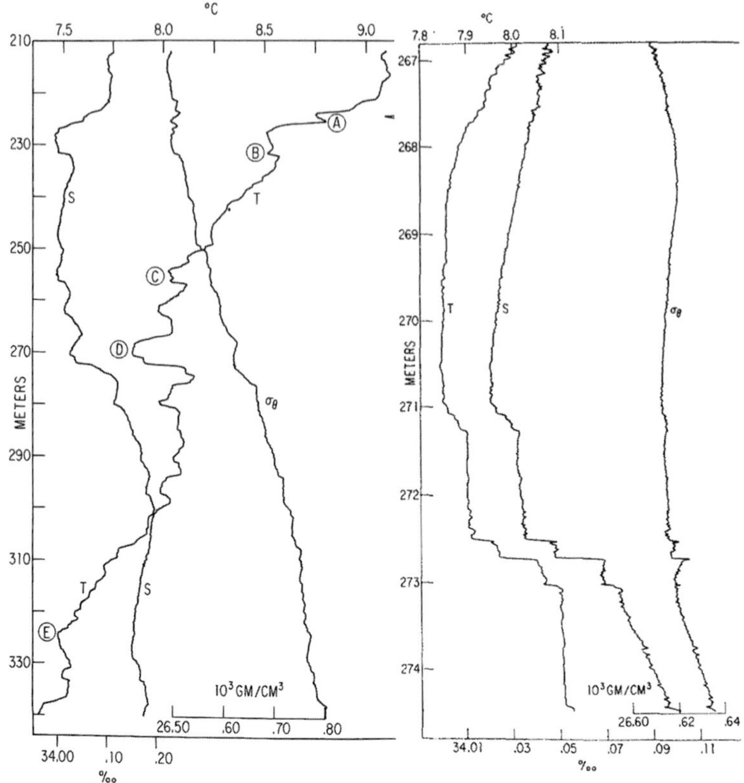

Figure 4.42 Microstructure profile through intrusions off San Diego. Diffusive layers below the temperature minimum at D are shown in more detail on the right. (Adapted from Gregg and Cox, 1972)

intrusions thickened rapidly as they spread (Figure 4.43). After the initial thickening, the advancing front formed alternating maxima and minima with diffusive layers and fingering on their boundaries. The similarity of intrusive structures in this experiment and those in the ocean is a strong reason for considering lateral diffusive instability as a process capable of producing intrusions across fronts.

Intrusion Density Ratios and Slopes

Depending on the background density ratio, $R_\rho(z)$, intrusions can also form from vertical perturbations to laterally homogenous profiles (Merryfield, 2000). The possibilities are shown schematically as slopes of $\beta S_z'$ versus $\alpha T_z'$ in Figure 4.44. Perturbations with slopes $\beta S_z'/\alpha T_z' < 1/R_\rho$ grow into fingering staircases, or into ones that are statically unstable. When the slope exceeds $1/R_\rho$, perturbations extend from temperature and salinity gradients unstable to fingering through diffusively stable gradients into gradients unstable to diffusive layering, the cases

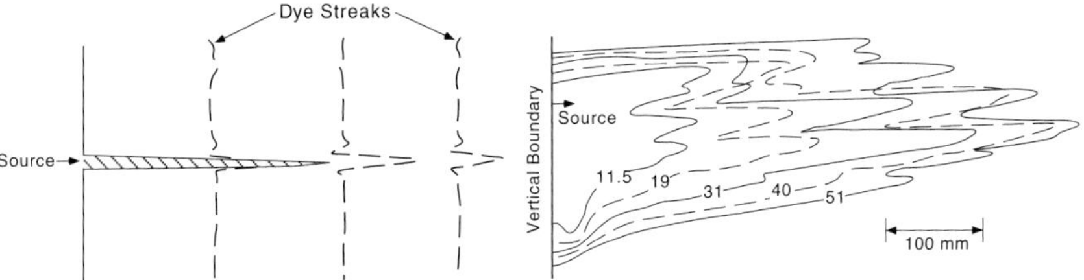

Figure 4.43 Evolution of diffusively stable and diffusively unstable intrusions. (Left) Salt water intruding at the same density into a salt-stratified tank remained on its original isopycnal, producing a nose current with return flows above and below. (Right) Sugar water with the same density as the salt intrusion entered a tank with the same salt stratification as the first. Having a different molecular diffusivity, the sugar intrusion expanded vertically into multiple intrusions with alternating diffusive and fingering interfaces. (From Turner, 1978)

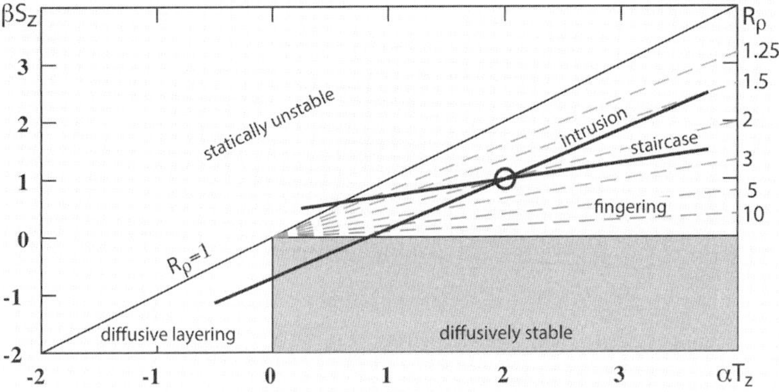

Figure 4.44 Schematic evolution of perturbations, growing in time outward along thick lines from the initial perturbation marked by the circle. Dashed lines have constant R_ρ. The space is divided into regimes that are statically, a.k.a. convectively, unstable, diffusively stable, and unstable to salt fingering and diffusive layering. The perturbation forming an intrusion grew into fingering and diffusive layering regimes separated by diffusively stable sections. The other perturbation, marked 'staircase', grew into fingering and convectively unstable regions, forming interfaces and layers in a staircase. (Based on Merryfield, 2000)

found in profiles with multiple intrusions. Equivalent relations also hold in profiles diffusively unstable to diffusive layering. Bebieva and Timmermans (2017) found this approach a good indicator of where diffusive staircases are found in the Arctic.

An alternate linear analysis finds a restrictive range of cross-frontal slopes for double-diffusivity intrusions,

$$0 < slope^{\mathrm{sf}}_{\mathrm{critical}} < \frac{\epsilon_z \overline{S_y}/\overline{S_z} + \overline{\sigma_y}/\overline{\sigma_z}}{\epsilon_z + 1}, \tag{4.36}$$

where $\epsilon_z \equiv (1 - R^{sf}_{flux})/(R_\rho - 1)$ (May and Kelley, 1997). The distribution of slopes at the subtropical front was much wider than this criterion, indicating that many intrusions may not have been formed by double diffusion (Shcherbina et al., 2009, 2010).

4.7.4 Formation Driven by Mesoscale Velocity Fluctuations

As suggested from the wide spread of intrusion slopes across the subtropical front, inference of diapycnal motion from even three-dimensional maps can be misleading unless intrusions can be followed in time. Ageostrophic flows within fronts deform temperature and salinity fields, often producing filaments laying across isopycnals with no diapycnal flow (Woods et al., 1986). This was demonstrated directly for the subtropical front by showing that an apparent intrusion in a salinity section was consistent with advection of a prior salinity section (Figure 4.45).

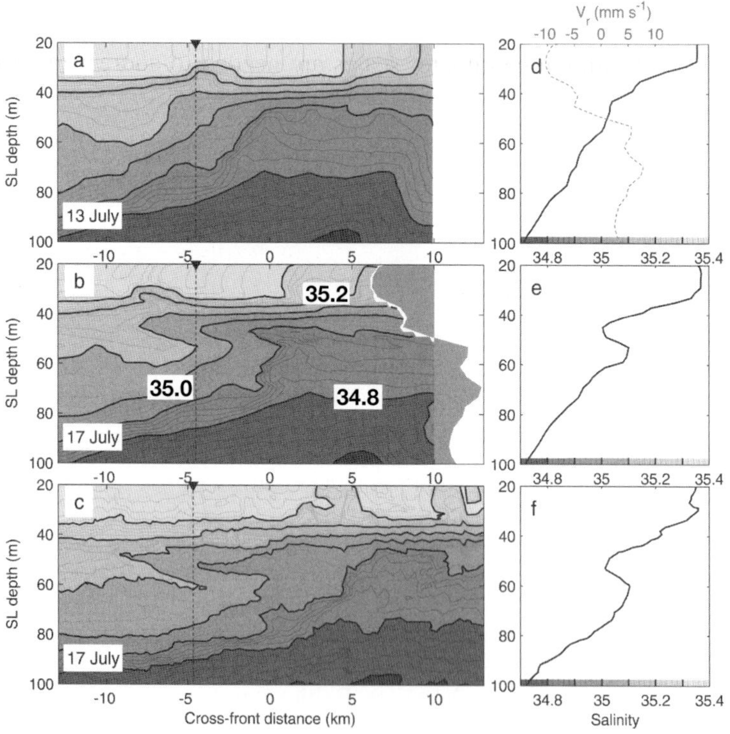

Figure 4.45 Simulation of an apparent intrusion beginning with observed salinity section (a). Simulated distortion by four days of cross-frontal advection by the observed velocity (d) produced a salinity inversion (b) in rough agreement with the section observed four days after the first, (c). Panels (d)–(f) have salinity profiles along the dashed line in the sections. (Adapted from Shcherbina et al., 2010. © American Meteorological Society. Used with permission)

This, however, does not explain why all intrusions originated on the cold, fresh side of the front. Both processes, double diffusion and differential advection, are likely involved, perhaps together on most or all intrusions, or possibly separately for different intrusions.

4.8 Perspectives

Fundamentals of double diffusion have been understood for 50 years, but many aspects of how it operates in the ocean remain unclear. Tall salt fingering staircases dominate diapycnal mixing at a few sites in the open ocean, and a tracer release in the Barbados staircase found a bulk diapycnal diffusivity of $\mathcal{O}(10^{-4})$ m^2 s^{-1}. How this flux relates to $\epsilon \sim 10^{-10}$ W kg^{-1} in the layers is not understood. Nor, is it clear how this relatively large bulk flux is produced in the thick interfaces of the staircase, where fingers, or sheets, appear strongly tilted by shear and are not thought to extend across the interface. Also, why do some interfaces contain homogenous sublayers, and how do they affect the fluxes? Observations of three-dimensional structure at centimeter scales are needed to address these issues, but close coupling with DNS will be needed to interpret them fully.

Diffusive staircases are more easily understood because fluxes result from molecular conduction across smooth centimeter-scale interfaces when $R_\rho^{-1} \gtrsim 2$, which is common in many staircases. These interfaces can be resolved by slow profiles with dual-needle conductivity and dynamically calibrated thermistors. These fluxes, however, have not been compared with simultaneous dissipation measurements to test $\epsilon = J_b$ balances in the layers. Larger, more sensitive, airfoils may be able to address this, as the fluxes are close to the noise levels of current airfoils. Overturns on the interfaces appear responsible for fluxes rising as $R_\rho \to 1$. Because the wavenumber bandwidth of the turbulence is likely to be too small for full isotropy, the enhanced fluxes will be difficult to estimate correctly from microstructure measurements and will need help from DNS.

Uncovering the role of double diffusion in the formation and evolution of intrusions will require measurements both more detailed and more extended in time than those to date. Microstructure is needed to test flux estimates, while meter-scale velocity, temperature, and salinity are required over tens of meters in the vertical and kilometers in the horizontal to follow the evolution of an intrusion. The apparent slow pace of intrusions requires tracking for weeks to months, beyond the duration of research vessels, but perhaps feasible with autonomous vehicles.

5

Sampling Mixing and Its Environment

5.1 Overview

Observations seek to estimate mixing from variables that can be measured, often in lieu of desired parameters, such as diapycnal buoyancy flux and diffusivity. One component of both the turbulent dissipation rate, ϵ, and the corresponding rate of thermal smoothing, χ_T, can be measured routinely with microstructure probes (Section 5.2). After assuming isotropy to estimate components that were not measured and applying finestructure measurements (Section 5.3) to obtain background gradients, models are used to infer diffusivity and buoyancy flux from the data. The results have been taken seriously since the inferences were tested by comparison with spreading rates of tracers injected into the pycnocline (Section 5.8).

In addition to probe characteristics, mixing observations are affected by how sensors move through the water. Most data were initially collected by free-fall and loosely tethered profilers (Section 5.4), but towed and self-propelled bodies are now widely used (Section 5.5), as are moorings and floats (Section 5.6). These give different views of mixing, but none by itself can adequately follow the evolution of turbulent patches. Backscatter of high-frequency acoustics provides the only observations continuous in time (Section 5.7), but this has been successful only where shallow gradients are strong near coasts or in estuaries.

5.2 Dissipation-Scale Sensors

Viscous, ϵ, and thermal, χ_T, dissipation rates are estimated from spectra. Taken over windows of one-half to several meters and corrected with probe response functions, spectra are integrated and scaled assuming the gradients are isotropic, i.e. the same in all directions. Often, response functions are not accurately known, owing to the difficulty of finding references that can resolve yet smaller scales, and to the need to calibrate each thermistor probe. In spite of these limitations, direct estimates of ϵ and χ_T are regularly obtained for low- and medium-turbulent intensities, although instrument vibration and noise sometimes contaminate spectra. In addition, very high dissipation rates require extrapolation of what can be measured using universal spectral forms. Estimates of χ_S and χ_ρ are also needed. Although a few measurements have been obtained under favorable circumstances, neither has generated enough data to be scientifically useful.

5.2.1 Velocity

The central 90% of shear variance forming ϵ occurs between 0.056 and 0.513 times the Kolmogorov wavenumber, k_η (Figure 3.7). More than half comes from wavenumbers greater than the spectral peak at $0.125\,k_\eta$. Measuring that variance for ϵ spanning $\lesssim 10^{-11}$ to $\gtrsim 10^{-2}$ W kg^{-1} requires resolving shear between 0.5 and 1,000 cpm (Figure 3.8). Fortunately, measurements are made easier by the increase in amplitude as ϵ rises.

Hot-Films

The hot-film (Figure 5.1) towed by Grant et al. (1962) to test Kolmogorov's spectrum was in an alternating current bridge that kept the film at a constant temperature. Overheat, limited to 5°C to minimize corrosion, gave a sensitivity of 0.2 volt per mm s^{-1}. Dissipation rates of $1.5 \times 10^{-9} \leq \epsilon \leq 1 \times 10^{-6}$ W kg^{-1} in the tidal channel provided strong signals, but the combination of low dissipation rates and strong thermal microstructure offshore has limited use of hot-films in the open-ocean thermocline. Hot-films, however, supplied the axial component in a study of spectral isotropy by Gargett et al. (1984) in Knight Inlet.

Airfoils

Osborn (1974) adapted airfoils measuring velocity in laboratory tanks to be the robust probes that are mainstays of oceanic turbulence measurements. Pointed

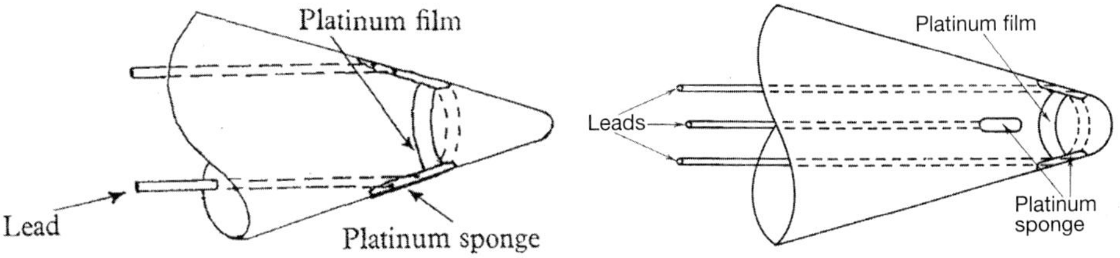

Figure 5.1 Schematics of platinum film turbulence sensors, both on the tips of glass-rods 4 mm in diameter. The films are 10 nm thick. (Left) hot-film velocity probe from Grant et al. (1962). (Right) cold-film temperature probe from Grant et al. (1968a).

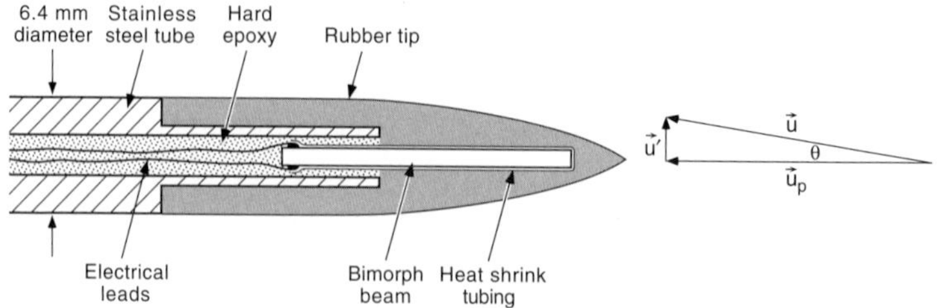

Figure 5.2 Airfoil schematic with $\vec{u}_p$ as the speed of the probe relative to the water and u' the cross-velocity generating a bending moment on the beam. (From Gregg, 1999. © American Meteorological Society. Used with permission)

into the mean flow, $\vec{u}_p$, airfoils detect velocity fluctuations, u', in a plane perpendicular to their axis (Figure 5.2), i.e. the normal component of a turbulent flow (Section 3.6.1). The sensing element is a piezoelectric crystal, a.k.a. a bimorph beam, encased in an axisymmetric rubber tip with a parabolic taper. Assuming potential flow at attack angle θ, Osborn and Crawford (1980) expressed the cross-force per unit probe length as

$$f_p = (1/2u_p^2)\, dA/dx \;\sin(2\theta) \quad [\mathrm{N\,m^{-1}}], \tag{5.1}$$

where dA/dx is the slope of the cross-sectional area, A, along the probe axis. Using $\sin 2\theta \approx 2u'/\overline{u}_p$, the net force on the beam is

$$F_p = \int f_p\, dx \approx \rho A u_p u' \quad [\mathrm{N}]. \tag{5.2}$$

For a given signal, u', the force is greater for thick probes moving rapidly. On profilers, signals are horizontal fluctuations, u' or v', whereas on tows and moorings they can be w' or v'. Moum et al. (1995) describe circuitry and processing.

Airfoil size determines its spatial response, which Oakey (1982) represents as a one-pole low-pass filter,

$$H^2_{\mathrm{airfoil}} = \frac{1}{1 + (k_1/k_{\mathrm{af}})^2}, \tag{5.3}$$

where k_1 is a one-dimensional wavenumber, and k_{af} characterizes the probe. Dynamic calibrations by Macoun and Lueck (2004) find $k_{\mathrm{af1}} = 88$ cpm for their standard airfoil and $k_{\mathrm{af2}} = 49$ cpm for their smaller probe. The af1 response reduces the signal by a factor of ~5 at 100 cpm (Figure 3.8), the middle of the dissipation range for $\epsilon = 10^{-5}\ \mathrm{W\,kg^{-1}}$. Rather than ignore stronger signals, larger dissipation rates can be estimated by extrapolating with universal curves (Wesson and Gregg, 1994, appendix).

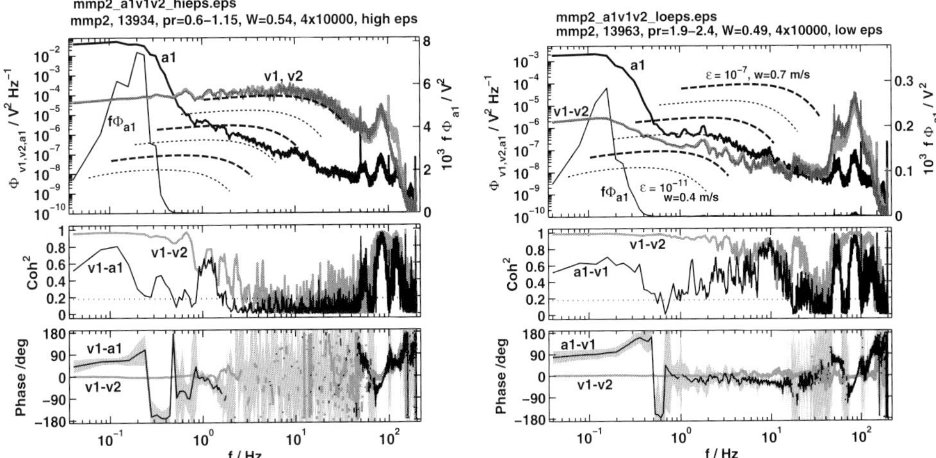

Figure 5.3 Spectra of recorded voltages of two airfoil channels, v1 and v2, and an accelerometer, a1, for moderately strong, $\epsilon = 6.6 \times 10^{-7}$ (left), and weak, 1.1×10^{-10} W kg^{-1} (right), turbulence. The variance-preserving accelerometer spectrum, fϕ_{a1}, shows the 0.2 Hz wobble peak. Dashed and dotted curves for ϵ at half-decade intervals were multiplied by airfoil electronic transfer functions. (From Gregg et al., 2012)

Airfoil signals are often contaminated by plankton impacts, vibration, and intermittent probe failure. Low dissipation rates can also be contaminated by airfoil temperature sensitivity and by vehicle wobble (Figure 5.3). The wobble peak at 0.2 Hz is prominent in the variance-preserving accelerometer spectra and contaminates a substantial portion of the dissipation range for $\epsilon \lesssim 10^{-10}$ W kg^{-1}. To some extent, contamination coherent between accelerometers and airfoils can be removed (Goodman et al., 2006; Gregg et al., 2012), but the degree of improvement depends on spatial separation.

Dissipation noise levels are often reported as the minimum value in ϵ distributions, but that underestimates the effect of noise on the low end of the distribution. Raw records show thin signals occupying varying fractions of spectral windows. This produces a second linear range on probability plots of $\log \epsilon$ (Figure 5.4), and the break-in-slope between them is the minimum dissipation rate not affected by noise, in this case 10 times the minimum estimate of 2.5×10^{-11} W kg^{-1}.

Electromagnetic Velocity Probe

In 1983, A. Arjannikov began developing small electromagnetic velocity sensors at the Central Research Institute (GRANIT) in Leningrad, USSR. Electrodes embedded between four permanent magnets detect voltages from tangential flows across the probe tip (Figure 5.5), with sums and differences translating into standard Cartesian velocities (Soloviev et al., 1999, appendix A). The probes are rugged;

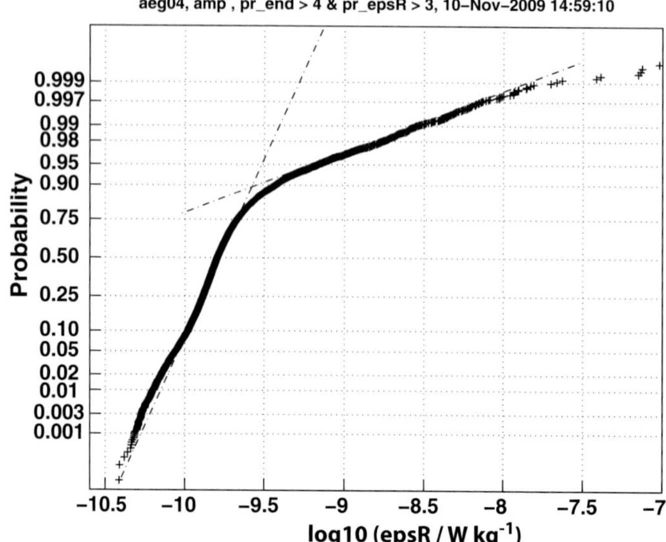

Figure 5.4 Normal probability plot of ϵ from the loosely tethered Modular Microstructure Profiler where $N^2 \lesssim 7 \times 10^{-6}$ W kg^{-1} in the Aegean. Dissipation rates are 5 m averages, analyzed in three sections to remove variance coherent with accelerometers. The break-in-slope at $\log_{10} \epsilon = -9.6$ ($\epsilon = 2.5 \times 10^{-10}$ W kg^{-1}) was taken as the noise level. (From Gregg et al., 2012)

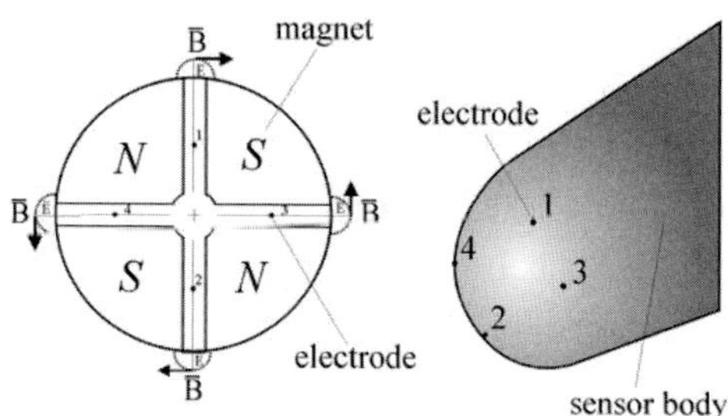

Figure 5.5 Schematic of 3-axis EM velocity probe. The nose has four permanent magnets with electrodes between them. The R16 version is 16 mm in diameter. (From Lai et al., 2000. © American Meteorological Society. Used with permission)

one version, 40 mm in diameter, has operated on bows of research ships and at high speeds on Russian submarines. The smaller, R16, sampled mixed-layer turbulence on a rising profiler (Soloviev et al., 1999).

Laboratory calibrations (Figure 5.6) yield the spatial response as

$$H^2_{\text{em}}(k) = \frac{1}{1 + (k/k_s)^n}, \tag{5.4}$$

with $k_s = 295/(2\pi)$ and $n = 3.3$ for the 16 mm-diameter probes (Lai et al., 2000). The spatial transfer function for the R16 is compared with airfoil responses in

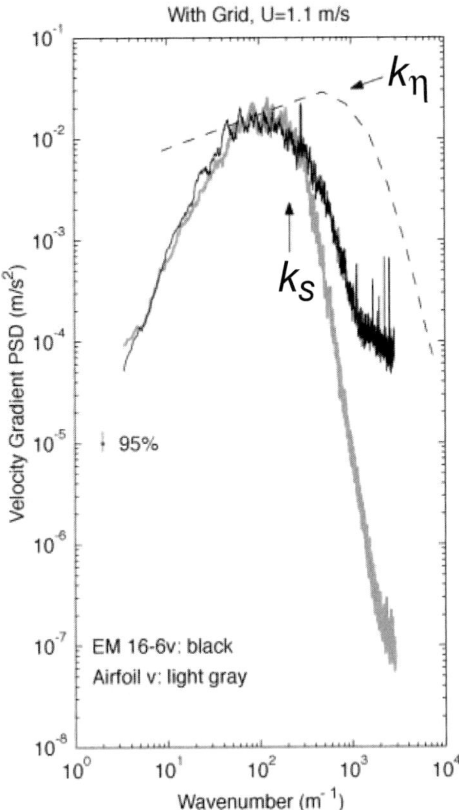

Figure 5.6 Shear spectra for the R16 electromagnetic probe and an airfoil in turbulence with $\epsilon = 7 \times 10^{-4}$ W kg^{-1}. k_s is the wavenumber scale of the probe, and k_η is the Kolmogorov wavenumber. Responses are comparable for $k < 300$ rad m^{-1}. The airfoil falls off more rapidly than the electromagnetic probe at higher wavenumbers, but the higher R16 noise level precludes a full comparison. (Adapted from Lai et al., 2000. © American Meteorological Society. Used with permission)

Figures 3.8 and 5.6. One or more accelerometers can be placed close to the tip, at least in the 40 mm-diameter version, allowing cancelation of noise induced by ship motion. Soloviev et al. (1999) give the useful bandwidth as 0.05–200 Hz and the root-mean-square (rms) noise as ~1 mm s^{-1}. In summary, the electromagnetic (EM) probe appears very useful for measuring turbulence from a wide variety of vessels, but its higher noise level restricts the range of signals detected in the pycnocline. Uniform construction, however, should insure that dynamic calibrations on one probe apply to all.

5.2.2 Temperature

Because heat diffuses more slowly than momentum, the main contributions to χ_T occur at length scales smaller than those of ϵ (Section 3.7.2). Across the range

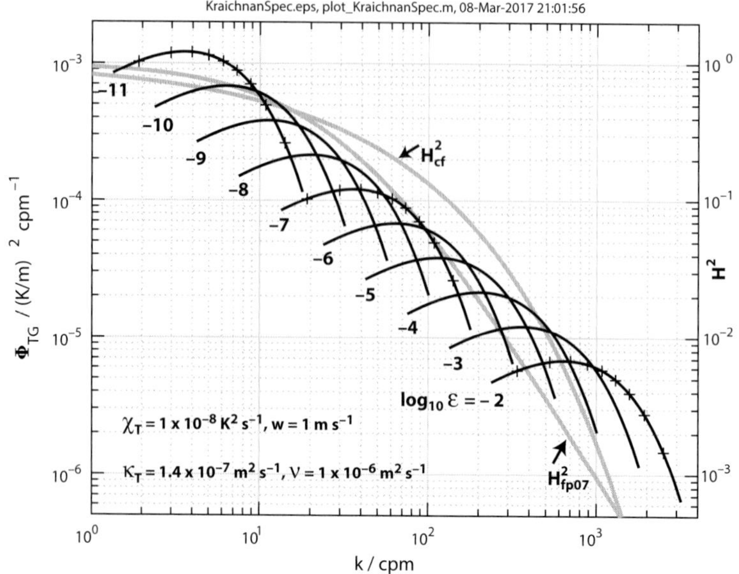

Figure 5.7 Kraichnan (1968) temperature gradient spectra for an arbitrary χ_T and $\log_{10}\epsilon$ spanning the oceanic range. (Keeping χ_T constant artifically forces spectral amplitudes down as ϵ increases.) Crosses on the spectra mark cumulative fractional variance from 0.1 to 0.9. Overlaid in gray are spatial response functions H^2_{cf}, for a cold-film (Fabula, 1968b), and H^2_{fp07}, for an FP07 thermistor with a 5 ms time constant (Gregg and Meagher, 1980).

of dissipation rates in the ocean, resolving χ_T requires measuring scales from 1 m to 0.3 mm (Figure 5.7). Temperature fluctuations are detected by measuring voltage changes, E_0, of known currents, i, passing through thermally sensitive resistors. Signal magnitude is proportional to the product of probe resistance, R_T, and sensitivity, $\alpha_{dRdT} \equiv (1/R_T)(dR_T/dT)$, as

$$\frac{dE_0}{dT} = i\,\alpha_{dRdT}\,R_T \quad [\mathrm{V}]. \tag{5.5}$$

Thermal diffusion attenuates signals in the probe as well as in the boundary layer surrounding it. Consequently, slower speeds are better, the opposite of the situation for airfoils. Resistance elements used in the ocean have principally been platinum films, a.k.a. cold-films, and thermistors, which are semiconductor material doped for high thermal sensitivity. Thermocouples have also proved useful in some applications.

Cold-Films

Grant et al. (1968b) constructed a cold-film by applying a thin film of platinum to the tapered tip of a 4 mm-diameter glass-rod (Figure 5.1). The platinum was electrically insulated from seawater by using a vacuum chamber to deposit a very

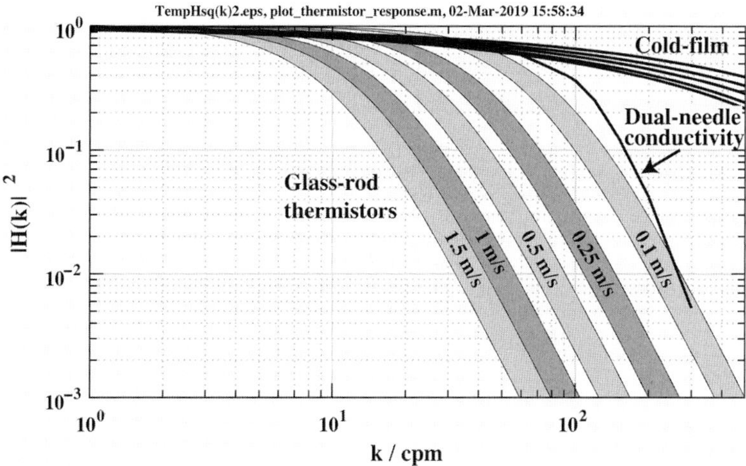

Figure 5.8 Power transfer functions for cold-films (5.6), glass-rod thermistors (5.7), and a dual-needle SBE-7 conductivity probe. Shading of the glass-rod responses at each speed span measured time constants of 8.42–11.3 ms. The dual-needle response combines the frequency term of Sommer et al. (2013b) and the spatial response of Hill and Woods (1988).

thin layer of quartz over the platinum. The third contact in the water nullifies apparent resistance changes due to variations in the electrical conductivity of seawater. Cold-films have good spatial resolution but poor temperature sensitivity and low resistance, $R_T \sim (10-20)\ \Omega$. Oakey (1977) gives $\sim$0.2 mK for typical noise levels, and Fabula (1968a) modeled the dynamic response as

$$H^2_{\text{cold-film}} = exp\left[-2\Delta(\pi f/\kappa_T)^{1/2}\right], \tag{5.6}$$

with $\Delta = (0.0023\kappa_T)^{1/2}(u/1.7)^{-0.32}$ for the boundary layer thickness. Attenuation is slight at 1 cpm but is nearly a factor of 10 at 100 cpm (Figs. 5.7 and 5.8).

Consequently, in a channel with $5 \times 10^{-8} \leq \epsilon \leq 5 \times 10^{-5}$ W kg^{-1}, Grant et al. (1968b) resolved the viscous-convective subrange but not much of the viscous-diffusive subrange before noise dominated the signal. Owing to their thermal sensitivity and short lives in seawater, cold-films are not widely used.

Thermocouples

Based on the Seabeck Effect, thermocouples operate on the electrical potential at a junction between two dissimilar metals. For a given pair of metals, the potential depends on the temperature of the junction, which does not require insulation from seawater. Surprisingly, thermocouples are sufficiently robust for routine use in the ocean. To maximize the signal, Nash et al. (1999) chose a chromel-constantan junction with a sensitivity of 5.8×10^{-5} V K^{-1} and rms noise of 6 mK over 10–100 Hz.

Another cold junction inside their profiler was a reference for the seawater junction, which was 65 μm in diameter and resolved temperature fluctuations one-tenth the size of those resolved with FP07 thermistors on the same profiler. Near 100 Hz, the dynamic response was near unity, but there was no reference for determining the response at higher frequencies. Thus, thermocouples can measure large-amplitude thermal turbulence at high dissipation rates, but a dynamic response function is needed for full utility. Nonetheless, Moum and Nash (2009) found them useful for calibrating dynamic responses of thermistors.

Thermistors

The source of nearly all χ_T measurements in the ocean, glass-rod thermistors can have high resistance, $R_T \sim (0.7-5)$ MΩ, and excellent sensitivity, $\alpha_{\text{dRdT}} \sim 0.04\ \text{K}^{-1}$, but their frequency response is poor. With $R_T \sim 5$ MΩ, rms noise levels of 3 μK are possible (Gregg et al., 1978). The most common microstructure probe, known as 'Fasttip' or FP07, is about 2 mm in diameter, with the thermistor in a small dimple on the nose of a glass-rod.

Gregg and Meagher (1980) represented dynamic calibrations as a two-pole low-pass filter,

$$H^2_{\text{FP07}} = \frac{1}{(1 + (2\pi\tau_{FP07} f)^2)^2}, \tag{5.7}$$

with $\tau_{FP07} = \tau_0\, u_p^{-0.32}$ and u_p being the probe speed (Figure 5.8). Owing to manufacturing differences, individual responses had $\tau_0 = 5-10$ ms. Subsequent studies found some probes with two-pole responses and others closer to one-pole (Nash and Moum, 1999, 2002). For speeds of 1.5–0.1 m s^{-1}, attenuation begins at 2–10 cpm. By 100 cpm, power attenuation is a factor of $\sim$5 at 0.1 m s^{-1} and 1,000 at 1 m s^{-1}, precluding most microstructure profilers from measuring major contributions to χ_T at the $\sim$1 m s^{-1} fall rates often used. The effect of slower speeds was examined by Sommer et al. (2013b), who sought to resolve centimeter-scale diffusive steps in Lake Kivu. Comparing responses at three speeds allowed them to infer responses for an FP07 as true interface thickness to measured thickness (Figure 5.9). The FP07 cannot resolve structures 30 mm thick and smaller at 0.89 m s^{-1}, but slowing to 0.19 m s^{-1} pushes resolution to 10 mm.

5.2.3 Salinity and Density

Salinity gradients must be forced to wavenumbers 10 times those of the thermal viscous-diffusive range before χ_S is large enough to dissipate turbulent fluctuations. This pushes the salt dissipation range to 10^1 to 4×10^4 cpm (Figure 5.10), requiring a resolution of 25 μm to determine χ_S at the highest dissipation rates.

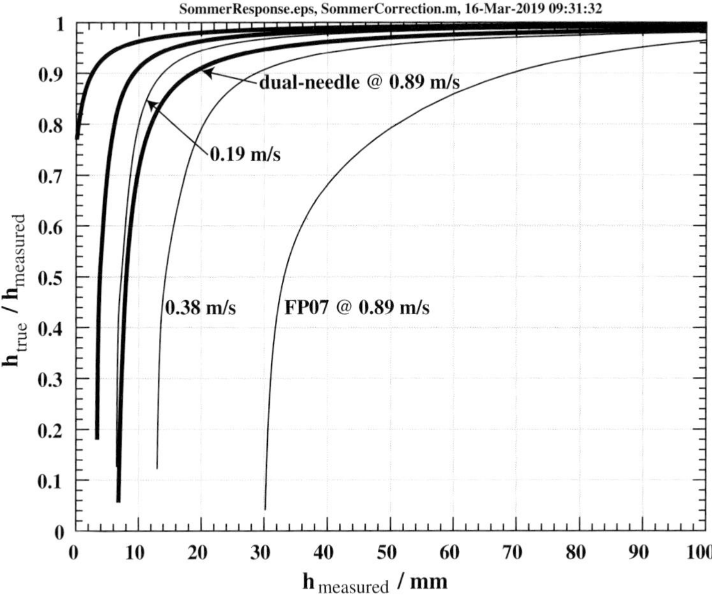

Figure 5.9 Responses of an FP07 thermistor (thin lines) and a dual-needle conductivity probe (thick lines) as the ratio of true thickness to measured thickness. The FP07 function (5.7) was evaluated with $\tau = 10$ ms. The dual-needle response combines spatial and frequency functions; the latter is (5.7) with $\tau = 2.2$ ms. (Based on Sommer et al., 2013b)

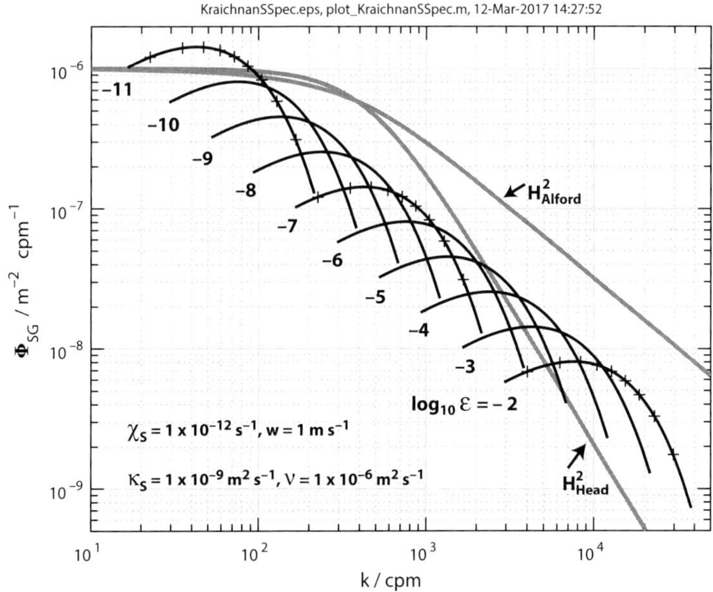

Figure 5.10 Kraichnan (1968) salinity gradient spectra corresponding to the temperature gradient spectra in Figure 5.7. Overlaid in gray are response functions for the Head (1983) 4-electrode conductivity probe and the Alford et al. (2006b) refractometer with $l_s = 2$ mm.

Two approaches have been used, measuring electrical conductivity and detecting optical shifts induced by index of refraction fluctuations. Both variables also respond to temperature, but those fluctuations are smoothed by diffusion at larger scales.

Electrical Conductivity

Owing to its strong dependence on temperature, electrical conductivity is an awkward parameter for obtaining salinity. Similar in principle to temperature measurements, dissipation-scale conductivity fluctuations are detected by the voltage change of a known current passing through seawater between electrodes. Several configurations concentrate electrical potential changes at dissipation scales, but electrochemical interactions make small electrodes quite noisy. Most successful approaches use multiple electrodes, often two in a low-impedance circuit to pass the current, and another pair in a high-impedance circuit to detect the voltage drop (Gregg and Cox, 1971).

Dual-Needle Probes

The viscous-convective subrange of temperature was first observed by Gibson and Schwarz (1963) using a single-needle conductivity probe in a saltwater tank containing temperature fluctuations. The sensing volume was concentrated around a 0.2 mm by 1 mm platinum film operated with a large remote ground plane. A larger dual-needle probe was developed by Gregg et al. (1981) as a reference for obtaining dynamic responses of finescale conductivity sensors in a two-layer tank stirred to have a sharp interface. This probe did not have the sensitivity to be useful in the ocean, but Meagher et al. (1982) constructed a larger version, later manufactured by Sea-Bird Electronics as the SBE-7 (Figure 5.11). The rms noise level from 1–100 Hz was 1×10^{-6} Siemen m^{-1}, equivalent to 10 μK at constant salinity.

The response of the SBE-7 to fluctuations is controlled by the spatial distribution of the electric field and by water flowing through a viscous boundary layer. Hill and Woods (1988) model the spatial response as $H(k) = \exp(-2\pi ka)$, where $a = 0.23$ mm is the half-distance between electrodes. Sommer et al. (2013b) measured the frequency response by comparing profiles at different speeds through thin diffusive steps in Lake Kivu. A double-pole filter with $\tau = 2.2$ ms fit the responses. The product of spatial and frequency functions, the net response is overlaid on temperature transfer functions in Figure 5.8. The SBE-7 response at 0.9 m s^{-1} is equivalent to one of the slower glass-rods at 0.2 m s^{-1} (Figs. 5.8, 5.9). At the slower speed, dual-needles can resolve a few millimeters.

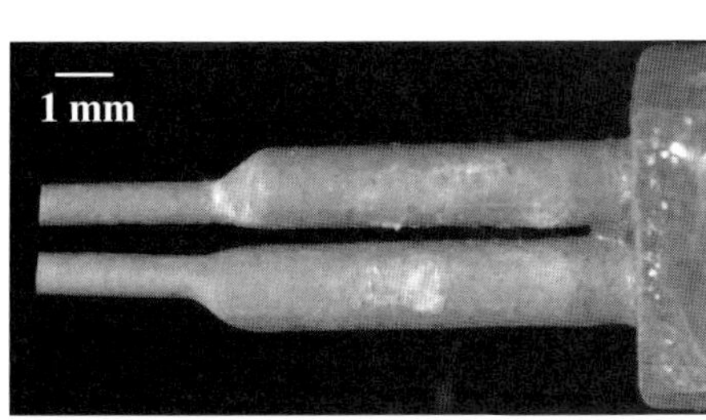

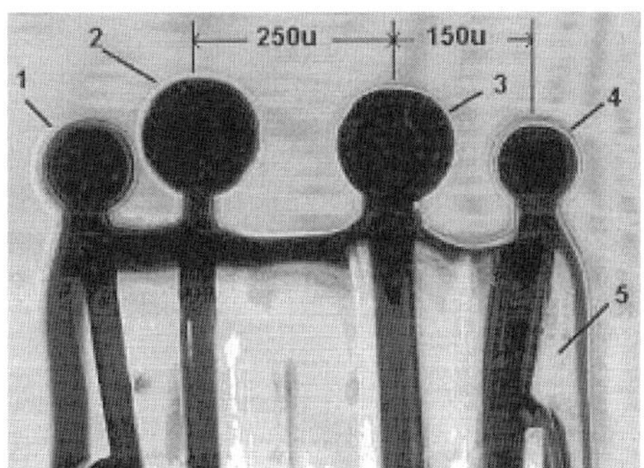

Figure 5.11 (Left) SBE-7 dual-needle conductivity probe, with electrodes at the tips. (Right) Head (1983) conductivity probe with four electrodes in a glass matrix (5); electrodes 2 and 3 supply current and 1 and 4 measure voltage. (From Sommer et al., 2013b [left] and Nash and Moum, 1999 [right], the latter © American Meteorological Society. Used with permission)

Head 4-Electrode Probe

By shrinking a 4-electrode assembly to about 0.5 mm (Figure 5.11, right), Head (1983) produced a probe capable of detecting part of the salinity dissipation range. Represented as

$$H^2_{\text{Head}}(k) = \frac{1}{1 + (k/k_c)^2}, \tag{5.8}$$

with $k_c = 455$ cpm, the response is attenuated by 3 db at 300 cpm (Figure 5.10). For a linearized dependence of conductivity on temperature and salinity, the conductivity gradient spectrum is

$$\Phi_{C_z}(k) = \left(\frac{\partial c}{\partial S}\right)^2 \Phi_{S_z}(k) + \frac{\partial c}{\partial S}\frac{\partial c}{\partial T}\Phi_{T_z S_z}(k) + \left(\frac{\partial c}{\partial T}\right)^2 \Phi_{T_z}(k) \tag{5.9}$$

(Washburn et al., 1996). At lower wavenumbers, the relative amplitudes of the three components vary with dS/dT. The cross-spectrum can be significant, but at high wavenumbers only the salinity contribution remains, as shown by Nash and Moum (1999) in Figure 5.12.

Density

Directly related to density, the index of refraction is an attractive parameter for measuring it. Variations of the index are small, less than 1% at a given pressure, but optical measurements can be very sensitive, leading Alford et al. (2006b) to fuse two single-mode optical fibers and suspend them across a 2 mm gap between support posts. The amount of light from one fiber coupling into the other, α_{fiber},

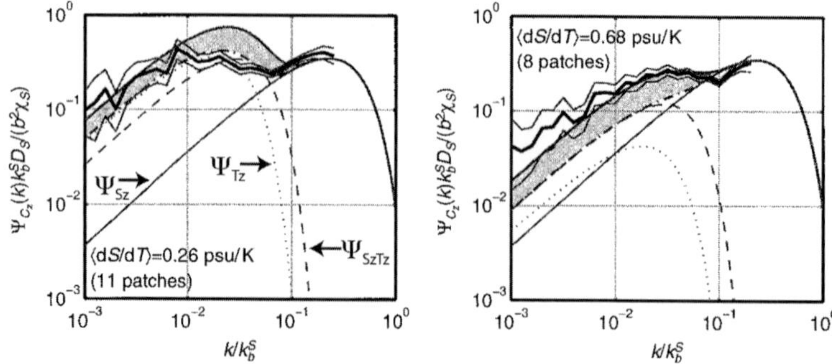

Figure 5.12 Normalized conductivity gradient spectra from the Head probe at small (left) and large (right) dS/dT. $k_b^S \equiv (\epsilon/\nu\kappa_S)^{1/4}$ is the salinity Batchelor scale, and spectra are divided by $\chi_S b^2/(k_b^S\kappa_S)$. Shading shows where coherence, $\gamma_{S_zT_z}(k)$, is significant. The thick gray line, e.g. the top line on the left, is the spectrum expected when S' and T' are fully coherent. Salinity variance dominates for $k/k_b^S > 0.1$ and is also the density variance at these scales. At lower wavenumbers, contributions from temperature and TS coherence decrease as dS/dT increases. (Adapted from Nash and Moum, 1999. © American Meteorological Society. Used with permission)

varies with the refractive index, n_{sw}, as $\Delta n_{sw} = (\partial\alpha_{\text{fiber}}/\partial n_{sw})\Delta\alpha_{\text{fiber}}$. The greatest sensitivity occurs when $\alpha_{\text{fiber}} \approx 0.5$ and corresponds to $\Delta\rho/\rho < 9 \times 10^{-9}$ kg m^{-3}.

The spatial transfer function is

$$H^2_{\text{Alford}} = (2/\pi)\tan^{-1}(1/l_s k), \tag{5.10}$$

where $l_s = 2$ mm characterizes the probe, and k is in cpm. For $k \lesssim 500$ cpm, the response follows the Head probe, but it falls off less rapidly at higher wavenumbers, not reaching 0.1 until 3×10^3 cpm (Figure 5.10). This yields a distinct salinity, and density, portion of the spectrum under moderate turbulence where salinity fluctuations dominate the TS relation (Figure 5.13). In addition to temperature, at lower wavenumbers the probe also responds to strumming induced by velocity turbulence. Minimization of this contamination is needed to make the technique suitable for routine use.

5.3 Energy-Scale and Finestructure Sensors

In addition to measuring dissipation scales, understanding turbulence requires resolving energy-containing scales as well as the finescale stratification background against which overturns work. Energy-containing scales vary from tens of centimeters, and possibly smaller, at background in the upper pycnocline to tens of meters in the abyss. Resolving the smaller ones requires in-situ sensors on

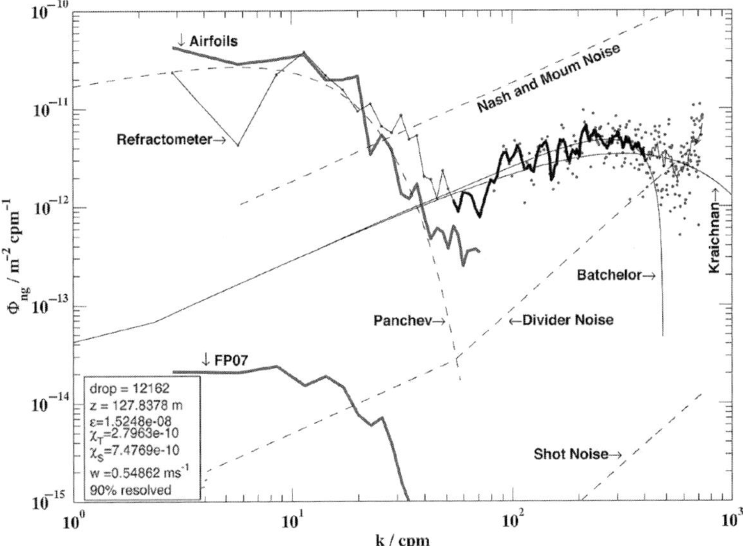

Figure 5.13 Spectrum of the vertical gradient of the refractive index in moderate turbulence in Puget Sound, where dS/dT is large, and salinity gradients dominate the index spectrum after the viscous cutoff. Vibrations induced by velocity turbulence dominate the spectrum for $k < 6$ cpm, but temperature fluctuations have negligible effect. (From Alford et al., 2006a. © American Meteorological Society. Used with permission)

autonomous platforms. Because the platforms are affected by shear, their motion must be monitored also (Section 5.3.1).

Fluctuations of temperature and salinity are important aspects of turbulence, and they determine the density stratification suppressing turbulence (Section 5.3.2). Where salinity gradients are weak, energy-containing scales are easily resolved with temperature. Elsewhere, salinity must be inferred from simultaneous measurements of temperature and electrical conductivity, but mismatches in resolution between temperature and conductivity probes often generate salinity spikes that obscure the stratification.

5.3.1 Velocity and Shear

Electromagnetic Velocity Profiling

Electric currents induced by seawater moving through the geomagnetic magnetic field, $\boldsymbol{F}^{\text{geo}}$, provide a unique means of profiling velocity in the ocean (Sanford et al., 1978). As a consequence of Ohm's Law, seawater, with electrical conductivity, σ, generates weak electric fields, $\boldsymbol{E}$, and currents

$$\boldsymbol{j} = \sigma(\boldsymbol{E} + \boldsymbol{u} \times \boldsymbol{F}^{\text{geo}}). \tag{5.11}$$

Typical electrical currents are 25 μA. Water velocities are calculated as,

$$u = -\frac{1}{F_z^{\text{geo}}}\left(E_y + \frac{j_y}{\sigma}\right), \quad v = \frac{1}{F_z^{\text{geo}}}\left(E_x + \frac{j_x}{\sigma}\right) + \left(\frac{F_h^{\text{geo}}}{F_z^{\text{geo}}}\right) w. \tag{5.12}$$

Because $F_h^{\text{geo}}/F_z^{\text{geo}} \approx 0.5$ at mid-latitudes and $w_{\text{rms}}^{\text{iw}} = 5(f/f_{30^o})$ mm s^{-1}, northward velocity proportional to vertical water velocity, w, can usually be ignored as a small error. Owing to the dependence on the northward component of Earth's magnetic field, useful signals cannot be found within 2° of the magnetic equator (Kennelly et al., 1986).

Because electric field components cannot be measured adequately from a profiler, this technique does not obtain the barotropic component of the flow, i.e. the currents are baroclinic, with an unknown constant-depth offset,

$$u - \langle u \rangle_z = \frac{1}{F_z^{\text{geo}}}\left(\frac{j_y}{\sigma} - \left\langle\frac{j_y}{\sigma}\right\rangle_z\right) \tag{5.13}$$

$$v - \langle v \rangle_z = -\frac{1}{F_z^{\text{geo}}}\left(\frac{j_x}{\sigma} - \left\langle\frac{j_x}{\sigma}\right\rangle_z\right). \tag{5.14}$$

Sanford et al. (1985) added acoustic Doppler to determine the offset by sensing profiler motion near the bottom. GPS coordinates before and after profiles provide offsets for drops turning around at mid-depth.

Voltages j_x/σ and j_y/σ are measured by electrodes on opposite sides of insulating shells surrounding profilers rotating under the influence of slanted vanes. Low-frequency electrode drift is determined by fitting sinusoids to the data. Noise levels are typically 10–15 mm s^{-1}. In the upper thermocline, with typical shears of 0.01 s^{-1}, vertical resolution is 5–10 m, sufficient to resolve the variance needed to estimate turbulent dissipation rates produced by breaking internal waves (Section 7.6.1). Energy-containing scales are rarely large enough to be detected electromagnetically, but the EM data are essential for determining the zero level for small-scale shear measured acoustically on the same profilers (Winkel et al., 1996). Combining data from the acoustic current meter, the electromagnetic probes, and airfoils gives a composite velocity profile over scales from the 1 km profile length to the ~10 mm airfoil resolution (Figs. 4.21 and 7.23).

An expendable version (Sanford et al., 1982) is sold commercially as the Expendable Current Profiler (XCP). It can be used from ships underway to observe the evolution of velocity and shear along transects, e.g. internal tide generation over Kaena Ridge (Figure 8.38). Owing to the smaller electrode separation, the noise level is higher, ~20 mm s^{-1}, than for standard profilers. When internal waves are at background, noise limits XCP spatial resolution to 0.02–0.03 cpm, but elevated spectra have been observed to 0.1 cpm.

Acoustic Differential Travel-Time Current Meters

Differences in travel times between MHz acoustic pulses transmitted in opposite directions along short paths measure flow relative to the host platform (Evans et al., 1979; Hayes et al., 1984). On profilers, two crossed paths can be configured to measure horizontal currents in front of a cluster of microstructure sensors (Figure 5.23). Typical characteristics are path lengths of 0.2 m and rms noise levels of 0.5 mm s^{-1} (Winkel et al., 1996).

Measurements on profilers also depend on vehicle stability and require correction for tilt and lateral displacement. Combining an acoustic current meter with electromagnetic sensors allows accurate reconstruction of horizontal currents $(u(z), v(z))$ with scales of hundreds of meters to 0.1 m (Winkel et al., 1996); without correction, acoustic current meter (ACM) spectra from the 4.2 m-long Multiscale Profiler (MSP) have a deep notch at 0.2 cpm due to the vehicle responding to shear. Adding airfoils extends spectra to $\approx$ 100 cpm, often resolving the dissipation range and allowing fully resolved composite spectra from one profiler (Gregg et al., 1996), instead of requiring simultaneous drops of multiple instruments (Gargett et al., 1981). Moreover, three-axis ACMs installed on a fixed frame can resolve energy-containing scales in bottom boundary layers (Gross et al., 1984; Williams, 2014), allowing estimation of dissipation rates using the inertial subrange.

Acoustic Doppler Velocimeter

Acoustic Doppler velocimeters (ADVs) detect energy backscattered from particles in small turbulent volumes. Typically, short MHz acoustic pulses are transmitted along an axis surrounded by receive transducers to obtain all three velocity components. Sensing volumes are small, 10 mm across or slightly less (Voulgaris and Trowbridge, 1998), and usually contain adequate particle densities only near ocean boundaries. To permit phase detection between successive pulses, ADVs must be mounted on stable platforms, e.g. a rigid mast beneath a ship (Holleman et al., 2016), a bottom-mounted frame (Agrawal and Belting, 1988), or a neutrally buoyant float, which produced the spectra in Figure 3.6. Noise levels are too high, $\approx$ 0.5 mm s^{-1} (Voulgaris and Trowbridge, 1998), to routinely detect the dissipation range, but, when turbulence is moderate or strong, ϵ can be estimated within a factor of 2 from the inertial subrange.

Pitot Tube

Pitot tubes have resolved inertial subranges of vertical velocity when installed on a profiler (Moum, 1990), and of horizontal velocity when clamped to a mooring (Moum, 2015). A differential pressure transducer (Figure 5.14) pointed into the

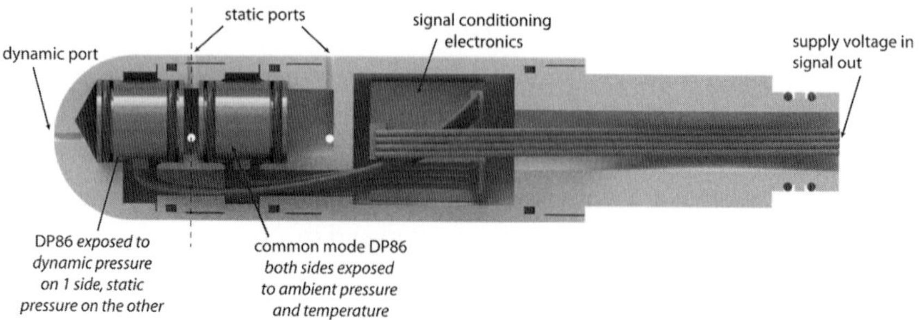

Figure 5.14 Cross-section of a pitot tube for sensing mean velocity and velocity fluctuations in the inertial subrange on a mooring. The unit is 156 mm long, 32 mm in diameter, and has duplicate internal differential pressure sensors to reduce sensitivity to temperature, acceleration, and static pressure. (From Moum, 2015. © American Meteorological Society. Used with permission)

flow senses the difference between dynamic pressure at the stagnation point on the nose and ambient static pressure on the side,

$$\Delta p = \rho u^2/2 - p_{\text{static}} \quad [\text{Pa}]. \tag{5.15}$$

Spectra exhibit well-defined inertial subranges, providing independent ϵ estimates on profilers from a sensor more robust than airfoils. On moorings, pitot tubes can provide long time series of u and ϵ, e.g. showing multiple cycles of fortnightly tidal modulation of near-bottom velocity and turbulence with noise levels $<10^{-9}$ W kg^{-1}.

Particle Image Velocimetry

Particle image velocimetry (PIV) determines three-dimensional distributions of velocity vectors by tracking successive positions of particles. Beginning with two-dimensional images every second (Bertuccioli et al., 1999; Doron et al., 2001), the technique now can sample a 0.2 m cube at 25 Hz (Nimmo Smith, 2008); Figure 5.15 is a two-dimensional cut. While removing large objects, such as swimming larvae, sophisticated algorithms track particles and form velocities that are interpolated and extrapolated using the equations of motion (Steele et al., 2013). Owing to needs for electrical power and scatterers, e.g. plankton and suspended sediment, PIV has been successful close to the bottom in shallow water, where it has identified intermittent 'hairpin' vortices and estimated all three components of the Reynolds stress.

Electromagnetic and Acoustic Vorticity Meters

By supplying a strong magnetic field, $\boldsymbol{B}$, in one direction, electromagnetic vorticity meters (Figure 5.16) allow estimation of that vorticity component from the Laplacian of electrical potential,

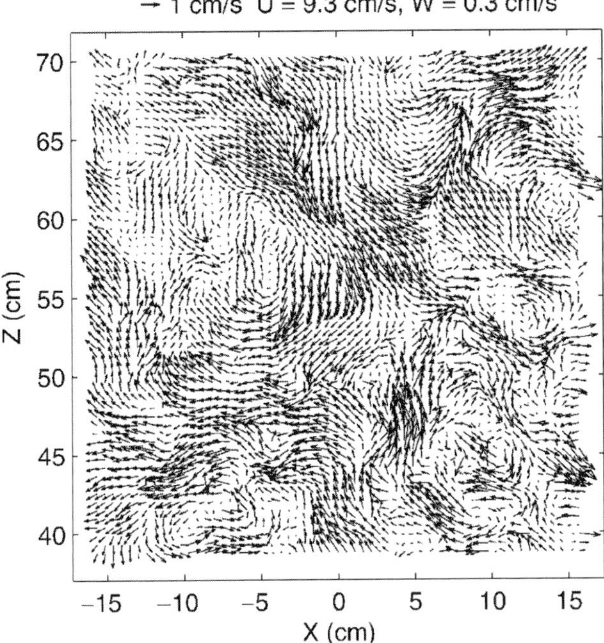

Figure 5.15 Instantaneous velocity vectors in a vertical-horizontal plane from a particle image velocimeter after removal of mean velocities at the top. (From Nimmo Smith et al., 2005. © American Meteorological Society. Used with permission)

$$\nabla^2 \phi \approx \boldsymbol{B} \cdot \boldsymbol{\omega} \quad [\mathrm{V\,m^{-2}}] \tag{5.16}$$

(Sanford et al., 1999). For instance, the electromagnetic vorticity meter in Figure 5.16 has electrodes 1–4 at the corners of a square with a diagonal distance $2l = 0.178$ m and electrode 0 in the center. When the flow-splitter plate is vertical and aligned with the mean flow, spanwise vorticity is

$$\omega_y \equiv \frac{\partial u}{\partial z} - \frac{\partial w}{\partial x} = \frac{e_1 + e_2 + e_3 + e_4}{Bl^2}, \tag{5.17}$$

where e_i is the potential difference between electrodes i and 0. The noise level is $\approx 10^{-3}\ \mathrm{s}^{-1}$. Because vorticity is concentrated near the Kolmogorov scale, the signal is attenuated as the dissipation rate increases: threefold for $\epsilon = 10^{-9}\ \mathrm{W\,kg^{-1}}$ and 10-fold for $\epsilon = 10^{-6}\ \mathrm{W\,kg^{-1}}$.

When combined with an ADV on the same slowly profiling platform (Sanford and Lien, 1999), the vorticity meter yielded profiles of enstrophy, $< \omega_y'^2(z) >$, and the vertical flux of spanwise vorticity, $< w'\omega_y' >$, through the bottom boundary layer in a turbulent tidal channel (Figure 3.2). In isotropic turbulence, each vorticity

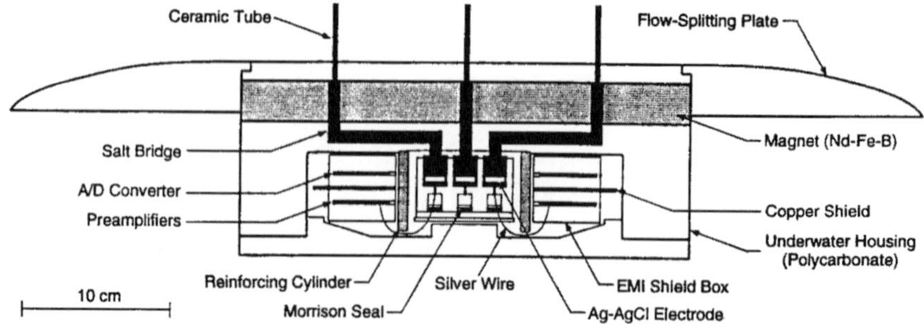

Figure 5.16 Side view of the electromagnetic vorticity meter showing three of the five electrodes; the other two are in line with the outer ones shown. Magnetic field lines are perpendicular to the flow-splitting plate. When the plate is vertical the electrodes measure ω_y. (From Sanford et al., 1999. © American Meteorological Society. Used with permission)

component equals $(\epsilon/3\nu)^{1/2}$, providing another test of isotropy when two vorticity components can be measured or when direct dissipation rates are available.

Forming closed paths with arrays of acoustic sensors measuring differential travel times offers another way of measuring vorticity. Thwaites et al. (1995) describe a family of three-axis current meters with individual path lengths of 0.15, 0.45, and 1.5 m. The paths are too long for measuring enstrophy, and flow-induced noise of 0.034 and 0.01 s^{-1} for the first two is also too high for measuring vorticity. The probes, however, measure vertical shear in the presence of surface waves and have been used to good effect in surface mixed layers and in shallow-water bottom boundary layers.

5.3.2 Temperature, Electrical Conductivity, and Salinity

Temperature

Measuring temperature requires balancing sensitivity, spatial resolution, accuracy, and sampling frequency to achieve the primary observational goals. At present, the Sea-Bird temperature probe SBE-3, packaged in several forms for different uses, is the standard for temperature finestructure from moving and fixed platforms. Developed at the Applied Physics Laboratory of the University of Washington (APL/UW) by A. Pederson as a sensor for the Self-Propelled Underwater Research Vehicle (SPURV), an early autonomous vehicle, the unit uses a 'bare' thermistor mounted in a thin stainless steel tube to protect it from ambient pressure (Figure 5.17). Thermistor resistance, R_T, is an element in a Wien Bridge oscillator with frequency $\omega^2 = 1/R_T R_2 R_1 C_2$, where R_2 is a balancing resistor and C_1 and C_2 are capacitances. Careful component selection and aging before first use produce initial accuracies of ± 1 mK. Period counting at 24 Hz gives a resolution of

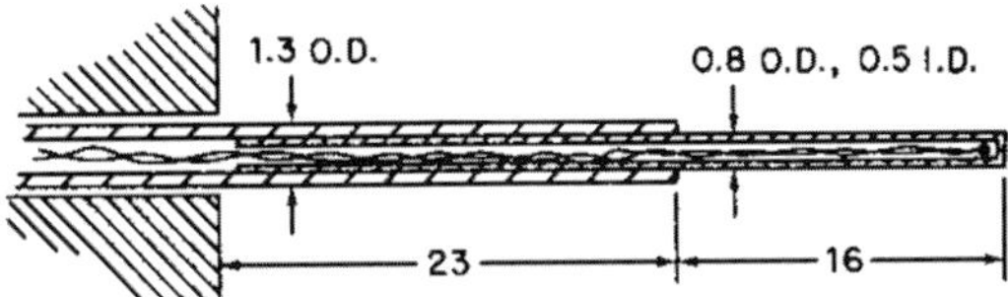

Figure 5.17 Sea-Bird temperature probe with dimensions in millimeters. (From Gregg and Hess, 1985. © American Meteorological Society. Used with permission)

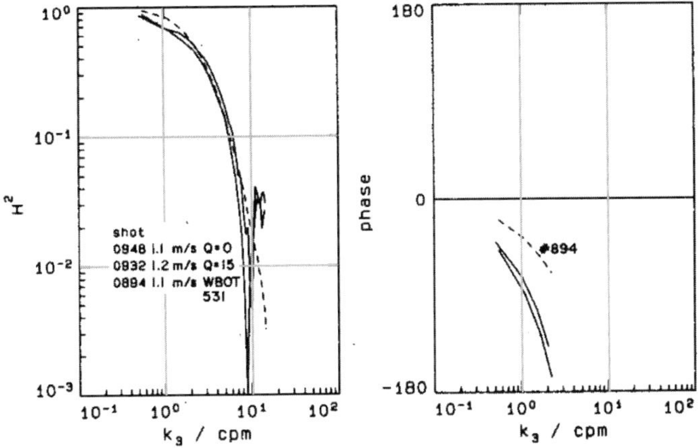

Figure 5.18 Amplitude and phase response of Sea-Bird temperature (dashed) and conductivity (solid). Q is the conductivity flow in $\mathrm{cm}^3\ \mathrm{s}^{-1}$. (Adapted from Gregg and Hess, 1985. © American Meteorological Society. Used with permission)

±0.3 mK. Dynamic response tests found half-power attenuation (Figure 5.18) at 5.6/3.9/2.2 cpm for speeds of 0.25/0.5/1.0 m s^{-1} (Gregg and Hess, 1985), precluding detection of energy-containing scales of thin overturns in the upper thermocline.

Electrical Conductivity

In addition to trade-offs similar to those encountered with temperature, measuring electrical conductivity faces the additional problem of being used primarily to obtain salinity. Depending on the ΘS relation, the salinity contribution to conductivity may be substantially less than the thermal component, requiring increased accuracy of T and C to be useful. Early conductivity-temperature-depth profilers (CTDs) measured conductivity with inductive coils near platinum thermometers and produced very spiky salinity from real-time analog computations. Resolution was at best ~10 m.

Subsequent approaches measure voltage drops from electric currents passing through control volumes and compute salinity after filtering conductivity and temperature. Gregg and Cox (1971) resolved centimeter-scale conductivity across a

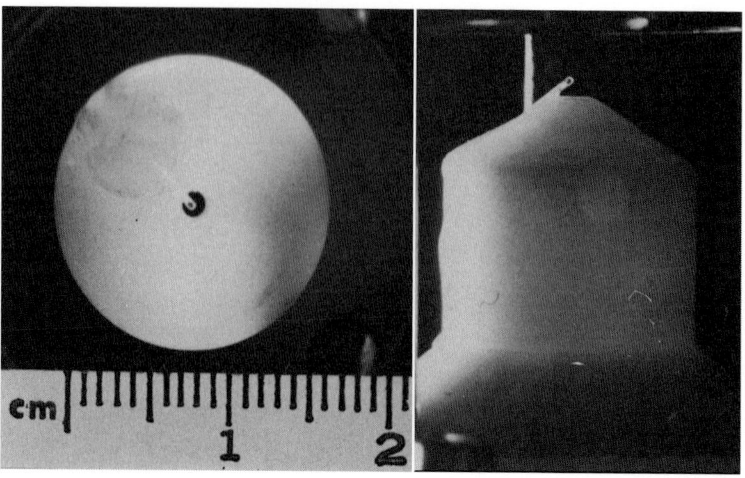

Figure 5.19 High-resolution conductivity/temperature probe with four electrodes (not shown) and a glass-rod thermistor in front of the sensing hole. (From Gregg and Cox, 1971)

1.1 mm-diameter hole behind a small glass-rod thermistor (Figure 5.19). Water was sucked through the hole by a spring-loaded piston forcing silicone oil through a small orifice. A pair of large electrodes, one inside and one outside, provided the electrical current through the hole, while another pair measured the voltage drop without the noise contamination inherent in passing current to seawater. Salinity spiking was minimal and sensitivity was good (Figs. 4.26, 4.42, and 8.28), but the probe required recharging between profiles.

Working at Woods Hole, Brown (1974) used four electrodes with a 3 cm duct open at both ends and near a platinum thermometer coupled with a glass-rod thermistor to greatly improve the quality of wire-lowered CTD data. The open conductivity cell design, however, produced a transient response much slower than expected for the size of the cell (Gregg et al., 1982), the cell constant drifted, and the offset from the temperature sensor resulted in significant salinity spiking. Consequently, a larger three-electrode cell (Pederson, 1973) that completely contains the electrical field (Figure 5.20) and can be pumped has become the oceanographic standard as Sea-Bird model SBE-4.

The electrical resistance of the Sea-Bird conductivity cell, R_2, is one side of a Wien Bridge oscillator with frequency $\omega^2 = 1/R_1R_2C_1C_2$. Conductivity is obtained as $c = R_1KC_1C_2\omega^2$, where $K = R_2\sigma$ is the cell constant. Sea-Bird gives the accuracy of a recently calibrated probe as $\pm 3 \times 10^{-3}$ S m^{-1} and the noise level as $\pm 3 \times 10^{-4}$ S m^{-1} when sampled at 24 Hz. Spatial resolution varies with the speed of the probe through the water and the pumping rate, Q (Gregg and

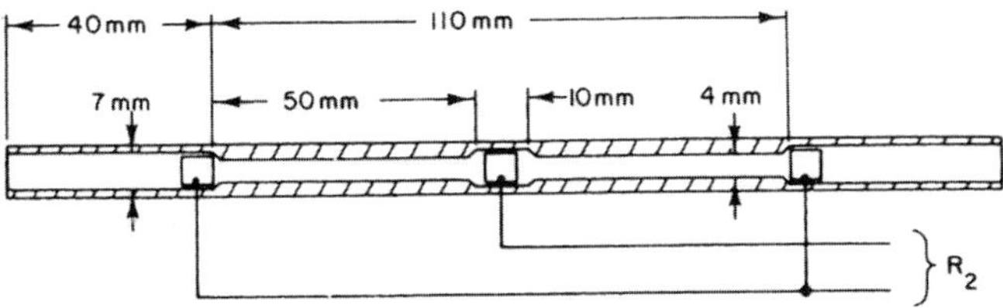

Figure 5.20 Schematic of a Sea-Bird conductivity cell (SBE-4) with three electrodes in a Pyrex tube. The two outer electrodes are at the same potential, completely containing the electric field in the cell. Cell resistance, R_2, between the inner electrode and the pair of outer electrodes, is one component of a Wien Bridge oscillator. (From Pederson and Gregg, 1979)

Hess, 1985). At 1 m s^{-1}, the half-power wavenumber is 1.7/2.0/2.2/3.7/5.4 cpm for $Q = 0/9/15/20/30$ cm^3 s^{-1}. Figure 5.18 shows the amplitude and phase response for $Q = 0$ and 15 cm^3 s^{-1}. Resolution improves as speed decreases, e.g. to a half-power response of 14 cpm at 0.25 m s^{-1} and $Q = 20$.

Calculating Salinity

As seen in Figure 5.18, at $\approx$ 1 m s^{-1}, Sea-Bird temperature and conductivity have nearly matched amplitude responses. The phase difference is larger, but it can be compensated by offsetting the records. Subsequent to the dynamic calibrations of Gregg and Hess (1985), Sea-Bird mounted the temperature probe perpendicular to the flow so it can be placed in front of the conductivity cell. This significantly improved salinity calculations, but likely altered the temperature response.

In spite of being addressed by several authors (Horne and Toole, 1980; Giles and McDougall, 1986), spiking often contaminates salinity and density over scales of $\lesssim$ 1 m (Figure 4.29). Severity varies with the ΘS relation and with the wavenumber/frequency content of the signals, both of which vary greatly along most records, the latter in response to the episodic nature of the turbulent dissipation rate, ϵ. Filtering to match amplitude responses and offsetting to minimize phase differences can ameliorate spiking, but at the cost of decreasing spatial resolution. Boosting the slower response to match the faster amplifies high-frequency noise where mean gradients are low.

5.4 Profilers

5.4.1 Wire-Lowered CTDs and LADCPs

Wire-lowered instruments are appealing, with ample electrical power supplying many sensors. Fall rates, however, depend on vertical heaving induced by the ship, as well as on winch speed, typically 1 m s^{-1}. Depending on sea state, instantaneous

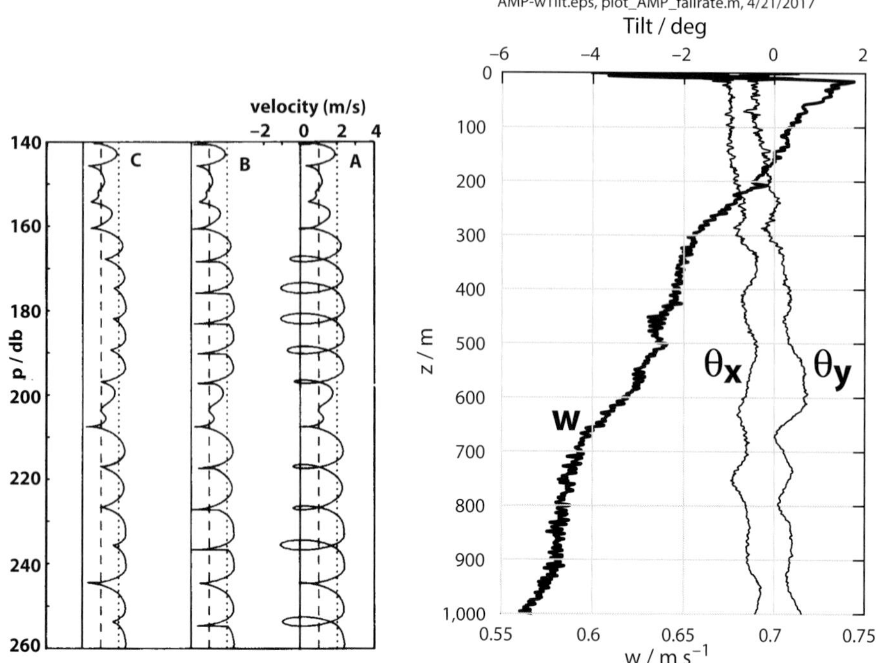

Figure 5.21 (Left) CTD fall rates for differing sea states. (Adapted from Giles and McDougall, 1986.) (Right) Fall rate, w, and tilts, θ_x and θ_y, for a loosely tethered microstructure profiler. Stratification decreased below 300 m, affecting w and tilt.

speeds can range from >2 m s^{-1} to <-1 m s^{-1} (Figure 5.21, left). Obtaining useful microstructure data requires smooth fall rates decoupled from ships, and, for χ_T, slower speeds. Typical microstructure profilers fall smoothly but too fast for optimum temperature measurements (Figure 5.21, right).

Owing to its stability, the Floating Instrument Platform (FLIP) is a unique vessel for wire-lowered profiling. When moored on Kaena Ridge, Pinkel et al. (2012) deployed two heavily weighted Sea-Bird CTDs in streamlined packages from a fast winch running at 3.6 m s^{-1}. One unit cycled between 20 and 420 m every four minutes, while the other sampled 370–800 m. After processing to match probe responses, the Sea-Bird data resolved structures of 2 m. Other deployments (Alford and Pinkel, 2000) included dual-needle conductivity probes to examine mixing directly. Obtained simultaneously with powerful Doppler current profilers, these FLIP data provide unique views of internal waves and their mixing.

Lowered Acoustic Doppler Current Profilers (LADCPs) attached to CTDs measure finescale shear and strain, allowing dissipation rates and diapycnal diffusivities to be estimated throughout the water column (Polzin et al., 2002). The LADCP transmits coded high-frequency acoustic pulses along pairs of narrow

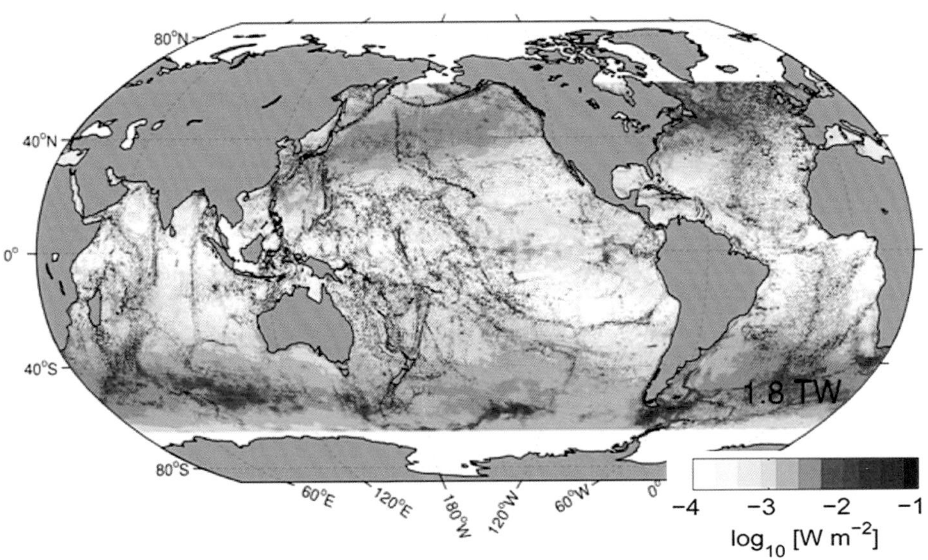

Figure 1.14 Power input to internal waves by internal tides and winds at near-inertial frequencies. Wind forcing varies smoothly over broad areas and is most intense at high latitudes. Internal tide forcing is strongest over ridges and seamounts and appears as lines of dots that are most intense at mid- and low latitudes in the western Pacific. (From Waterhouse et al., 2014. © American Meteorological Society. Used with permission.)

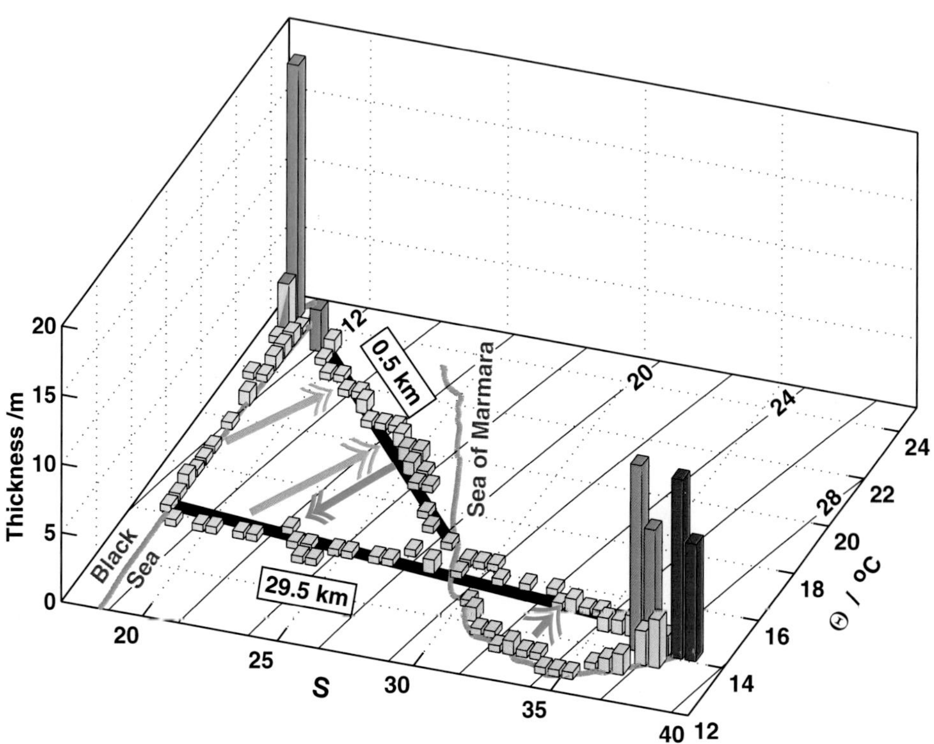

Figure 2.20 Volumetric θS diagram for the two-layer exchange flow through the Bosphorus. Numbered gray contours are σ_θ levels, and the heights of the columns are proportional to the thickness of water binned with $\Delta\theta = 0.5°C$ and $\Delta S = 0.5$ psu. (From Gregg and Özsoy, 2002.)

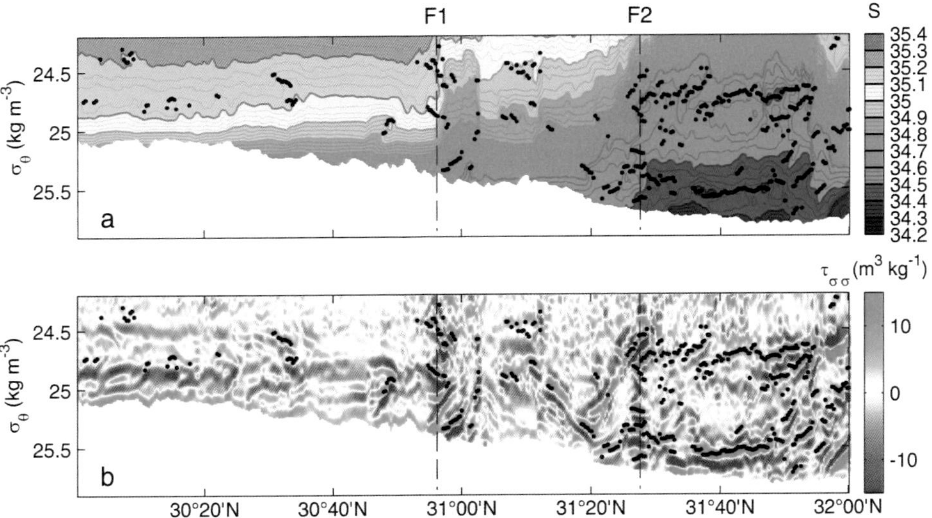

Figure 4.38 Salinity (a) and diapycnal spiciness curvature (b) along 158°W across two branches, F1 and F2, of the Subtropical Front. Salinity minima, black dots, are coherent over parts of $\tau_{\theta\theta}$ maxima and minima (shading). (Adapted from Shcherbina et al., 2009. © American Meteorological Society. Used with permission.)

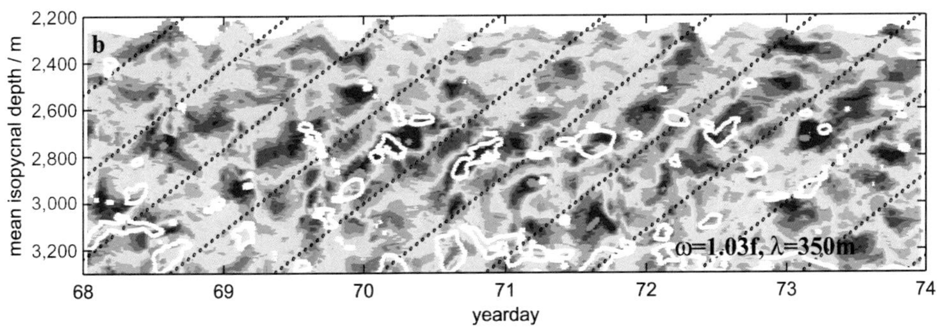

Figure 6.4 Zonal shear near the Mendocino Escarpment spanning ±0.001 s^{-1} versus time on a semi-Lagrangian vertical reference. The observations are consistent with $\omega = 1.03\,f$ and $\lambda = 350\ m$. (From Alford, 2010. © American Meteorological Society. Used with permission.)

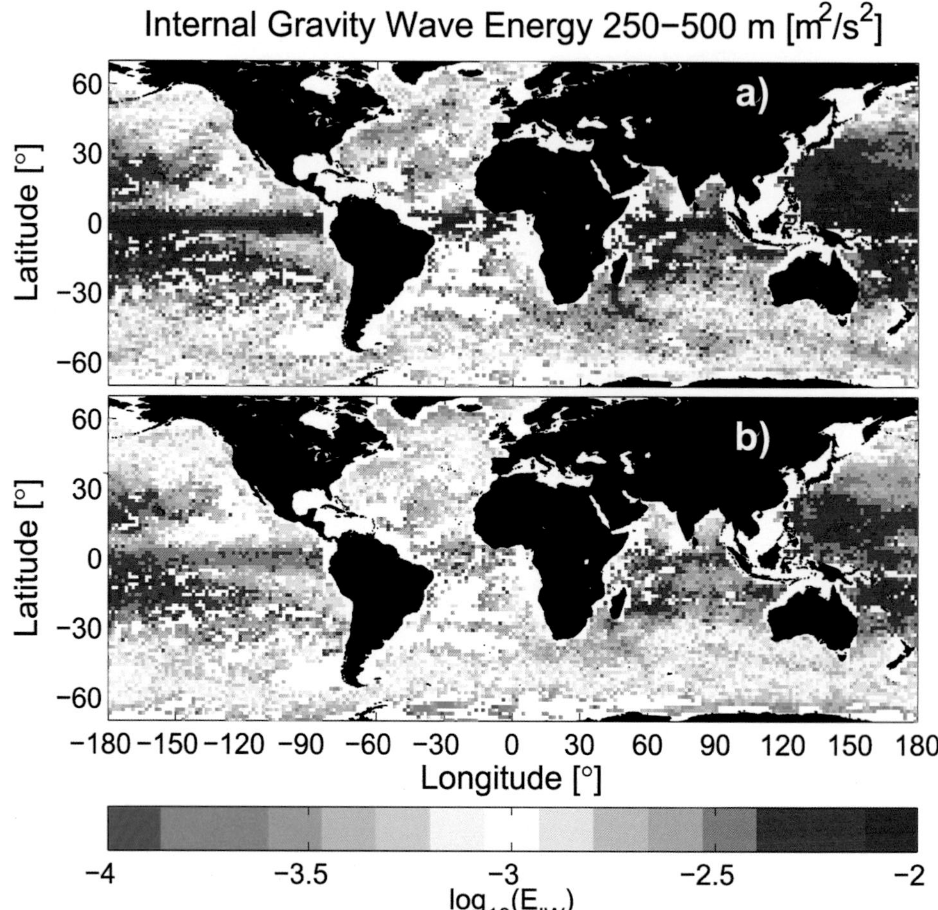

Figure 6.25 Internal wave energy from Argo profiles between 2006 and 2015, using density displacement (a) and strain (b). Intensity is highest in the western North Pacific and along the equator and lowest in the Southern Ocean. (From Pollman et al., 2017. © American Meteorological Society. Used with permission.)

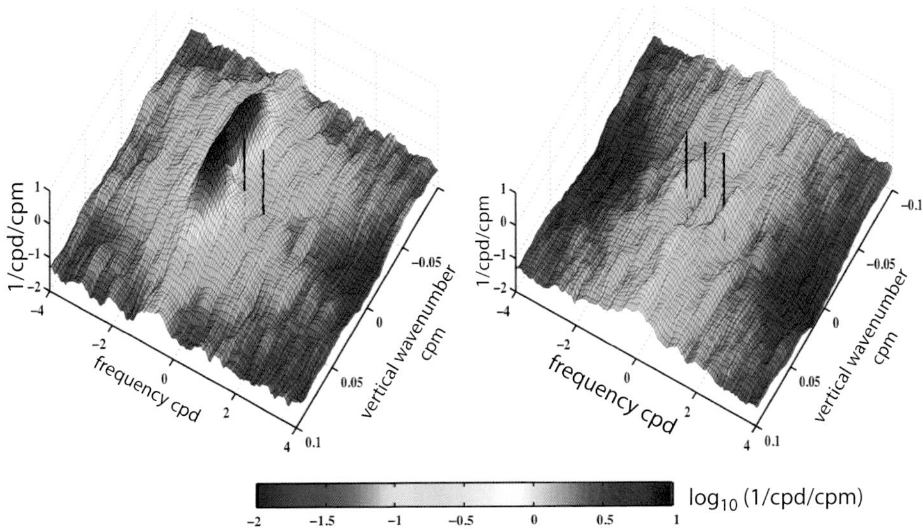

Figure 6.28 Two-dimensonal spectra of normalized shear, $\Phi_{Fr}(\omega, k_z)$ with $Fr \equiv Shear(z)/N(z)$ (left), and strain (right), both on semi-Lagrangian references. Vertical black lines mark frequencies of $-f, 0, f$. (Adapted from Pinkel, 2014. © American Meteorological Society. Used with permission.)

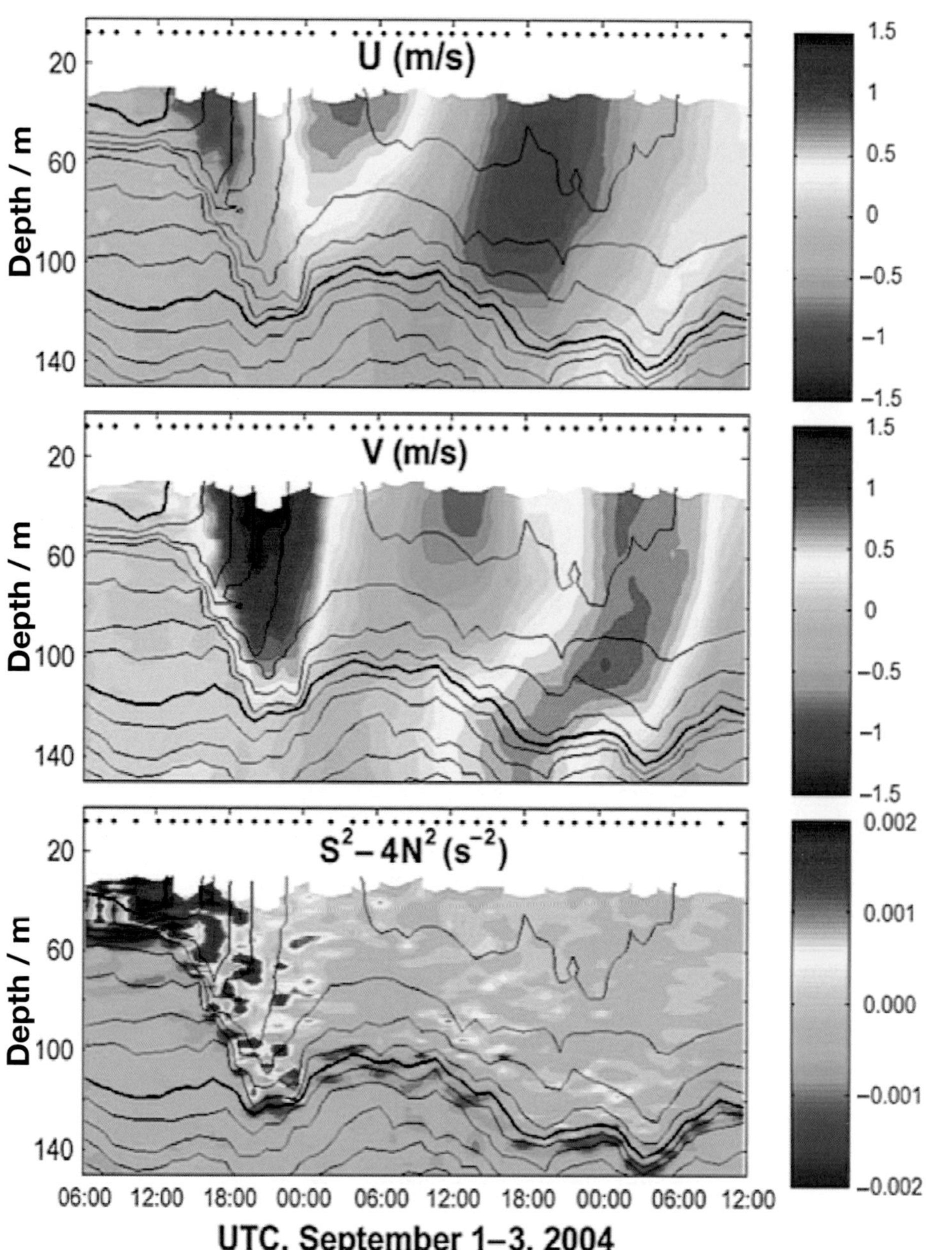

Figure 7.2 Mixed-layer deepening and near-inertial generation measured by an EM-APEX float 50 km to the right of Hurricane Frances. Isotherms are at 0.5°C intervals. The surface mixed layer deepened 80 m in six hours, cooling the surface by 2.2°C. Due to the intense dissipation and mixing, near-inertial currents of 1 m s^{-1} and larger in the mixed layer quickly decayed as waves propagated downward. (Adapted from Sanford et al., 2011. © American Meteorological Society. Used with permission.)

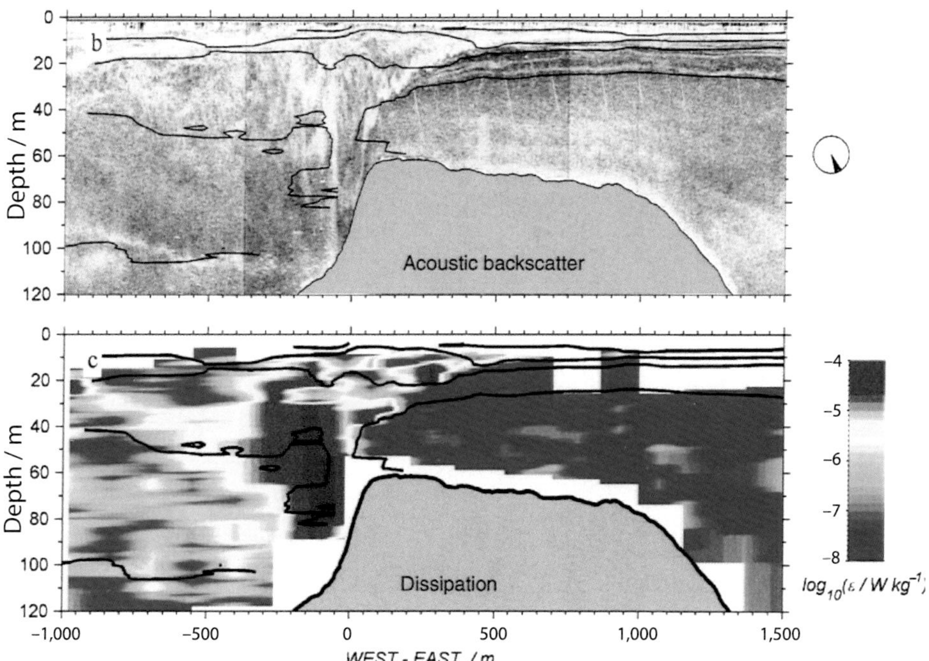

Figure 7.8 (b) Backscatter intensity from a narrow beam of high-frequency acoustics over the seaward side of the Knight Inlet Sill during ebb tide. Flow was hydraulically controlled at the sill crest, forming an hydraulic jump downstream. (c) Intense turbulent dissipation in the plunging flow and rotor. (Adapted from Klymak and Gregg, 2004. © American Meteorological Society. Used with permission.)

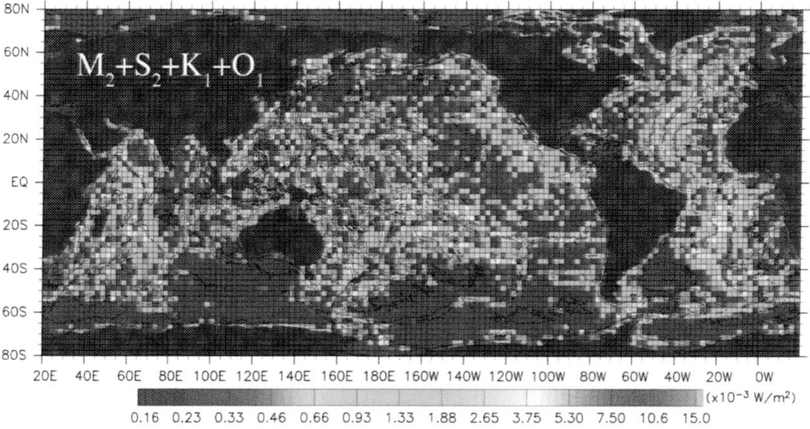

Figure 7.10 Baroclinic tidal energy conversion rate for four tidal components extrapolated to zero horizontal grid spacing. Estimates are averages over 2.5° by 2.5° intervals. (Adapted from Niwa and Hibiya, 2014.)

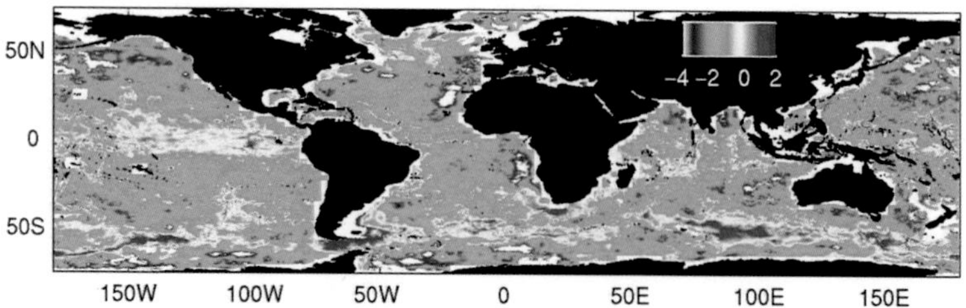

Figure 7.12 Energy flux into internal lee waves ($\log_{10} J_E/\text{mWm}^{-2}$) estimated using global descriptions of bottom stratification, velocity, and roughness. Large fluxes occur in the Southern Ocean and near the equator in the eastern Pacific. (From Nikurashin and Ferrari, 2011.)

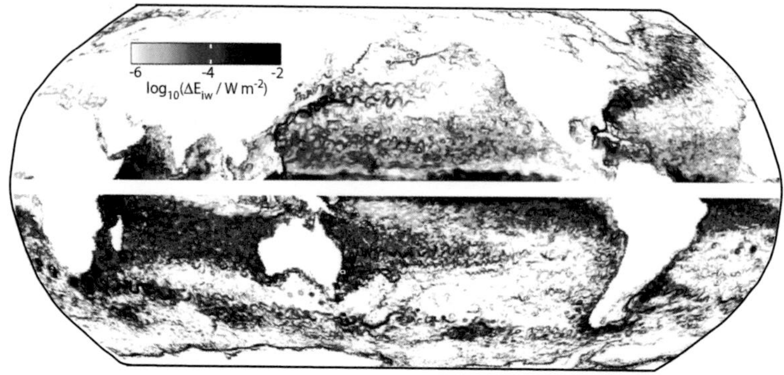

Figure 7.14 Internal wave power gain from balanced geostrophic flows, except within 3° of the equator. (Adapted from Nagai et al., 2015.)

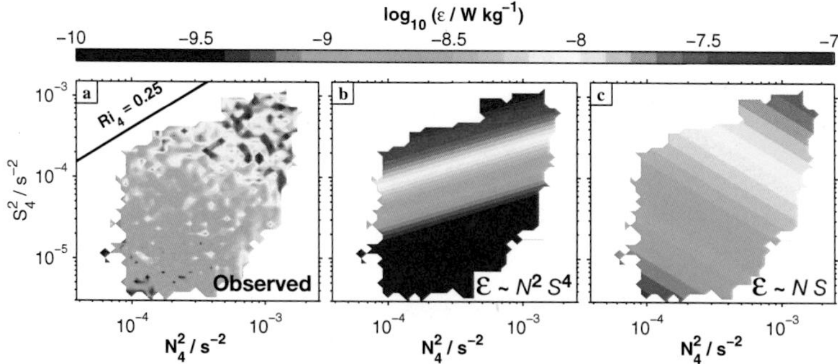

Figure 7.33 Observed (a) and scaled (b,c) dissipation rates in 70 m of water over the New England shelf. Stratification and shear were over 4 m intervals. Using them in (7.16) produces an ϵ pattern nearly orthogonal to observations, in contrast to applying (7.20), which was constructed to agree with the data. (Adapted from MacKinnon and Gregg, 2003. © American Meteorological Society. Used with permission.)

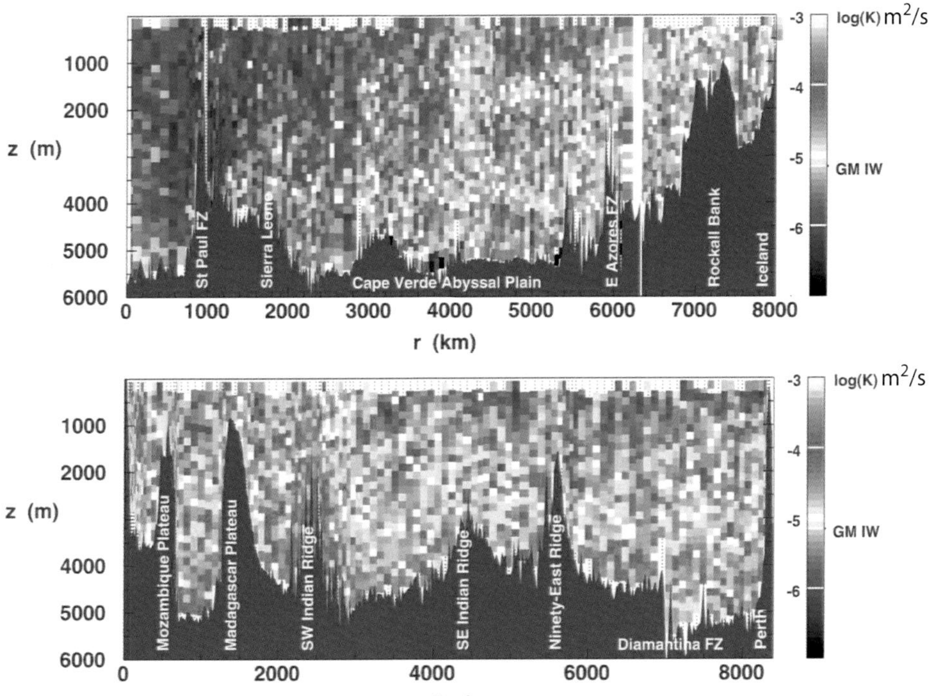

Figure 8.15 Vertically integrated diapycnal diffusivity from finescale shear and strain in the Atlantic from 7°S to Iceland (upper) and in the Indian Ocean along 32°S from Madagascar to Perth (lower). (Adapted from Kunze et al., 2006a. © American Meteorological Society. Used with permission.)

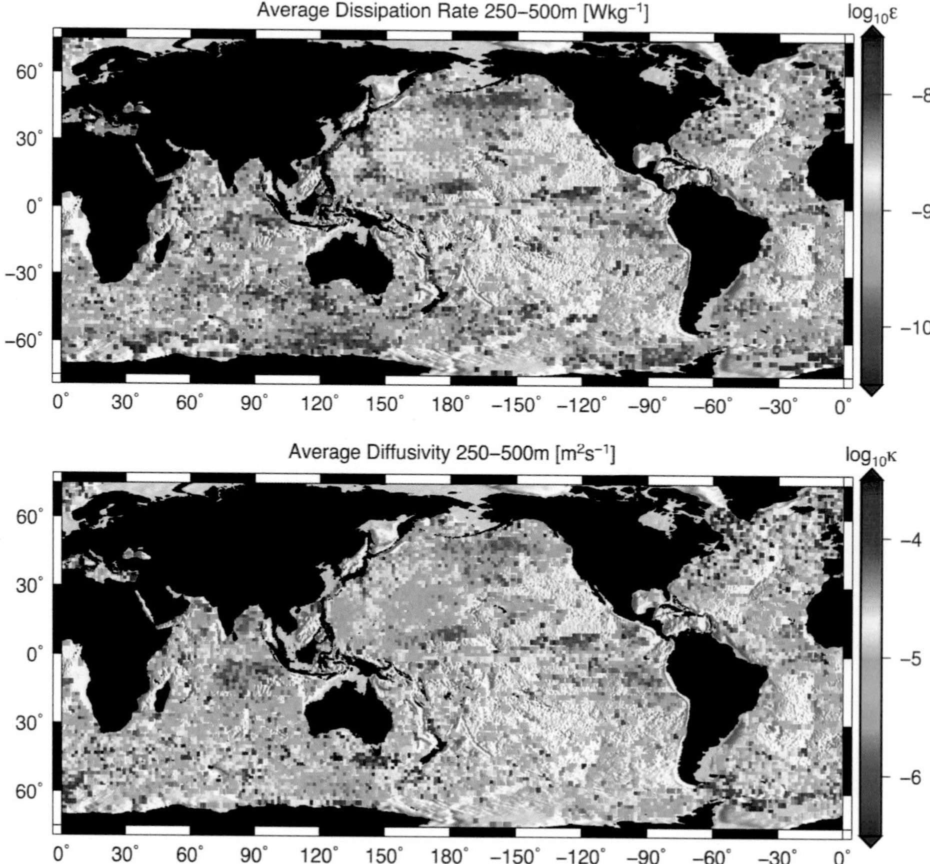

Figure 8.16 Average dissipation rates and diapycnal diffusivities from strain in Argo profiles between 2006 and 2011. (Adapted from Whalen et al., 2012.)

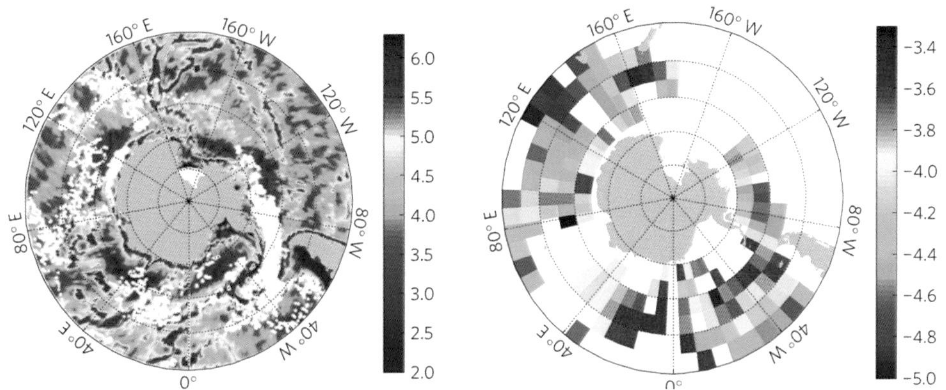

Figure 8.29 (Left) Bottom character as $\log_{10} Roughness(\mathrm{m}^2)$. (Right) $\log_{10}$ $(K_\rho(\mathrm{m}^2\mathrm{s}^{-1})$ over 300–1,800 m. (Adapted from Wu et al., 2011.)

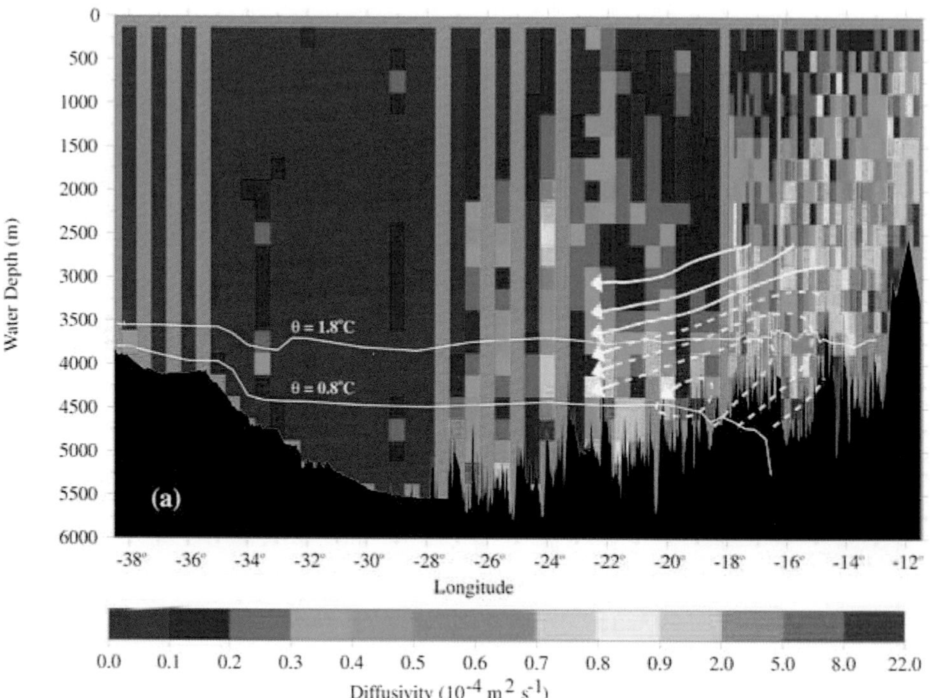

Figure 8.35 Diapycnal diffusivity across the Brazil Basin from velocity microstructure (Polzin et al., 1997; Ledwell et al., 2000). White lines are stream functions inferred from an inverse calculation (St. Laurent et al., 2001). (Adapted from Mauritzen et al., 2002.)

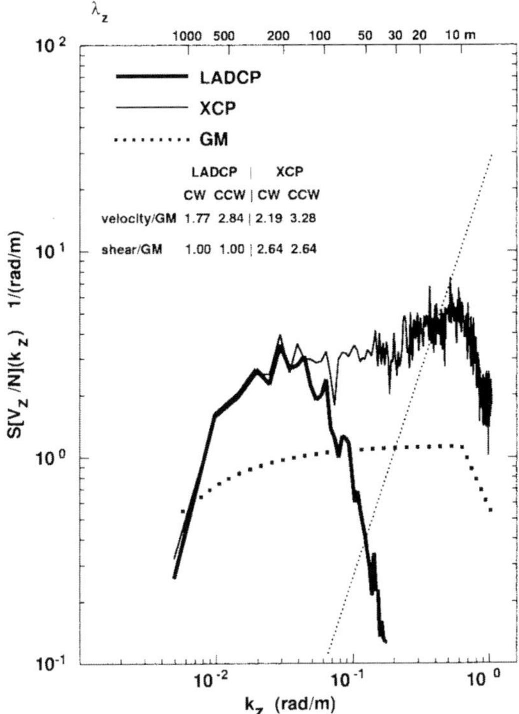

Figure 5.22 Spectra of normalized vertical shear observed with an LADCP and an Expendable Current Profiler (XCP) compared to the GM internal wave model for $\overline{N} = 0.0017\ \mathrm{s}^{-1}$. Noise dominates XCP spectra to the right of the thin dotted diagonal line. (From Polzin et al., 2002. © American Meteorological Society. Used with permission)

beams inclined 20° to 30° from vertical. Horizontal velocities for each range bin are calculated from Doppler shifts along the beams. After correction for tilting, inferred velocities are averages along horizontal paths between range beams, which lengthen with range. Comparing with Expendable Current Profilers (XCPs) shows identical spectra to ~0.03 cpm (Figure 5.22). Though the LADCP spectra roll off steeply at higher wavenumbers, Polzin et al. (2002) argue that K_ρ can be estimated within a factor of 3–4 from the shear variance that can be measured.

5.4.2 Free-Fall and Loosely Tethered Profilers

Recognizing that solid-state electronics permitted deploying robust electronics in the ocean, and inspired by Grant et al. (1962), in the mid-1960s C.S. Cox at Scripps developed the free-fall Microstructure Recorder (MSR) to measure temperature microstructure in the open ocean. Ballasted several kilograms negative to minimize the relative increase in buoyancy with increasing depth, the MSR was slowed to $\lesssim 0.1$ m s^{-1} by 1.5 m-long wing blades generating lift as they rotated the tube. At that speed, a glass-rod thermistor could resolve temperature gradients in the thermocline off San Diego (Cox et al., 1969). Until the electronics became reliable, the MSR survived many failures owing to a mechanical release that dropped ballast at pressures beyond a preset depth. Gregg and Cox (1971) added centimeter-scale conductivity, and Osborn and Cox (1972) used the temperature microstructure to estimate diapycnal diffusivity by approximating the $\overline{\Theta'^2}$ equation.

In 1975 the Fine and Microstructure Experiment (FAME) brought the MSR together with free-fall profilers carrying an optical shadowgraph, airfoils (Osborn, 1974), and sensors for measuring velocity electromagnetically (Sanford, 1975) and acoustically (Rossby, 1969). The comparisons related mixing to finescale features (Gregg and Sanford, 1980) and provided composite shear spectra (Gargett et al., 1981). Even though the profilers were launched within a few minutes, differences in fine and microstructure between profilers demonstrated the need to put sensors on the same vehicle. This realization led to the Multi-Scale Profiler (MSP) in the early 1980s and then the High-Resolution Profiler (HRP).

In addition to a Sea-Bird CTD and temperature and velocity microstructure, the MSP fell at 0.35 m s^{-1}, measuring velocity finestructure using electromagnetic induction and a travel-time acoustic current meter. After modeling the MSP response to horizontal currents (Figure 5.23), corrected shear from the acoustic current meter resolved fluctuations over scales of 100 m to 1 m. Subsequently, the HRP, carrying an ACM and airfoils, extended measurements to full depth (Schmitt et al., 1988). Its successor (HRP II) and later commercial profilers by Rockland Scientific have included electromagnetic velocity sensors to correct ACM data. Requiring several hours to launch, recover, and rearm, these large profilers can barely resolve daily cycles at one site. They have proved more effective in discovering low-frequency and large-amplitude changes in mixing intensity over large areas, e.g. across the Brazil Basin (Polzin et al., 1997).

To permit operations in estuaries and near coasts, as well as more frequent profiling of the upper ocean, Elliott and Oakey (1975) and Caldwell et al. (1975) developed loosely tethered microstructure profilers with data links in strong, flexible cables. These became workhorses, obtaining several thousand profiles during two to three weeks. Later versions tracked the diurnal cycle of convectively forced turbulence in the surface mixed layer (Brainerd and Gregg, 1992) and in billow trains of breaking internal waves and solitons in the pycnocline (Moum et al., 2003; MacKinnon and Gregg, 2003). Tethers, however, increase drag, slow descent, and add vibrations contaminating dissipation-range signals. Tethers can also energize the first bending mode of profilers, producing some of the largest spikes in dissipation spectra (Figure 5.24). The frequency of the bending mode increases as

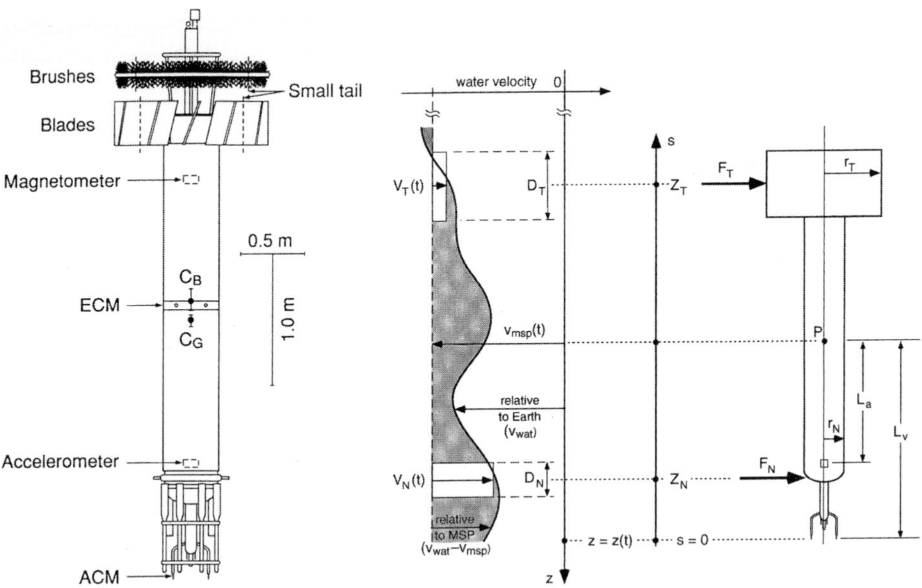

Figure 5.23 Schematics of the Multi-Scale Profiler (MSP, left) and its response to horizontal currents (right). C_B and C_G are centers of buoyancy and of mass (gravity). ECM indicates electrodes for detecting motionally induced velocity, and ACM is the acoustic current meter. Horizontal drag is represented by tail, F_T, and nose, F_N, forces. Shading in the right panel shows water velocity relative to the MSP. (From Winkel et al., 1996. © American Meterological Society. Used with permission)

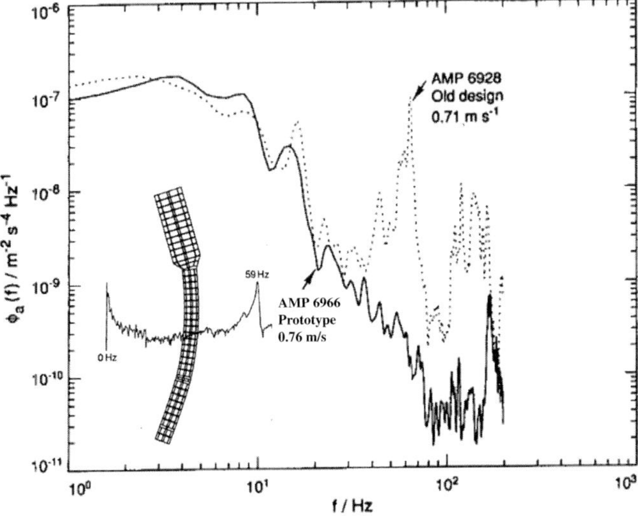

Figure 5.24 Spectra of different profilers with first bending modes at lower and higher frequencies. The insert shows the mode and a spectrum on a linear axis. (From Miller et al., 1989. © American Meteorological Society. Used with permission)

profilers are made thicker and shorter (Miller et al., 1989). Since the mode cannot be eliminated, our next profilers were designed with the bending-mode frequency exceeding the 100 Hz upper limit of airfoils.

5.4.3 Wire-Crawlers and Wire-Walkers

McLane Profilers and wire-walkers enable long time series by moving instruments up and down wires suspended from floats or ice. McLane profilers use pinch-rollers to crawl up and down mooring wires at ≈ 0.25 m s^{-1} and can operate throughout the water column (Morrison et al., 2000). Sensors include CTDs and acoustic current meters, but vibrations from motion and wire strumming set the velocity noise level (Thwaites et al., 2011), so far precluding turbulence measurements.

Powered solely by vertical displacements of surface waves, wire-walkers move along weighted wires suspended from surface floats. As the surface rises, the frame is allowed to slide down the wire, but cams lock rubber wheels in contact with the wire when the surface falls (Pinkel et al., 2011). Release of the cams at the bottom of the wire allows the positively buoyant frame to rise freely and smoothly at 0.24–0.50 m s^{-1}, permitting successful microstructure measurements (R. Pinkel, personal communication, 2017).

5.4.4 Floats and Gliders

Initially developed to monitor the general circulation (Davis et al., 1992), profiling floats proved so useful that CTD sensors were added (Davis et al., 2001). Subsequently, these were joined by many specialized probes (Figure 5.25, right), beginning with temperature microstructure for the North Atlantic Tracer Release Experiment (NATRE) (Sherman and Davis, 1995) and electromagnetic velocity for process studies (Sanford et al., 2005). Temperature and salinity from Autonomous Profiling Explorer (APEX) floats, widely deployed as the Argo float network, led to the first global maps of mixing intensity (Section 8.4). Some recent floats carry airfoils as well as fast thermistors.

After launch, Argo floats descend to 1,000 m, where they become neutrally buoyant by pumping hydraulic fluid from an internal reservoir to an external bladder. They then 'sleep' for 10 days before descending to 2,000 m to begin collecting CTD data while rising at $\sim$0.1 m s^{-1}. Data are transmitted while on the surface, before the floats return to 1,000 m for another snooze. Floats usually last four years and collect 150 profiles. During process studies, specialized floats are often cycled more rapidly to shallower depths.

D'Asaro et al. (1996) followed a different approach in designing a float (Figure 5.25, left) that is more truly Lagrangian by closely approximating the

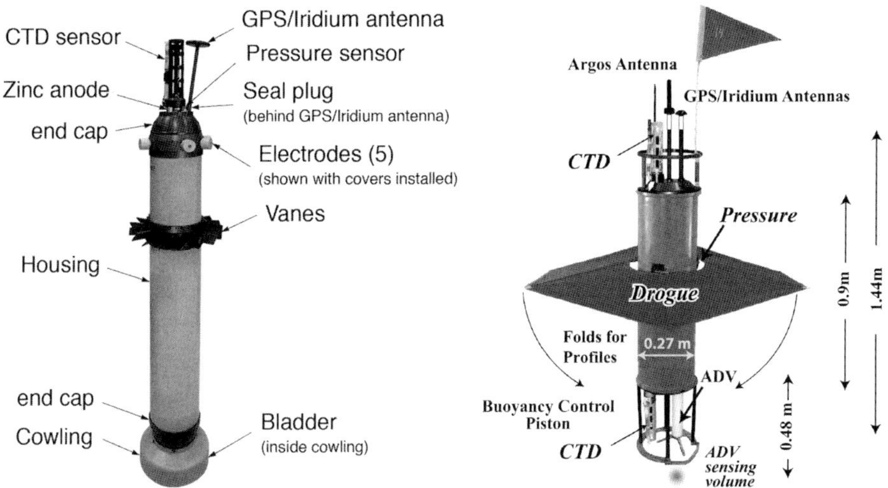

Figure 5.25 EM-APEX float (left from T. Sanford, personal communication, 2019) and a neutrally buoyant Lagrangian float (right from Lien and D'Asaro, 2006. © American Meteorological Society. Used with permission)

compressibility of seawater; the compressibility of standard neutrally buoyant floats differs from that of seawater.[1] A collapsible screen provides high vertical drag while deployed, making the float Lagrangian for $\omega \gtrsim N/30$. One version also measures turbulence with an ADV (Section 3.7.1). Pressure and temperature sensors have also yielded turbulent kinetic energy and vertical heat flux, $\overline{w'\Theta'}$, when convective eddies cycled Lagrangian floats through the depth of a convecting mixed layer (D'Asaro, 2001; Steffen and D'Asaro, 2002). In the pycnocline, adjusting ballast allows floats to 'tag' temperature or salinity surfaces for extended sampling of small-scale features, such as thermohaline intrusions (Alford et al., 2005; Shcherbina et al., 2010).

Unlike floats, gliders can point microstructure probes into the flow while ascending (Figure 5.26) and descending by moving the battery to shift the center of mass relative to the center of buoyancy. Controlling rudder and climb/dive angles allows gliders to maintain stations or follow tracks against weak flows (Eriksen et al., 2001). Adding airfoils gives normal components of velocity microstructure along the glide path. Dissipation noise levels, $\lesssim 10^{-10}\ \mathrm{W\,kg^{-1}}$, are comparable to those on the quietest profilers. Detection of overturns (Smyth and Thorpe, 2012) and the effects of anisotropy along slant paths differ from those along profiles, but, with the suite of available sensors, floats can survey and monitor mixing (Beaird et al., 2012; Fer et al., 2014; Palmer et al., 2015) much more cheaply than can ships.

[1] Otherwise, floats gradually migrate off density surfaces under the cumulative effect of internal wave displacements.

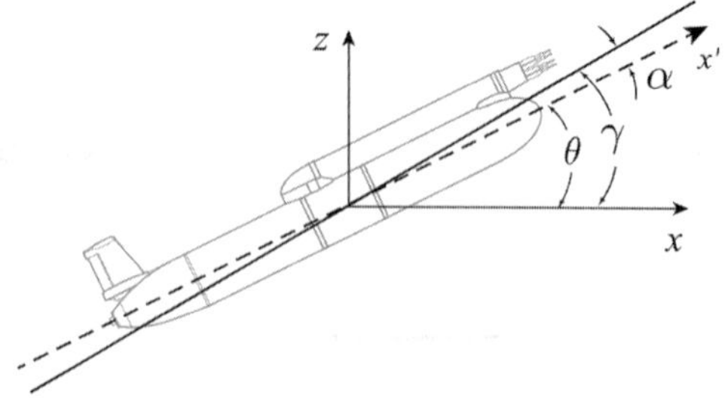

Figure 5.26 Schematic of a Slocum glider ascending with a Rockland Microstructure Rider on top. The glide angle, $\gamma = \theta + sgn(\alpha)$, is typically $+24^o$ or -34°, where θ is the pitch angle and $\alpha \sim 2$–4° is the angle of attack. (From Fer et al., 2014)

5.5 Towed Bodies and Self-Propelled Vehicles

Constant-depth horizontal transects supplement profiles to provide a second dimension necessary for understanding internal waves and mixing in general, and salt fingers in particular. Tight depth-cycling runs map horizontal structures, e.g. thermohaline intrusions, but the additional lift and drag increase vibrations, contaminating microscale measurements.

5.5.1 Tows to Measure Turbulence

Horizontal turbulence measurements with dissipation noise levels comparable to free-fall profilers have been achieved by towing slowly, $\lesssim 1$ m s^{-1}, and mounting probes on streamlined bodies attached to submerged tow points (Lueck, 1987; Moum et al., 2002). This was accomplished by depressing the tow cable from the ship with a heavy weight or depressor and using a long flexible line that pulled the body along a nearly horizontal path behind and below the tow point (Figure 5.27). Motion was further stabilized by a high-drag tail of chimney flue brushes. In one case, pitch was less than 2° and roll less than 1° (Lueck, 1987). Orienting the sensitive axis of one airfoil horizontally and another vertically tested isotropy, and including a fast thermistor enabled calculation of heat fluxes, $\overline{w'\Theta'}$, from wavenumbers of 1–40 cpm (Fleury and Lueck, 1994); lower wavenumbers were contaminated by body motion.

5.5.2 Depth-Cycling Tows

Depth-cycling tows address the need to map finescale features with one set of probes in lieu of adjudicating calibration offsets between probes on multiple profilers. Line drag, however, forces a trade-off between ship speed and the depth and tightness of ‘sawtooth’ trajectories. Accompanying vibrations often contaminate

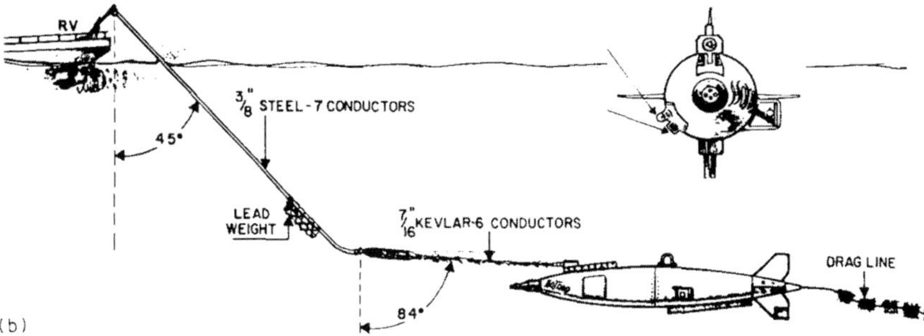

Figure 5.27 Configuration to decouple towed turbulence measurements from ship motion. As seen in the head-on insert, horizontal and vertical vanes stabilized body motion. Turbulence probes were on the nose, and temperature and conductivity probes on the lower side. (From Lueck, 1987)

microstructure measurements, but useful dissipation rates can be inferred from overturns along depth-cycling runs (Klymak and Gregg, 2004). Although individual ϵs from overturns are less accurate than microstructure measurements, the much greater spatial coverage with tows provides more accurate area averages. Care is needed, however, to avoid spurious overturns when tows cut through horizontal gradients.

5.5.3 Submersibles and Submarines

Using a hot-film and two airfoils on a Pisces IV submersible in energetic inland waters, Gargett et al. (1984) measured three components of dissipation-scale velocity to examine isotropy as a function of the buoyancy Reynolds number, a measurement that could not be made with profilers. The submersible's 11 Hz propellor frequency was a problem, but, with care, the dissipation noise level was as low as 1×10^{-9} W kg^{-1}. Separately, Osborn and Lueck (1985) used two airfoils on a research submarine in coastal waters to obtain useful shear spectra and $\overline{v'w'}$ for frequencies < 40 Hz. They also identified salt fingers from narrowband spectra of temperature gradients from an FP07 thermistor. Although these observations demonstrated the capabilities of submersibles and research submarines, horizontal tows and autonomous vehicles usually do as well at much less cost.

5.5.4 Autonomous Underwater Vehicles

Turbulence sensors were installed on autonomous underwater vehicles (AUVs) in the 1970s (Irish and Nodland, 1978), but, owing to propeller-induced vibrations, obtaining good data took several decades (Levine and Lueck, 1999) and coherent processing with accelerometers (Goodman et al., 2006). Some vehicles were large,

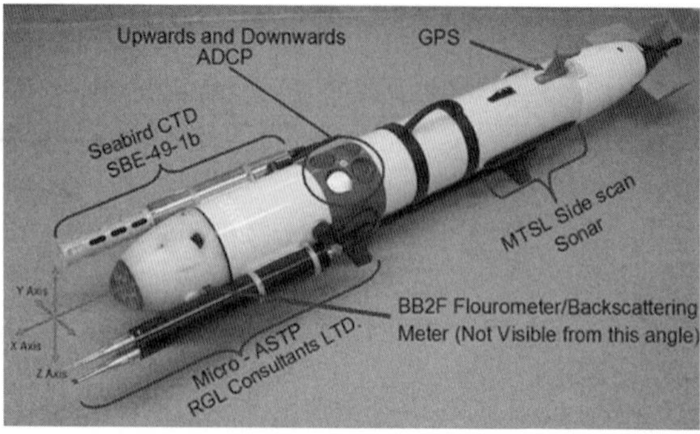

Figure 5.28 T-REMUS, 2 m long, carrying a microstructure package, upward and downward 1.2 MHz ADCPs, a CTD, a flurometer, and a side-scan sonar. (From Goodman and Wang, 2009)

e.g. Autosub (7 m long and 0.9 m in diameter) and capable of two-day missions. Carrying side scan sonars while cruising 2–10 m below the surface, Autosub related turbulence to Langmuir cells and bubble clouds under moderate winds (Thorpe et al., 2002). For one-second averages, the noise level of dissipation rates was 5×10^{-9} W kg^{-1}. Other AUVs, e.g. T-REMUS (2 m long and 0.3 m in diameter), are much smaller (Figure 5.28) and limited to runs of several hours, sufficient to map turbulence during sawtooth runs across a narrow continental shelf. In spite of broad vibrations between 10 and 20 Hz and a cluster of others near 70 Hz, dissipation noise was $\sim 1 \times 10^{-9}$ W kg^{-1}.

5.6 Moorings and Fixed Platforms

Microstructure and finestructure sensors have been mounted on moorings and bottom-mounted towers to obtain times series and absolute velocity.

5.6.1 Moorings

To observe mixing through several tidal cycles in a turbulent channel, Lueck et al. (1997) mounted airfoils, FP07 thermistors, and two ducted rotors on the nose of an aircraft wing tank attached to a taut mooring. A compass and accelerometers monitored package motion, and dissipation rates greater than 4×10^{-10} W kg^{-1} could be estimated when flow exceeded 0.1 m s^{-1}.

Much longer series of temperature microstructure were obtained using 'χpods' pointing Fastip thermistors into flows past moorings (Moum and Nash, 2009). Before deployment, each thermistor was mounted on a profiler with a thermocouple to obtain its dynamic response. Initial measurements were made in a very difficult

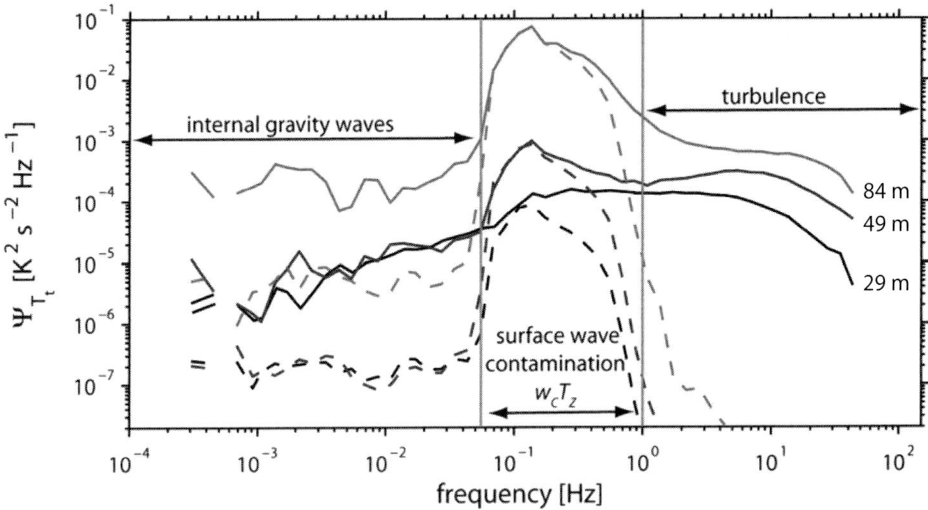

Figure 5.29 Spectra of temperature derivatives from χpods moored on the equator. (Adapted from Moum and Nash, 2009. © American Meteorological Society. Used with permission)

environment: just below an equatorial mixed layer. Surface waves contaminated frequencies between 0.05 and 1 Hz, but the dissipation range was detectable at higher frequencies (Figure 5.29), in spite of ± 1 m displacements and 0.5 m s^{-1} currents from surface waves. Vibrations of the mooring cable were not a significant problem; Perlin and Moum (2012) compared measurements on two χpods 9 km apart with microstructure profiles between them. Averages over 15 days were within 95% confidence limits for 14 of the 17 comparisons. Much of the variability was attributed to horizontal differences in mixing intensity rather than to the uncertainty in the measurements.

5.6.2 Fixed Platforms

The stability obtained with sensors on fixed platforms enables measurements of inertial and dissipation subranges. For instance, fast thermistors and clusters of three small ducted rotors mounted orthogonally in horizontal planes beneath Arctic ice yielded the vertical turbulent heat flux, $\overline{w'T'}$ (Figure 5.30).

Lorke and Wuest (2005) observed the inertial subrange in the turbulent bottom boundary layer of a shallow lake using a tripod-mounted ADCP that detected phase shifts between successive backscattered pulses (Figure 5.31). Averaging over $k^{-5/3}$ portions of the spectra yields

$$\epsilon = \exp\left\langle \ln\left(\phi_{\mathrm{vel}}/0.51(w/u)^{-5/3}\right)^{3/2}\right\rangle \quad [\mathrm{W\,kg^{-1}}]. \tag{5.18}$$

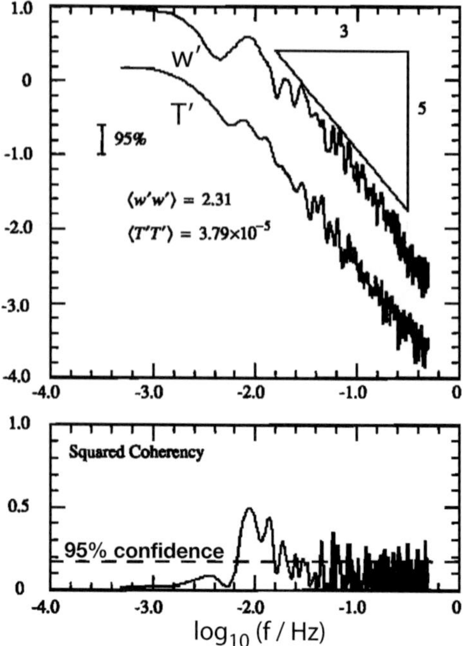

Figure 5.30 Frequency spectra and squared coherence of vertical velocity, w', and temperature, T', in a turbulent boundary layer under Arctic ice. Variances are in cm^2 s^{-2} and K^2. The temperature spectrum was shifted down one decade. Contributions to the vertical heat flux, $\overline{w'T'}$, occur at the energy-containing scales with $f \sim 10^{-2}$ Hz, slightly less than the beginning of the $f^{-5/3}$ inertial subrange. A similar relation is expected for the buoyancy flux in stratified turbulence. (Adapted from McPhee, 1992)

The technique works with three-beam ADCPs; four beams also yield momentum fluxes by the variance method,

$$\overline{u'w'} = \frac{\overline{u_3'^2} - \overline{u_4'^2}}{4 \sin\theta \cos\theta}, \quad \overline{v'w'} = \frac{\overline{u_1'^2} - \overline{u_2'^2}}{4 \sin\theta \cos\theta} \quad \left[\mathrm{m^2 s^{-2}}\right], \tag{5.19}$$

where u_i are beam variances, with beams 1 and 3 opposite beams 2 and 4, and θ as the inclination of the beams from vertical (Stacey et al., 1999).

5.7 Remote Sensing

5.7.1 High-Frequency Acoustic Backscatter

Continuous images of turbulence at shallow depths can be obtained from backscattered high-frequency acoustic energy when the turbulence occurs in strong temperature and salinity stratification (Proni and Apel, 1975). Particularly striking

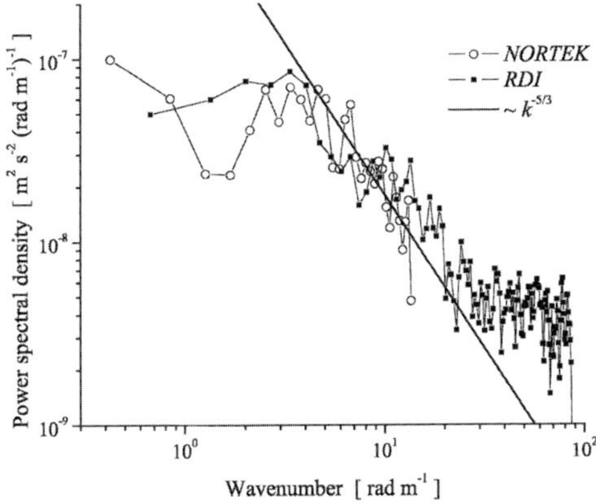

Figure 5.31 Spectra of along-beam velocity in a turbulent boundary layer of a shallow lake. Mounted on a tripod, the 1.5 MHz NORTEK ADCP faced downward. The 614 kHz RDI ADCP sat nearby on the bottom. Both exhibit portions of $k^{-5/3}$ slopes characteristic of inertial subranges. (From Lorke and Wuest, 2005. © American Meteorological Society. Used with permission)

images of overturns and solitary waves have been obtained during tidal flows over the sill in Knight Inlet (Farmer and Smith, 1980; Armi and Farmer, 2002). By comparing returns at multiple frequencies, Orr (1981) determined that most came from biology, but the lack of simultaneous biological and physical sampling limited the interpretation. Thorpe and Brubaker (1983), however, demonstrated backscatter from thermal microstructure by towing a sphere in a lake. Bragg backscatter from microstructure was subsequently quantified by Seim et al. (1995), who compared returns from a calibrated transducer with simultaneous profiles of ϵ and χ_{T} in overturning billows in a tidal channel (Figs. 1.2 and 1.4).

Received acoustic levels, RL, in decibels (dB), are corrected for source strength, SL, volume spreading, V, and transmission loss, TL, using the sonar equation,

$$S_v = RL - SL - 10\log_{10} V + TL \quad [\mathrm{dB}]. \tag{5.20}$$

The observed scattering cross-section, $\sigma_{\mathrm{obs}} = 10(S_v/10)\ \mathrm{m}^{-1}$, is predicted for Bragg scattering from sound speed fluctuations as

$$\sigma_{\mathrm{est}}(k_{\mathrm{Bragg}}) = -\frac{k_{\mathrm{Bragg}}^3}{32}\frac{d}{dk}\Phi_{c'}(k_{\mathrm{Bragg}}) \quad [\mathrm{m}^{-1}], \tag{5.21}$$

where the Bragg wavenumber is twice the acoustic wavenumber, $k_{\mathrm{Bragg}} = 2k$. Following Goodman (1990), Seim (1999) expressed the spectrum of fractional sound speed fluctuations, c', as

$$\Phi_{c'}(k) = a^2\Phi_T(k) + b^2\Phi_S(k) + 2ab\Phi_{ST}(k) \quad [\mathrm{m}^{-1}], \tag{5.22}$$

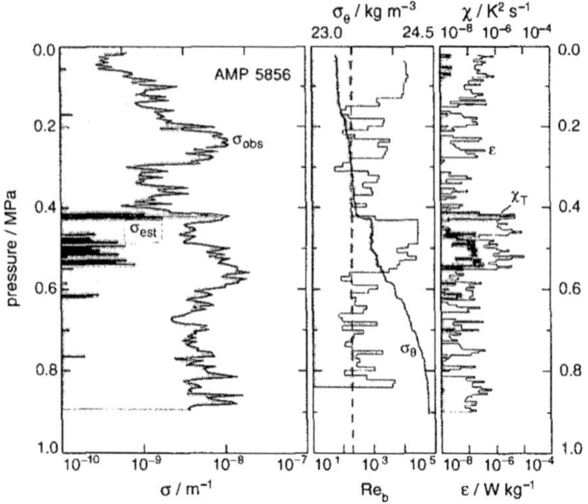

Figure 5.32 (Left) Observed backcattering cross-section, σ_{obs}, compared with the estimated cross-section for 200 kHz using only a^2 and b^2 terms, evaluated using ϵ and χ_T (right) from AMP 5856 through a large billow. (Middle) Buoyancy Reynolds number, Re_b, and potential density, σ_θ. (From Seim et al., 1995. © American Meteorological Society. Used with permission)

where $a \equiv (1/c)\partial c/\partial T|_{S,p}$, $b \equiv (1/c)\partial c/\partial S|_{T,p}$, $\Phi_{ST}(k)$ is the cospectrum of temperature and salinity, and Φ_T and Φ_S are one-dimensional Batchelor spectra (3.51).

Neglecting the cospectrum and considering contributions only from Φ_T and Φ_S, Seim et al. (1995) evaluated σ_{est} using ϵ and χ_{T} though large overturns in Admiralty Inlet (Figs 1.2 and 1.4), assuming that χ_S is proportional to χ_T consistent with the $\partial S/\partial\Theta$ relation. Interpolated onto the trajectory of the microstructure profiler, σ_{obs} had several local maxima, some of which were within 10 dB of σ_{est} (Fig 5.32), which was significant only in the overturns. Observed cross-sections were consistently several dB smaller than σ_{obs}, a bias attributed to the tendency of small averages of lognormal variables to be less than the mean and to the tiny volume sampled by microstructure probes compared to the backscattering volume.

Subsequently, Seim (1999) evaluated the full expression for σ_{est} using microstructure on the thermally stratified New England shelf and in salt-stratified Admiralty Inlet (Fig. 5.33). On the shelf, the peak cross-section maintained nearly the same magnitude from the inertial subrange to the thermal diffusive cutoff; the salinity contribution at $f > f(k_{\kappa T})$ extended the response beyond 10^3 kHz, but with less than one-tenth the amplitude. In Admiralty Inlet, the largest cross-section occurred in the salinity range, beyond $f(k_{\kappa T})$, while the cospectral terms suppressed scattering in the inertial subrange. Because the cospectrum has yet to be

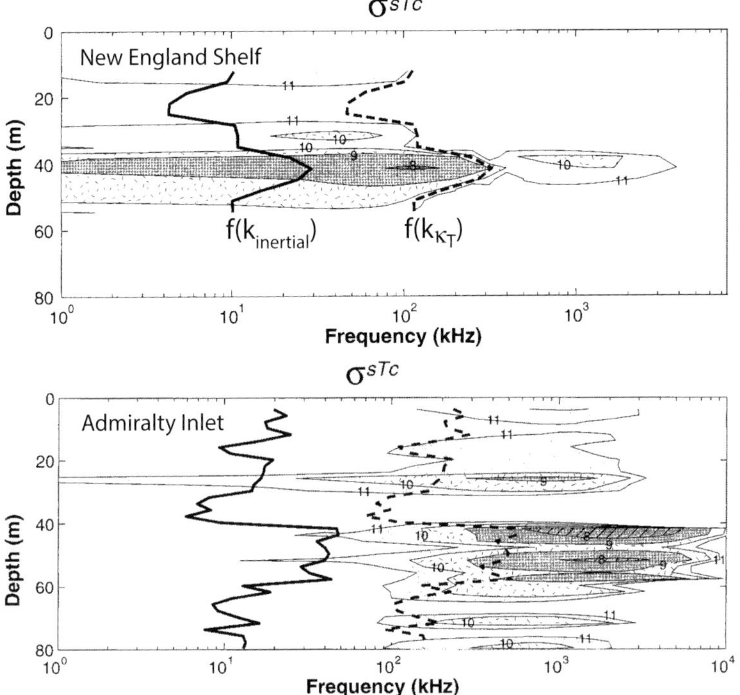

Figure 5.33 Contours of $\log_{10}\sigma^{STc}$, which is $\log_{10}\sigma_{\text{est}}$ including the cospectrum, Φ_{TS}. Thick solid and dashed lines mark $f(k_{\text{inertial}})$, separating the inertial subrange from the viscous dissipation range, and $f(k_{\kappa_T})$, the thermal Batchelor scale. On the thermally stratified shelf, the relatively weak salinity gradient extended the patch of strong thermal backscatter to 100–1,000 kHz, but with an amplitude reduced at least 10-fold. In the salt-stratified inlet, the strongest scattering occurred in the salinity frequency range at 100 kHz and above, while the cospectrum strongly suppressed scattering below $f(k_{\text{inertial}})$. (Adapted from Seim, 1999. © American Meteorological Society. Used with permission)

measured, these estimates are tentative and demonstrate what is needed to utilize this technique fully. So far, high-frequency backscatter has been useful only in estuaries and at some coastal sites.

5.7.2 Low-Frequency Acoustic Backscatter

Large arrays of low-frequency, 10–100 Hz, acoustics transmitted through the water column to profile sediments detect returns from index of refraction fluctuations in the water column as well as from below the seafloor. Water column reflections are strongest in the thermocline, e.g. the upper 1 km, where acoustic wavelengths of 15–75 m can resolve structures as small as 4 m (Holbrook et al., 2003). Comparisons with Expendable Bathythermograph (XBT) and Expendable Current

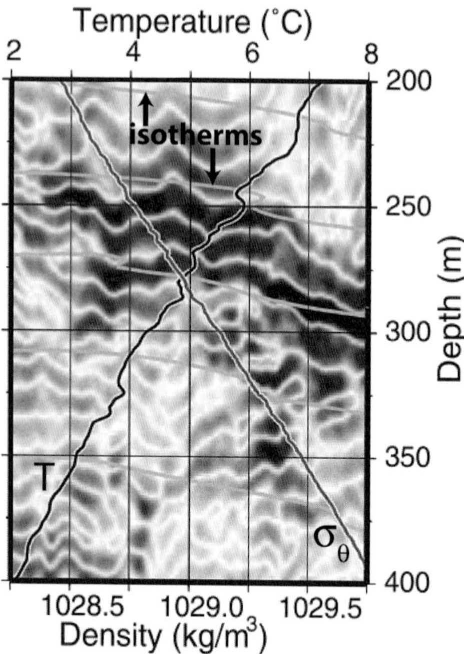

Figure 5.34 Low-frequency acoustic backscatter, greyscale with white contours, from a water mass boundary in the Norwegian Sea, including a density-compensated temperature inversion at 250 m. Profiles of temperature and potential density are overlaid, as well as temperature contours from a simultaneous survey with expendable CTDs. Reflections generally follow isotherms perturbed by internal waves. (Adapted from Nandi et al., 2004)

Profiler (XCP) profiles demonstrate strong correspondence between bright reflectors and temperature gradients enhanced by thermohaline intrusions and internal wave strain. For example, in Figure 5.34 10 m vertical corrugations in the finestructure have horizontal wavelengths of $\approx$700 m, suggesting internal wave displacements. Spectra of these displacements versus horizontal wavenumber match the GM internal wave model in the open ocean and rise above GM within 3 km of continental slopes (Holbrook and Fer, 2005). Thus, low-frequency acoustics are a unique means of mapping the horizontal structure of internal waves and their relation to boundary mixing.

5.8 Tracers

Thickening rates of tracers injected into the pycnocline provide estimates of cumulative mixing for comparison with the instantaneous mixing rates from microstructure. Injections to study diapycnal mixing began with Ewart and Bendiner (1981) releasing rhodamine dye from a neutrally buoyant float, first at 300 m, then at

1,000 m in the Pacific. Horizontal and vertical spreading were mapped for up to 66 hours with the SPURV climbing and diving at 45° under acoustic control. The SPURV carried prototypes of what became Sea-Bird temperature and conductivity probes, as well as a fluorometer that could detect dye concentrations to 10^{-9} kg m^{-3}. Simultaneous microstructure profiles were taken for comparison, but the MSR malfunctioned. Applying an exponential solution to the vertical diffusion equation to the growing separation between upper and lower edges of the patch yielded $K_\rho = (\Delta z)^2/5.69\,t = 7 \times 10^{-7}$ m^2 s^{-1}, barely above the molecular diffusivity of heat.

Subsequent discussions examined how best to extend diapycnal studies to much longer time scales to test diffusivities inferred from microstructure. These led to sulfur hexafluoride (SF_6), an artificial chemical very stable in the environment and detectable at $\sim 1 \times 10^{-17}$ mol kg^{-2} (Watson and Ledwell, 2000). Owing to concerns about energy deposited by self-propelled vehicles, towed sleds were used for sampling and release. Between 1985 and 1991 two releases were made in basins off Southern California to develop techniques (Ledwell et al., 1986; Ledwell and Watson, 1991; Ledwell and Bratkovich, 1995; Ledwell and Hickey, 1995). Shear and strain from XCPs and CTDs accompanying a release in Santa Monica Basin were consistent with elevated diffusivities inferred from the tracer (Gregg and Kunze, 1991).

NATRE began in May 1992 by releasing 139 kg of SF_6 in nine streaks within 1 m of a reference density surface at 300 m (Figure 5.35). Microstructure was sampled at intervals (Toole et al., 1994; Sherman and Davis, 1995), and, during the 30 months the tracer was tracked, the reference density surface was subducted 50–60 m, while the tracer patch was advected southward. The center of mass of the

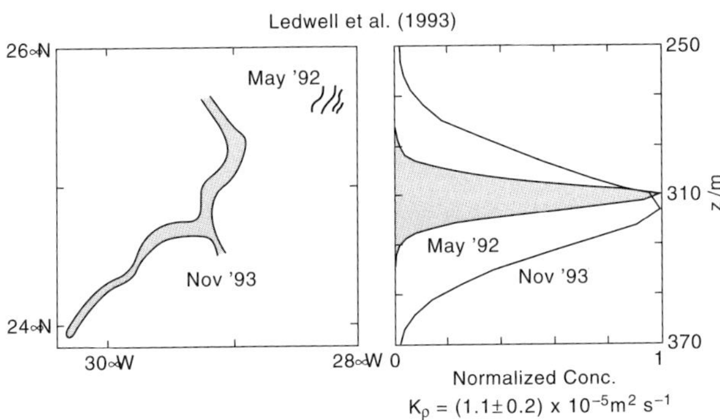

Figure 5.35 Evolution of SF_6 streaks injected into the thermocline at 310 m during NATRE. Tracer concentrations were normalized to the same peak value for visual comparison. (Adapted from Ledwell et al., 1993)

tracer also descended relative to the reference isopycnal. This velocity, computed between sampling periods, was represented as a diapycnal velocity, w_ρ.

For long tracer releases, concentration changes produced by diapycnal diffusivity must be separated from those resulting from vertical divergence of vertical tracer transport. The corresponding advection-diffusion equation for average tracer concentration, $\overline{c}$, is

$$\frac{\partial \overline{c}}{\partial t} + \frac{\partial \overline{wc}}{\partial z} = \kappa_T \frac{\partial^2 \overline{c}}{\partial z^2}. \tag{5.23}$$

Horizontal averaging is represented by an overbar, and

$$\overline{wc} = \overline{w}\,\overline{c} - K_\rho \frac{\partial \overline{c}}{\partial z}. \tag{5.24}$$

Mean vertical velocity, $\overline{w}$, is the sum of the velocity of the reference density surface, σ, and $\overline{w}_\rho$ through the surface, i.e. $\overline{w} = (\partial \overline{z}/\partial t)_\sigma + \overline{w_\rho}$ with $\overline{w}_\rho = (\partial \overline{z}/\partial \sigma)_t (d\overline{\sigma}/dt)$. The latter was twice the divergence needed to conserve potential vorticity as the tracer moved south.

When referenced to height, h, above the reference density surface, the vertical advection-diffusion equation becomes

$$\frac{\partial \overline{c}}{\partial t} + \left(\overline{w}_\rho - \frac{\partial K_\rho}{\partial h}\right)\frac{\partial \overline{c}}{\partial h} + \frac{\partial}{\partial h}\left[\left(\frac{\partial h}{\partial t}\right)_\sigma \overline{c}\right] = K_\rho \frac{\partial^2 \overline{c}}{\partial h^2}, \tag{5.25}$$

with $\overline{w}_\rho$ assumed independent of height. Low-frequency straining of the water column moves isopycnals relative to the reference density surface, as represented by $(\partial h/\partial t)_\sigma$. Least squares analysis yielded good estimates of K_ρ and less accurate ones of $\partial K/\partial h$ and w_ρ. Subsequent studies applied similar techniques to continental shelves (Ledwell et al., 2004), to deep convection (Watson et al., 1999), and to the abyss (Ledwell et al., 2000). Owing to the increasing use of SF_6 in the atmosphere, recent long-term releases have used SF_5CF_3 (Ho et al., 2008; Holtermann et al., 2012; Ledwell et al., 2016).

5.9 Perspectives

By comparison with the rate of instrument development in other fields, reliance on the same microstructure probes for nearly 50 years is striking! The probes are widely deployed on many types of vehicles, exploring mixing in many regimes, but serious deficiencies remain: inadequate spatial resolution of dissipation-scale gradients, uncertainty about dissipation-scale isotropy, and salinity spiking.

Airfoils resolve gradients produced by dissipation rates between 10^{-9} and 10^{-7} W kg^{-1}, adequate for most of the pycnocline, but not for hot spots believed to produce the most significant mixing, or for large parts of the abyss with very weak or no mixing. Now, larger dissipation rates must be estimated by extrapolating what can be measured using probe response functions and universal spectral forms. Both approaches likely produce significant errors. Smaller dissipation rates are ill defined, owing to uncertainties about probe noise and contamination by vehicle motion. Extending direct resolution of gradients at both ends of the intensity range is needed. Smaller airfoils have been made, but the change is modest relative to length scales of the highest dissipation rates. Looking ahead, developments in microelectromechanical systems (MEMS) may lead to new transducers. A sustained effort, however, will be needed to develop durable probes capable of repeated cycling to high pressure in seawater. Extending resolution to lower dissipation rates should be easier because airfoils can be made significantly larger to increase their sensitivity. Also, there would be more space for installing accelerometers closer to them to monitor and reduce motion effects.

Glass-rod thermistors measuring centimeter-scale temperature gradients are not made uniformly, differing in the shapes and sizes of the semiconductor material, the amount of protective glass, and the number and size of air bubbles in the glass. To combine the sensitivity of thermistors with the speed of cold-films, a sol gel was developed to allow depositing thin films of MΩ semiconductor material on surfaces such as glass (Kukuruznyak et al., 2001). Insulating, a.k.a. passivating, the film from seawater, however, has not been solved. It appears possible but requires a larger investment than has been made. In addition to resolving much smaller gradients, such a 'thinistor' could be replicated in identical copies, so one careful dynamic calibration would suffice for all.

Development of smaller microstructure probes would also allow routine measurements of isotropy. Cox's original microstructure profiler (MSR) had a thermistor on the tip of one of the rotating wing blades. Some use was made of the data, but, in most cases, the wing-tip thermistor responded too slowly to resolve structures at the speed it was moving. An array of thinistors at varied spacings along a rotating strut to produce varying attack angles should be able to assess isotropy and identify salt fingers.

Although the probes partially resolving dissipation-scale salinity and density gradients are intriguing, they have not been followed by routine measurements. The demonstrations, however, justify the engineering necessary to make probes suitable for widespread use. It has not been shown, however, that these approaches can lead to salinity and density profiles without salinity spiking. In view of the minimum spiking with the 'small hole' conductivity and thermistor probe, an attempt should be made to replace the silicone oil and spring mechanism with a peristaltic, a.k.a. roller, pump that pinches the tube so tightly that the electrical resistance is very high around the back path.

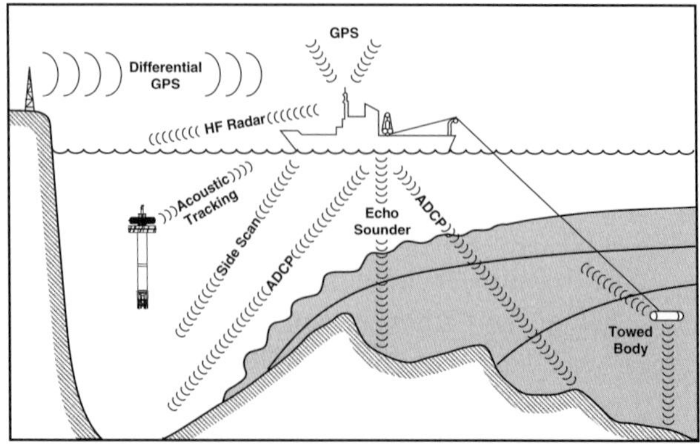

Figure 5.36 Sampling flow over a sill with two vehicles, one free-fall and the other towed. ADCPs on the vessel and on the towed body measure velocity and high-frequency acoustic backscatter displayed on the ship to guide sampling.

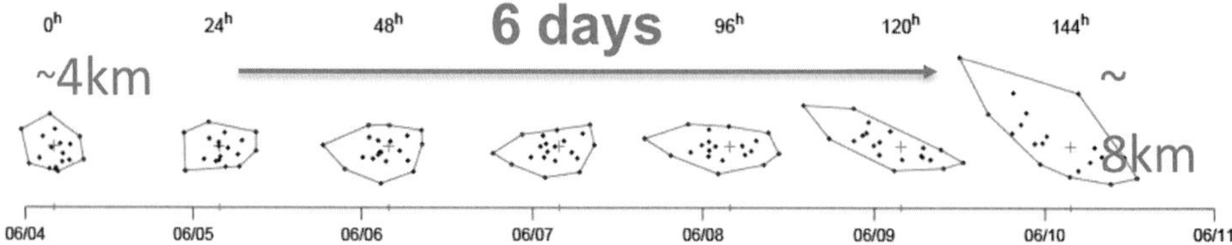

Figure 5.37 Positions of 18 EM-APEX floats during the 2011 LatMix program. The floats profiled synchronously, and half had sensors for χ_T. (Figure courtesy of Tom Sanford. Results are in Lien and Sanford, 2019)

Improved probes will address some outstanding issues, but others require many simultaneous measurements. A series of programs of this sort at the sill in Knight Inlet, British Columbia greatly advanced understanding of hydraulically controlled flows. The most elaborate included three vessels like the one in Figure 5.36 mapping channel-wide flow structures in parallel. More sensors and better acoustic tracking are needed to adapt this approach to small-scale questions, such as the three-dimensional structure of the mixing. Recently, one such possibility was explored during the Lateral Mixing program, when 18 EM-APEX floats were released in three concentric circles with radii of 0.5, 1, and 2 km. To examine the evolution of relative vorticity and mixing, the floats profiled synchronously (Figure 5.37).

6

Internal Waves and the Vortical Mode

6.1 Overview

Ubiquitous throughout the ocean, internal waves are oscillations produced by buoyancy forces restoring equilibrium after stratification is disrupted, e.g. by fluctuating wind stress (Figure 6.1). Internal wave fundamentals are discussed by Sutherland (2010). The emphasis here is on aspects related to mixing in the ocean.

Possible frequencies, ω_{iw}, are bounded by the effective Coriolis frequency, f_{eff}, and by the buoyancy frequency, N. That is, $f_{eff} \leq \omega_{iw} \leq N$, where

$$f_{eff} \approx f + (1/2)(\partial v/\partial x - \partial u/\partial y) \quad \left[\mathrm{s}^{-1}\right] \tag{6.1}$$

is the sum of planetary vorticity from the vertical component of Earth's rotation rate, $f \equiv 2\omega_{earth} \sin(\text{latitude})$ (a.k.a. the Coriolis frequency), and the relative vorticity from horizontal shear of mesoscale and submesoscale flows (Kunze, 1985). The corresponding frequency bandwidth, N/f, is typically $\sim$90 but can become very large on the equator, where $f = 0$, and as small as 7 in the deep Arctic. Wavelengths span meters to hundreds of meters in the vertical and can be tens of kilometers in the horizontal.

The internal wave field and its interactions are complex, making observations and their limitations crucial to understanding. Section 6.2 explores the main types of measurements describing internal waves and their degenerate relatives, vortical motions. Ensemble averages of several wave properties over a few kilometers in space and hours to days in time from diverse locations are the basis for the Garrett and Munk internal wave models (Section 6.6.2, Appendix B).

Normal mode solutions of the linear equations describing internal waves lead to the dispersion relation and to the vortical mode (Section 6.3). The dispersion equation leads to propagation characteristics, scalings with stratification, and reflections from the bottom and turning latitudes. Adding horizontal shear produces minima and maxima in f_{eff} at mesoscale fronts and eddies that reflect or trap incident

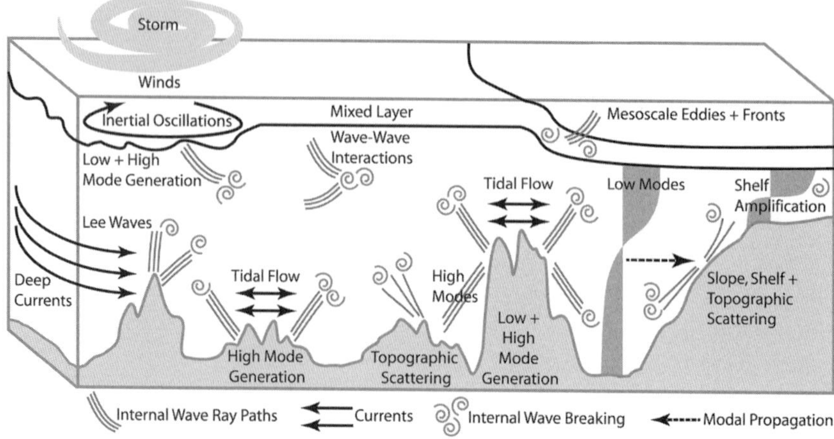

Figure 6.1 Schematic of internal wave generation, propagation, and dissipation. (From MacKinnon et al., 2017. © American Meteorological Society. Used with permission)

internal waves (Section 6.4). By enhancing shear, some of these interactions lead to breaking and mixing, as is also the case where vertical shear is elevated (Section 6.5).

Much has been learned about internal waves using spectra in Eulerian frames (Section 6.6). Averages from disparate locations led Garrett and Munk (1972) to formulate a common spectral description. In addition to being a reference for subsequent observations, their model, GM, is a basis for calculations of wave–wave interactions and for estimating mixing rates from variances of internal wave shear and strain. In particular, variances obtained by integrating spectra from conductivity-temperature-depth (CTD) casts and Argo floats have yielded global maps of internal wave intensity and revealed seasonal cycles.

A vortical mode is a lens of water, homogenized by mixing and rotating slowly in geostrophic or nonlinear cyclostrophic balance (Section 6.7). Simulations of stratified turbulence show decaying turbulence evolving into lenses of alternating vertical vorticity (Figure 6.2). Vortical modes naturally separate from internal waves when spectra are computed in semi-Lagrangian reference frames. In addition, they can be distinguished by applying consistency relations to Eulerian spectra. These show vortical motions contributing a significant fraction of the total shear variance in the pycnocline. Spatial scales of vortical modes extend from geostrophy to turbulence, overlapping internal waves. Because vortical motions do not propagate, their frequency comes from straining and can be represented as $f R_o$, where $R_o \equiv u/fL$ is the Rossby number. When $R_o > 1$, vortical mode frequencies are in the internal wave band.

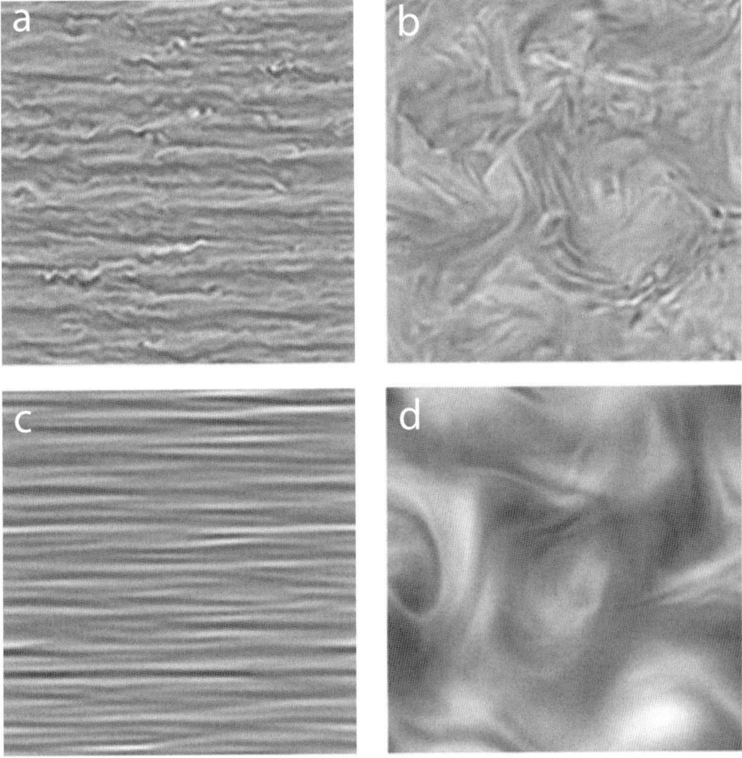

Figure 6.2 Vertical sections of horizontal vorticity, Ω_y (left), and horizontal sections of vertical vorticity, Ω_z (right), from a simulation of stratified turbulence dominated by vortical motions. In (a) and (b) stratification suppressed large overturns but not small ones. Increased stratification in (c) and (d) suppressed all overturns, increasing the prominence of vorticity. (From Waite and Bartello, 2004)

6.2 Observations

Full understanding of internal wave dynamics requires observing the three-dimensional evolution of the field under varying forcing and dissipation. Since this has not been done, our knowledge has been pieced together from diverse measurements capturing a few aspects of internal waves at one time and place. Measurements come from sensors on moorings, wire-lowered, wire-crawling, and free-fall profilers, towed chains and bodies, and Doppler acoustic profiling. In addition to being mostly 'one-shot' realizations, these observations are often Eulerian, fixed in space, rather than semi-Lagrangian, on density surfaces.

Serious measurements started when solid-state sensors became available in the 1960s and initially focused on time series of temperature and velocity on moorings. Power and coherence spectra showed distributions of kinetic and potential energy

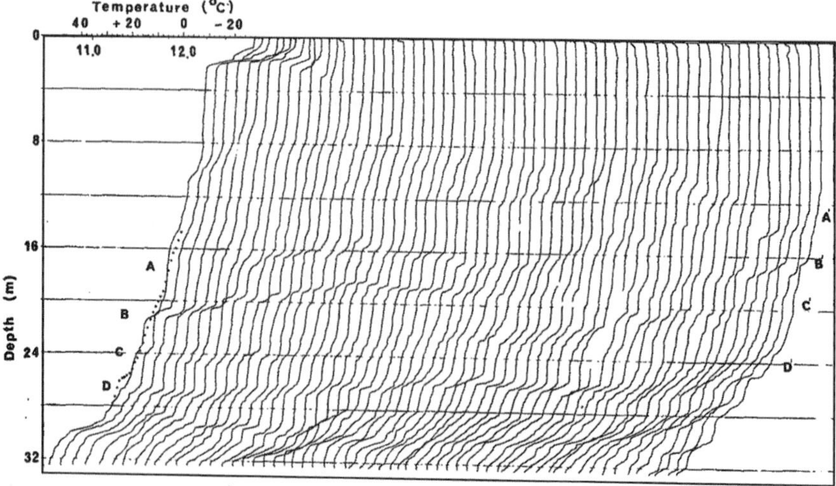

Figure 6.3 Temperature, offset 0.1°C, every 11 minutes from a barge moored in a Welsh lake. Persistent gradients produced by internal wave strain migrate in temperature as well as in depth. Letters A to D on opposite sides mark end points of four high-gradient features. (From Lazier, 1973. © American Meteorological Society. Used with permission)

versus frequency and wavenumber consistent with linear internal wave dynamics (Fofonoff, 1969). Later measurements on the Internal Waves Experiment (IWEX) tri-mooring included current meters and temperature probes distributed between 4 m and 400 m from the apex to provide varied horizontal and vertical spacings for six weeks (Briscoe, 1975).

CTD recorders lowered on hydrographic wires have produced most in-situ vertical measurements. They are often limited to meter-scale resolution by vibration and ship roll. Contamination can be avoided with free-fall expendables and by profiling from the Floating Instrument Platform (FLIP), ice floes, or barges anchored in lakes. In addition to heaving isotherms, the example in Figure 6.3 shows 'kinks' migrating through the temperature field, demonstrating that the finestructure was produced by transient internal wave strain rather than by mixing. As components of hydrographic surveys and of process studies, CTDs/LADCPs (Lowered Acoustic Doppler Current Profilers) are the major sources of abyssal internal wave data.

Remote acoustic probing of internal wave velocities developed relatively early (Pinkel, 1981) and remains a mainstay. Acoustic Doppler current profilers (ADCPs) infer water velocity from differences in Doppler shifts of high-frequency sound backscattered from beams at different angles. When mounted on surface platforms, ADCPs operating at frequencies of tens of kilohertz measure velocity as deep

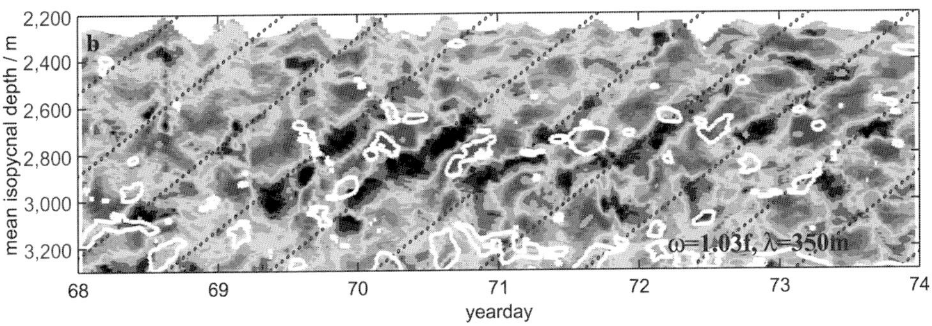

Figure 6.4 Zonal shear near the Mendocino Escarpment spanning $\pm 0.001\ \mathrm{s}^{-1}$ versus time on a semi-Lagrangian vertical reference. The observations are consistent with $\omega = 1.03\,f$ and $\lambda = 350\ m$. (From Alford, 2010. © American Meteorological Society. Used with permission.) (A black and white version of this figure will appear in some formats. For the color version, please refer to the plate section)

as a kilometer. Using eight beams from FLIP, four upward from 400 m and four downward, gave a nominal resolution of 4.5 m to depths of 800 m (Pinkel et al., 2012). Better resolutions can be obtained using higher frequencies over shorter ranges and by mounting ADCPs on submerged vehicles or lowering them on wires (LADCPs). Sound pulses transmitted from a moving vessel and backscattered from small particles drifting with the water yield lines of two-dimensional velocity, but rapid scanning from a fixed location similar to atmospheric radars is not possible, owing to the slow speed of sound in water and the averaging required.

Sensors on packages crawling up and down taut moorings provide extended time series showing internal wave propagation (Figure 6.4). In many cases, these measurements also resolve density overturns of a meter and smaller that cannot be detected with profilers attached to rolling ships. This allows estimation of mixing rates from overturn characteristics. In addition, direct turbulence measurements are also possible.

Avoiding contamination by ship heave, free-fall profilers are excellent platforms for high-resolution, full-depth profiling of voltages induced by currents carrying ionized seawater through Earth's magnetic field (Figure 6.5). Sensitive electrodes coupled with small acoustic current meters and microstructure probes fully resolve velocity profiles, relating changes in internal wave parameters to variations of turbulent dissipation rates.

The horizontal structure of the internal wave field is investigated by depth-cycling runs from autonomous underwater vehicles and shipboard CTDs, as well as arrays of sensors on towed chains (Figure 6.6). Though slower, survey lines of autonomous gliders can measure patterns of internal wave activity.

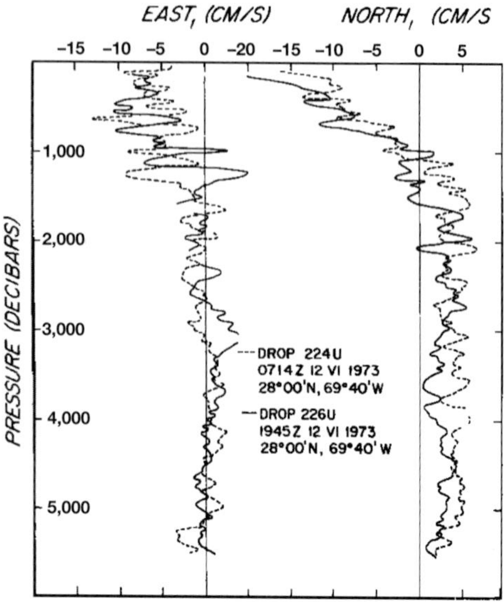

Figure 6.5 Two records from the Electromagnetic Velocity Profiler (EMVP) half an inertial period, 12.5 hours, apart. Near-inertial internal waves produced the mirror imaging over scales of several hundred meters, shorter than the structure of the first baroclinic mode dominating velocity above 1,000 decibars. Data are from the Mid-Ocean Dynamics Experiment (MODE). (Adapted from Sanford, 1975)

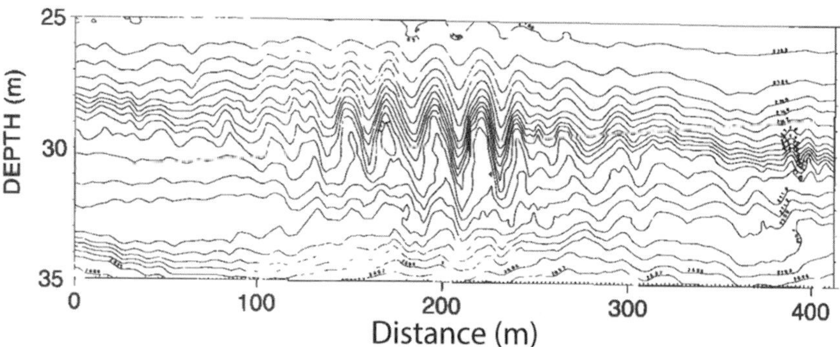

Figure 6.6 A packet of 20–30 m waves on a high-gradient layer in the Sargasso Sea observed with temperature sensors 0.5 m apart on a chain towed at 2.7 m s^{-1}. Isotherms are contoured at 0.04°C intervals. (From Marmorino et al., 1987)

6.2.1 Lagrangian and Semi-Lagrangian Measurements

Length scales of internal waves and turbulence overlap in Eulerian measurements, but their frequency content naturally separates on Lagrangian frames (Section 5.4.4), with internal waves cutting off sharply at N (Figure 6.7). Also, semi-Lagrangian frames from rapidly repeated CTD profiling reveal velocity,

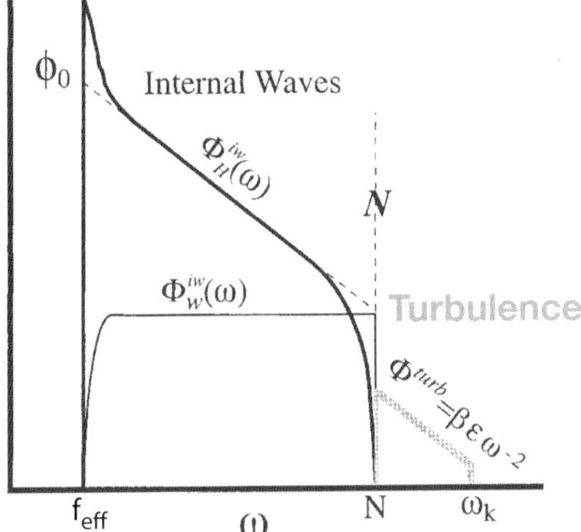

Figure 6.7 Schematic Lagrangian frequency spectra of internal waves and three-dimensional turbulence illustrate their natural separation at the buoyancy frequency, N. $\Phi_{\mathrm{H}}^{\mathrm{iw}}$ and $\Phi_{\mathrm{W}}^{\mathrm{iw}}$ are spectra of horizontal and vertical internal wave velocities. The Lagrangian form of the Kolmogorov frequency spectrum, Φ^{turb}, extends to $\omega_{\mathrm{K}} = (\epsilon/\nu)^{1/2}$. (Adapted from D'Asaro and Lien, 2000b. © American Meteorological Society. Used with permission)

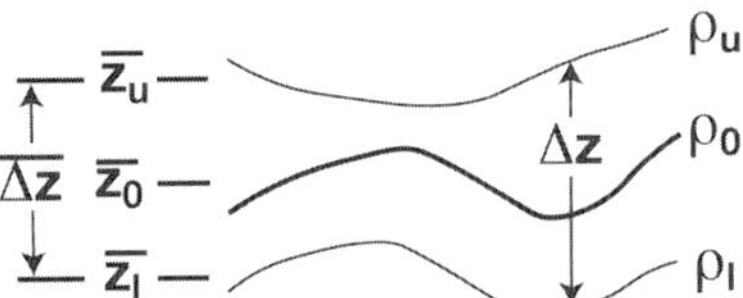

Figure 6.8 Schematic strain, $\Delta z/\overline{\Delta z}$, between isopycnals ρ_u and ρ_l, which are centered about ρ_0 at equilibrium. Strain is referenced to $\rho_0 \equiv \overline{\rho(z_0, t)}$ and to semi-Lagrangian depth $\overline{z_0} = \overline{z(\rho_0(t))}$.

shear, and strain changing much more slowly in time along density surfaces than at fixed depths (Sherman and Pinkel, 1991). This approach is particularly useful when examining vertical displacement and strain.

Vertical displacement, $\zeta_i(t) \equiv z_i(t) - \overline{z_i(t)}$, is the separation between the instantaneous position, $z_i(t)$, of the ith isopycnal from its average depth, $\overline{z_i(t)}$ (Figure 6.8). Strain, the gradient of displacement, ζ_z, can be computed in several ways. One approach defines strain between isopycnals ρ_u and ρ_l as the normalized difference in their depths,

$$\zeta_z\left(\overline{z_0(t)}, t\right) \equiv \frac{z_u(t) - z_l(t)}{\overline{z_u(t)} - \overline{z_l(t)}} = \frac{\Delta z(t)}{\overline{\Delta z(t)}} = \frac{\zeta_u(t) - \zeta_l(t)}{\overline{\Delta z(t)}}. \tag{6.2}$$

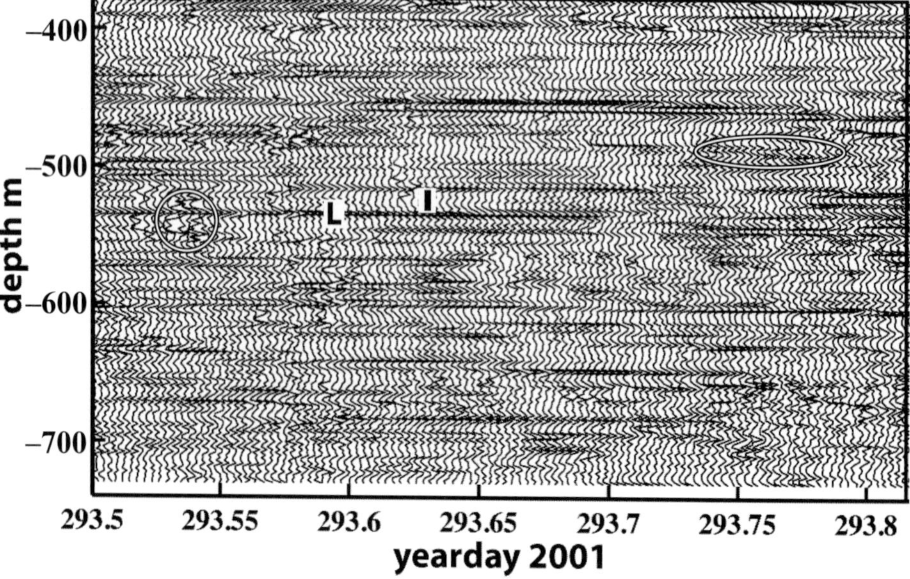

Figure 6.9 Isopycnal separation as $\log_{10}(\Delta z/\overline{\Delta z})$ for $\overline{\Delta z} = 2$ m versus semi-Lagrangian depth. 'L' is a weakly stratified layer and 'I' is a highly stratified interface in the original profiles. The circle and oval enclose an overturn and high-frequency internal waves. (Adapted from Pinkel, 2014. © American Meteorological Society. Used with permission)

Centering $\overline{z_u}$ and $\overline{z_l}$ about $\overline{z_0}$ makes $\overline{z_0}$ the semi-Lagrangian reference depth for strain between ρ_u and ρ_l. An equivalent approach expresses strain in terms of buoyancy frequency squared,

$$\zeta_z\left(\overline{z_0(t)}, t\right) = \frac{\Delta z/\Delta\rho}{\overline{\Delta z}/\Delta\rho} = \frac{\overline{N^2(z_0)}}{N^2(\rho_u, \rho_l, t)}, \tag{6.3}$$

where the numerator is the background stratification, and the denominator is the instantaneous stratification between ρ_u and ρ_l. Because isopycnals cannot cross without overturning, strain under this definition is positive, > 1 where the water column is stretched to be weakly stratified and < 1 where it is compressed to form strong stratification.[1] Other strain definitions are inverse to these (Kunze et al., 1990; Polzin et al., 2003), but calculations with sinusoidal strain fields show that only (6.2) is immune to the self-distortion produced by reversible fine structure (R. Pinkel, personal communication, 2013).

[1] Some authors subtract 1 so compressed isopycnals correspond to negative strain.

Plotting the logarithm of strain versus semi-Lagrangian depth (Figure 6.9), causes high-gradient compressive strain-forming interfaces in observed profiles to have small positive values and appear as homogenous layers in a semi-Lagrangian frame. Thus, the example marked 'I' corresponds to an observed interface. Likewise, low-gradient layers in original profiles result from high strain and appear in semi-Lagrangian frames as spikes marked 'L'. These weakly stratified layers are often associated with vortical motions. Many interfaces and layers persist for hours to days, unlike the wiggles of high-frequency internal waves propagating across density surfaces and the chaotic inversions resulting from density overturns and lateral intrusions.

Following demonstration by Pinkel and Anderson (1997) that velocity difference, $\Delta u(t)$, and separation, $\Delta z(t)$, between isopycnals are independent, the finite-difference form of the gradient Richardson number can be written as

$$Ri_g = \frac{\Delta\rho/\Delta z(t)}{\Delta u(t)^2/\Delta z(t)^2} = \frac{\Delta\rho}{|\Delta u(t)^2|}\Delta z(t). \tag{6.4}$$

Consequently, low values are most likely in low-gradient regions, where $\Delta z(t)$ is small.

6.3 Linear Waves in a Steady Unsheared Flow

Properties of single waves in steady unsheared backgrounds include the dispersion relation linking wavenumber and frequency, the inclination of $\boldsymbol{k}$ from horizontal, propagation, and reflection from boundaries.

6.3.1 *Frequency and Wavenumber*

An internal wave is characterized by amplitude, A, frequency, ω, and vector wavenumber, $\boldsymbol{k}$ (Figure 6.10). Wavenumber squared has equivalent forms of

$$\boldsymbol{k}\cdot\boldsymbol{k} = k^2 = k_i^2 = k_x^2 + k_y^2 + k_z^2 = k_h^2 + k_z^2 \quad [\mathrm{m}^{-2}], \tag{6.5}$$

where $i = 1,2,3$. Radian wavenumber is used for theoretical expressions, while most data are in cyclic units with 1 rad/m (1 m^{-1}) equivalent to $1/(2\pi)$ cycles per meter ($1/(2\pi)$ cpm). When the internal wave field is horizontally isotropic, it is convenient to use horizontal wavenumber, k_h, which is unchanged when the background does not change laterally. Vertical wavenumber is constant when flow and stratification are uniform vertically.

Wave frequency relative to low-frequency flow is termed the 'Lagrangian' or 'intrinsic' frequency, ω_{L}. When waves propagate in still water with varying

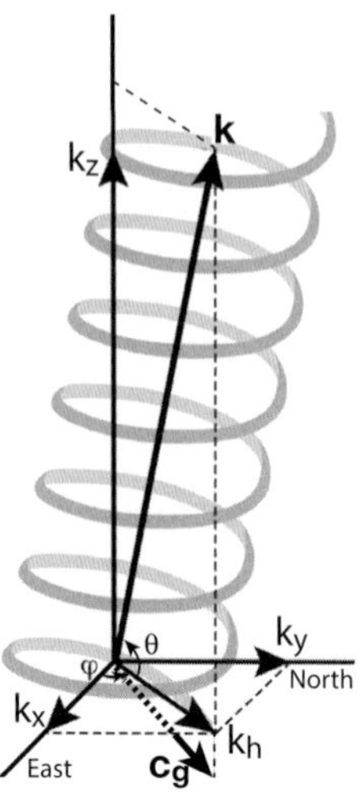

Figure 6.10 Path of a phasor rotating anticlockwise upward for a single internal wave in the northern hemisphere having vector $\boldsymbol{k}$ inclined θ above horizontal and rotated φ anti-clockwise from east (x). All water parcels in planes perpendicular to $\boldsymbol{k}$ move with the same magnitude and direction. In general, $k_x \neq k_y$, and the phasor follows an elliptical helix. The group velocity vector is perpendicular to $\boldsymbol{k}$, inclined downward by $\pi/2 - \theta$. (Adapted from Meyer et al., 2015a. © American Meteorological Society. Used with permission)

stratification, $\omega_{\rm L}$ is constant, maintained by k_z and k_h changing in response to variations in N. If the medium changes with time or is sheared, $\omega_{\rm L}$ is not normally constant. At a position fixed relative to the flow, the Eulerian frequency, $\omega_{\rm E}$, is the sum of the Lagrangian frequency and a Doppler shift,

$$\omega_{\rm E} = \omega_{\rm L} + \boldsymbol{k} \cdot \boldsymbol{u}. \tag{6.6}$$

When the background flow is steady, $\omega_{\rm E}$ is constant.

The speed of wave crests relative to the mean flow is the Lagrangian phase velocity, $\boldsymbol{c}_{\rm phase}$, aligned with the unit wave vector, $\hat{\boldsymbol{k}}$,

$$\boldsymbol{c}_{\rm phase} = (\omega_L/k)\hat{\boldsymbol{k}}, \quad \boldsymbol{c}_{g_{\rm L}} = \nabla_k \omega_{\rm L}, \quad \boldsymbol{c}_{g_{\rm E}} = \boldsymbol{c}_{g_{\rm L}} + \boldsymbol{u}. \tag{6.7}$$

The velocity of the wave's energy relative to the fluid is given by the wavenumber gradient, ∇_k, of Lagrangian frequency. Termed the 'Lagrangian group velocity', $\boldsymbol{c}_{g_{\rm L}}$ is perpendicular to $\boldsymbol{k}$, and the path followed along the group velocity is termed a 'ray'. Phase propagates upward when energy propagates downward, and vice versa. Relative to a fixed observer, the Eulerian group velocity, $\boldsymbol{c}_{gE}$, sometimes simply $\boldsymbol{c}_g$, is the vector sum of $\boldsymbol{c}_{g_{\rm L}}$ and $\boldsymbol{u}$.

6.3.2 Energy and Action

As the sum of potential and kinetic terms, internal wave energy density is

$$E_{\rm iw} = \frac{N^2\zeta^2}{2} + \frac{u^2+v^2+w^2}{2} \quad \left[\frac{\rm J}{\rm kg} = \frac{\rm m^2}{\rm s^2}\right], \tag{6.8}$$

with a corresponding energy flux of

$$\boldsymbol{J}_{\rm iw} = \rho E_{\rm iw}\boldsymbol{c}_g \quad [\rm W\,m^{-2}], \tag{6.9}$$

where the group velocity can be Lagrangian or Eulerian. During propagation through inhomogenous stratification and shear, $E_{\rm iw}$ and $\omega_{\rm L}$ change with the background, owing to wave–mean flow interactions. However, their ratio, wave action density,

$$\mathcal{A} \equiv E_{\rm iw}/\omega_{\rm L} \quad [\rm J\,s\,kg^{-1} = m^2\,s^{-1}], \tag{6.10}$$

is conserved (Bretherton and Garrett, 1968) as

$$\partial\mathcal{A}/\partial t + \nabla\cdot(\mathcal{A}\boldsymbol{c}_g) = 0 \quad [\rm J\,kg^{-1}], \tag{6.11}$$

where $\mathcal{A}\boldsymbol{c}_g$ is action flux (Gill, 1982, section 8.12). At steady state, action density varies as $\mathcal{A} \sim |c_g|^{-1}$.

6.3.3 The Dispersion Relation

Steady unsheared linear flow is described by

$$\partial u/\partial t - f_{\rm eff}v + (1/\rho_0)\partial p/\partial x = F_1 \tag{6.12}$$

$$\partial v/\partial t + f_{\rm eff}u + (1/\rho_0)\partial p/\partial y = F_2 \tag{6.13}$$

$$\partial w/\partial t + N^2\zeta + (1/\rho_0)\partial p/\partial z = F_3 \tag{6.14}$$

$$\partial\zeta/\partial t - w = F_4 \tag{6.15}$$

$$\partial u/\partial x + \partial v/\partial y + \partial w/\partial z = 0, \tag{6.16}$$

where F_i are forces, ζ is vertical displacement, p is perturbation pressure, and $\rho \equiv (\rho_0/g)N^2\zeta$ is perturbation density (Lien and Müller, 1992). Expressing plane wave solutions for unforced motions as products of slowly varying amplitudes and fast oscillations, vertical velocity is

$$w = w_0 e^{i(k_x x + k_y y + k_z z - \omega_{\rm E} t)}. \tag{6.17}$$

There are two sets of normal mode solutions,

$$\omega_{\pm} = \pm\left(\frac{f_{\text{eff}}^2 k_z^2 + N^2 k_h^2}{k_z^2 + k_h^2}\right)^{1/2} \quad \text{with} \quad f_{\text{eff}} \le \omega \le N \tag{6.18}$$

$$\omega_0 = 0, \tag{6.19}$$

with eigenvectors for wavenumber k,

$$\begin{pmatrix} u_+ \\ u_- \\ N\zeta \end{pmatrix} = \underbrace{\frac{1}{\sqrt{2}k|\omega_{\pm}|}\begin{pmatrix} \pm i f_{\text{eff}} k_z \\ |\omega_{\pm}|k \\ \mp N k_h \end{pmatrix}}_{\text{internal waves}}, \quad \underbrace{\frac{1}{k|\omega_{\pm}|}\begin{pmatrix} N k_h \\ 0 \\ -i f_{\text{eff}} k_z \end{pmatrix}}_{\text{vortical mode}} \tag{6.20}$$

(Riley and Lelong, 2000). The first set, the internal wave dispersion relation, has one component rotating anticlockwise (+) in time and the other clockwise (–). In the northern hemisphere, components rotating clockwise in time have upward phase and downward group velocities; wave components rotate clockwise with increasing depth. Likewise, anticlockwise rotations have downward phase and upward group velocities. These reverse in the southern hemisphere, e.g. clockwise motions have downward phase velocity. When $k_h = 0$, wave frequency equals the effective Coriolis frequency, $\omega_{\pm} = \pm f_{\text{eff}}$, and the motions are termed 'inertial waves'. When $k_z = 0$, wave frequency equals the buoyancy frequency, $\omega_{\pm} = \pm N$.

The vortical mode, a.k.a. the balanced mode or the degenerate solution, does not propagate and carries all of the potential vorticity of these solutions. Potential vorticity,

$$PV \equiv \frac{(\boldsymbol{\omega} + f\hat{z}) \cdot \nabla\rho}{\rho} \quad [\text{s}^{-1}\ \text{m}^{-1}], \tag{6.21}$$

is the component of total vorticity (planetary plus relative) projected on the density gradient, and $\boldsymbol{\omega}$ is relative to the rotating system (Riley and Lelong, 2000). A more relevant frequency for these motions is $\omega_0 = u/l$, from velocity and length scales (Vanneste, 2012).

6.3.4 Scaling for Changes in Stratification

Exact solutions of the equations for linear flow, (6.12–6.16), are possible over flat bottoms where N and f_{eff} are constant. Elsewhere, the asymptotic Wentzel-Krammers-Brillouin (WKB) approach (Bender and Orszag, 1978) is used, e.g. to examine vertical variability of internal waves in pycnoclines where stratification changes over vertical scales longer than those of the waves. Asymptotic solutions to the internal wave equations by Phillips (1966) show $w \propto N(z)^{-1/2}$, $k_z \propto N(z)$,

$\zeta \propto N(z)^{-1/2}$, and $u \propto N(z)^{1/2}$, dependencies used by Briscoe (1975) to compare observations at different depths on the IWEX tri-mooring.

With $k_z \propto N$, length scales must be inversely proportional to N. Therefore, scaling to remove the N dependence of vertical scale changes is

$$dz_{\text{wkb}} = dz(z)\left(\overline{N}(z)/N_{\text{ref}}\right), \quad z_{\text{wkb}} = z_{\text{ref}} + \int_{z'_l}^{z'_u} \left(\overline{N}(z')/N_{\text{ref}}\right) dz', \tag{6.22}$$

where $z' \equiv z - z_{\text{ref}}$ centers integration on z_{ref}. The scaling expands observed vertical length scales at $N(z) < N_{\text{ref}}$ and compresses those at greater stratification (Figure 6.11). WKB internal wave normalizations useful for mixing include vertical displacement, ζ, horizontal velocity, u, vertical strain, $d\zeta/dz$, and vertical shear, du/dz:

$$\zeta_{\text{wkb}} = \zeta(z)\left(\overline{N}(z)/N_{\text{ref}}\right)^{1/2}, \frac{d\zeta_{\text{wkb}}(z_{\text{wkb}})}{dz_{\text{wkb}}} = \left(\frac{d\zeta(z)}{dz}\right)\left(\frac{\overline{N(z)}}{N_{\text{ref}}}\right)^{-1/2} \tag{6.23}$$

$$u_{\text{wkb}} = u(z)\left(\overline{N}(z)/N_{\text{ref}}\right)^{-1/2}, \frac{du_{\text{wkb}}(z_{\text{wkb}})}{dz_{\text{wkb}}} = \left(\frac{du(z)}{dz}\right)\left(\frac{\overline{N(z)}}{N_{\text{ref}}}\right)^{-3/2}. \tag{6.24}$$

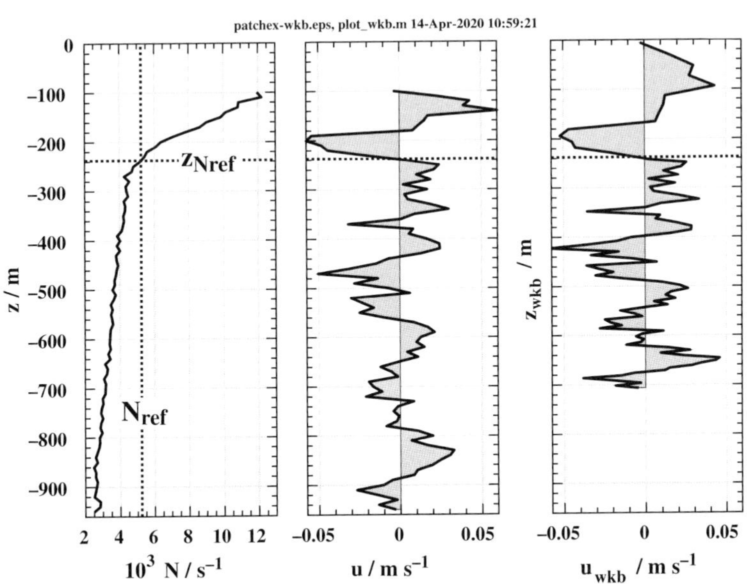

Figure 6.11 WKB-normalized velocity referenced to $N_{\text{ref}} = N_0 = 0.0052\ \text{s}^{-1}$, which is at $z_{\text{Nref}} = -238$ m in $\overline{N}(z)$, at the left. Scaling expands vertical length scales above z_{Nref} and compresses those below it, unlike velocity magnitude, which is diminished above and increased below.

The expressions for displacement and horizontal velocity are consistent with GM scalings (B.31) and (B.20). If strain and shear are computed against observed depth, they will also follow GM scalings (B.41) and (B.25). These expressions for strain and shear, however, differ from GM scalings because they are for gradients computed along WKB-normalized depths, in contrast to GM scalings for gradients computed versus observed depths. Working with WKB-normalized depth is useful when analyzing data across significant changes in stratification.

6.3.5 Aspect Ratio, Inclination, and Polarization

Rearranging the dispersion relation yields the aspect ratio,

$$|k_z|/|k_h| = |\lambda_h|/|\lambda_z| = \sqrt{(N^2 - \omega^2)/(\omega^2 - f_{\text{eff}}^2)}. \tag{6.25}$$

Near-inertial waves have large aspect ratios, e.g. waves with frequencies less than twice the inertial frequency have horizontal wavelengths 30 or more times their vertical wavelengths (Figure 6.12). By contrast, buoyancy waves, with $\omega \sim N$, have aspect ratios of unity and smaller. The ratio ω/f_{eff} also determines θ_w, the inclination of $\boldsymbol{k}$ from horizontal, as seen by writing the dispersion equation as $\omega^2 = N^2 \cos^2\theta_w + f_{\text{eff}}^2 \sin^2\theta_w$ or

$$\theta_w = \tan^{-1}(k_z/k_h) = \tan^{-1}\left[\sqrt{(N^2 - \omega^2)/(\omega^2 - f_{\text{eff}}^2)}\right]. \tag{6.26}$$

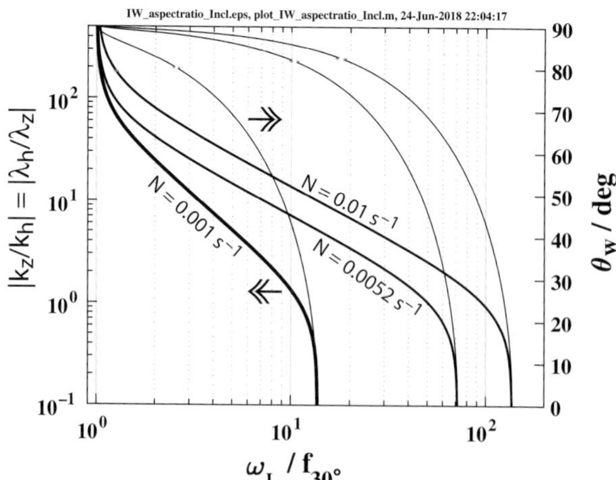

Figure 6.12 Aspect ratio (left y-axis) and inclination (right y-axis) of a single internal wave versus frequency normalized by $f_{30°}$. Curves are for strong, $2N_0$, moderate ($N_0 = 0.0052$ s$^{-1} = 3$ cph), and weak, $N_0/5$, stratification at 30° latitude. Intersections with the lower x-axis equal the frequency bandwidth, N/f_{eff}.

Particle motions are polarized, forming elliptical orbits in planes perpendicular to $\boldsymbol{k}$. Owing to uncertainty in absolute velocity, it is convenient to express the polarization of profiles in terms of shear and strain,

$$\frac{\partial u}{\partial z} = \frac{(N^2 - \omega^2)}{(\omega^2 - f_{\text{eff}}^2)} \left(-ik_x + \frac{k_y f_{\text{eff}}}{\omega} \right) w \tag{6.27}$$

$$\frac{\partial v}{\partial z} = \frac{(N^2 - \omega^2)}{(\omega^2 - f_{\text{eff}}^2)} \left(-\frac{k_x f_{\text{eff}}}{\omega} - ik_y \right) w \tag{6.28}$$

$$\frac{\partial \zeta}{\partial z} = -\frac{k_z}{\omega} w. \tag{6.29}$$

Ratios of horizontal kinetic energy, $hke \equiv (1/2)(u^2 + v^2)$, to potential energy and of shear to strain follow as

$$Shear^2 = \left(\frac{\partial u}{\partial z}\right)^2 + \left(\frac{\partial v}{\partial z}\right)^2 = \left(\frac{N^2 - \omega^2}{\omega^2 - f_{\text{eff}}^2}\right)^2 k_h^2 \left(1 + \frac{f_{\text{eff}}^2}{\omega^2}\right) \langle w^2 \rangle \tag{6.30}$$

$$N^2 \left(\frac{\partial \zeta}{\partial z}\right)^2 = \frac{N^2}{\omega^2} k_z^2 \langle w^2 \rangle \tag{6.31}$$

$$\frac{hke}{pe} = \frac{Shear^2}{N^2 \zeta z^2} = \frac{\omega^2}{N^2} \left(1 + \frac{f_{\text{eff}}^2}{\omega^2}\right). \tag{6.32}$$

6.3.6 Rotary Decomposition and Horizontal Orientation

Viewed from above, particle orbits are circular when $\omega = f_{\text{eff}}$ and become more elliptical as frequency increases, until they are linear at $\omega = N$. The angle, φ, of $\boldsymbol{k}$ relative to east can be determined from alignment of the major axis of the ellipse or from cross-spectra between u and v,

$$\Phi_{uv} = CSpec_{uv} - i\,QSpec_{uv}, \tag{6.33}$$

with real and complex parts termed 'co-' and 'quad-spectra'. Alternatively, rotary decomposition yields clockwise and anticlockwise components,

$$u_{\text{acw}} = \left(1/\sqrt{2}\right)(u + iv), \quad u_{\text{cw}} = \left(1/\sqrt{2}\right)(u - iv). \tag{6.34}$$

Horizontal orientation can be determined by combining cross-spectra,

$$\tan(2\varphi) = \frac{2CSpec_{uv}}{CSpec_{uu} - CSpec_{vv}}, \quad \tan(2\varphi) = -\frac{QSpec_{u_{\text{cw}} v_{\text{acw}}}}{CSpec_{u_{\text{cw}} v_{\text{acw}}}} \tag{6.35}$$

$$\tan(\varphi) = -\frac{CSpec_{\zeta u}}{CSpec_{\zeta v}}, \quad \tan(\varphi) = \frac{CSpec_{\zeta u_{\text{cw}}}}{QSpec_{\zeta u_{\text{cw}}}} \tag{6.36}$$

$$\tan(\varphi) = \frac{QSpec_{\zeta v}}{QSpec_{\zeta u}}, \quad \tan(\varphi) = -\frac{CSpec_{\zeta u_{\text{acw}}}}{QSpec_{\zeta u_{\text{acw}}}}, \tag{6.37}$$

as functions of frequency or wavenumber (Lien and Müller, 1992, table 3). In the northern hemisphere, clockwise rotation in time corresponds to eastward velocity, u, leading northward velocity, v, with increasing depth.

6.3.7 Turning Depths, Turning Latitudes, and Bottom Reflection

With frequencies slightly greater than the local effective Coriolis frequency, near-inertial waves are free to propagate poleward until their frequency no longer exceeds the local inertial frequency, which increases with latitude. After reaching this turning latitude, waves are reflected toward the equator. Equatorward paths steepen as f_{eff} decreases and ω/f_{eff} increases. If the wave's frequency exceeds the local N as it descends, it will be reflected upward from the turning depth. Otherwise, it will reflect from the bottom (Figure 6.13). Garrett (2001) estimates that near-inertial waves travel 400–1,000 km before reaching bottom.

If the bottom is flat and horizontal over the scale of the wave, internal waves reflect specularly, preserving their frequency and retaining their inclination, θ_{w}, but with the opposite sign (Figure 6.14). When the slope of the bottom is less than the ray slope, $\theta_{\text{topo}} < \theta_{\text{w}}$, reflection is termed 'supercritical', and the reflected wave continues in the same direction, upslope in the case shown. During subcritical reflection, $\theta_{\text{topo}} > \theta_{\text{w}}$, waves are reflected back. Reflection, supercritical and subcritical, increases wavenumber, i.e. crests and troughs are closer, increasing c_g and the likelihood of shear instability. Critical reflection, $\theta_{\text{w}} = \theta_{\text{topo}}$, produces the largest changes and occurs when the bottom slope is

$$\theta_{\text{topo}}^{\text{critical}} = \sin^{-1}\left(\sqrt{(\omega^2 - f_{\text{eff}}^2)/(N^2 - f_{\text{eff}}^2)}\right), \tag{6.38}$$

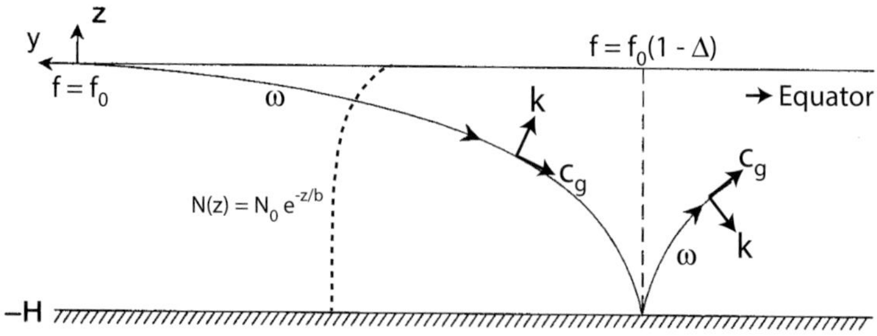

Figure 6.13 Path of a near-inertial wave generated at the surface and propagating downward and toward the equator. The wave frequency, ω_{L}, will not change, but it will encounter increasingly lower inertial frequencies as latitude decreases. (Adapted from Garrett, 2001, © American Meteorological Society. Used with permission)

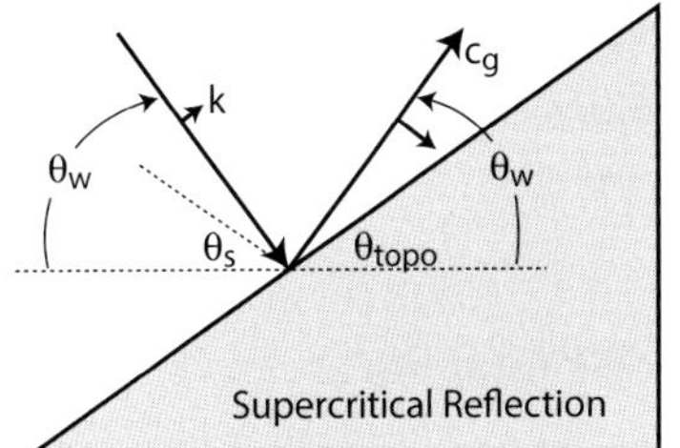

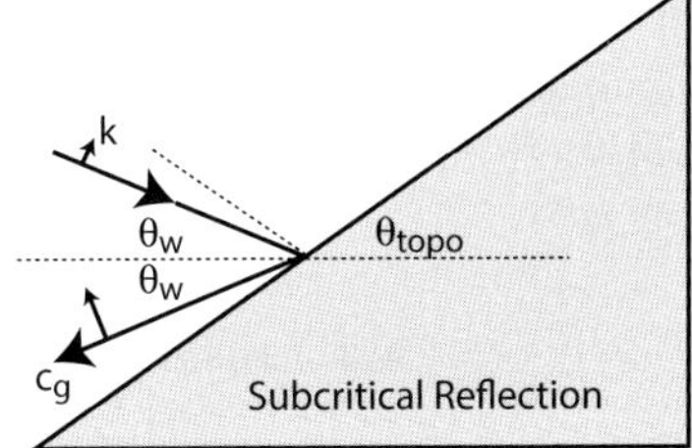

Figure 6.14 Plane internal waves at normal incidence reflecting from a bottom inclined θ_{topo} from horizontal. Because frequency is preserved, incident and reflected waves have the same horizontal inclination, $|\theta_w|$. When $\theta_w > \theta_{topo}$ (supercritical reflection); the group velocity continues in the same direction, unlike the situation for subcritical reflection, when $\theta_w < \theta_{topo}$. Wavenumber magnitude increases in both cases, making shear instability more likely.

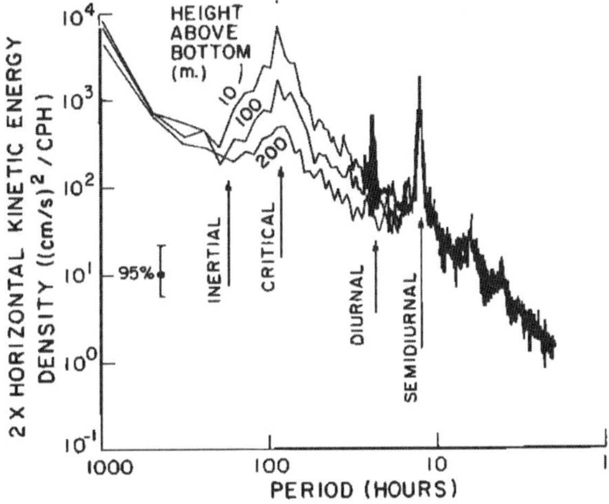

Figure 6.15 Spectra of horizontal kinetic energy above Muir Seamount have a broad peak at the critical period, $2\pi/\omega^{\text{critical}}$. Peak amplitude decreases with height. (From Eriksen, 1982)

or when the wave frequency is

$$\omega^{\text{critical}} = \sqrt{N^2 \sin^2 \theta_{\text{topo}} + f_{\text{eff}}^2 \cos^2 \theta_{\text{topo}}}. \tag{6.39}$$

As they reflect, waves propagating upslope and incident at oblique angles turn in azimuth toward the direction of maximum bottom slope (Eriksen, 1982), and waves propagating downslope turn away.

Laboratory experiments (Thorpe, 1987) show waves breaking when reflection is critical, and spectra of near-bottom ocean temperatures exhibit peaks at critical frequencies (Figure 6.15). With average inclinations of 2–4°, most continental slopes are close to critical for the M_2 internal tide, a situation Cacchione et al. (2002) attribute to erosion of the sediment under the sustained stress of incident internal tides.

6.4 Linear Waves in Horizontal Shear

Relative vorticity accompanying horizontal shear forms minima and maxima of f_{eff} at the edges of mesoscale and submesoscale features. Having the same length scales as near-inertial waves, these changes in f_{eff} trap and amplify the waves (Kunze, 1985). Typical changes of f_{eff} are illustrated in Figure 6.16 for a baroclinic jet, with velocity varying as stratification across the jet changes vertically. Because v is negative (out of the page), $\partial v/\partial x$ is negative on the warm (left) side and positive on the cold (right) side, producing a minimum in f_{eff} on the warm side.

For a barotropic front,[2] f_{eff} is constant with depth (Figure 6.17). Lateral variations in f_{eff} across the front trap waves and/or reflect them at horizontal turning points. In the left panel, waves originating at the surface west of the front and propagating eastward penetrate the trough of negative vorticity to be reflected in the middle, where they encounter $f_{\text{eff}} > f$ at the ridge of positive vorticity. Near-inertial waves formed atop the warm side are trapped because their frequency exceeds f_{eff} in the trough. Waves formed over the east side and propagating westward (right panel) are reflected at the ridge of positive vorticity on the east side, where f_{eff} exceeds their frequency of f.

Kunze and Sanford (1984) invoked trapping to explain the distribution of near-inertial energy across the subtropical front north of Hawaii (Figure 6.18). The structure of f_{eff} was similar to that in Figure 6.16, but the front, stretching from west to east across the Pacific, was locally distorted to be more nearly north/south

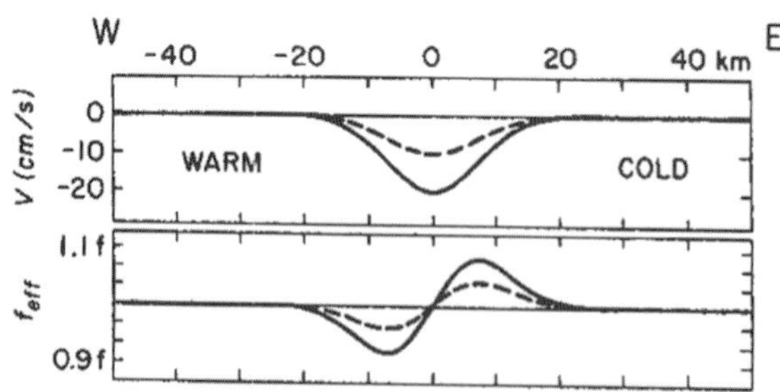

Figure 6.16 Velocity and effective Coriolis frequency in a southward baroclinic jet (out of the page) at the surface (solid) and 100 m (dashed). (From Kunze, 1985. © American Meteorological Society. Used with permission)

[2] Produced by sloping sea surfaces, barotropic fronts have horizontal velocities that are constant with depth.

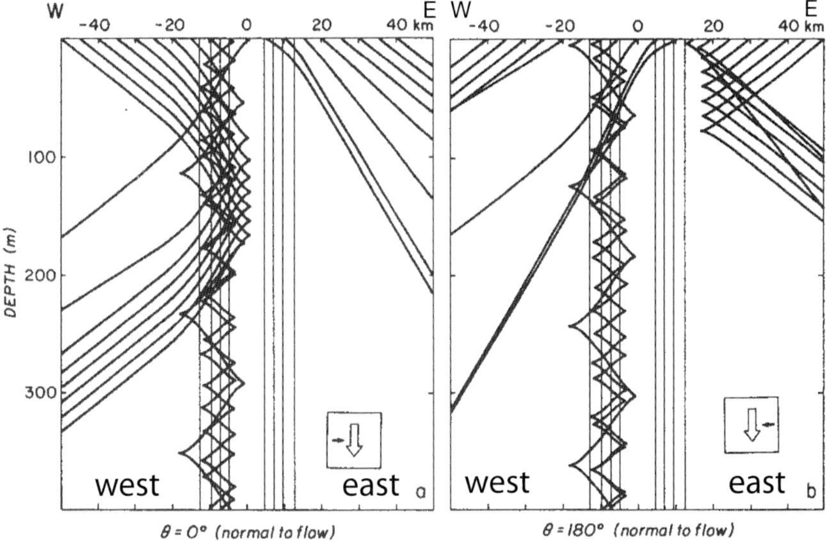

Figure 6.17 Schematic of near-inertial rays (thick lines) formed at the surface and propagating downward and eastward/westward (left panel/right panel) into a barotropic jet, marked by thin isotachs. Flow is toward the reader with the same horizontal pattern as the surface velocity in the previous figure. (Adapted from Kunze, 1985. © American Meteorological Society. Used with permission)

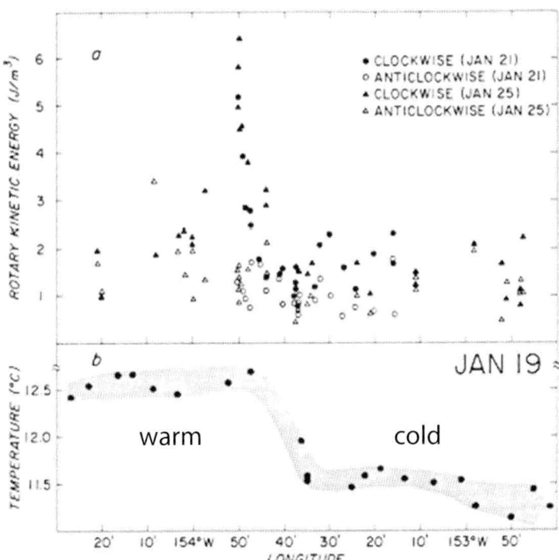

Figure 6.18 XCP surveys during Fronts 80 found elevated near-inertial energy (filled) at the warm side of the Pacific subtropical front, which was southerly to where the survey was made. Mean along-front flow (out of the page) decreased relative vorticity on the warm side of the front where clockwise energy spiked. (Adapted from Kunze and Sanford, 1984. © American Meteorological Society. Used with permission)

where Expendable Current Profiler (XCP) sections found clockwise (downward) near-inertial energy concentrated in a local minimum of $f_{\rm eff}$ on the west (warm) side (Figure 6.16). Similar concentrations have subsequently been found in other locations.

6.5 Linear Waves in Vertical Shear

Near-inertial waves propagating in geostrophic shear have the approximate dispersion relation

$$\omega_{\rm L} = \omega_E - \boldsymbol{k} \cdot \boldsymbol{u} \approx f_{\rm eff} + \frac{N^2 k_h^2}{2 f k_z^2} + \frac{1}{k_z}\left(\frac{\partial \overline{u}}{\partial z} k_y - \frac{\partial \overline{v}}{\partial z} k_x\right) \tag{6.40}$$

(Kunze, 1985, eq. 14). The first two terms dominate; the third is usually less than 5% of $\omega_{\rm L}$, consistent with geostrophic shear variances significantly less than N^2. Using (6.40), Kunze (1985) traced near-inertial rays propagating into a baroclinic jet (Figure 6.19). Owing to the strength of the jet, Doppler shifts dominate the effect of $f_{\rm eff}$. Waves from behind (left) experience positive shifts ($k_y < 0, v < 0$), reducing $\omega_{\rm L}$ to make the waves more inertial. As a result, they reflect from the jet at horizontal turning points where $\omega_L = f_{\rm eff}$. Waves from ahead of the jet (right panel) experience negative Doppler shifts, allowing them to penetrate the jet, where they are trapped, enhancing shear and mixing, as observed by Lueck and Osborn (1986).

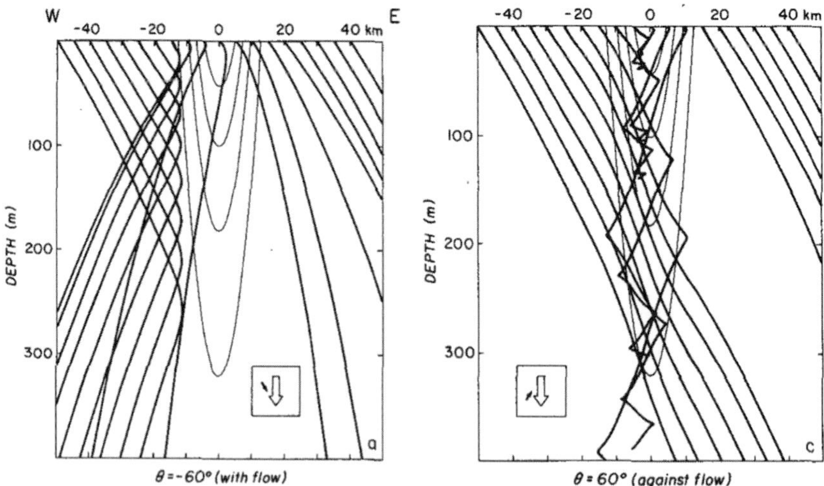

Figure 6.19 Near-inertial rays (thick lines) propagating downward and obliquely into a baroclinic jet (thin lines) from behind (left) and ahead (right). The jet is flowing southward (out of the page), and east is to the right. (Adapted from Kunze, 1985)

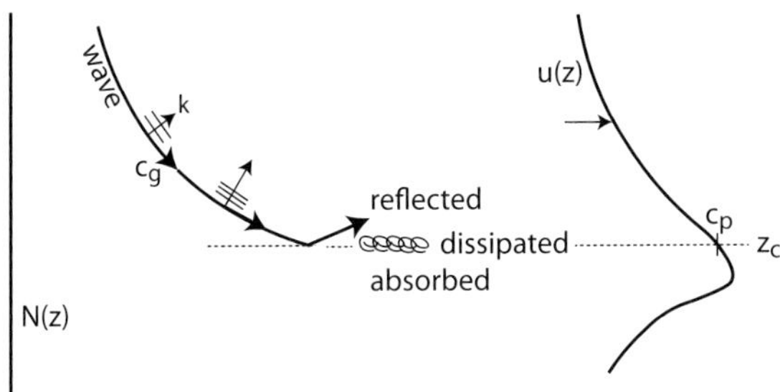

Figure 6.20 A positive Doppler shift in a steady shear flow stops propagation relative to the water, forming a critical layer at z_c, where $\omega_E/k = u(z_c)$. Compressing phase planes enhances dissipation. Energy is also reflected and/or absorbed.

Critical Layers

When internal waves propagate in steady, vertically sheared flow (Figure 6.20), Eulerian frequency, ω_E, and horizontal wavenumber, k_h, are constant (Bretherton, 1966), but Doppler shifts change the Lagrangian frequency, ω_L. Positive shifts decrease ω_L, rotating $\boldsymbol{c}_g$ toward horizontal and $\boldsymbol{k}$ toward vertical, while increasing k_z. If at some depth, z_c, the Doppler shift reduces ω_L to f_{eff}, the horizontal phase speed will equal the speed of the mean current, $\omega_L/k = u(z_c)$, and the wave will cease propagating relative to the fluid. The wave is said to have a critical layer at critical depth z_c. Approaching z_c, the principal effects on the wave are

$$\omega_L \rightarrow f_{\text{eff}},\ k_z \rightarrow \infty,\ c_{gz} \equiv \partial\omega_E/\partial k_z \rightarrow 0,\ c_{ph} \equiv \omega_L/k_h \rightarrow u(z_c). \qquad (6.41)$$

To maintain constant action flux, $c_{g_z} E_{\text{iw}}/\omega_L$, as the vertical group velocity goes to zero and the Lagrangian frequency approaches f_{eff}, E_{iw} increases toward the critical layer. Because the approach is slow, interactions can deposit significant wave energy in the mean flow without breaking. Breaking, however, can occur, as well as reflection, preventing significant energy transmission through the critical layer. Laboratory experiments and numerical simulations find several outcomes at critical layers: acceleration of the mean flow by absorbing the wave without mixing, acceleration of the mean flow with turbulent mixing, and generation of new internal waves without accelerating the mean flow. Three-dimensional direct numerical simulations of a single internal wave find two-dimensional waves breaking at critical layers in three dimensions with characteristics of both advective overturning and shear instability (Winters and D'Asaro, 1994). For $Ri = 1/2$ at the critical layer, about $1/3$ of the energy of large-amplitude waves was reflected and $\sim 1/4$ was dissipated. The remainder accelerated the mean flow.

Unambiguous observations of critical layers in the ocean have not been reported, but they have been inferred from patterns of dissipation and shear variance. At the

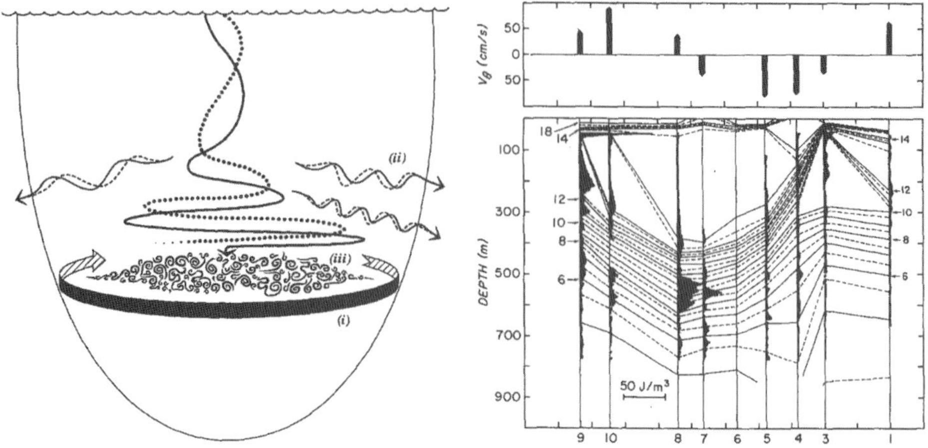

Figure 6.21 (Left) Schematic of a near-inertial wave (eastward solid, northward dotted) propagating into a critical layer in a warm-core ring rotating clockwise (anticyclonically) relative to 800–900 m. Wave energy is lost to (i) the mean flow, (ii) untrapped waves, and (iii) turbulence. (From Kunze et al., 1995.) (Right) Observed azimuthal velocity (upper panel) and internal wave horizontal kinetic energy overlaid on isotherms (lower panel). The concentration of kinetic energy deep in the ring was attributed to critical layer trapping. (Adapted from Kunze and Lueck, 1986. © American Meteorological Society. Used with permission)

base of a warm-core ring, Lueck and Osborn (1986) found dissipation rates 60 times those at similar depths outside the ring. The distribution of turbulence in the ring was consistent with a critical layer where near-inertial waves tried to propagate out of a minimum in f_{eff}. The situation is shown schematically in Figure 6.21, along with observations of enhanced near-inertial shear where a critical layer is expected. Critical layers also play important roles within internal wave fields; ray-tracing of small-amplitude test waves propagating through a GM background found that within a few inertial periods small waves encountered critical layers in the flow fields of larger waves (Henyey et al., 1986).

6.6 Eulerian Spectra

Frequency spectra of horizontal velocity can be in Cartesian or rotary forms,

$$\Phi_{vel} = \Phi_u + \Phi_v = \Phi_{vel}^{cw} + \Phi_{vel}^{acw}, \tag{6.42}$$

where Φ_{vel}^{cw} and Φ_{vel}^{acw} are clockwise (CW) and anticlockwise (ACW) components, in time or space, corresponding to anticyclonic and cyclonic motions in the northern hemisphere. Peaks of the M_2 internal tide typically appear in both rotary components, corresponding to up- and downgoing waves. The inertial peak, $\omega \sim f_{\text{eff}}$,

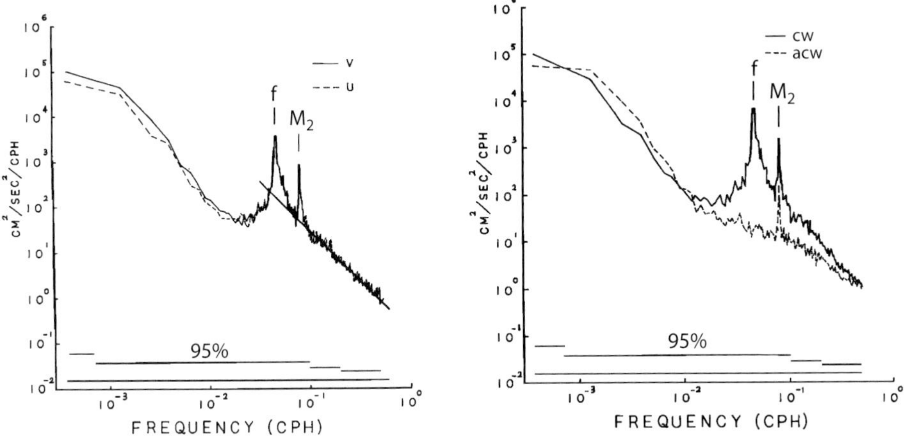

Figure 6.22 Cartesian (left) and rotary (right) velocity spectra at 600 m and 28°N have peaks at inertial, $\omega \sim f$, and M_2 tidal frequencies, in cycles per hour (CPH) superimposed on a smooth continuum. M_2 peaks are in both rotary components, but the inertial peak is only clockwise. The straight line on the left has a slope of –2.23. A 'spectral gap' separates internal waves from low-frequency energy. (Adapted from Fu, 1981)

is anticyclonic (CW) in the northern hemisphere and contains most of the internal wave kinetic energy (Figure 6.22). For example, Alford et al. (2017) found that ~60% of root-mean-square (rms) shear at scales greater than 20 m and ~80% greater than 80 m is near-inertial. Moreover, near-inertial shear at moorings persists for about seven days in most locations and 25 days at the M_2-PSI (Parametric Subharmonic Instability) critical latitude of 28.8°N. Formation of the high-frequency 'continuum spectrum' is in part due to Doppler smearing of low-frequency peaks (Pinkel, 2008b) and in part to interactions between near-inertial waves (Polzin and Lvov, 2011).

Observed spectra are often expressed as horizontal kinetic energy (hke), vertical kinetic energy (vke), and potential energy (pe),

$$\Phi_{hke} = (1/2)\Phi_{vel}, \quad \Phi_{vke} = (\omega^2/2)\Phi_\zeta, \quad \Phi_{pe} = (N^2/2)\Phi_\zeta, \tag{6.43}$$

where ζ is vertical displacement from the equilibrium depth. Spectra of total kinetic energy (tke) and total energy (E_{iw}) follow as

$$\Phi_{tke} = \Phi_{hke} + \Phi_{vke}, \quad \Phi_{E_{\text{iw}}} = \Phi_{tke} + \Phi_{pe}. \tag{6.44}$$

Of particular importance to mixing, spectra of shear and strain are formed by multiplying velocity and displacement spectra by k_z^2,

$$\Phi_{shear}(k_z) = k_z^2 \Phi_{vel}(k_z) \left[\frac{\mathrm{s}^{-2}}{\mathrm{m}^{-1}}\right], \quad \Phi_{strain}(k_z) = k_z^2 \Phi_{\zeta}(k_z) \left[\frac{1}{\mathrm{m}^{-1}}\right], \tag{6.45}$$

shifting variance toward a high wavenumber.

6.6.1 Internal Wave Consistency Relations

To examine whether fluctuations with frequencies between f and N were indeed signatures of internal waves, Fofonoff (1969) took Fourier transforms of the linear equations of motion (6.16) to develop consistency relations between one-dimensional spectra. Müller and Siedler (1976) and Lien and Müller (1992) extended the approach to include rotary as well as Cartesian spectra. For instance, the ratio of potential energy to horizontal kinetic energy for internal waves varies with frequency as

$$\frac{\Phi_{pe}^{\mathrm{iw}}}{\Phi_{hke}^{\mathrm{iw}}} = \frac{N^2 \Phi_{\zeta}^{\mathrm{iw}}}{\Phi_{vel}^{\mathrm{iw}}} = \frac{N^2 \Phi_{strain}^{\mathrm{iw}}}{\Phi_{shear}^{\mathrm{iw}}} = \frac{N^2(\omega^2 - f^2)}{(N^2 - \omega^2)(\omega^2 + f^2)}, \tag{6.46}$$

demonstrated over the internal wave band in Figure 6.23. Other tests apply to rotary, quad-, and co-spectra, e.g.

$$\frac{\Phi_{vel}^{\mathrm{iw}}(\omega)}{\Phi_{uv-quadspec}^{\mathrm{iw}}(\omega)} = -\frac{(\omega^2 + f_{\mathrm{eff}}^2)}{\omega f_{\mathrm{eff}}}, \quad \frac{\Phi_{vel-acw}^{\mathrm{iw}}(\omega)}{\Phi_{vel-cw}^{\mathrm{iw}}(\omega)} = \left(\frac{\omega - f_{\mathrm{eff}}}{\omega + f_{\mathrm{eff}}}\right)^2. \tag{6.47}$$

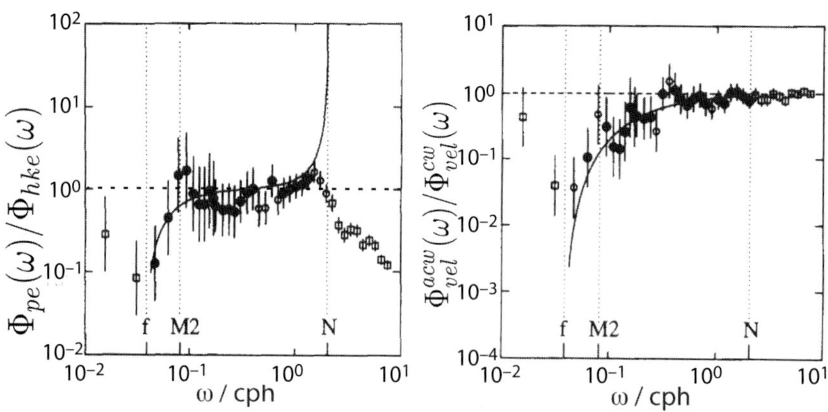

Figure 6.23 Observations (symbols with 95% confidence limits) compared to consistency relations (solid lines) for linear internal waves. (Left) Potential to horizontal kinetic energy (6.46). (Right) Anticlockwise to clockwise horizontal kinetic energy (6.47). Solid circles mark agreement and open circles disagreement; open squares lie outside the internal wave band. (Adapted from Lien and Müller, 1992)

6.6.2 The Garrett and Munk Spectrum

Garrett and Munk (1972) demonstrated that spectra of one-dimensional Eulerian measurements – towed, dropped, and moored – are consistent with a common distribution of internal wave energy in three-dimensional wavenumber and frequency. Their model (GM) is kinematic rather than dynamic, representing averages over several days and a few kilometers, agreeing with all consistency relations, but not derived from equations of motion. The model was subsequently adjusted by Garrett and Munk (1975) (GM75), Cairns and Williams (1976) (a.k.a. GM76), and Munk (1981) (a.k.a. GM79). By providing a reference for observations (Section 6.6.3) and for modeling wave–wave interactions (Section 7.5), GM transformed approaches to internal waves. Although GM did not pass rigorous internal wave consistency tests during IWEX analysis (Müller et al., 1978), and in spite of additional inconsistencies, the model remains the standard description of internal waves and the basis for scaling and predicting dissipation rates in most of the ocean (Section 7.5). Appendix B describes the final version, including one-dimensional spectra relevant to mixing. Major features include:

Energy Density and Modal Content

GM is formulated with a dimensionless constant, E_0, proportional to energy density. Examining additional observations, however, demonstrated significant variability (Wunsch and Webb, 1979), including correlations of shear and strain variability with the magnitudes of turbulent dissipation rates (Gregg, 1989b; Wijesekera et al., 1993; Polzin et al., 1995; Gregg et al., 2003). To accommodate these variations, in (6.8) the energy density, $E_{\rm iw}$, is proportional to $(E/E_0)E_0$. In addition to spectral slope, wavenumber content is specified by a dimensionless mode number, j_*, as the upper bound of the energy-containing range. The GM value is $j_* = 3$, but observations yield other values, often larger.

Latitudinal Dependence and the Inertial Peak

In GM79, horizontal velocity (B.19) and vertical shear (B.24) are proportional to f, yielding integrals constant with latitude (Figure 6.24). Owing to the increase in frequency bandwidth toward the equator, constant integrals imply decreasing spectral amplitude. Observed frequency spectra, however, do not decrease rapidly near the equator; rather, spectral amplitudes appear independent of f, and their integrals increase. GM79 approached this discrepancy by replacing f with the Coriolis parameter at 30° (Figure 6.24). Alternatively, Levine (2002) split the frequency dependence at ω_{S_2}, the frequency of the twice-daily solar tide. A more complicated function for $\omega < \omega_{S_2}$ better represents the inertial peak off the equator and yields frequency integrals increasing toward the equator.

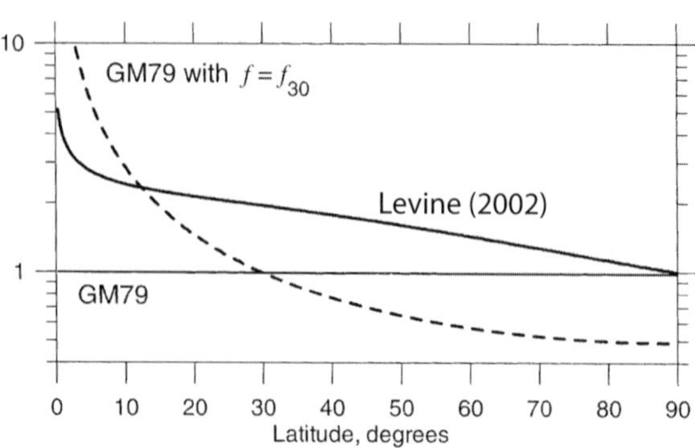

Figure 6.24 Integrals from f to N of the internal wave frequency function versus latitude for Levine (2002), for GM79, and for GM79 with $f = f_{30°}$. (Adapted from Levine, 2002. © American Meteorological Society. Used with permission)

Separability

For analytic convenience, GM is separable, i.e. the product of one function dependent only on frequency and another dependent only on wavenumber. That is, all wavenumbers have the same frequency dependence and all frequencies have the same wavenumber dependence. This has long been recognized as wrong (Sherman and Pinkel, 1991). Subsequent analyses of velocity on density surfaces, a.k.a. a semi-Lagrangian reference, give a more accurate description (Section 6.2.1).

Waveguide Thickness

Based on typical profiles deeper than 1 km, GM parameterizes stratification as an exponential density decrease with a scale-depth of $b = 1,300$ m. Most observations, however, are from the upper 1 km, where shape and scale-depth change. To maintain a uniform reference, these inconsistencies are often ignored, but that is not possible in shallow water. Levine (2002) addressed this by defining a wave guide between upper and lower turning depths and calculating its thickness from a normalized integral of $N(z)$.

Vertical Symmetry

Differentiating the dispersion relation (6.18) to obtain the vertical group velocity,

$$c_{gz} = \frac{\partial \omega}{\partial k_z} = -\frac{N^2 k_h^2}{k_z^3 f} \quad \left[\frac{\text{m}}{\text{s}}\right], \tag{6.48}$$

yields vertical energy fluxes if k_h can be estimated from lateral velocity variations. GM assumes equal upward and downward energy fluxes, and they are often vertically symmetric (Waterhouse et al., 2014; Alford et al., 2016), but not always. At Station P in the North Pacific, downward energy is 2–3 times upward in summer and 3–7 times in winter (Alford et al., 2012).

Shear-to-Strain Ratio

For a single internal wave, the ratio of normalized shear-to-strain variances, R_ω, varies with frequency from > 10 when $\omega \sim f$ to 1 at the buoyancy frequency, N (Figure B.5). For GM79, the ratio is 3 (Eq. B.42). Because the rate at which internal waves dissipate energy by breaking depends on this ratio, it is estimated from observations by integrating shear and strain spectra as functions of wavenumber or frequency. To improve dissipation estimates, Ijichi and Hibiya (2015) reformulated R_ω to distinguish the broadband GM field from the narrowband near-inertial field.

High-Wavenumber Cutoff

Garrett and Munk (1972) chose a vertical scale of 2 m to cut off the internal wave spectrum, citing the beginning of the turbulence range observed by Cox et al. (1969). In view of observations showing slopes of strain (Gregg et al., 1973; Gregg, 1977a) and shear (Gargett et al., 1984) spectra changing from nearly flat to k_z^{-1} at 0.1 cpm, Munk (1981) added an ad hoc fix incorporating that cutoff. This accords with observations, but reasons for the cutoff remain uncertain. Internal wave scaling predicts that the cutoff should change with stratification, which has not been observed. Moreover, Holloway (1980) argues that internal waves become nonlinear near vertical scales of 60 m. The cutoff and the k_z^{-1} range are discussed in more detail in Section 7.8.

6.6.3 Spatial and Temporal Variability

Across the globe, internal wave energy inferred from strain in Argo profiles varies 100-fold, principally between high and low latitudes (Figure 6.25). The highest levels occur broadly throughout the northwest Pacific, driven by storms and tidal currents across submarine ridges. Surprisingly in view of the strong winds, the Southern Ocean has the lowest energies outside the Arctic. High values at low latitude are consistent with energy propagating equatorward building up to high intensity at steady state, owing to reduced mixing efficiency near the equator (Section 7.6.2).

Under Arctic ice, internal waves are less energetic than in the open ocean. Horizontal kinetic energy is 40–50 times less, and, owing to differences in spectral

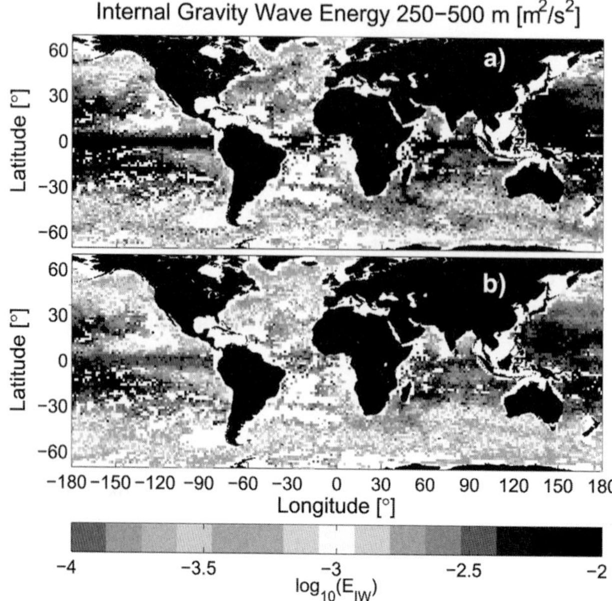

Figure 6.25 Internal wave energy from Argo profiles between 2006 and 2015, using density displacement (a) and strain (b). Intensity is highest in the western North Pacific and along the equator and lowest in the Southern Ocean. (From Pollman et al., 2017. © American Meteorological Society. Used with permission.) (A black and white version of this figure will appear in some formats. For the color version, please refer to the plate section)

slope, WKB-normalized shear variance is 10 times less (Figure 6.26). There is also a large seasonal cycle, low beneath the ice pack in winter and high under open water during summer (Pinkel, 2008b). Upward and downward fluxes are nearly equal, surprisingly so, in view of weak tidal and mesoscale flows capable of generating internal waves at the bottom. Pinkel (personal communication, 2018) believes that the upward waves likely were generated at the surface and returned to shallow depths after encountering turning depths where $N^2 < f^2$. Moreover, dissipation rates are so low that waves can propagate long distances with little attenuation.

Long-term observations of high-frequency internal waves at 200–500 m in the North Atlantic (34°N, 70°W) also reveal a seasonal cycle by factors of 2–3 about the annual mean, but the phasing differs from that in the Arctic: high under winter storms and low during the weak winds of late autumn (Briscoe and Weller, 1984). Subsequent analysis of current meters moored around the globe found kinetic energy above 4,500 m enhanced by factors of 4–5 between 25–45°N, mostly in the western Atlantic and Pacific (Alford and Whitmont, 2007). The amplitude seems to be only twofold in the southern hemisphere, but statistics are less robust.

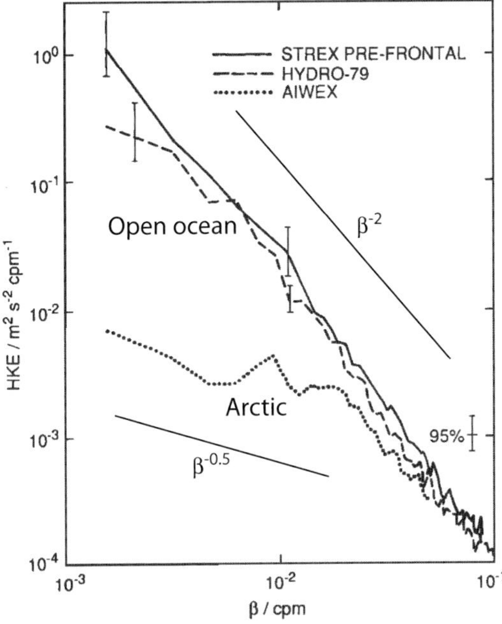

Figure 6.26 Spectra of horizontal kinetic energy in the Arctic and in the open ocean, WKB-normalized to $N_{\text{ref}} = N_0 = 0.0052\ \text{s}^{-1}$. At low vertical wavenumber, β, the Arctic spectrum is lower than open-ocean spectra by factors of 30–100. At $\beta \geq 0.03$ cpm, the Arctic spectrum steepens and is only slightly below the others. Low and high wavenumbers are best fit with $j_* = 60$ and 20. (Adapted from D'Asaro and Morehead, 1991)

6.7 The Vortical Mode

The vortical mode has long been recognized as the degenerate solution ($\omega \sim 0$) of the linear internal wave equations (6.12–6.16). Initially termed 'pancake eddies' or 'blini' (Russian pancakes), vortical motions are thin, rotating lenses of homogenous or partially mixed water. They have been observed in laboratory tanks, growing slowly in late wakes of submerged bodies (Figure 6.27). Numerical simulations reveal similar structures developing as stratified turbulence decays (Figure 6.2). Vortical motions can be in geostrophic balance, with Rossby numbers of zero ($Ro = 0$), or in cyclostrophic balance ($Ro \gtrsim 1$), consistent with strongly stratified turbulence (Riley et al., 1981).

6.7.1 In Semi-Lagrangian Reference Frames

In contrast to smooth Eulerian frequency spectra, semi-Lagrangian spectra of shear and strain have energy concentrated in a few constant-frequency ridges parallel to

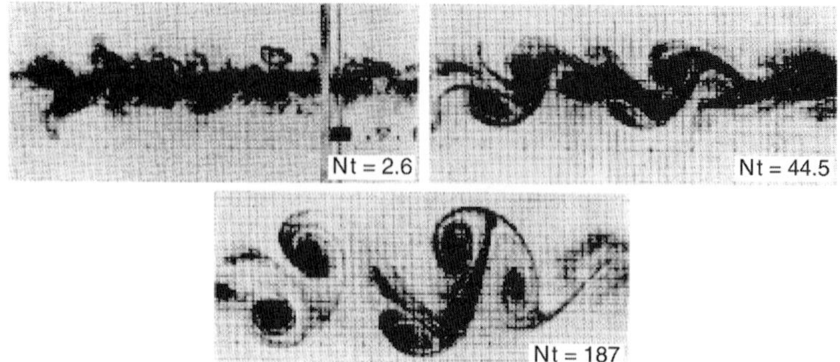

Figure 6.27 Looking down on the dyed wake of a sphere towed leftward through stratified water. The aging wake organized into thin counter-rotating vortices, a.k.a. vortical motions. (From Kao and Pao, 1978. Reprinted by permission of Taylor and Francis Ltd)

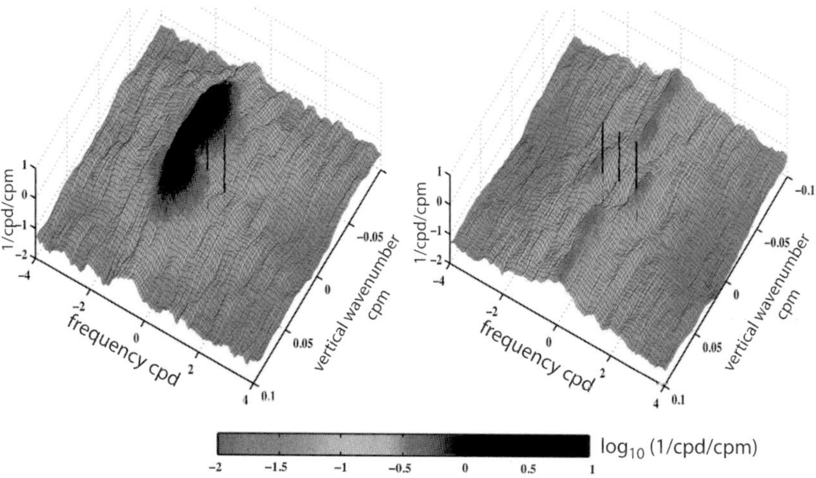

Figure 6.28 Two-dimensonal spectra of normalized shear, $\Phi_{Fr}(\omega, k_z)$ with $Fr \equiv Shear(z)/N(z)$ (left), and strain (right), both on semi-Lagrangian references. Vertical black lines mark frequencies of $-f, 0, f$. (Adapted from Pinkel, 2014. © American Meteorological Society. Used with permission.) (A black and white version of this figure will appear in some formats. For the color version, please refer to the plate section)

the wavenumber axis (Figure 6.28). At low wavenumber, i.e. $|k_z < 0.05|$, a spectral gap separates vortical energy at zero frequency from internal wave near-inertial peaks at $\omega \sim \pm f$. These ridges dominate shear and taper off as wavenumber magnitude increases. At intermediate wavenumber, Doppler shifting smears anticyclonic shear variance to form much smaller ridges at positive, cyclonic, frequencies. These obscure any vortical energy that may be present. The strain spectrum has a

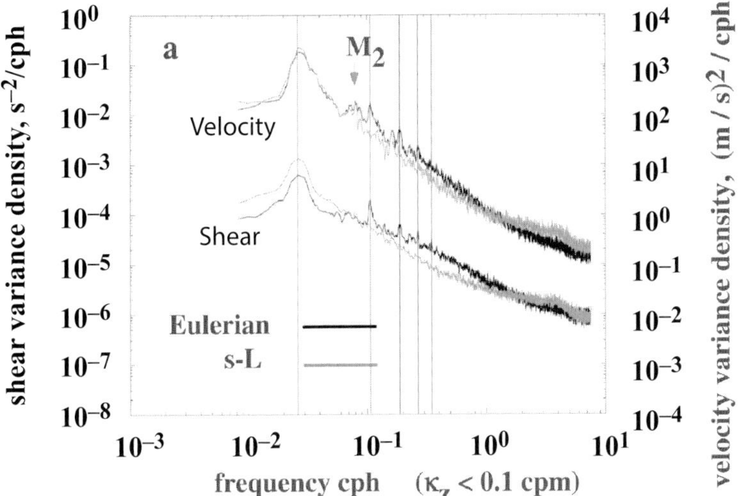

Figure 6.29 Spectra of anticyclonic (CW) horizontal velocity and vertical shear on Eulerian and semi-Lagrangian (s-L) references. Vertical lines mark harmonics in Eulerian, but not semi-Lagrangian, spectra produced by Doppler shifting. (Adapted from Pinkel, 2008a. © American Meteorological Society. Used with permission)

vortical ridge at $\omega = 0$ that has low amplitude at small wavenumbers and is highest between $|k_z| \sim 0.05$ and 0.1 cpm.

Integrating two-dimensional velocity and shear spectra from -0.05 cpm to $+0.05$ cpm yields frequency spectra with strong inertial peaks and a continuum that decreases in amplitude with increasing frequency (Figure 6.29). The M_2 tide produces a peak in Eulerian and semi-Lagrangian velocity spectra but not in shear. Peaks at frequencies greater than $M_2 + f$ in Eulerian velocity are attributed to vertical Doppler shifting because they are absent in the semi-Lagrangian spectrum. The consequences of the inherent frequency spectrum being discrete rather than continuous have not been seriously addressed.

6.7.2 In Eulerian reference frames

In addition to a strong vortical mode peak at the origin, wavenumber-frequency spectra from the Arctic demonstrate varying wavenumber content at different frequencies, in contrast to GM (Figure 6.30). Vortical motions can only be formed by mixing, e.g. by breaking internal waves (Section 7.5) and/or by a horizontal cascade of stratified turbulence (Section 3.8). Once vortical motions exist, whether they act as an additional source of dissipation as they break down is not understood. Such a source could be partly responsible for scatter in dissipation rates now attributed solely to breaking internal waves.

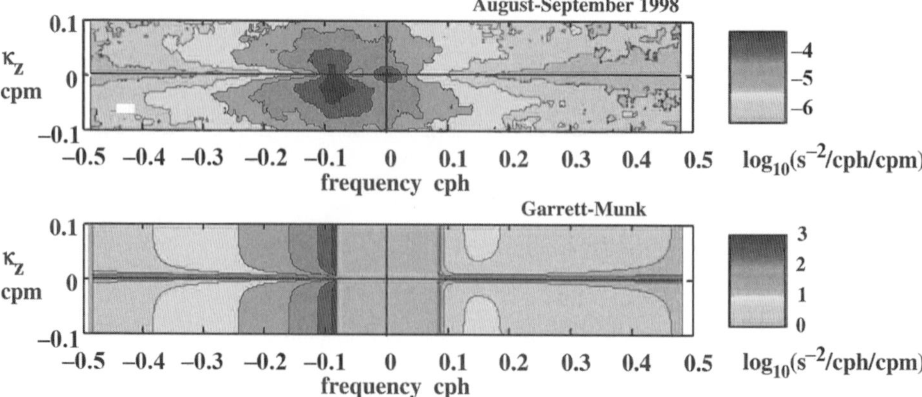

Figure 6.30 Spectra of $\log_{10} Shear^2$ in the Arctic (top) and in GM (bottom) showing how observed spectra do not have the same wavenumber behavior at all frequencies. The Arctic spectrum also has a distinct vortical peak at zero frequency. (Adapted from Pinkel, 2008b. © American Meteorological Society. Used with permission)

Distinguishing structures produced irreversibly by mixing from those resulting from transient strain is difficult, even with closely spaced profiles. There are, however, consistency relations for spectra of the linear vortical mode,

$$\frac{\Phi_{vel}}{N^2} = \left(\frac{N^2 k_h^2}{f^2 k_z^2}\right) \Phi_{displ} \tag{6.49}$$

Rotary , Cartesian

$$\Phi_{vel}^{acw} = \Phi_{vel}^{cw}, \ \Phi_{uv}^{quadspec} = 0 \tag{6.50}$$

$$\Phi_{\zeta u}^{cospec} = \Phi_{\zeta v}^{cospec}, \ \Phi_{\zeta v}^{quadspec} = 0 \tag{6.51}$$

$$\Phi_{\zeta u}^{quadspec} = \Phi_{\zeta v}^{quadspec}, \ \Phi_{\zeta u}^{quadspec} = 0, \tag{6.52}$$

to compare with (6.46) and (6.47) for linear internal waves (Lien and Müller, 1992). Applying the consistency relations allowed the velocity spectrum in Figure 6.31 to be decomposed into internal waves, vortical motions, and noise (Willebrand et al., 1977, p. 98). At low frequency, the vortical mode spectrum is over a decade below the internal wave spectrum, but near N it is slightly higher. Ratios of horizontal kinetic energy and potential energy in spectra also led D'Asaro and Morehead (1991) and Kunze (1993) to conclude that some spectra they observed had contributions from vortical motions. In particular, Polzin et al. (2003) argued that vortical motions are a major component of spectra in the k_z^{-1} 'saturated' range of shear spectra, discussed more thoroughly in Section 7.8.

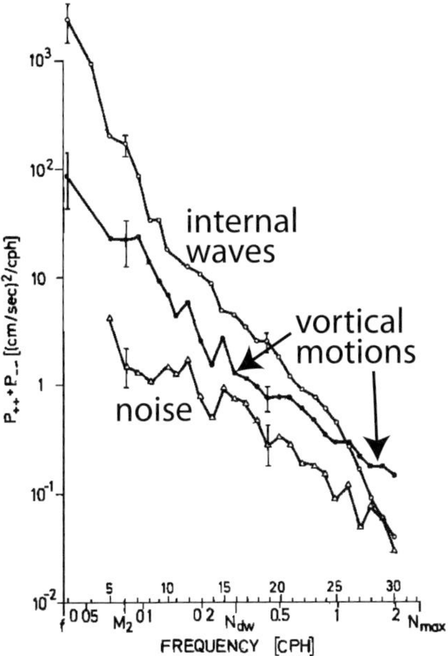

Figure 6.31 Decomposition of an observed velocity spectrum, $\Phi_{vel}(\omega)$, into internal waves, vortical motions (initially termed 'current finestructure'), and noise, based on applying consistency relations to spectra and cross-spectra. The vortical mode contribution increases with frequency and eventually dominates. (Adapted from Müller et al., 1978)

6.8 Perspectives

Walter Munk often joked that the success of a scientific contribution is measured by how long it impedes progress, meaning that few contributions resolve all issues, but some do so well that other investigators put off the hard work of finding the next step. By Walter's measure, and many others, GM succeeded brilliantly. In spite of accumulating discrepancies, particularly in higher-order moments and in horizontal spectra, GM remains the primary description of the internal wave field. Its replacement should be a high priority.

Based mostly on measurements of vertical displacement and horizontal velocity, GM describes the energy of the wave field averaged over days and tens of kilometers, assuming it is composed of a random collection of individual waves. Interactions driving mixing, however, occur over much shorter intervals in instantaneous wave fields often dominated by one or more near-inertial waves plus a few high-frequency packets. Detailed time series encompassing individual wave fields are needed to formulate a physics-based model of the internal wave field as it generates mixing.

7

Interactions and Dissipation of Internal Waves and the Vortical Mode

7.1 Overview

Internal waves generated by winds and flows over rough bottoms (Figure 7.1) subsequently break to mix the pycnocline. Global maps of dissipation rates in the breaking waves have been made by applying models of wave–wave interactions to finescale shear and strain from conductivity-temperature-depth (CTD) surveys and Argo profiling floats. Assuming a constant mixing efficiency of 0.2 allows these maps to be rescaled in terms of diapycnal diffusivity. Comparisons with thickening rates of artificial tracers confirm these diffusivities within a factor of 2. Improving on them, however, will require deeper understanding of how the turbulence develops. Examining these links is a major theme of this chapter.

By imparting anticyclonic energy at inertial frequencies, winds spin up surface mixed layers, and lateral differences in horizontal velocity within these layers pump underlying stratified water to generate internal waves with frequencies slightly above inertial (Section 7.2). Although the basic mechanism is known, owing to the varying resolution and accuracy of wind inputs, estimates of global production differ fivefold.

Driven by tides and deep mesoscale eddies, the flow of stratified water over irregular bottoms is the other major source of internal waves (Section 7.3). One aspect, internal tide generation by surface tides flowing over ridges and seamounts, is well understood, as is their propagation across ocean basins. Less understood, the generation of freely propagating, high-frequency internal waves and lee waves by deep flows over rough bottoms has been inferred from observations of shear and dissipation increasing exponentially near the bottom. Owing to the sparsity of detailed observations, however, assessing abyssal generation is at an early stage. Generation also occurs within the ocean's interior if geostrophic flows lose balance when meanders or mesoscale eddies impinge on them (Section 7.4). This mechanism has been inferred from serendipitous observations, and much remains to be learned.

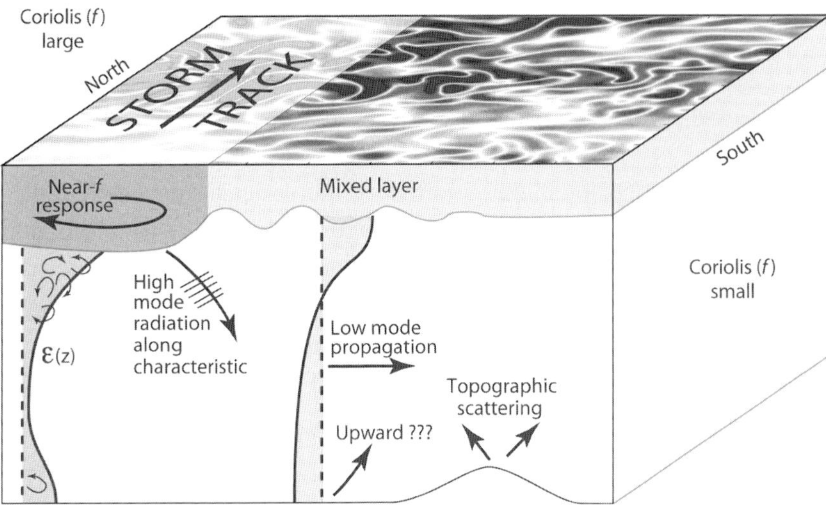

Figure 7.1 Schematic of wind generation of near-inertial waves and their subsequent propagation, bottom interaction, and dissipation, shallow and deep. (Adapted from Simmons and Alford, 2012)

For 'GM-like' (i.e. like the Garrett and Munk spectrum) internal wave fields, analytic theory, ray tracing, and direct simulations demonstrate resonant interactions between triads of waves driving net energy fluxes toward small scales and ultimate breaking (Section 7.5). The flux varies primarily with stratification, N^2, and internal wave energy density, E_{iw}, which is proportional to shear variance. These relationships have been tested extensively and modified to meet conditions of wave field variability (Section 7.6). In particular, adaptation to scaling with strain permits global estimation of mixing levels using data from Argo floats and CTD surveys. Accumulating discrepancies when internal waves differ significantly from GM illuminate the need to understand dissipation in solitary waves, over continental shelves, and in stratified boundary layers and submarine canyons.

As wave–wave interactions move energy to smaller scales, interactions become increasingly nonlinear until the waves break (Section 7.7). Kelvin–Helmholtz (KH) shear instabilities are considered the most common breaking mechanism, but Holmboe instabilities and advective overturning are also likely. Observations are not yet adequate to sort out the relative importance of these mechanisms.

Vertical wavenumbers separating internal waves and turbulence are saturated (Section 7.8). That is, changes of internal wave energy density do not affect spectral levels between the internal wave cutoff, k_z^c, and the largest overturns. Rather, the saturated range shifts to lower wavenumbers as spectral amplitudes increase in the internal wave and the turbulent ranges bounding it. Of the models advanced for the saturated range, the vortical mode and a horizontal turbulent cascade are the most convincing. Because this scale range links internal waves to turbulence, it must be included in a full understanding of mixing in the pycnocline.

7.2 Generation by Wind Stress

Wind stress with anticyclonic energy often excites resonant responses at the local inertial frequency in surface mixed layers (Figure 7.1), clockwise in the northern hemisphere and the opposite south of the equator. The strongest near-inertial forcing is generated by hurricanes and typhoons (Figure 7.2). With typical frequencies of $\omega \leq 1.2f$, near-inertial internal waves are observed in discrete packets having $\lambda_z \sim 100-400$ m and $\lambda_h \sim 10-100$ km (Alford et al., 2016). Waves generated

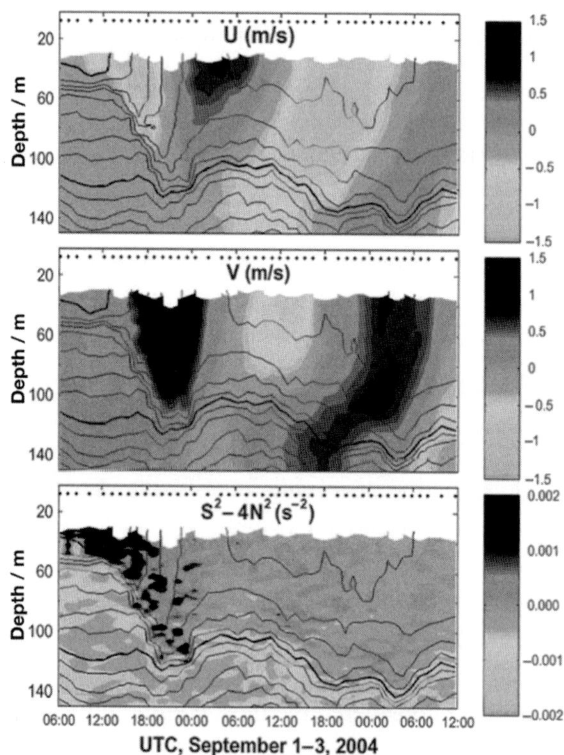

Figure 7.2 Mixed-layer deepening and near-inertial generation measured by an EM-APEX float 50 km to the right of Hurricane Frances. Isotherms are at 0.5°C intervals. The surface mixed layer deepened 80 m in six hours, cooling the surface by 2.2°C. Due to the intense dissipation and mixing, near-inertial currents of 1 m s^{-1} and larger in the mixed layer quickly decayed as waves propagated downward. (Adapted from Sanford et al., 2011. © American Meteorological Society. Used with permission.) (A black and white version of this figure will appear in some formats. For the color version, please refer to the plate section.)

at the surface are characterized by anticyclonic rotation in depth as they propagate downward, and by upward phase lines in time series at one location. High internal modes tend to dissipate locally, while low modes propagate far with little loss. As seen by the vertical group velocity at near-inertial frequencies (6.48), vertical propagation depends quadratically on horizontal wavenumber, which can be affected by inhomogeneities in winds or lateral processes in the surface layer. Latitudinal differences in the Coriolis frequency, i.e. $\beta \equiv df/dlat$, represent one of these processes (D'Asaro, 1989), and interaction with mesoscale eddies is likely another. Models of wind forcing, e.g. Pollard and Millard (1970), compare well with surface drifters, justifying their use for global estimates. These lateral differences in forcing produce convergences and divergences in the mixed layer that are projected onto internal modes, principally the lowest, producing a red spectrum peaking at f (Gill, 1984).

Global models show inputs of near-inertial energy concentrated on the western sides of major ocean basins, under mid-latitude storm tracks, and in the Southern Ocean (Figure 7.3). Other models show similar patterns, but there is wide disagreement about net inputs; estimated global averages range from 0.3 TW to 1.5 TW, depending on grid scales and wind products. Using 0.1° resolution, Rimac and von Storch (2016) obtain 0.35 TW. Much of the energy enters the ocean with frequencies slightly above the local inertial frequency and thus can propagate in any

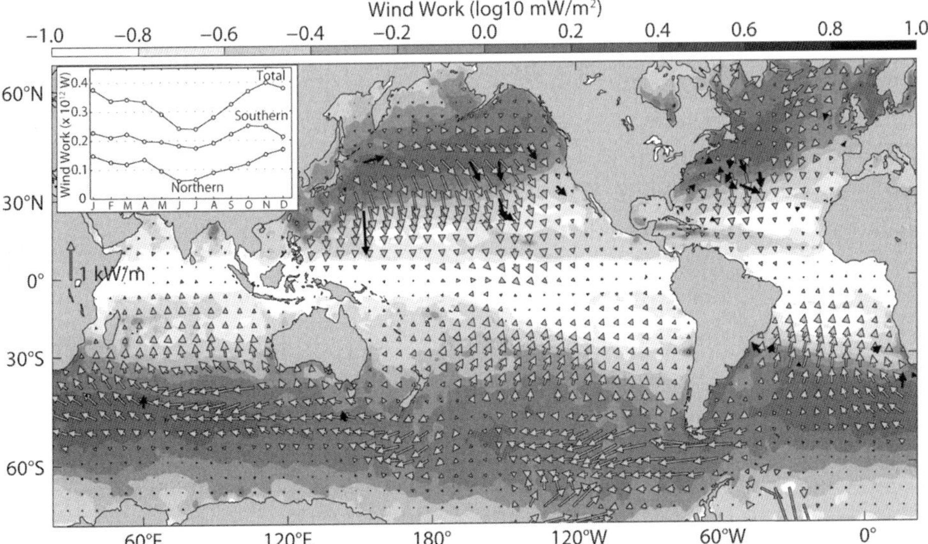

Figure 7.3 Horizontal energy fluxes of the first vertical internal mode (arrows) superimposed on annual mean wind work on the sea surface (shading). (Adapted from Simmons and Alford, 2012)

direction, including poleward. Poleward energy, however, soon reaches a turning latitude, where its frequency equals the local Coriolis frequency, causing the wave to be reflected equatorward. Consequently, near-inertial fluxes under the storm track are strongly equatorward, carrying 3%–16% of the energy input (Simmons and Alford, 2012).

7.3 Topographic Generation

Stratified flows, u_0, over irregular topography characterized by k_h^{topo} generate internal waves when

$$f \leq u_0 k_h^{\text{topo}} \leq N \quad \left[\mathrm{s}^{-1}\right]. \tag{7.1}$$

The most energetic topographic waves are found where surface, i.e. barotropic, tides are forced over tall ridges to produce baroclinic waves at frequencies of the major tidal components (Table 7.1). Topographic internal waves have also been observed above continental shelf breaks, seamounts, and rough sections of abyssal bottoms. Low baroclinic modes at tidal frequencies propagate long distances with minimal attenuation. Tidal frequencies are in the internal wave band at low latitude, allowing poleward propagation until the wave's frequency equals the local inertial frequency, where propagation is redirected equatorward. These turning latitudes vary from 75.3° for M_2 to 27.6° for O_1. Internal tides also displace the sea surface. Though only centimeters high, stationary components are phase-locked to astronomical forcing, allowing extensive averaging to extract signals from satellite altimetry (Ray and Mitchum, 1996). Averaging 20 years of observations provides unprecedented detail (Figure 7.4), including M_2 modes-2 and 3 (Zhao, 2018) and other major tidal components, e.g. O_1 and K_1.

Though not as energetic, topographic waves are also generated at abyssal depths, particularly by geostrophic flows. Not all topographic waves propagate freely; some have frequencies outside the internal wave band; others become stationary lee waves behind obstacles if their phase velocity, $c_{\text{phase}} \equiv \omega_L/k = -u$, is equal

Table 7.1 *Major tidal components, turning latitudes (where $\omega_{\text{tide}} = f$), and PSI-critical latitudes ($\omega_{\text{tide}}/2 = f$). Amplitudes are relative to M_2 as 100.*

Name/Symbol/Amplitude	Period hours	Frequency rad s^{-1}	Turning/PSI-critical latitudes, deg
Principal lunar/M_2/100	12.42	1.41×10^{-4}	75.3/28.8
Principal solar/S_2/46.6	12.00	1.45×10^{-4}	84.0/29.8
Lunisolar diurnal/K_1/58.4	23.93	7.29×10^{-5}	30.0/14.5
Lunar diurnal/O_1/41.5	25.82	6.76×10^{-5}	27.6/13.4

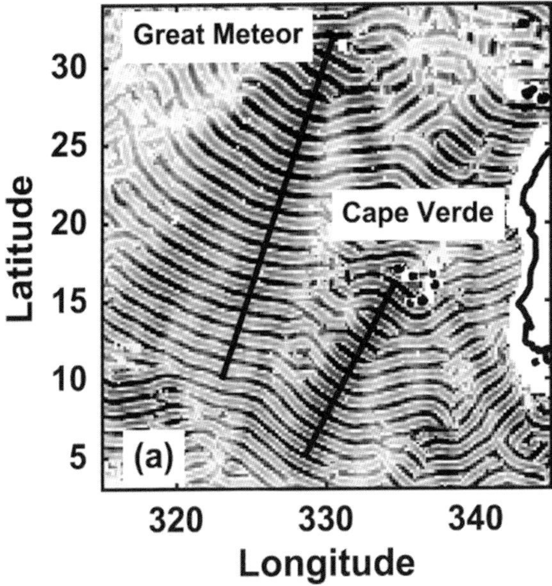

Figure 7.4 Stationary components of surface displacements (~0.1 m) of mode-1 M_2 internal tides propagating from the Great Meteor seamount and the Cape Verde Islands in the North Atlantic; the coast of Africa is outlined on the right. In deep water $\lambda_{M_2} \sim 150$ km. The displacements were computed using 20 years of altimetry data. (From Zhao, 2016)

and opposite to the background flow. If the bottom slope is sufficiently steep, the flow is squeezed beneath an overturning 'rotor' until hydraulic control is lost downstream as the current mixes with slower and deeper water, causing the flow to rebound upward. Some of the most intense turbulence in the ocean occurs in hydraulic controls and rotors where strong tidal currents flow over sills. As the tide slackens, waves trapped at a sill or shelf break are released upstream, often beginning as an internal bore, or solibore, and evolving into a train of internal solitary waves.

7.3.1 Parameters Controlling Topographic Generation

Topographic generation involves seven dimensional variables:

Forcing: velocity, u_0, and frequency, ω_0
Topography: vertical, H, and horizontal, L, length scales
Environment: stratification, N, Coriolis frequency, f, and water depth, D

(Legg and Huijts, 2006). Forcing is principally by the major tidal components and geostrophic eddies (Nikurashin and Ferrari, 2010a). The velocity of barotropic tides, u_0, is uniform with depth, and some geostrophic eddies have strong near-bottom flows. Topographic length scales are typically characterized by root-mean-square (rms) values along the flow, but some bottom roughness must be specified with two dimensions.

By the Buckingham π theorem (Kundu and Cohen, 2004), seven dimensional variables containing two fundamental units, in this case length and time, can be represented by up to five independent nondimensional parameters. Important dimensionless parameters include:

Fractional height, i.e. height relative to water depth,

$$F_H \equiv H/D, \tag{7.2}$$

affects generation when height is a significant fraction of water depth, and flow must be accelerated to pass over.

Steepness, St, is the ratio of topographic slope, H/L, to wave slope,

$$St \equiv (H/L)/\sqrt{(\omega^2 - f^2)/(N^2 - \omega^2)}. \tag{7.3}$$

Discussed in Section 6.3.7, steepness separates into subcritical ($St < 1$) regimes with linear dynamics (Bell, 1975b,a) and supercritical regimes ($St > 1$), which are increasingly nonlinear as slopes steepen.

Excursion, Ex, is the ratio of horizontal displacement ($\Delta x = u_0/\omega_0$) to the lateral topographic scale, L,

$$Ex \equiv \Delta x/L \sim u_0/(\omega_0 L) = u_0 k_h^{\text{topo}}/\omega_0. \tag{7.4}$$

Short excursions ($Ex < 1$) energize waves at the forcing frequency, ω_0, by moving water up- and downslope without crossing the crest. When not limited by stratification, maximum vertical displacements are $\Delta z = u_0 H/\omega_0 L$. Long excursions, $Ex > 1$, produce stationary lee waves (Nikurashin and Ferrari, 2010a) with first harmonics of $\omega_{\text{topo}} \equiv u_0/L$ related to bottom curvature and higher harmonics to higher derivates of topographic height (Simmons, 2008). Other mechanisms producing harmonics include triad interactions (Section 7.5.1), such as the parametric subharmonic instability (PSI) between M_2 and $M_2/2$, which yields half-integer multiples, e.g. $(3/2)M_2$.

Stratification Froude number, Fr_N (3.6), measures how much topography impedes flow. Weak forcing, $Fr_N = u/(HN) \ll 1$, blocks flows, while strongly forced flow, $Fr_N \gg 1$, is relatively unaffected. Intermediate forcing, $Fr_N \sim 1$, is characterized by prominent hydraulic responses at the crest of the topography. Many responses are transient, changing with the tidal cycle.

Table 7.2 *Regimes of simulated M_2 tidal forcing of stratified flow over isolated Gaussian topography by Legg and Huijts (2006).*

	Short	Tall
	• $H/D = 0.04$ • $0.125 \leq Fr_N \leq 2$	• $H/D = 0.5$ • $0.01 \leq Fr_N \leq 0.17$ (nonlinear)
Narrow	• $St = 2$ (supercritical) • $0.34 \leq Ex \leq 5.5$ • upward beams at ω_{M_2} plus higher harmonics as Fr_N increases • rough abyssal seafloors	• $St = 8$ (supercritical) • $0.12 \leq Ex \leq 1.9$ • upward & downward beams • hydraulic jumps, high dissipation • Knight Inlet Sill
Wide	• $St = 0.08$ (subcritical) • $0.014 \leq Ex \leq 0.23$ • mode-1 at ω_{M_2} • low dissipation • Mid-Atlantic Ridge	• $St = 2$ (supercritical) • $0.0029 \leq Ex \leq 0.47$ • upward & downward beams become disorganized at high G • Hawaiian Ridge

7.3.2 Internal Tide Generation

Penetrating the water column with undiminished amplitude, surface tides generate internal tides over irregular abyssal bottoms, as well as over seamounts and ocean ridges. To explore the responses, Legg and Huijts (2006) ran two-dimensional nonhydrostatic simulations for M_2 surface tides across isolated Gaussian shapes that were wide or narrow and tall or short, four cases in all. Each case had five current speeds from $u_0 = 0.02$ to 0.32 m s^{-1}, and was classified by excursion, steepness, and stratification Froude number (Table 7.2).

The short, wide bump, a prototype of the Mid-Atlantic Ridge, had a barotropic-to-baroclinic conversion rate, C_∞, matching the linear prediction of Bell (1975b,a) for an infinitely deep ocean,

$$C_\infty = \frac{\pi}{8}\rho_0 \frac{[(N^2 - \omega_0^2)(\omega_0^2 - f^2)]^{1/2}}{\omega_0} u_0^2 H^2 \quad \left[\frac{\mathrm{W}}{\mathrm{m}}\right]. \tag{7.5}$$

For a given frequency, C_∞ has a quadratic dependence on forcing speed and topographic height.

Weak flow over the short, narrow bump produced upward ω_{M_2} beams (Figure 7.5) and a conversion rate nearly twice that of the short, wide bump, consistent with predictions for nonlinear conversion at a knife-edge (St. Laurent et al., 2003; Llewellyn Smith and Young, 2002), even though the bump had significantly gentler slopes. Faster flow, u_0, increased Fr_N and generated stronger

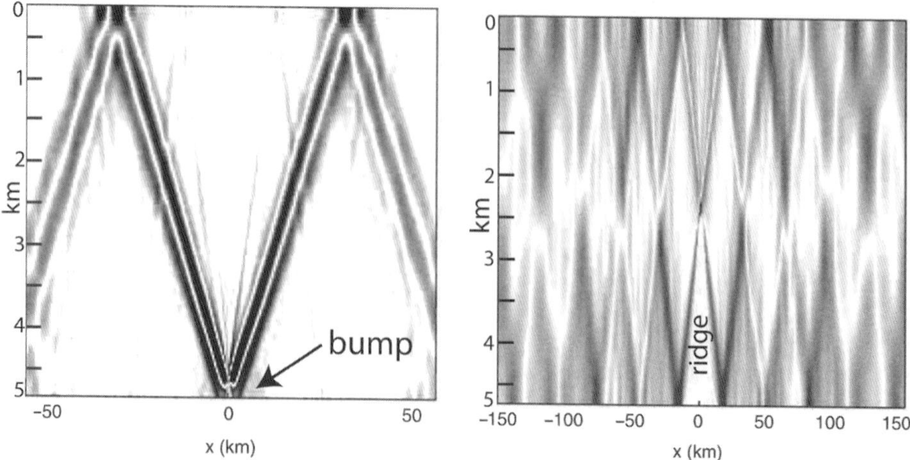

Figure 7.5 Simulations of M_2 baroclinic velocities over a short, narrow bump (left) and a tall, wide ridge (right), both with minimal forcing, $u = 0.02$ m s^{-1}. With $Fr_N = 0.125$, the bump radiated narrow upward beams with weak harmonics. With $Fr_N = 0.01$, the ridge generated a mode-1 response. (Adapted from Legg and Huijts, 2006)

harmonics, which equaled the magnitude of primary response when forcing was strongest. The enhanced conversion rate during weak forcing mostly disappeared as excursions increased with stronger forcing, likely due to destructive interference in an increasingly complex wave field. This case represents generation over the rough topography found on the sides of the Mid-Atlantic Ridge, a site of elevated mixing. Initially, the elevated mixing was attributed solely to internal waves generated by tidal flows over the rough bottom (Toole et al., 1994), but reanalysis suggests that much of the turbulence was produced by hydraulic responses to low-frequency flows along fracture zones perpendicular to the ridge (Thurnherr et al., 2005).

Both tall cases, $H/D = 0.5$, are strongly nonlinear, producing upward and downward beams and hydraulic responses during peak tides. Acceleration over tall, steep ridges with large H/D enhances conversion by $\sim$15% relative to similar features with small H/D (St. Laurent et al., 2003). As with the short cases, harmonics increased with higher forcing, and enhanced mixing would be expected to follow. The tall, wide configuration is a prototype for the Hawaiian Ridge, where intensive measurements documented beams emanating from Kaena Ridge in the Kauai Channel (Figure 7.6). The cross-ridge M_2 energy flux in the channel, 17 ± 2.5 kW m^{-1} (Lee et al., 2006), agreed with linear calculations using the Princeton Ocean Model (Rudnick et al., 2003). This was accompanied by increased dissipation, 3 ± 1.5 TW within 60 km of the ridge, about 15% of the energy converted from surface to internal tides (Klymak et al., 2006a).

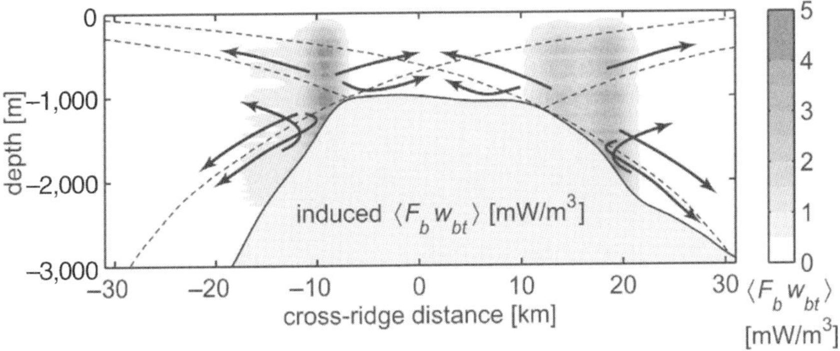

Figure 7.6 Barotropic forcing power across Kaena Ridge, Hawaii computed from barotropic currents derived from satellite altimetry. Arrows show energy paths along critical internal tide rays (dashed). (Adapted from Nash et al., 2006. © American Meteorological Society. Used with permission)

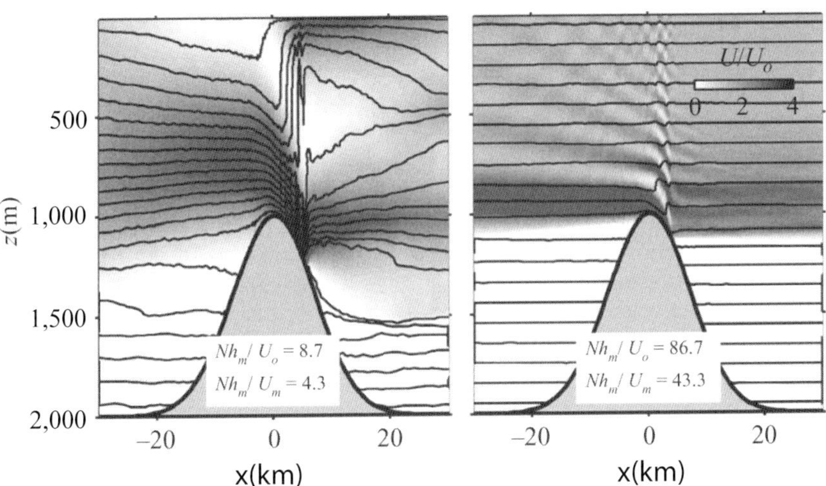

Figure 7.7 Simulations of lee waves in steady flow over a ridge that is blocking flow from upstream. Moderate $G \equiv Nh_m/U_0$ ($Fr_N = 0.11, 0.23$) yields a low-mode response, and large G ($Fr_N = 0.012, 0.023$) a high-mode response. h_m is the peak height of the ridge. (From Klymak et al., 2010)

Flows across tall, narrow features are often partially blocked and hydraulically controlled at the crest. That is, some internal modes have phase velocities slower than the forcing flow and cannot propagate upstream against it. Thin supercritical overflows continue downslope beneath a turbulent overturning 'rotor' until control is lost as depth increases and the plume thickens by mixing into water on the lee side. After control is lost, the flow rebounds (Figure 7.7, left) in a turbulent

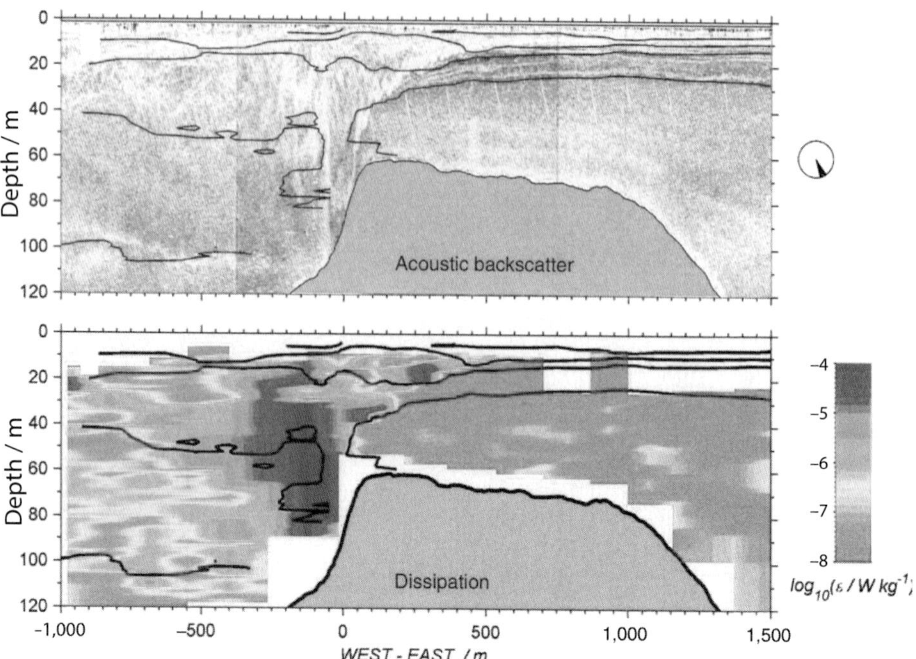

Figure 7.8 Backscatter intensity from a narrow beam of high-frequency acoustics over the seaward side of the Knight Inlet Sill during ebb tide. Flow was hydraulically controlled at the sill crest, forming an hydraulic jump downstream. Intense turbulent dissipation in the plunging flow and rotor. (Adapted from Klymak and Gregg, 2004. © American Meteorological Society. Used with permission.) (A black and white version of this figure will appear in some formats. For the color version, please refer to the plate section.)

hydraulic jump, dissipating excess energy as it joins the slower background flow. Partially blocked controlled flows in the Strait of Gibraltar, Knight Inlet, and Luzon Strait have been studied in detail, revealing intense dissipation in hydraulic jumps downstream of controlled flows (Figure 7.8). As the tide slackens, waves arrested during maximum flow are released to propagate upstream, often forming trains of internal solitary waves, discussed in more detail below. Released solitons are also strongly turbulent.

7.3.3 Global Distribution of Internal Tides

Unlike surface-generated inertial currents, twice-daily internal tides have frequencies above the local Coriolis frequency and can propagate freely, except in the Arctic, where they encounter turning latitudes. Daily tides, however, are limited to latitudes less than 30.0° for K_1 and 27.6° for O_1. Within these ranges, low modes propagate as beams with barely detectable energy loss until they encounter shoaling

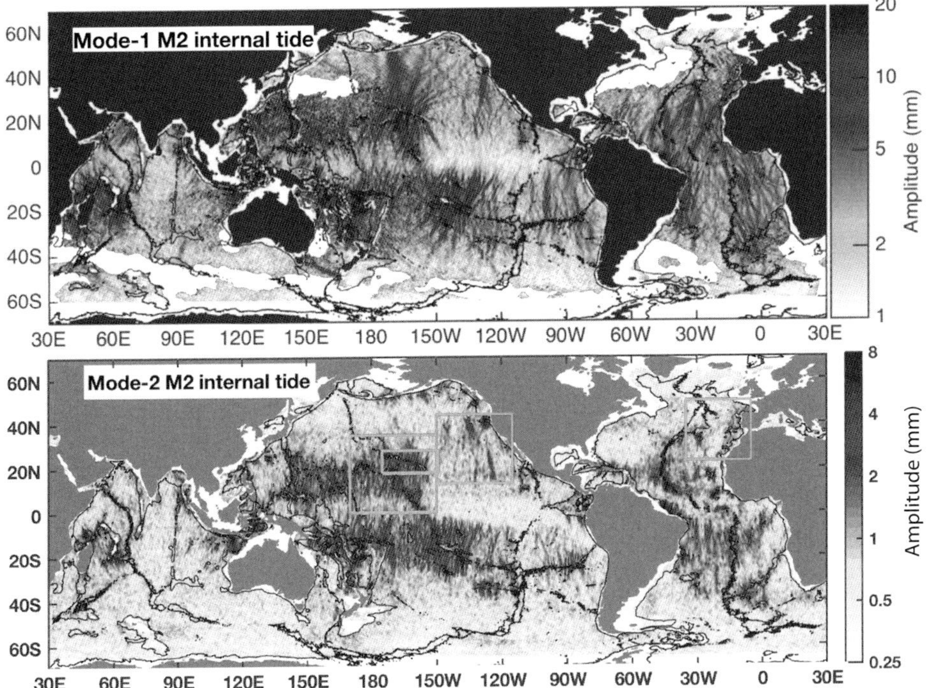

Figure 7.9 Surface displacements of stationary components of mode-1 and mode-2 M_2 internal tides from 20 years of altimetry. High mesoscale activity masks mode-1 signatures in the northwestern Atlantic and Pacific. (Adapted from Zhao et al., 2016 and Zhao, 2018)

bottoms. The primary energy loss during propagation seems to occur where PSI is most likely, i.e. latitudes where the Coriolis frequency is half the frequency of the internal tide, $f = \omega_{\text{tide}}/2$ (Table 7.1). As a result of PSI the tidal wave decays into two inertial waves (Section 7.5).

Observed by satellite altimetry, displacements of the sea surface are the most direct global signature of the M_2 internal tide (Figure 7.9). Amplitudes are only a few centimeters, 20 at most, but they form stationary patterns around generation sites, allowing averaging of repeat passes by multiple satellites over many years to extract signals from noise. The most intense generation occurs in the western Pacific at ocean ridges and island chains such as the Aleutians, Hawaii, and, largest of all, the Solomon Islands. In some cases, source variability along ridges creates interference patterns concentrating the energy into beams. These are most evident for mode-1, e.g. those radiating for thousands of kilometers from Hawaii, some refracted by the mean circulation (Figure 8.41). By comparison, mode-2 displacements are smaller and decay closer to the ridge.

Internal Tides in Numerical Models

Many global estimates have been made of internal wave production by linear dynamics. As an example, using twice-daily (M_2 and S_2) and daily (O_1 and K_1) tidal components, Niwa and Hibiya (2014) used a hydrostatic expression to compute the vertically integrated conversion rate,

$$J_{\rm iw}(x,y) = \overline{\int_{-H}^{0} g\rho' \,\overline{w}\, dz}^{\,\rm time} \qquad \left[\frac{\rm W}{\rm m^2}\right], \tag{7.6}$$

where $\overline{w} = \frac{z-\zeta}{H+\zeta}\left(\overline{u}\frac{\partial H}{\partial x} + \overline{v}\frac{\partial H}{\partial y}\right)$ is the vertical velocity induced by barotropic flow over topography, H is bottom depth, and ζ is sea surface displacement. Runs with topographic grids of $1/5°$–$1/20°$ revealed an exponential increase with decreasing grid size of the flux of energy into internal waves. To estimate production with fully resolved topography, Niwa and Hibiya extrapolated the calculated trend to zero grid spacing. Their calculated energy conversion has patterns similar to those inferred from altimetry, with 'hot spots' mostly in the western Pacific and along the Mid-Atlantic Ridge (Figure 7.10). Net conversion with zero grid spacing is 1.2 TW, 35% larger than with the best available topography, and within the 1 ± 0.25 TW estimate of deep-ocean tidal dissipation by Egbert and Ray (2000). With large H/D, tall, steep ridges enhance conversion by $\sim$15% relative to similar features with low H/D (St. Laurent et al., 2003).

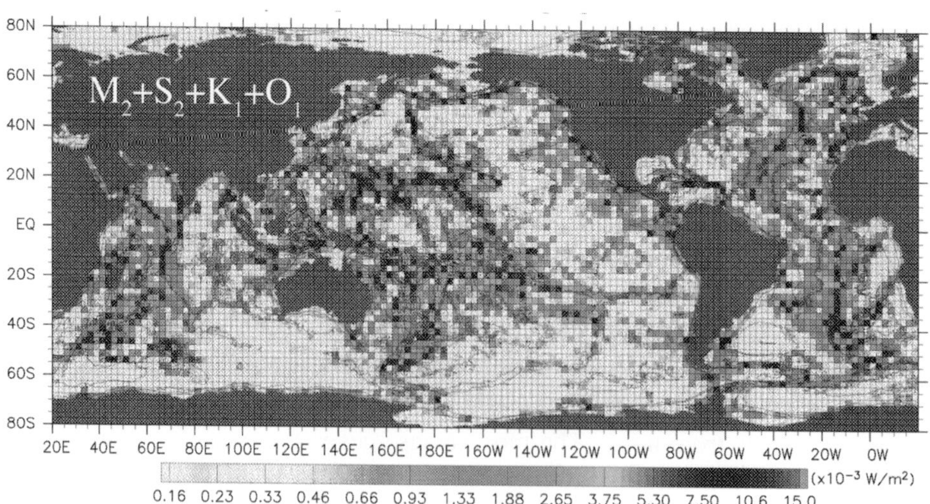

Figure 7.10 Baroclinic tidal energy conversion rate for four tidal components extrapolated to zero horizontal grid spacing. Estimates are averages over 2.5° by 2.5° intervals. (Adapted from Niwa and Hibiya, 2014.) (A black and white version of this figure will appear in some formats. For the color version, please refer to the plate section.)

7.3.4 Generation by Geostrophic Flows over Deep Bottom Roughness

Although not detectable from satellites, internal waves are also generated in the abyss when strong currents, mostly deep geostrophic eddies, flow over irregular topography. The evidence comes from elevated near-bottom shear and strain in Lowered Acoustic Doppler Current Profilers (LADCPs)/CTDs in the Southern Ocean (Naveira Garabato et al., 2004; Kunze et al., 2006b). Geostrophic eddies in the Drake Passage dominate the deep flow with $u_0 \sim 0.1$ m s^{-1}, and high-resolution bathymetry reveals a rough bottom with $H = 305$ m and $L = 5\text{–}10$ km. Using weakly nonlinear theory and fully nonlinear simulations, Nikurashin and Ferrari (2010b,a) demonstrated that waves generated at the topographic frequency, $\omega_{\text{topo}} = u_0/H$, break while propagating upward (Figure 7.11, upper), consistent with nonlinear forcing. Breaking slows the geostrophic flow and energizes near-inertial and higher frequencies, creating a complex flow varying in time. In response to the breaking, wave amplitude decreases 10-fold in the bottom kilometer. For comparison, the calculation was repeated with the same forcing and topography one-third as high to simulate the seafloor west of the Drake Passage, where LADCPs/CTDs find low background levels of shear and strain. Consistent with linear forcing, the wave field is stationary with weak harmonics, and waves propagating with little change in amplitude (Figure 7.11, lower).

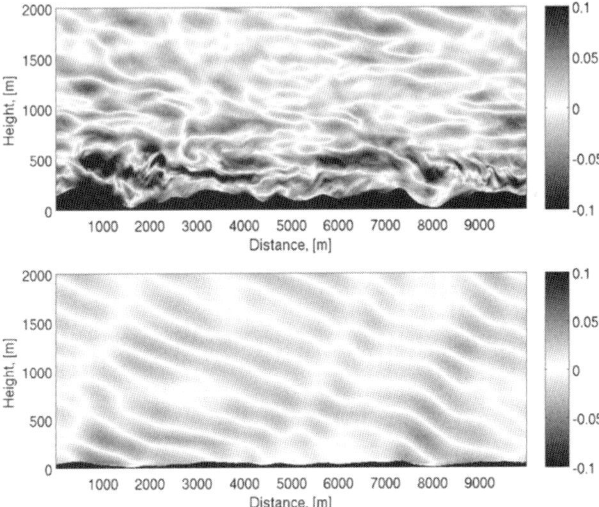

Figure 7.11 Simulated zonal wave velocities (m s^{-1}) in the Southern Ocean produced by mean flows of 0.1 m s^{-1}. (Upper) over rough topography ($St = 0.3 - 0.5$) in the Drake Passage. (Lower) over gentler topography to the west with $St = 0.2\text{–}0.3$ and 1/3 the amplitude. (From Nikurashin and Ferrari, 2010a. © American Meteorological Society. Used with permission)

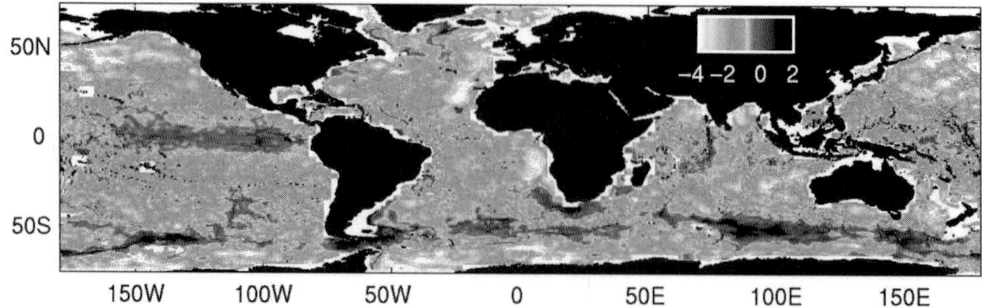

Figure 7.12 Energy flux into internal lee waves ($\log_{10} J_E/\mathrm{mWm}^{-2}$) estimated using global descriptions of bottom stratification, velocity, and roughness. Large fluxes occur in the Southern Ocean and near the equator in the eastern Pacific. (From Nikurashin and Ferrari, 2011.) (A black and white version of this figure will appear in some formats. For the color version, please refer to the plate section.)

Global Estimate from Geostrophic Flow into Internal Waves

Using linear theory, roughness from single-beam bathymetry, abyssal stratification from an atlas, and velocity from a global model, Nikurashin and Ferrari (2011) estimated internal wave production in the deep ocean. Net conversion was 0.2 TW, 16% of the estimate by Niwa and Hibiya (2014) for conversion from surface to internal tides. Half is in the Southern Ocean (Figure 7.12), where deep geostrophic currents cross rough topography. Conversion is also high in the eastern equatorial Pacific, where currents are not as fast but the topography is steeper. The estimates may be biased high because the effect of bottom boundary layers in reducing conversion was not considered.

7.4 Generation by Geostrophic Adjustment of Balanced Flows

Several observations have found strong near-inertial waves in and beneath prominent geostrophic fronts. Using a powerful shipboard Acoustic Doppler Current Profiler (ADCP), Rainville and Pinkel (2004) discovered spatially coherent near-inertial waves beneath and shoreward of the Kuroshio off southern Japan. Vertical and horizontal scales were 10–50 m and 10–60 km, and turbulent dissipation rates predicted using observed shear variances in (7.14) were 40 times those at GM background. Subsequently, Alford et al. (2013) observed intense near-inertial waves propagating downward and toward to the south with $\lambda_z \sim 250$ m from the Subtropical Front north of Hawaii (Figure 7.13). Frontogenesis and wind generation were considered less likely than creation by geostrophic adjustment of the front, which had a Rossby number of $Ro \equiv u/fl_h = 0.2 - 0.3$.

These and other observations have been attributed to radiation of internal waves from fronts in quasi-geostrophic balance adjusting to meanders or perturbations

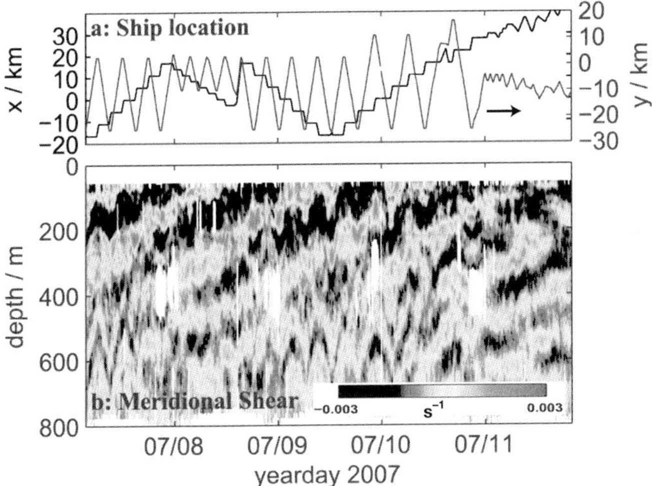

Figure 7.13 Ship position (a) and shear (b) beneath the Subtropical Front north of Hawaii along a rectangular grid with short east–west (x) and long north–south (y) legs. The shear was coherent across the track, with upward phase and downward energy propagation. (Adapted from Alford et al., 2013)

induced by large-scale flows. As they evolve over weeks, balanced flows can shed excess energy at internal wave frequencies. Referred to as spontaneous generation, energy transfer comes through interactions like $\overline{u'u'}\,\partial\overline{u}/\partial x$ and $\overline{u'v'}\,\partial\overline{u}/\partial y$. Simulations indicate that the same interactions can also absorb internal wave energy (Nagai et al., 2015).

Nagai et al. (2015) estimated the global rate of spontaneous generation as 0.36 TW, within the range of 0.1 TW–1.5 TW from other estimates. Although poorly constrained, the magnitude is significant relative to wind and topographic generation. Conversion is most intense near major fronts, including western boundary currents, equatorial currents, and the Antarctic Circumpolar Front (Figure 7.14). The largest area of strong conversion is in the Indian Ocean, attributed to monsoons in the north and the Agulhas Current in the south. The major uncertainty is whether the internal waves are reabsorbed or dissipated.

7.5 Wave–Wave Interactions

The GM spectrum provides a framework for adapting techniques developed for surface gravity waves to internal waves. These efforts underlie much ongoing work on internal waves and mixing, including the first global maps of diapycnal turbulent diffusivity inferred for GM-like internal waves (Section 8.4). Understanding of the mechanisms producing the fluxes, however, remains speculative, owing to the

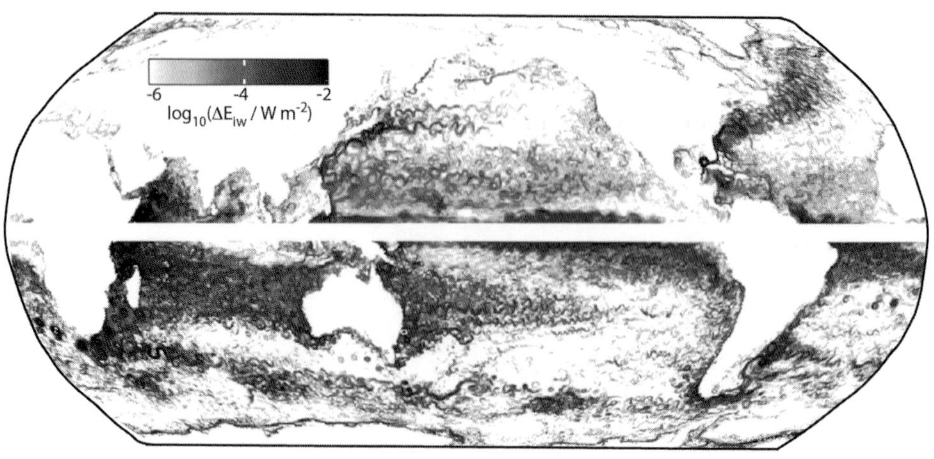

Figure 7.14 Internal wave power gain from balanced geostrophic flows, except within 3° of the equator. (Adapted from Nagai et al., 2015.) (A black and white version of this figure will appear in some formats. For the color version, please refer to the plate section.)

difficulty of observing wave–wave interactions. After considering resonant interactions and caveats about them, ray-tracing, a.k.a. eikonal calculations, is considered as an alternate approach.

7.5.1 Resonant Interactions

Using numerical simulations applying Hasselman (1966) weak interaction theory, McComas and Bretherton (1977) determined that most interactions in a GM internal wave field result from resonant triads characterized by

$$\mathbf{k}_1 + \mathbf{k}_2 = \mathbf{k}_3 \quad \text{and} \quad \omega_1 + \omega_2 = \omega_3. \tag{7.7}$$

Shown schematically in Figure 7.15, three triad interactions dominate:

Elastic scattering (ES): High-frequency wave k_1 scatters off near-inertial wave k_2 propagating in the same direction with twice the wavenumber. The interaction produces high-frequency wave k_3 with the same wavenumber magnitude as k_1 but propagating in the opposite direction. Similar to Bragg scattering, the interaction equalizes upward and downward fluxes using near-inertial waves as passive catalysts.

Induced diffusion (ID): High-frequency, high-wavenumber wave k_3 scatters off near-inertial wave k_2 that has low wavenumber. The interaction transfers energy to k_1 having high frequency and high wavenumber. By diffusing energy across vertical wavenumbers, shear of the near-inertial wave smoothes the spectrum, removing spectral gaps, and produces a net flux to high wavenumber (McComas and Müller, 1981a).

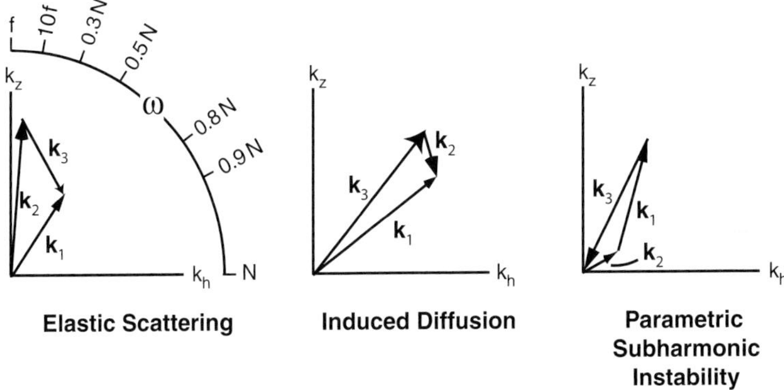

Figure 7.15 The three primary resonant interactions versus vertical, k_z, and horizontal, k_h, wavenumber with inclinations appropriate to their frequency, ω. (Based on McComas and Müller, 1981b)

Parametric subharmonic instability (PSI): Low-wavenumber wave k_2 decays into k_1 and k_3 propagating in opposite directions from each other, both with higher wavenumber and half the frequency of k_1. PSI can transfer energy from internal tides to internal waves at latitudes where the tidal frequency is at least twice the inertial frequency. Simulations show the interaction to be most effective where the Coriolis frequency is half the tidal frequency, e.g. $f = \omega_{M_2}/2$. Table 7.1 gives these critical latitudes for the most important tidal components, and Figure 7.16 presents observational evidence for PSI at the M_2 critical latitude. Enhanced near-inertial shear was also found at K_1 and O_1 critical latitudes in the South China Sea (Alford, 2008) and over rough topography at mid- but not high latitudes (Nagasawa et al., 2002; Carter and Gregg, 2006). By moving energy toward f, PSI is at least partly responsible for the inertial peak (McComas and Müller, 1981a).

7.5.2 Dissipation Rates from Interaction Time Scales

Induced diffusion and parametric subharmonic instability move energy across the equilibrium range separating energy-containing scales, $\beta < \beta_*$, from the high-wavenumber cutoff at β_c (Figure 7.17).[1] (Elastic scattering does not affect the spectrum when the wave field is vertically symmetric.) McComas and Müller (1981a,b) computed time scales for energy transfer of

$$\tau_{\mathrm{id}} = \beta_*^{-2} f^{-1} N^2 E_0^{-1} \quad [\mathrm{s}], \tag{7.8}$$

$$\tau_{\mathrm{psi}} = (32\sqrt{10}/27\pi)\beta_*^{-2} f^{-1} N^2 E_0^{-1} = 1.19\tau_{\mathrm{id}} \quad [\mathrm{s}], \tag{7.9}$$

[1] This discussion retains the notation in the original papers, e.g. β instead of k_z.

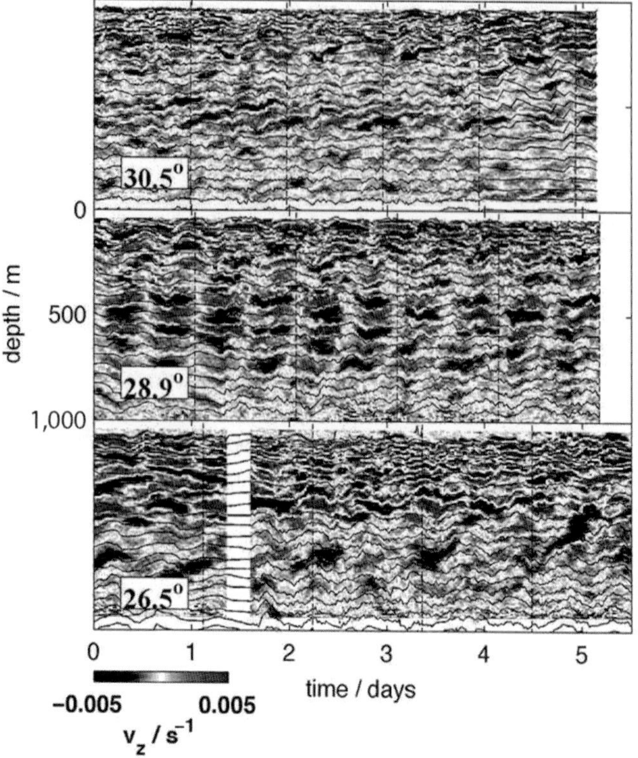

Figure 7.16 Meridional shear with isopycnals at three latitudes along the northward M_2 internal tide beam from the French Frigate Shoals. Strong near-inertial shear between 500 and 750 m at the critical latitude (28.9°) was attributed to PSI because it did not propagate vertically, unlike near-inertial shear at the other latitudes. (Adapted from Alford et al., 2007)

with E_0 as the GM energy level (Table B.1) (McComas and Müller, 1981a). At steady state, the internal wave dissipation rate predicted by McComas and Müller is the energy level divided by the net time scale,

$$\epsilon_{\mathrm{iw}}^{\mathrm{MM}} \equiv \frac{E_{\mathrm{GM}}}{\tau_{\mathrm{id}} + \tau_{\mathrm{psi}}} = \left(\frac{27\pi}{32\sqrt{10}} + 1\right)\pi^2 j_*^2 b^2 N^2 E_0^2 f \quad [\mathrm{W\,kg^{-1}}], \tag{7.10}$$

with $b = 1,300$ m as the GM scale depth of the thermocline. When internal wave intensity is uniform vertically, $K_\rho = 0.2\epsilon/N^2$ is constant with depth. A bulk average, depending only on large-scale parameters of the internal wave field and the stratification, $\epsilon_{\mathrm{iw}}^{\mathrm{MM}}$ is not affected by how internal waves break.

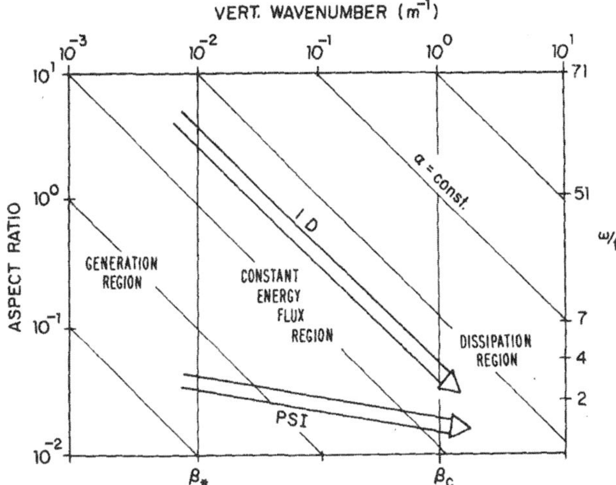

Figure 7.17 Schematic of PSI and ID transporting energy to high wavenumber and low frequency. The left y-axis is the aspect ratio of vertical to horizontal wavenumber. (From McComas and Müller, 1981a. © American Meteorological Society. Used with permission)

7.5.3 Reservations about Resonant Interactions

Holloway (1980) criticized using weak interaction theory to estimate interaction time scales, arguing that internal waves are too energetic by two orders of magnitude to be treated as a collection of weakly interacting waves. Specifically, interaction time scales are less than a wave period over much of the frequency-wavenumber domain. In response to McComas and Müller (1981b), Holloway (1982) clarified his arguments, focusing on contradictions between observations and assumptions. As an example, small waves, $\lambda_z < 70-100$ m, span velocity differences of larger GM background waves that exceed their phase speed. Drastically altering the small waves in less than one period, the encounters are not weak.

Where weak interaction approximations should apply and fail is displayed in Figure 7.18. Developed by Hirst (1991), the region of likely failure on the right reflects Holloway's prediction of problems where interaction times are shorter than wave periods. Failure at low wavenumber results from waves hitting bottom before having time to interact. The resonant interaction assumption may be useful between these two regions, but additional problems at critical layers could occur within the two dashed regions. The first (lower left) results when a near-inertial wave propagates into a region where baroclinic vorticity produces an effective inertial

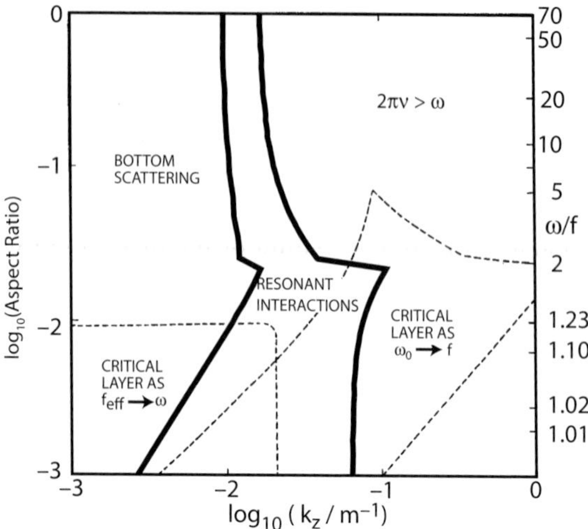

Figure 7.18 The resonant interaction approximation should apply between the solid lines and likely fails outside them. Failure is expected from bottom scattering (left), wave–wave interaction times shorter than one wave period (right), and in some critical layers. (Adapted from Hirst, 1991)

frequency, f_{eff}, greater than the wave's intrinsic frequency, causing velocity to approach zero and k_z to become infinite. Hirst assumes that this condition could be found in strong baroclinic shear, such as mesoscale eddies. The second type of critical layer could occur when spatial variability in a geostrophic current shifts a wave's intrinsic frequency to f or f_{eff}.

7.5.4 Dissipation Rates from Ray-Tracing

To avoid weak interaction theory, Henyey et al. (1986) applied the eikonal technique, a.k.a. ray-tracing, to calculate energy and action fluxes using the Wentzel-Kramers-Brillouin (WKB) approximation to the wave equation. Following Henyey and Pomphrey (1983) and Flatté et al. (1985), they examined transport in space and wavenumber of small test waves refracted by larger waves having a GM spectrum. Considering waves to break when their vertical wavenumber is twice $k_z^c = \beta_c/2\pi$, the high-wavenumber cutoff in cycles per meter gives the dissipation rate as

$$\epsilon_{\text{iw}}^{\text{HWF}} = \frac{1-r}{1+r}\int_{\omega_i}^{N} \Phi_{\text{flux}}(2k_z^c, \omega)\, d\omega \quad [\text{W kg}^{-1}], \tag{7.11}$$

where r is the ratio at $2k_z^c \equiv \beta_c/2\pi$ of fluxes toward lower and higher wavenumbers.

To test their model, Henyey et al. (1986) developed an analytic approximation representing the flux spectrum entering the saturation range as $\Phi_E(k_z, \omega) \mid dk_z(\omega)/dt \mid$. After several approximations,

$$\epsilon_{\rm iw}^{\rm HWF} = (1.67/\pi) j_*^2 b^2 N^2 E_0^2 f \cosh^{-1}(N/f) \quad [\mathrm{W\,kg^{-1}}]. \tag{7.12}$$

One power of N comes from WKB scaling of the dissipating waves and the other from background shear being cut off at a universal Richardson number proportional to N. The f dependence reflects decreasing interaction efficiency and wave aspect ratio as waves approach the equator. Monte Carlo simulations verified both N and $E_{\rm GM}$ dependencies.

Although the theoretical approaches of McComas and Müller (1981b) and of Henyey et al. (1986) differ greatly, their models have the same functional form, except for $\cosh^{-1}(N/f)$, which is significant only at low latitude. Numerical coefficients, however, differ considerably: 8.2 for $\epsilon_{\rm iw}^{\rm MM}$ and 0.53 for $\epsilon_{\rm iw}^{\rm HWF}$. Part of this is due to Henyey et al. (1986) using GM79 versus McComas and Müller (1981a) using GM76.

7.5.5 Direct Simulations

To test the dependence of ϵ on energy level and stratification, Winters and D'Asaro (1997) ran nonhydrostatic simulations using the first 40 vertical modes in a domain 2 km deep and 80 by 10 km horizontally. Energy levels were $E/E_0 = 1$, 1/2, 2, and 1/100 at the six observational sites of Gregg (1989a). A hyperviscosity operator damped vertical modes greater than 40 and horizontal scales less than 1 km. After an inertial period, the dissipation rate, $\epsilon = dE/dt$, was insensitive to the damping. Dissipation rates averaged over three and five inertial periods after initial transients produced the comparisons versus E and N in Figure 7.19; least squares fits found $\epsilon \propto N^{1.9}$ for simulations I–III. Dissipation for simulation IV, with $E = E_0/100$, was much higher than expected from the trend of the other cases, presumably due to numerical artifacts.

7.5.6 Perspectives

In view of the different approaches of McComas and Müller (1981b) and of Henyey et al. (1986) to internal wave dissipation, it is curious that both obtain the same functional dependence, $\epsilon \propto E_0^2 N^2 f$, save for the additional $\cosh^{-1}(N/f)$ in the latter. Numerical simulations are consistent with $\epsilon_{\rm iw} \propto E_0^2 N^2$, and, as seen in the next section, so are observations. Nonetheless, criticisms of resonant interaction theory, upon which (7.10) is based, seem valid. One possibility is that resonant interaction theory captures the wave–wave energy flux over the limited wavenumber range

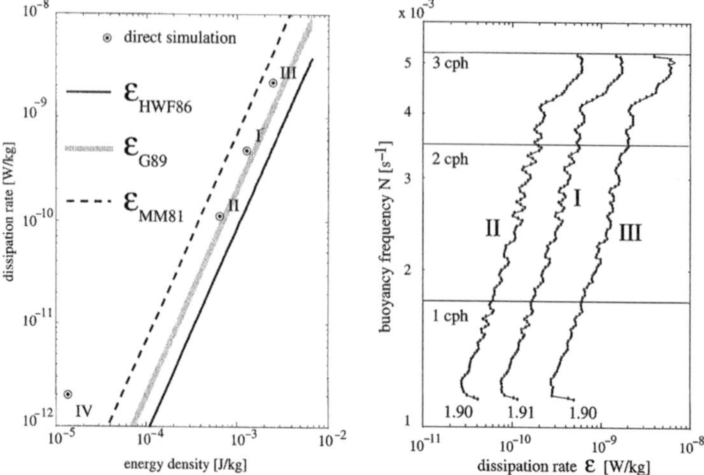

Figure 7.19 (Left) Dissipation rates versus internal wave energy density. Observations by Gregg (1989b), ϵ_{G89}, are twice ϵ_{iw}^{HWF} because the model used GM79 versus GM76 used by Gregg. (Right) N versus ϵ for simulations with energy levels 1 (I), 1/2 (II), and 2 (III) times GM. Simulation IV (left) had $E = E_{GM}/100$ and apparently was strongly affected by numerical noise owing to the low energy level. (From Winters and D'Asaro, 1997. © American Meteorological Society. Used with permission)

shown by Hirst (1991), and these are the dominant fluxes. In any event, the subject is not closed.

7.6 Testing Dissipation Scaling Based on Shear and Strain

Agreement within a factor of 2 between diapycnal diffusivities from ϵ and χ_T with those from tracers (Section 3.12.1). is the principal justification for K_ρ from Osborn (1980) and K_T from Osborn and Cox (1972). Here, the focus is on predicting ϵ from finescale shear and strain. Model predictions by McComas and Müller (1981a,b) and Henyey et al. (1986) are in terms of energy, E, but half the velocity variance is at vertical scales longer than 500 m (Figure 2.2). To avoid being limited to a small fraction of velocity variance across large changes in stratification, using ~100 m sections obtains 90% of shear and strain variance, often with nearly uniform N.

7.6.1 Scaling Dissipation with Shear

Comparisons began with Multi-Scale Profiler (MSP) observations in the northeastern Pacific during the Patches Experiment (PATCHEX), where internal wave fields were very close to GM and $\epsilon_{iw} \propto N^2$ (Gregg and Sanford, 1988). Estimating internal wave decay time scales as

$$\tau_{iw-decay} = \frac{E_0}{<\epsilon>} \frac{<Shear_{10}^2>}{Shear_{GM}^2} \quad [s] \tag{7.13}$$

gave 40 days in the seasonal pycnocline and 100 days in the main pycnocline, consistent with the view that low dissipation rates allow many distant sources to maintain a global field with modest temporal and spatial variability. Other sites were compared with PATCHEX by adapting (7.12) to

$$\epsilon_{iw} = 7 \times 10^{-10} \langle N^2/N_0^2 \rangle \langle Shear_{10}^4 / Shear_{GM}^4 \rangle \quad [\mathrm{W\,kg^{-1}}] \tag{7.14}$$

and ignoring small variations in latitude (Gregg, 1989a). Stratification differences were also minor, but $Shear_{10}^4$ varied by two decades (Figure 7.20). On average, (7.14) reduced ϵ variability to about twofold. Scaled dissipation rates were twice those predicted by Henyey et al. (1986), owing to a discrepancy in shear variance between GM76 and Munk (1981). Consequently, the scaling constant, 7×10^{-10}, is empirical. Twofold accuracy was also found by Lee et al. (2006) over a wide range of dissipation rates along the Hawaiian Ridge (Figure 7.21). The largest discrepancies occurred at the ends of the dissipation distribution, where the number of observations was small and the averages suspect.

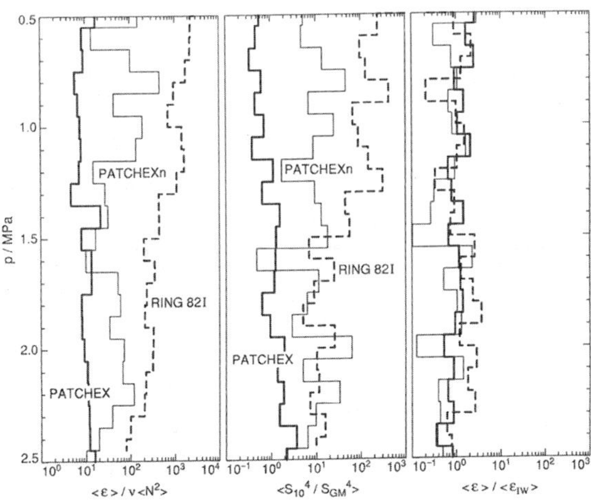

Figure 7.20 (Left) Dissipation rates normalized by the buoyancy Reynolds number. (Middle) 10 m $Shear_{10}^4$ normalized by the equivalent GM shear. (Right) ϵ normalized by (7.14). Shear variances were multiplied by 2.11 to correct for the first-difference filter. PATCHEX data are from MSP, RING data from XCPs. (From Gregg, 1989a)

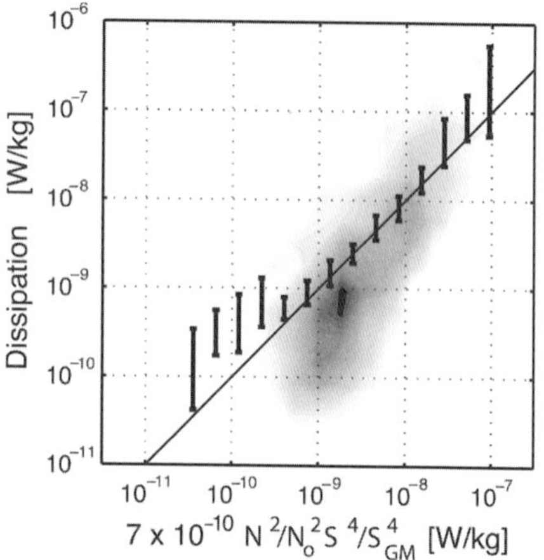

Figure 7.21 Survey-averaged binned dissipation rates from Absolute Velocity Profiler (AVP) along the Hawaiian Ridge, as bars between 95% confidence limits, versus prediction using (7.14). Shading shows the probability density of 5 m dissipation rates. (Adapted from Lee et al., 2006. © American Meteorological Society. Used with permission)

7.6.2 Scaling Dissipation with Shear and Strain

Subsequent comparisons with spectra departing from GM find better agreement: 1) if the integration for shear variance is cut off where the accumulated variance matches GM, i.e.

$$\int_{k_0}^{k_z^c} \Phi_{\text{shear}}(k_z)\, dk_z = \int_{k_0}^{0.1} \Phi_{\text{shear}}^{\text{GM76}}(k_z)\, dk_z = 0.66 N^2 \quad [\text{s}^{-2}] \tag{7.15}$$

(Gargett, 1990); k_0 is chosen so there is little variance at smaller wavenumbers, 2) if strain variances over the same interval are used to estimate average wave field frequency (Polzin et al., 1995), and 3) if the latitude dependence is included near the equator (Gregg et al., 2003). Then, the scaling using shear and strain variances is

$$\epsilon_{\text{iw}}^{\text{shear}} = \epsilon_0 \underbrace{\left(\frac{N}{N_0}\right)^2}_{\text{I}} \underbrace{\left(\frac{0.1}{k_z^c}\right)^2}_{\text{II}} \underbrace{\frac{3}{2\sqrt{2}} \frac{(R_\omega + 1)}{R_\omega \sqrt{R_\omega - 1}}}_{\text{III}} \underbrace{\frac{f \cosh^{-1}(N/f)}{f_{30^\circ} \cosh^{-1}(N_0/f_{30^\circ})}}_{\text{IV}} \tag{7.16}$$

Term II represents the principal variation of dissipation with internal wave intensity. It is the square of the ratio of observed and GM energy densities, or of shear variances, and is estimated by integrating spectra to reduce statistical

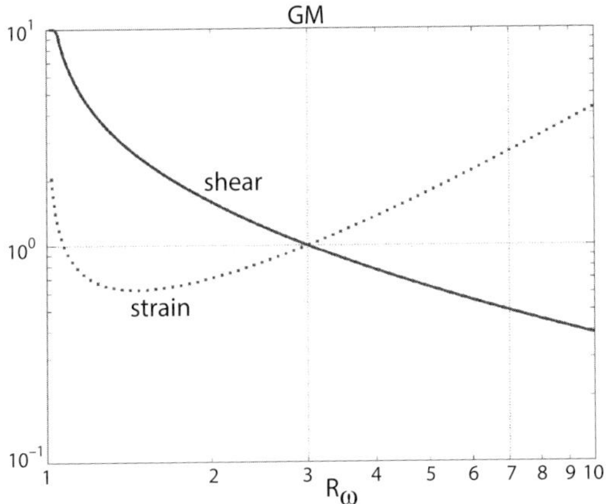

Figure 7.22 Term III in shear scaling (7.16) and in strain scaling (7.22). When $R_\omega = 3$, the shear-to-strain value for GM76, term III is unity in both expressions. $R_\omega = 7$ in the data of Kunze et al. (2006b). (Adapted from Kunze et al., 2006a. © American Meteorological Society. Used with permission)

uncertainty. Shear variance is preferred, owing to its concentration at high wavenumber. A spectrum elevated above GM cutoffs at a lower k_z^c (Figure 7.23) such that its integral returns the same normalized shear variance as GM integrated to 0.1 cpm (Gargett et al., 1981; Munk, 1981). Consequently, the ratio of shear variances in the internal wave band is $0.1/k_z^c$. Term III (Figure 7.22) corrects for variations with average wave frequency, estimated from R_ω (B.43), in the ratio of total energy to horizontal kinetic energy. Term IV arises from the rate of Doppler shifting being proportional to the ratio of vertical and horizontal length scales of shear and the ratio varying with latitude. Collectively, (7.16) produces tight fits to ϵ scaled as diapycnal diffusivity (Figure 7.24). Outliers were near rough bottoms, where the wave field may have been altered too much for GM to apply.

The product of terms I to III predicts the dissipation rate at 30° latitude. Comparing $\epsilon/\epsilon_{30^\circ}$ demonstrates that term IV accurately predicts dissipation variations with latitude (Figure 7.25); observations and term IV both drop steeply to 0.01 on the equator using low-latitude data from the central (Tropic Heat 2) and western (COARE 3) equatorial Pacific. Normalized shear variance, term II, varied from 1,000 to 0.4 and was largest on the equator. The simplest interpretation is that steady state at low latitudes requires higher shear variances than at mid-latitudes owing to the reduced efficiency of wave–wave interactions in dissipating energy. Consequently, the reduced efficiency keeps K_ρ low on and near the equator, but the reduction is partially compensated by the buildup of internal wave energy coming from higher latitudes (Figure 7.26).

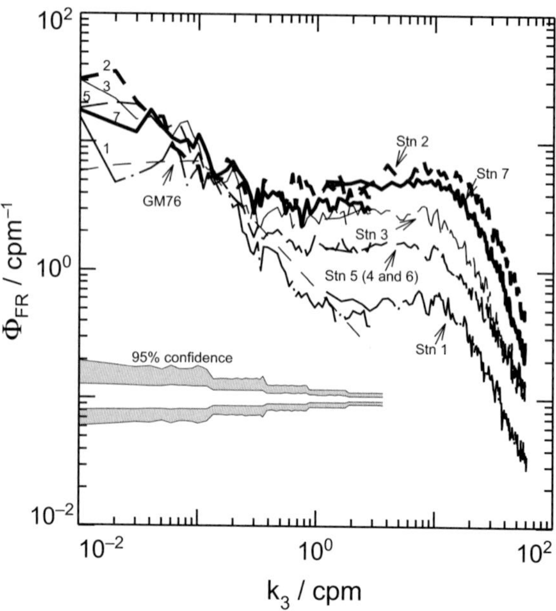

Figure 7.23 Average Froude spectra, $\Phi_{Fr} \equiv \Phi_{vshear}/N^2$, across the Florida Strait exhibit three spectral ranges whose boundaries vary with activity level. Constant amplitudes and k_z^{-1} slopes characterize the saturated range separating internal waves and turbulence. The turbulence range responds proportionately but nonlinearly to levels in the internal wave range, e.g. compare Stns. 1 and 2. Station positions are in Figure 7.28. (Adapted from Gregg et al., 1996)

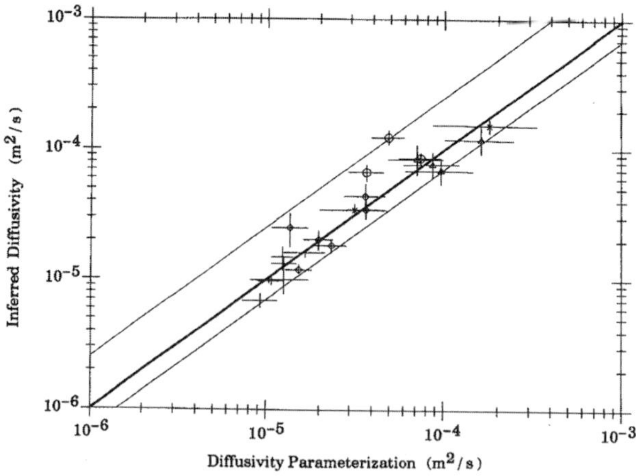

Figure 7.24 Bin averages of K_ρ using (7.16) for ϵ in four sets of HRP profiles. High outliers were near rough bottoms. Crosses are 95% confidence levels. (From Polzin et al., 1995. © American Meteorological Society. Used with permission)

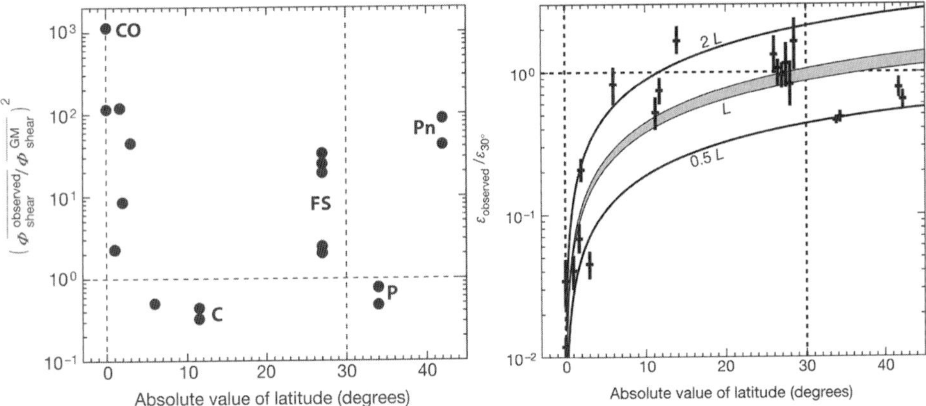

Figure 7.25 (Left) Latitudinal dependence of shear variance squared, and (right) term IV compared with $\epsilon/\epsilon_{30^\circ}$ computed using terms I to III. MSP data are Pn (PATCHEX north), P (PATCHEX), FS (Florida Strait, Stns. 5 and 6), C (CSALT) outside the thermohaline staircase, and CO (COARE 3). Unmarked symbols are from Tropic Heat 2. L is term IV, with shading for the range of N^2. Dots are averages of 3 to 69 estimates, each 125 m thick. Equatorial data were not in the undercurrent or deep jets. (Adapted from Gregg et al., 2003)

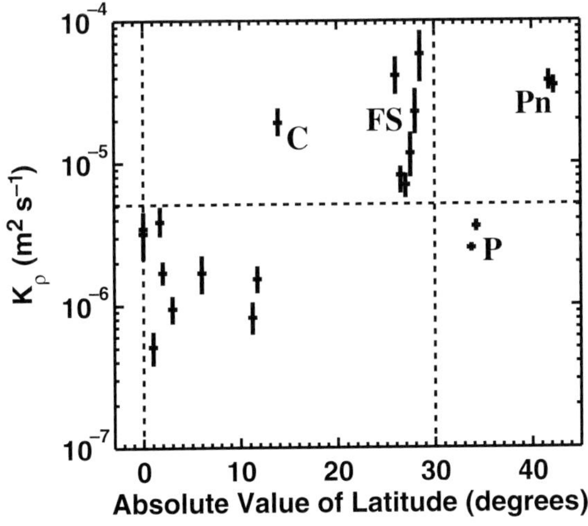

Figure 7.26 Diapycnal diffusivities for the averages in Figure 7.25. Elevated shear on and close to the equator compensates for some of the decrease in mixing represented by term IV. The horizontal dashed line is K_ρ at 30° latitude when internal waves are at GM. (Adapted from Gregg et al., 2003)

7.6.3 Discrepancies

Shear and strain scaling (7.16) predict dissipation rates within a factor of 2 when internal wave spectra are 'GM-like', with $R_\omega \sim 3$ and flat shear spectra, implying a broadband field of many uncorrelated waves. In addition, GM assumes negligible mean shear and stratification varying gradually with depth. A major goal of subsequent work has been to find and understand discrepancies. Many have been found, mostly near the seafloor, in the open ocean, as well as over continental shelves. Even though these regimes differ greatly from those assumed for GM, the dissipation scalings are the only references available, and the goal is to use the discrepancies to develop a better model of the internal wave field and its interactions. This is a high priority, owing to the focus on near-bottom mixing as a major component of meridional overturning circulation (Section 1.3).

Underestimates in the Florida Strait

A wide channel with strong geostrophic shear, the Florida Strait has dissipation rates 2–8 times predictions beneath the core of the Florida Current and also near the sidewalls (Figure 7.27, left). Diffusivity was less than 10^{-5} m^2 s^{-1} in the current core and not much stronger throughout much of the interior (Figure 7.28), where the scaling was within a factor of 2 of observations. The strongest dissipation, 10^{-7} W kg^{-1} in a stratified layer just above the homogenous bottom boundary layer, was eight times the scaling. Half of the velocity contrast across the Florida Current occurred across this layer, producing $K_\rho > 10^{-4}$ m^2 s^{-1}. Similar rates and underpredictions occurred over the steeply sloping sidewalls. At these locations,

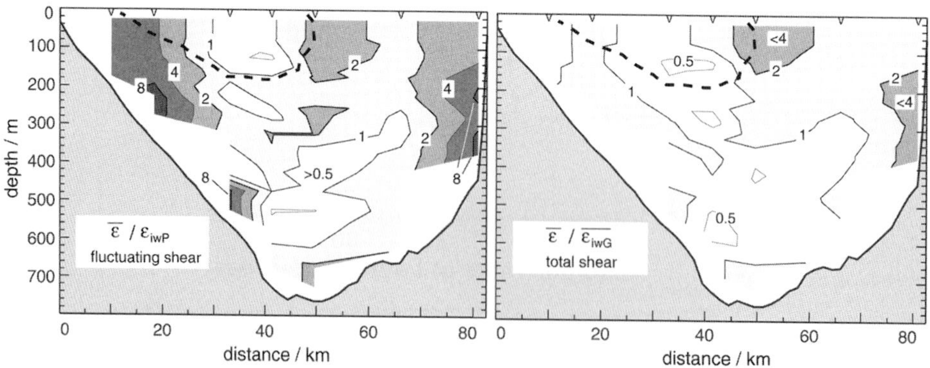

Figure 7.27 Scaled dissipation rates across the Florida Strait. (Left) Using (7.16) underestimates ϵ in regions of high mean shear, mostly near the boundaries. (Right) Using (7.14) with the total shear variance, i.e. fluctuating variance plus the square of the mean shear, gives smaller errors. (Adapted from Winkel et al., 2002. © American Meteorological Society. Used with permission)

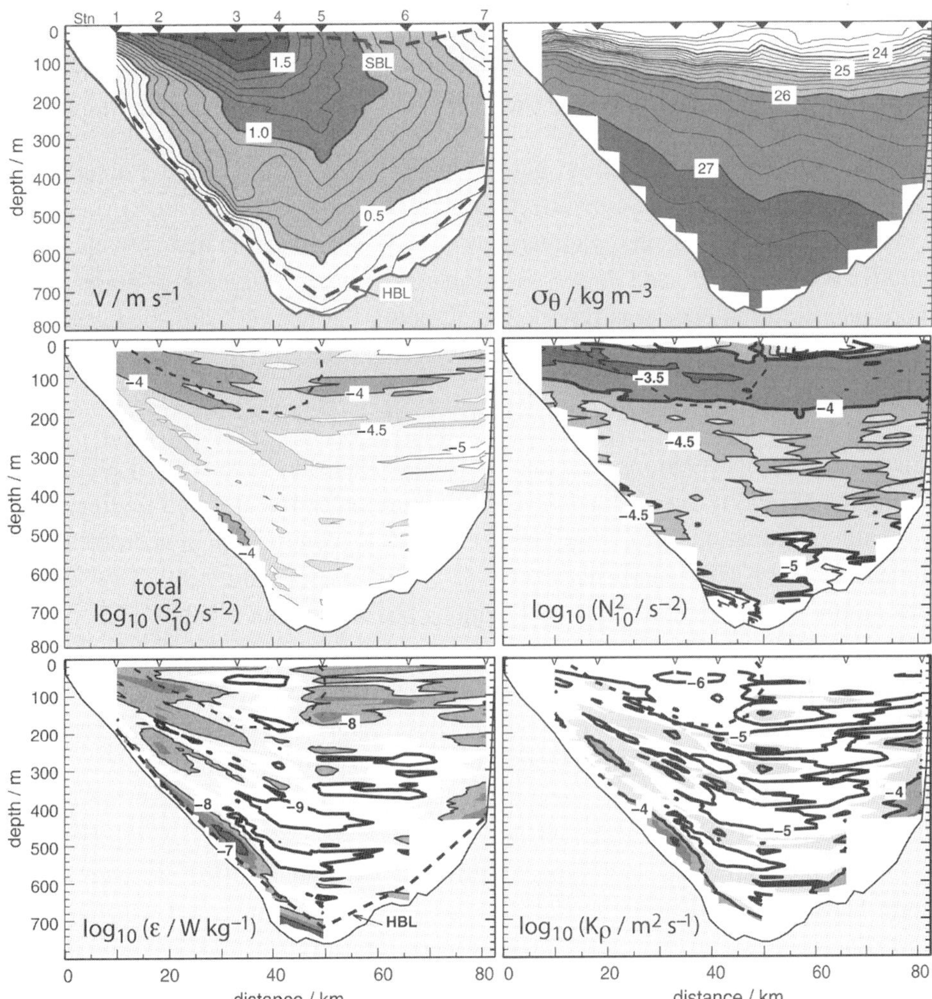

Figure 7.28 Averages at seven stations along 27°N across the Florida Strait, each sampled 10–14 times. Dashed lines mark homogenous boundary layers and the core, $V \leq 1.2$ m s^{-1}, of the Florida Current. (Adapted from Winkel et al., 2002. © American Meteorological Society. Used with permission)

internal waves were strongly polarized and dominantly across channel. Reasoning that mean shear adds to the Doppler shift moving small-scale waves toward dissipation, an ad hoc fix replaced fluctuating shear variance with total shear squared, a change that removed most of the discrepancies (Figure 7.27, right).

Underestimates in Monterey Submarine Canyon

Upper Monterey Submarine Canyon contains even larger discrepancies than the Florida Strait. Using (7.16), Kunze et al. (2002) found ϵ underestimated up to a factor of 30 where internal waves were an order of magnitude above GM. In some places, the shear-to-strain ratio was close to $R_\omega = 2.13$, the value for semi-diurnal internal tides, while deeper stations had $R_\omega < 1$ near the bottom. Tidal fluxes into the canyon and the production of lee waves by bottom currents were considered as sources of additional dissipation, but the analysis was inconclusive. Polzin et al. (2014) argued that the parameterization should not be expected to work where wave generation inserts energy at wavenumbers greater than the high-wavenumber cutoff, i.e. at $k_z > k_z^c$, and suggested Monterey Canyon as a case in point. Even more intense turbulence was reported in Kaoping Submarine Canyon (Lee et al., 2009), though dissipation scaling was not tested.

Overestimates in the Southern Ocean

Microstructure observations over the Kerguelen Plateau (Waterman et al., 2013) during the Southern Ocean Finestructure (SOFine) experiment found scaling overestimating dissipation rates. Waterman et al. (2014) identified eight stations where ϵ measured with HRP markedly exceeded predictions (Figure 7.29). The scaling was mildly exceeded at several other stations but was within a factor of 2 at most locations. Common anomaly characteristics included: rough bottoms, narrowband spectra centered at 0.01 cpm, low shear-to-strain ratios (R_ω), upward propagation, and strong mean shear with $Fr_N \geq 0.2$. These point to interactions between the mean flow and internal waves. Consequently, overprediction may have resulted from waves produced by these interactions propagating away to dissipate at remote sites or being absorbed at critical layers.

During the Diapycnal and Isopycnal Mixin Experiment (DIMES), overpredictions were also found in the Drake Passage where flows interacted with the bottom (Sheen et al., 2013). Overestimates were typically twofold, but some were eight times observations. Estimates within 2 km of the bottom were considered unreliable due to LADCP noise and weak velocity signals in low stratification. Owing to observations of waves propagating up from the bottom, wave–wave and wave–mean flow interactions were possible reasons for discrepancies at shallower depths.

Rather similar to the situation in the Drake Passage, a line of microstructure and LADCP observations south of Australia from 40°S to 60°S found over-predictions up to a factor of 3 in the upper water column. North of the Polar Front, measurements were close to background, with $\epsilon \sim 10^{-10}$ W kg^{-1} and $K_\rho \sim 10^{-5}$ m^2 s^{-1} (Takahashi and Hibiya, 2019). The discrepancies were in energetic regions farther south, exemplified by the spectrum in Figure 7.30. The overestimates were attributed to spectral anisotropy of downward near-inertial waves and upward lee waves. The spectrum also highlights the difficulty in estimating shear variance when shapes differ greatly from GM. The scaling assumes that variance at the

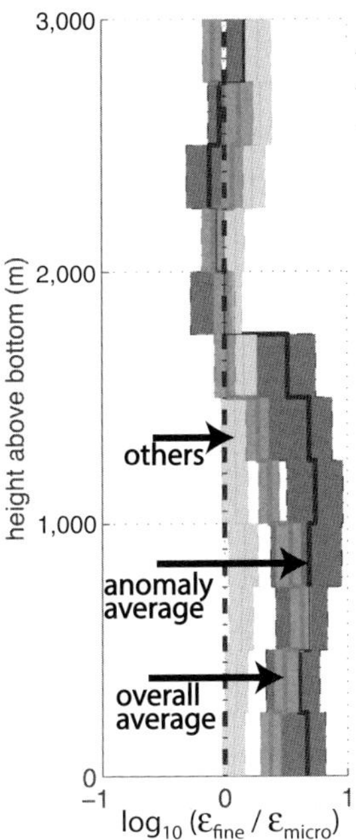

Figure 7.29 Averages of the ratios of dissipation rates predicted from finestructure, ϵ_{fine}, to observations, ϵ_{micro}, on the north side of the Kerguelen Plateau during SOFine. Averages are lines (dark, medium gray, light gray), and shading gives 90% confidence limits. The anomaly average was formed from eight profiles with observed dissipation rates significantly less than those predicted using (7.22). (Adapted from Waterman et al., 2014. © American Meteorological Society. Used with permission)

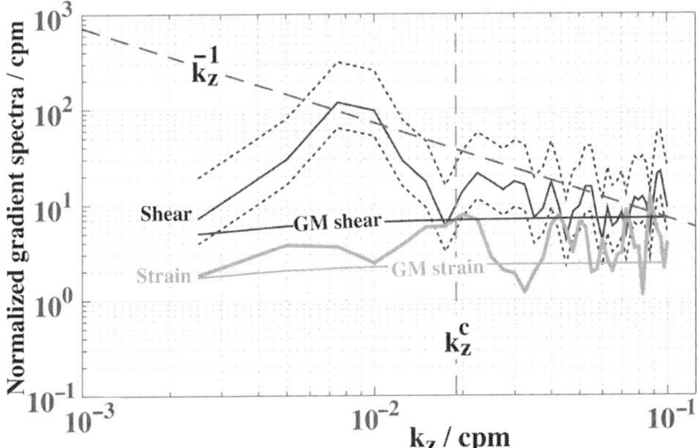

Figure 7.30 Normalized spectra from 1,100 to 2,100 m in a jet in the subAntarctic zone (45 °S, 110°W). (Adapted from Takahashi and Hibiya, 2019. © American Meteorological Society. Used with permission)

cutoff wavenumber, k_z^c, drives the flux toward higher wavenumbers and ultimate dissipation. Assuming flat spectra, $\propto k_z^0$, statistical confidence is increased by integrating observed and GM spectra from the record length to k_z^c, rather than simply taking spectral amplitudes at k_z^c. This is problematic for irregularly shaped spectra, and using an ADCP with limited vertical resolution can amplify errors.

Overestimates in Fields Dominated by Near-Inertial Waves

Analyzing observations near the Aleutians and the Izu-Ogasawara Ridge (Figure 7.31), Ijichi and Hibiya (2015) also observed dissipation rates lower than predicted. Noting discrepancies increasing as R_ω exceeded the GM shear-to-strain ratio of 3, Ijichi and Hibiya attributed the problem to (7.16) characterizing a single wave in a broadband field, in contrast to their observations of near-inertial waves dominating the field. To reduce discrepancies, Ijichi and Hibiya replaced the

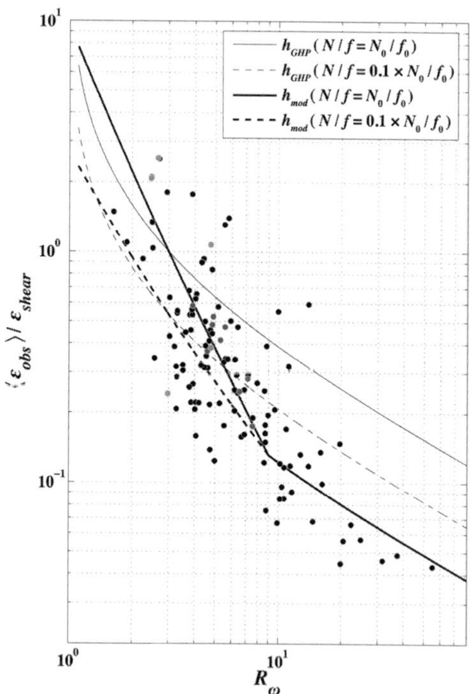

Figure 7.31 Observed dissipation rates (dots) normalized with $\epsilon_0 = 2.24 \times 10^{-10}$ W kg^{-1} and terms I and II of (7.16) versus the shear-to-strain ratio, R_ω. Thin lines show scaling with terms III and IV, the frequency part of (7.16) for two values of N/f. Thick lines are the revised frequency dependence in (7.17) proposed by Ijichi and Hibiya (2015). The revision is a better fit for $R_\omega \gtrsim 3$, particularly for near-inertial frequencies, when $R_\omega \gtrsim 10$. (Adapted from Ijichi and Hibiya, 2015. © American Meteorological Society. Used with permission)

frequency-dependent terms (III and IV) of (7.16) with a function for narrowband and broadband wave fields,

$$h_{IH}(R_\omega, N/f) = \frac{1 + 1/R_\omega}{4/3} \begin{cases} (L_1/L_0)R_\omega^{-L_2} & R_\omega < 9 \\ (1/L_0)\sqrt{2/(R_\omega - 1)} & R_\omega > 9 \end{cases} \tag{7.17}$$

where the coefficients include μ_{GM} for Doppler shifting in a GM field and

$$L_0 \equiv 2\pi^{-1}\cosh(N_0/f_0),\ \mu_{GM} \equiv 2\pi^{-1}\cosh(N/f) \tag{7.18}$$

$$L_1 \equiv 2\mu_{GM}^2,\ L_2 \equiv \log_{10}(2\mu_{GM}),\ \epsilon_0 = 2.24 \times 10^{-10}\ \mathrm{W\,kg^{-1}}. \tag{7.19}$$

Comparison with the observations shows the revision to be comparable to the original scaling at low R_ω but much closer to observations at high R_ω (Figure 7.31). Individual profiles show significant improvement in predictions at some depths, presumably where R_ω was large (Figure 7.32). This revision, however, does not resolve discrepancies with all near-inertial fields. Both (7.16) and (7.17) overestimate dissipation by a factor of 3 for the spectrum in Figure 7.30. This spectrum is from a region in the Southern Ocean surrounded by cyclonic and anticyclonic eddies. Corresponding vertical rotary spectra showed much larger anticlockwise

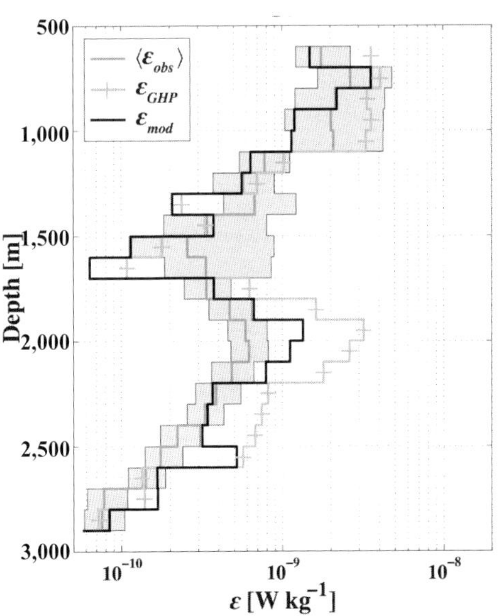

Figure 7.32 Observed dissipation rates near the Izu-Ogasawara Ridge in the North Pacific, compared with predictions with (7.16), labeled ϵ_{GHP}, and with (7.17), labeled ϵ_{mod}. The revision significantly improves the prediction in the lower half of the profile, but less so in the upper half. (Adapted from Ijichi and Hibiya, 2015. © American Meteorological Society. Used with permission)

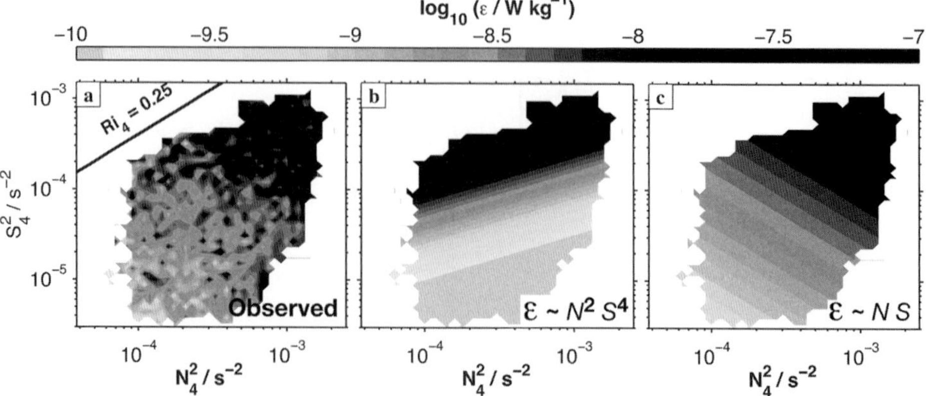

Figure 7.33 Observed (a) and scaled (b,c) dissipation rates in 70 m of water over the New England shelf. Stratification and shear were over 4 m intervals. Using them in (7.16) produces an ϵ pattern nearly orthogonal to observations, in contrast to applying (7.20), which was constructed to agree with the data. (Adapted from MacKinnon and Gregg, 2003. © American Meteorological Society. Used with permission.) (A black and white version of this figure will appear in some formats. For the color version, please refer to the plate section.)

than clockwise energy, indicating downward propagation from shallow generation. $R_\omega = 27$ demonstrates that the energy is mostly near-inertial. The authors attributed the discrepancy to spatial anisotropy similar to that found previously in the Drake Passage and near the Kerguelen Plateau.

Over Continental Shelves with Solibores

Over smooth continental shelves, ϵ follows a different dependence on shear and stratification than in deep water (Figure 7.33), even when data are selected to avoid internal solitary waves. Making plausible arguments and assuming that small waves leading to breaking are affected by large-scale, low-frequency waves, primarily near-inertial, MacKinnon and Gregg (2003) described the shallow water dependence as

$$\epsilon_{\mathrm{iw}}^{\mathrm{shelf}} = 6.9 \times 10^{-10}(N/N_0)(S_{\mathrm{lowfreq}}/S_0) \quad [\mathrm{W\,kg^{-1}}], \tag{7.20}$$

where $S_0 = N_0$ for convenience. The scaling accurately represents observations on the New England (Figure 7.34) and Monterey Bay shelves (Carter et al., 2005), the latter after adjustment of the scaling constant. Curiously, open-ocean scaling (7.16) worked well for dissipation in solibores crossing the shelf, which accounted for half of the net dissipation on the New England shelf during late summer. Dissipation in solibores is also examined in (Section 7.7.4).

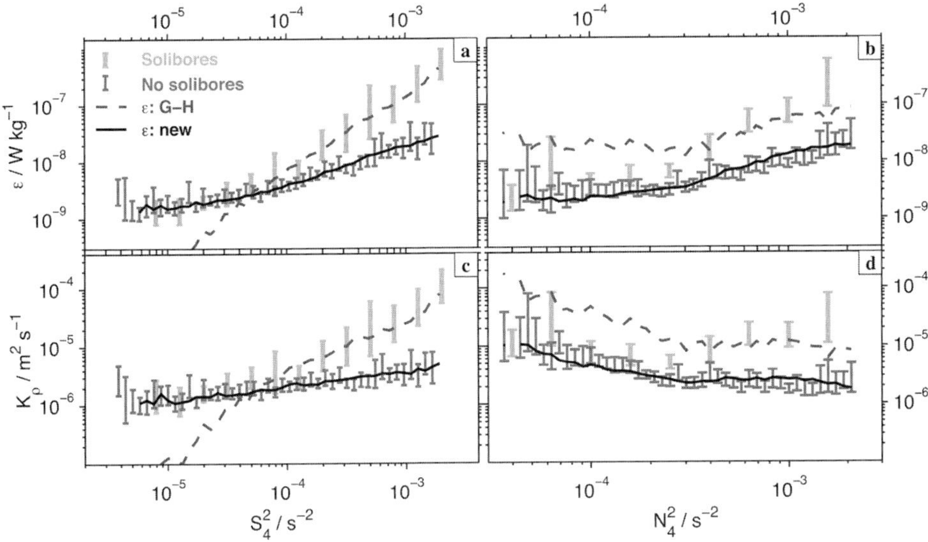

Figure 7.34 Average dissipation rates and diapycnal diffusivities on the New England shelf, with and without solibores. G-H scaling refers to (7.16), and new to (7.20). Data bars span 95% bootstrap confidence intervals. (From MacKinnon and Gregg, 2003. © American Meteorological Society. Used with permission.)

7.6.4 Scaling Dissipation with Strain

To estimate ϵ using density profiles from a drifting ice camp, Wijesekera et al. (1993) computed $Strain_{10} \equiv (\zeta_i - \zeta_{i+1})/10$, where ζ_i are vertical displacements of isopycnals 10 m apart on average. Strain in lieu of shear in (7.14) yields

$$\epsilon = 7 \times 10^{-10} \langle N^2/N_0^2 \rangle \langle Strain_{10}^4 / Strain_{\rm GM}^4 \rangle \quad [\text{W kg}^{-1}], \tag{7.21}$$

which predicted average dissipation rates within 1%. Subsequently, Kunze et al. (2006a) adapted (7.16) to its equivalent for strain,

$$\epsilon_{\rm iw}^{\rm strain} = \epsilon_0 \underbrace{\left(\frac{N}{N_0}\right)^2}_{\rm I} \underbrace{\left(\frac{0.1}{k_z^c}\right)^2}_{\rm II} \underbrace{\frac{1}{6\sqrt{2}} \frac{R_\omega(R_\omega + 1)}{\sqrt{R_\omega - 1}}}_{\rm III} \underbrace{\frac{f \cosh^{-1}(N/f)}{f_{30^\circ} \cosh^{-1}(N_0/f_{30^\circ})}}_{\rm IV}. \tag{7.22}$$

Except for different coefficients in III, the expression is the same as for shear-strain scaling, and the cutoff wavenumber, k_z^c, is computed in a similar way,

$$Strain_{\rm GM}^2 \equiv \int_{k_0}^{0.1} \Phi_{\rm GM-strain}(k_z)\, dk_z = 0.5 = \int_{k_0}^{k_z^c} \Phi_{\rm strain}(k_z)\, dk_z. \tag{7.23}$$

Strain is useful if salinity spiking does not corrupt density or if temperature can be used in lieu of density for profiles free of salt-stabilized temperature inversions.

Strain-based estimates applied to the large archive of Argo profiles provided the first global maps of ϵ (Section 8.4).

7.6.5 Perspectives

Analytical, numerical, and observational evidence suggest that the various forms for scaling ϵ capture essential elements of wave–wave interactions, making them useful for first-order estimates of global mixing rates in the large number of places where internal waves are close to GM. These, however, are not where mixing is most important, and much remains to be learned about the mixing of waves interacting with strong, low-frequency shear and of waves generated by flows over topography.

7.7 Nonlinearity

Photographs by Woods (1968) of dyed interfaces in shallow water (Figure 8.5) showed internal waves on thin interfaces rolling up and breaking in sequences similar to KH instabilities in tanks (Figure 7.35) and in numerical models. Subsequently, KH instabilities have been assumed to be the most common mechanism of shear instability in the pycnocline.

The gradient Richardson number, Ri_g (3.6), indicates the likelihood of waves breaking in stratified shear flows, and its reciprocal function, the stratification Froude number, Fr_N, measures nonlinearity as the ratio of inertial and buoyancy forces. Steady parallel shear flows are stable to small disturbances when $Ri_g > 1/4 \equiv Ri_g^{\text{critical}}$ (Miles, 1961; Howard, 1961), or equivalently when $Fr_N > 4$. Although necessary for shear instability, these criteria are not sufficient (Thorpe, 2005, Chapter 3). Nor are most oceanic shear flows steady and parallel. Consequently, in complex oceanic shear fields, gradient Richardson numbers indicate where shear instability is likely, rather than providing exact criteria.

In spite of Ri_g^{crit} not being clearly defined for instabilities in the pycnocline, observations show relatively few gradient Richardson numbers smaller than 1/4

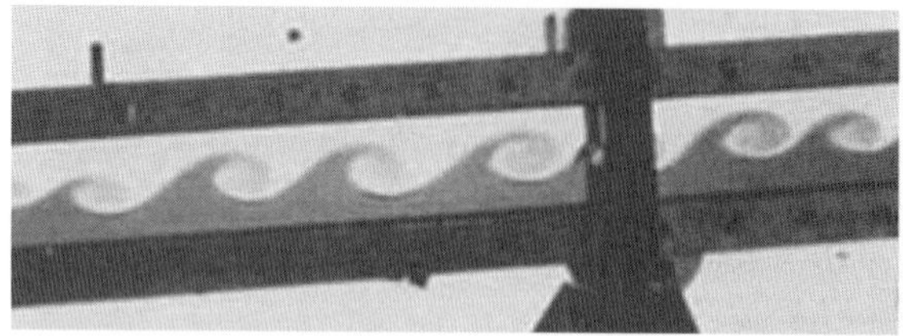

Figure 7.35 KH billows across a thin interface between clear and dyed layers in a tilt tank. (From Thorpe, 1971)

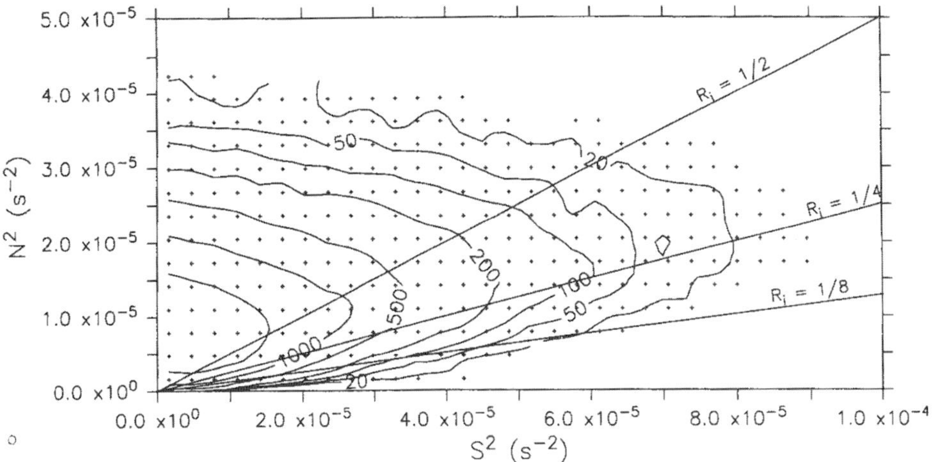

Figure 7.36 Probability density of Ri_g over $\Delta z = 4$ m as functions of stratification, N^2, and shear, S^2, between 400 and 800 m in the eastern North Atlantic during the North Atlantic Tracer Release Experiment. The probability density falls sharply as Ri_g decreases below 1/4. Internal waves were near GM. (From Polzin, 1996. © American Meteorological Society. Used with permission)

(Eriksen, 1982). The example in Figure 7.36 shows the probability density of Ri over 4 m scales in a pycnocline where most overturns are considerably smaller. The cutoff as Ri_g decreases below 1/4 is consistent with the conclusion of Hazel (1972) that 'Miles' necessary condition for instability is quite a good ad hoc sufficient criterion in the field'.

Finestructure produces minima in the gradient Richardson number at interfaces, and the nature of the instability varies with interface thickness, δ, relative to wavelength, k (Fringer and Street, 2003). Instabilities develop when $k\delta < 0.56$, but they lack the energy to form the two-dimensional convective instabilities necessary for strong turbulence. Instabilities with $0.56 < k\delta < 2.33$ and $Ri_0 < 0.13$ generate vigorous secondary convective instabilities and turbulence, mixing most effectively when $k\delta = 1$. Internal waves with $k\delta > 2.33$ are too thick to form KH billows and instead produce convective, or Rayleigh-Taylor, billows. Smyth and Winters (2003) found Holmboe waves developing from KH instabilities grown large. Holmboe waves become the primary instability on interfaces having velocity length scales more than 2.4 times density length scales.

7.7.1 Kelvin–Helmholtz Instabilities

KH instabilities develop when Ri_g is significantly below $1/4$. Using Hazel (1972), Kunze et al. (1990) expressed σ_{KH}, the maximum growth rate, as

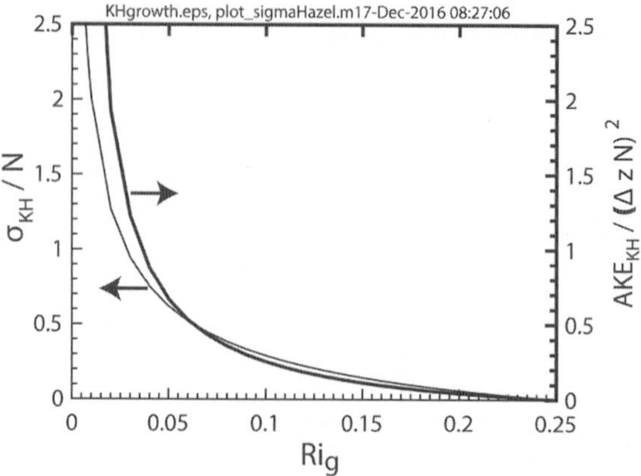

Figure 7.37 (Left axis) KH growth rate, σ, normalized by N. (Right axis) Available kinetic energy in a KH instability, normalized by $(\Delta z N)^2$. (Plotted from Kunze et al., 1990)

$$\frac{\sigma_{\mathrm{KH}}}{N} = \frac{\tau_{buoy}}{\tau_{\mathrm{KH}}} = \frac{1/\sqrt{Ri_g} - 2}{4}. \tag{7.24}$$

Only when $Ri_g \lesssim 0.03$ do KH instabilities grow large in less than a stability period (Figure 7.37). When the growth rate is large, small perturbations generate spanwise vortices, or billows. Horizontal wavelengths are typically 6–11 times interface thickness (Smyth and Moum, 2012). Initial growth is two-dimensional, confining kinetic energy to vertical and streamwise components. Using $\Delta u = N\Delta z/\sqrt{Ri_g}$, an overturn initiated with gradient Richardson number Ri_g will have an initial Reynolds number

$$Re_0 = \frac{\Delta u\, \Delta z}{\nu} = \frac{N(\Delta z)^2}{\nu\sqrt{Ri_g}}. \tag{7.25}$$

Initial Reynolds numbers are of $\mathcal{O}(10^3)$ where stratification exceeds the GM reference, $N \gtrsim N_0$, and overturns are tens of centimeters high (Figure 7.38). The small number of meter-scale overturns that produce much of the net dissipation have Re_0 of $\mathcal{O}(10^5)$. Abyssal overturns from tens of meters to more than 100 m, e.g. Nash et al. (2007), have Reynolds numbers of $10^6 - 10^8$ (Figure 7.38, right).

When the initial Reynolds number, Re_0, is small or moderate, billows grow by pairing until suppressed by stratification. Within gravitationally unstable regions of the core, shear-aligned convective secondary instabilities produce small streamwise vortices that rapidly transition to turbulence. Velocity is nearly isotropic after the transition (shaded on the top axis of Figure 7.39), but, as decay proceeds, vertical and spanwise components are increasingly weaker than streamwise flow.

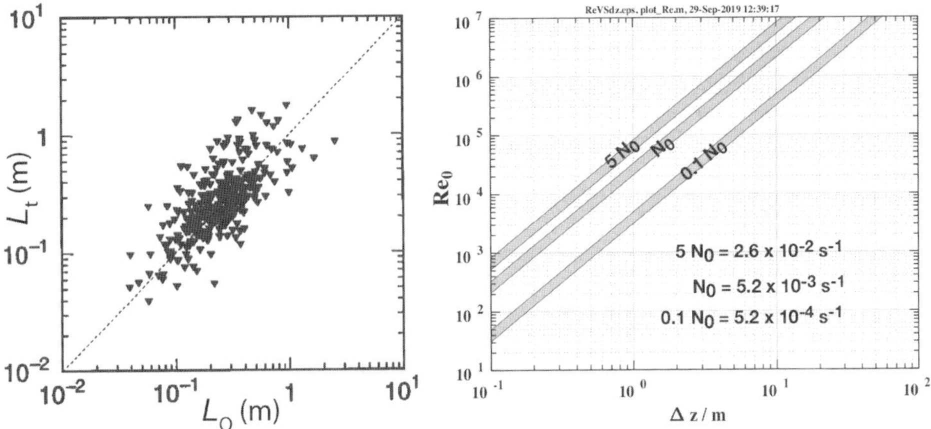

Figure 7.38 (Left) rms overturning scale (Thorpe scale) versus Ozmidov scale for selected overturns in the upper pycnocline (adapted from Moum, 1996). (Right) Initial Reynolds numbers of overturns as a function of overturn height, Δz, evaluated with (7.25).

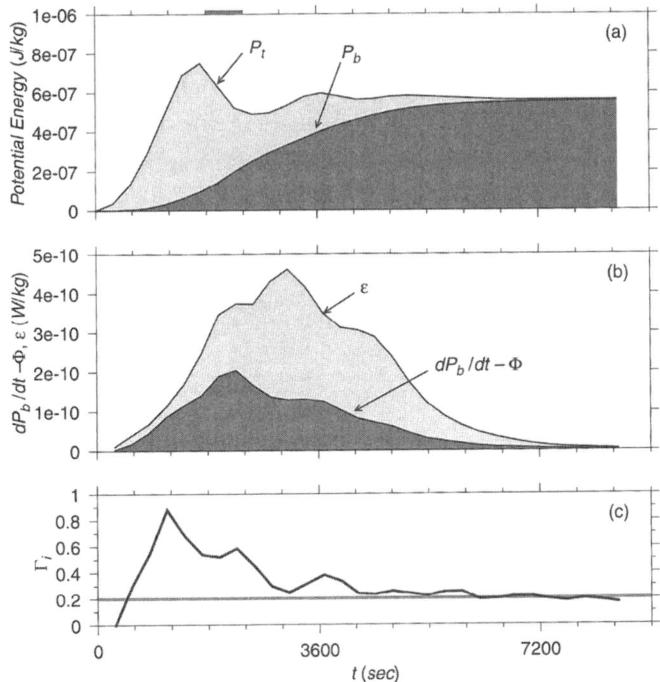

Figure 7.39 DNS of a KH instability: (a) total (P_t) and background (P_b) potential energy, (b) ϵ over the entire domain and the rate of change of background potential energy, and (c) mixing efficiency as $dgpe_r/\epsilon_{\text{turbulent}}$. (Adapted from Smyth et al., 2001. © American Meteorological Society. Used with permission)

With increasing Re_0, this sequence develops more rapidly, as vortex pairing is suppressed and additional secondary instabilities develop (Mashayek and Peltier, 2011; Salehipour et al., 2015). When normalized by $\Delta z N$, the available kinetic energy,

$$\frac{ake_{\mathrm{kh}}}{(\Delta z N)^2} = \frac{1 - 4Ri_g}{Ri_g}, \tag{7.26}$$

should be in large instabilities for given stratification and Ri_g (Fig 7.37).

7.7.2 Holmboe Instabilities

Because viscosity is much larger than thermal and haline diffusivity, velocity interfaces thicken more rapidly than density interfaces when flow is laminar. When the velocity interface is significantly thicker than the density interface, Holmboe instabilities can develop when waves traveling in opposite directions on upper and lower boundaries of the interface interfere to produce instabilities (Figure 7.40). The nature of the instabilities varies from cusps to vortices with the ratio of interface thicknesses, the initial Reynolds number, and the gradient Richardson number. The vortices move relative to each other and can combine to overturn the central interface when vertically aligned.

Mixing develops more slowly in Holmboe instabilities than it does in KH billows, but it follows a similar pattern. Potential energy increases rapidly, while dissipation remains low (Figure 7.41), initially producing a large mixing coefficient, Γ_i. The rate of increase in reference, a.k.a. background, potential energy slows after the transition to turbulence, and Γ_i decreases rapidly to $\sim$0.2.

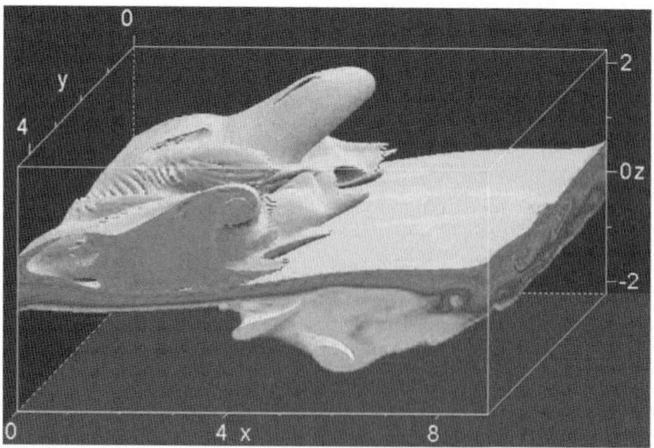

Figure 7.40 Turbulence-like vortices on an interface formed by a Holmboe instability in a DNS with $Ri_g^0 = 0.45$, $Re_0 = 1{,}200$, $Pr = 9$, and a velocity interface three times thicker than the density interface. (From Smyth et al., 2007. © American Meteorological Society. Used with permission)

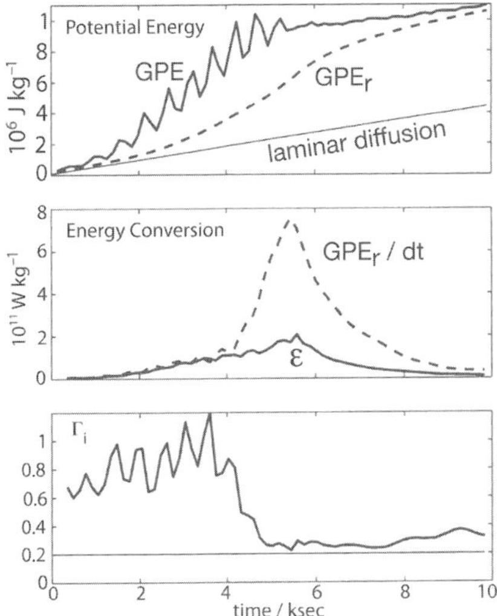

Figure 7.41 Mixing in a Holmboe instability with $Pr = 9$, $Re_0 = 1,200$, and $Ri_v^0 = 0.15$. The velocity interface was three times thicker than the density interface. Mixing efficiency, measured by Γ_i, is high when potential energy increases rapidly and dissipation is low. It drops rapidly to ~ 0.2 after the transition to turbulence. Owing to the low Reynolds number, laminar diffusion across sharpened nonturbulent gradients contributed $\sim 40\%$ of the potential energy increase. (Adapted from Smyth and Winters, 2003. © American Meteorological Society. Used with permission)

The mixing coefficient, Γ_i, is insensitive to Re_0, the initial Reynolds number. Net mixing, however, increases with increasing stratification, which delays or prevents the transition to turbulence, prolonging the laminar phase, when Γ_i is greatest. For similar conditions, Holmboe mixing is significantly weaker than KH mixing (Smyth et al., 2007).

7.7.3 Advective Overturning

In addition to shear instability, internal waves break when they are so steep that dense water hangs over less dense water. Termed 'convective', 'advective', or 'Rayleigh-Taylor overturning', this mode of instability requires a wave steepness greater than one,

$$Steepness \equiv u'/c_p = u'/(\omega/k) > 1, \tag{7.27}$$

corresponding to the Orlanski and Bryan (1969) criterion (7.33).

For single internal waves, simulations of near-inertial waves show shear instability dominating advective overturning and occurring simultaneously throughout the wave field (Dunkerton, 1997; Lelong and Dunkerton, 1998a,b). At intermediate

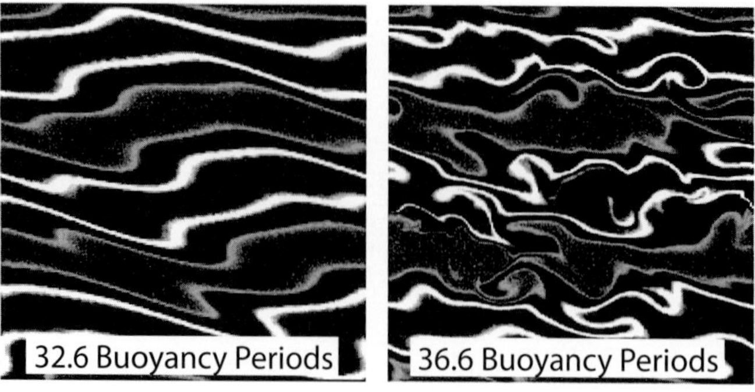

Figure 7.42 Simulated evolution of density in the vertical plane for an internal wave with $\omega = N/\sqrt{2}$, visualized with repeating greyscale bands. The instability began with vorticity contours inclined 20° from horizontal and evolved to overturn density with patterns very different from KH instabilities. (Adapted from Bouruet-Aubertot et al., 2001)

frequencies, shear instability initiates transverse overturning when the wave is convectively unstable; otherwise overturning occurs diagonally. At high frequencies, near N, convective instability modifies the transverse shear instability as waves grow large. Waves are susceptible to shear instability before convective instability begins, except when amplitude increases rapidly approaching a critical layer.

Advective overturns are not the only alternative to KH instabilities. Simulations of a single high-frequency wave by Bouruet-Aubertot et al. (2001) demonstrate instability to triad interactions, producing breaking patterns differing greatly from KH instabilities and advective overturns in parallel shear flows (Figure 7.42). The variety in these simulations demonstrates that much remains to be learned about overturning in the pycnocline.

7.7.4 Solitary Waves and Solibores

Tall waves in shallow water can balance nonlinear steepening and dispersive widening to propagate long distances without changing form. Termed 'solitons' when observed as isolated bumps, solitary waves frequently occur in packets ordered by descending amplitude and speed. For moderately long waves in a weakly dispersive medium, the dynamic balance is approximated by the Korteweg–deVries (KdV) equation (Lighthill, 1979) as

$$\underbrace{\frac{\partial u}{\partial t} + c_0 \frac{\partial u}{\partial x}}_{\text{linear nondispersive}} + \underbrace{u \frac{\partial u}{\partial x}}_{\text{nonlinear}} + \underbrace{d \frac{\partial^3 u}{\partial x^3}}_{\text{weak dispersion}} = 0 \quad \left[\mathrm{m\,s^{-2}}\right], \tag{7.28}$$

where $c_0 = \sqrt{gH}$ is the linear nondispersive speed, H the water depth, and $d = (c_0/6)H^2$. When the ratio of the nonlinear to the dispersive term is $\gtrsim 16$, nonlinear effects steepen the forward face, producing an internal bore or hydraulic jump. For smaller ratios, two solutions are possible. One gives a single hump or depression, and the other yields a periodic form known as cnoidal waves. To describe packets more accurately, another KdV solution gives dnoidal waves (Gurevich and Pitaevskii, 1973; Apel, 2002).

Solitary waves are commonly generated when tidal currents form hydraulic jumps at shelf breaks and over sills in straits and channels. The simulation in Figure 7.43 shows a sequence over a sill, but one over shelf breaks would be similar. Initially, flow in the simulation was northward (rightward) at maximum

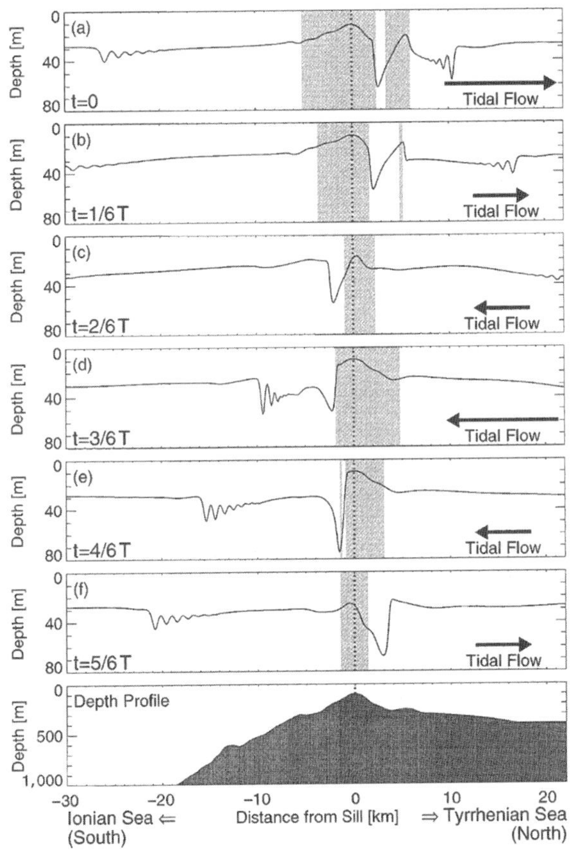

Figure 7.43 Simulated evolution of an interface representing the seasonal thermocline during a tidal cycle over a sill in the Strait of Messina. Shading shows where flow was supercritical, i.e. $G^2 \equiv u_1^2/g'h_1 + u_2^2/g'h_2 > 1$, where h_1 and h_2 are water depths above and below the interface. Soliton trains formed on both sides of the sill near maximum flow in that direction. T is the tidal period, $t = 0$ the time of maximum northward flow, and $g' \equiv g\Delta\rho/\rho_0 = 0.015$ across the interface. (From Brandt et al., 1997. © American Meteorological Society. Used with permission)

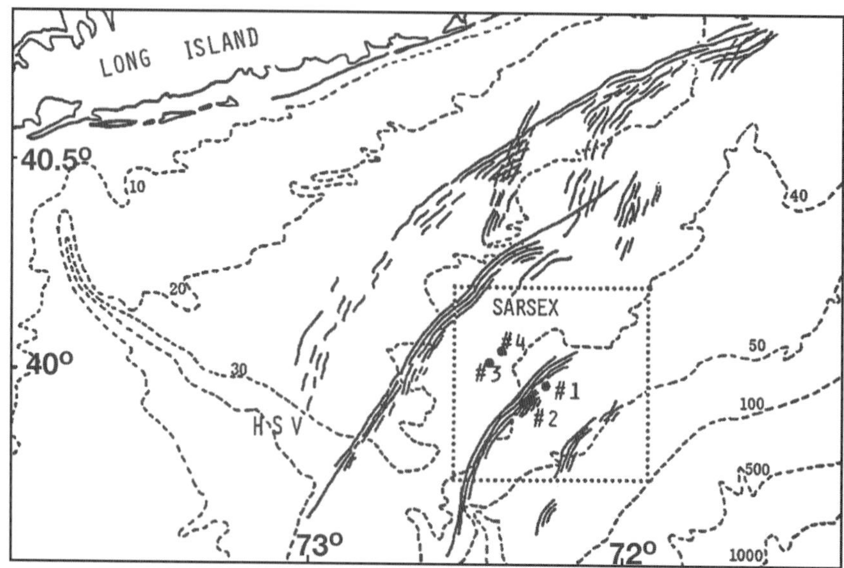

Figure 7.44 Surface signatures of soliton packets (solid) and bathymetry (dashed) of the New York Bight during the Synthetic Aperature Radar Experiment (SARSEX). Spacing between packets is the wavelength of the twice-daily internal tide. (Adapted from Liu, 1988)

amplitude, with the interface sloping upward toward the sill from the south. Flow was hydraulically controlled near the sill crest, producing an hydraulic jump where control was lost on the north side. The soliton train formed downstream of the jump and propagated northward as flow slackened. After the tide reversed, a similar sequence developed on the south side. In both cases, the waves were rank ordered, decreasing in amplitude behind a steep drop resembling an internal bore. Henyey and Hoering (1997) termed this stage a 'solibore', reflecting its dual nature as an internal bore and a packet of solitons. Solibores proceed with intense turbulent dissipation (Wesson and Gregg, 1988) until the borelike nature diminishes and the packet evolves into a train of rank-ordered solitary waves and then into nondissipative, nearly sinusoidal waves.

Whether and how solibores form varies with tidal strength, interface depth, and mean currents. Shallow solibores ruffle the sea surface, creating visual and radar signatures when winds are light (Figure 7.44). These allow wave properties to be determined by combining shipboard radar returns with simultaneous profiling (Chang et al., 2008). Solitary waves have been found over most continental shelves, as well as offshore (Pinkel, 2000), but generation and detection are episodic: winter homogenization of shelf water precludes generation, solibores are usually not formed during neap tides, and moderate seas can obscure surface signatures.

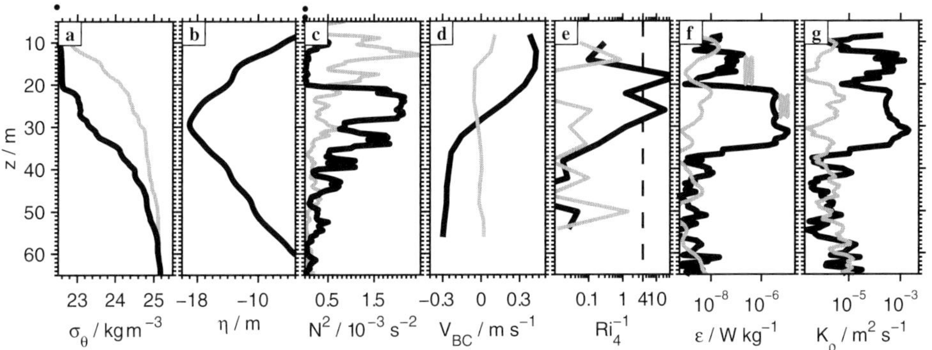

Figure 7.45 Averages of three troughs in one soliton packet (black) compared to profiles before their arrival (gray): (a) potential density, (b) vertical displacement, (c) 4 m stratification, (d) baroclinic cross-shelf velocity, (e) 4 m inverse gradient Richardson number, (f) ϵ, and (g) K_ρ. The dashed line in (e) is the minimum level expected for shear instability. Gray 'x's in panel (f) are estimates using (7.29). (From MacKinnon and Gregg, 2003. © American Meteorological Society. Used with permission)

Until they evolve into more sinusoidal dispersive waves, even moderate solitons are highly dissipative. For example, in late summer strong solitons over the 70 m-deep New England shelf depressed the pycnocline 18 m while increasing stratification and shear (Figure 7.45). In the upper half of the soliton, $Ri_4^{-1} > 4$, and ϵ and K_ρ were 100 times background, mixing more intensely than GM internal waves with the same stratification and shear (Figure 7.34). Exceeding 10^{-6} W kg^{-1}, dissipation rates were only slightly less than predictions by Kunze et al. (1990) for the energy release needed to reduce the observed supercritical value of Ri_4^{-1} to the critical value of 4,

$$\epsilon_{\mathrm{KWB}} = (\Delta z)^2 \overline{\left(\frac{S^2 - N^2}{24}\right)\left(\frac{S - 2N}{4}\right)} \quad \left[\mathrm{W\,kg^{-1}}\right]. \tag{7.29}$$

K_ρ in the solitons was 10^{-4} to 10^{-3} m^2 s^{-1}, contributing half of the net diffusivity. Similar results have been found on other shelves (Moum et al., 2003; Shroyer et al., 2010).

At some places, e.g. the Andaman, Sulu, and South China Seas, solitons generated at offshore ridges and straits approach continental shelves with forms consistent with KdV solutions for waves of depression (Klymak et al., 2006b). They become dissipative en route only if they encounter background shear. As water depth decreases, the waves steepen and become more nonlinear until the depth of the upper layer, H_1, equals that of the lower layer, H_2, triggering a very dissipative transition to waves of elevation (Figure 7.46). To describe this sequence, Liu et al.

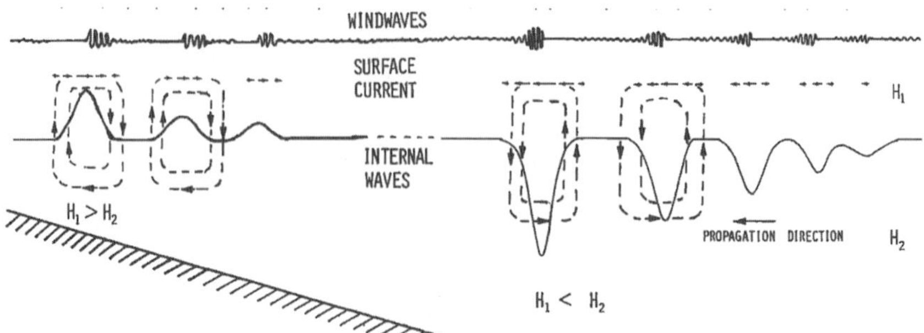

Figure 7.46 A train of solitary waves transitioning from deep-water waves of depression to shallow-water waves of elevation. The transition produces intense dissipation. Under light winds, orbital velocities induce surface ripples and foam detectable from satellites. (Adapted from Liu et al., 1998)

(1998) developed a single expression for soliton packets propagating from deep to shallow water,

$$(1 + c_0 + \underbrace{c_1\zeta + c_2\zeta^2)}_{\text{nonlinear}}\frac{\partial \zeta}{\partial x} + \underbrace{c_3\frac{\partial^3 \zeta}{\partial x^3}}_{\text{dispersion}} + \underbrace{c_4\lambda}_{\text{shoaling}} - \underbrace{c_5\frac{\partial^2 \zeta}{\partial x^2}}_{\text{dissipation}} = 0, \tag{7.30}$$

with $c_0 = [\Delta\rho g H_1 H_2/\rho_0(H_1 + H_2)]^{1/2}$ for the linear wave speed, and c_1 to c_5 as additional coefficients. Where the interface is at mid-depth, $H_1 = H_2$, the first-order nonlinear coefficient, c_1, changes sign in keeping with the transition between waves of depression and elevation, and c_2 becomes larger. The transition from deep to shallow water is violent, as shown by strong backscatter from beams of high-frequency acoustics (Figure 7.47). Direct turbulence measurements found vertically integrated dissipation levels of 50 mW/m^2, an order of magnitude larger than frictional dissipation of the internal tide or levels in the open ocean (St. Laurent, 2008).

Bogucki and Garrett (1993) modeled soliton mixing assuming that dissipation is fed by the wave's decay and can be estimated from interface thickening and the gradient Richardson number. As restated by Henyey and Hoering (1997), integrated dissipation is a very strong function of u_{sb}, the solibore speed,

$$\int \rho\epsilon_{\mathrm{solibore}}\, dA = Ri^2 \frac{\rho^2 u_{\mathrm{sb}}^5}{24 g \Gamma_{\mathrm{mix}} \Delta\rho} \quad \left[\frac{\mathrm{W}}{\mathrm{m}}\right]. \tag{7.31}$$

Addressing solibores rather than solitons, they assumed that dissipation is balanced by energy released as the interface rises at the end of the wave train (Figure 7.43).

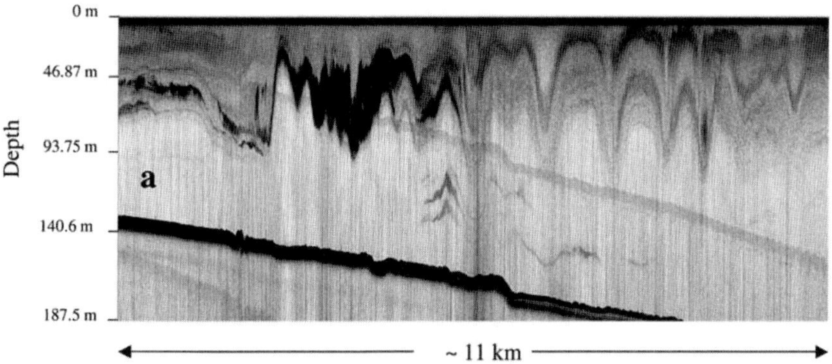

Figure 7.47 Backscatter intensity from a beam of high-frequency acoustics through a soliton packet propagating onto the Chinese shelf. Waves were transitioning from depression (right) to elevation (left). (Adapted from Orr and Mignerey, 2003)

Tracking a solibore packet across the Oregon shelf, Moum et al. (2007a) found wavelength and spacing decreasing as depth shoaled. Inshore of the 50 m isobath, the packet catastrophically lost its wavelike characteristics. The rate of energy loss, $dE_{\text{solibore}} = -(14 \pm 7)$ W m^{-1}, was close to the observed net dissipation, $\rho\epsilon_{\text{sb}} = (10 \pm 3)$W m^{-1}, but was not consistent with Henyey and Hoering (1997), possibly because the packet had evolved from a solibore to solitons.

Owing to the intensity of mixing produced by solibores and solitons and its spatial and temporal distribution, considerably more understanding is needed to parameterize coastal mixing adequately. The efficiency of solibore mixing is among the important issues, as the waves can break by shear instability in trains of KH instabilities, as well as by advective overturning when waves form closed cores (Lamb et al., 2019).

7.8 The Saturated Range

When internal waves are close to GM, spectra of vertical shear and strain change slope at $k_z^c \sim 0.1$ cpm from $k_z^{\sim 0}$ to $k_z^{\sim -1}$ (Figure 7.48). The k_z^{-1} range ends near 1 cpm at the wavenumber of the largest overturns and the beginning of the turbulent range. The rms overturning scale is approximated by the Ozmidov length scale (3.5). At GM levels, internal wave dissipation scaling (7.16) gives the Ozmidov length scale as

$$l_O \equiv \left(\frac{\epsilon}{N^3}\right)^{1/2} = \left(\frac{7 \times 10^{-10}(N/N_0)^2}{N^3}\right)^{1/2} = 0.07\left(\frac{N}{N_0}\right)^{1/2} \quad [\text{m}]. \qquad (7.32)$$

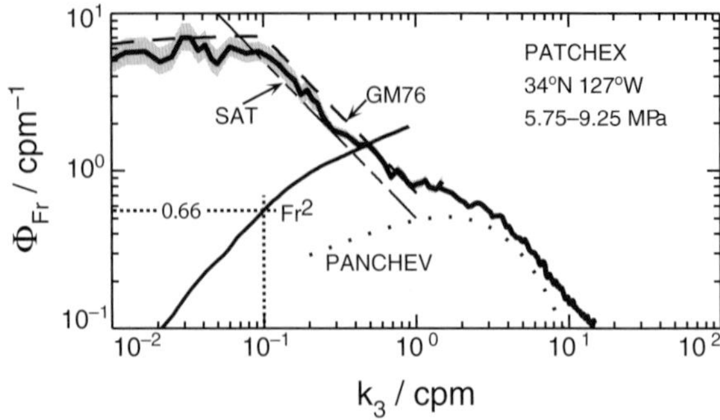

Figure 7.48 Froude spectrum, $\Phi_{\text{Fr}} \equiv \Phi_{vshear}/N^2$, and its cumulative integral, Fr^2, with GM76, the Panchev and Kesich (1969) turbulent spectrum, and the spectrum of saturated internal waves (SAT). (Adapted from Gregg et al., 1996)

For the spectrum in Figure 7.48, the Ozmidov scale is 0.093 m, corresponding to $k_z^O = 10.8$ cpm, much larger than the wavenumber of the largest overturns.

GM models initially missed the change in slope at k_z^c, but included it after it was found by high-resolution profiling of strain (Gregg et al., 1973; Gregg, 1977a) and shear (Gargett et al., 1981). Characteristics include:

- *Saturated amplitudes of vertical wavenumber spectra*: Vertical spectra of intense internal waves rise above GM at low wavenumbers, $k_z < k_z^c$, often with irregular shapes. Spectral amplitudes, however, do not increase beyond the cutoff, i.e. amplitudes are saturated between k_z^c and the dissipation range. Rather, the increased energy passes through and elevates spectra in the dissipation range (Figure 7.23).
- *Slopes of vertical spectra*: Strain and shear spectra change slope at k_z^c from nominally flat, k_z^0, to k_z^{-1} in the saturated range. Slope, however, is steeper than –1 within a few degrees of the equator, likely due to the reduced efficiency of wave–wave energy transfer (Gregg et al., 1996). Also, some strain spectra are flatter than –1, e.g. potential density in Figure 7.49. These are possibly contaminated by salinity spiking, which varies with the TS relation and with dissipation rate. Consequently, high-wavenumber slopes of potential density spectra should be viewed with caution until spiking is eliminated.
- *Variable wavenumber bandwidth*: When internal waves are at GM, the saturated bandwidth spans a decade, from $k_z^c = 0.1$ cpm to the beginning of the dissipation range at 1 cpm. The transition in slope at k_z^c, however, is not always apparent (Figs. 7.23, 7.48, 7.49), perhaps owing to noise and variability in small data ensembles. The bandwidth remains about one decade as the saturation range shifts to lower wavenumber when energy exceeds GM (Figure 7.23).

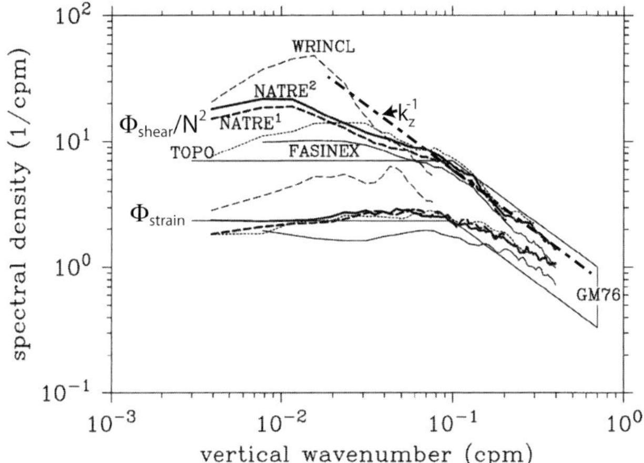

Figure 7.49 Average Froude spectra, $\Phi_{\mathrm{Fr}} \equiv \Phi_{\mathrm{shear}}/N^2$, and strain spectra from the HRP. The Froude spectra coalesce near 0.1 cpm with a slope slightly steeper than –1 in the saturated range, where the strain spectra have a shallower slope. (Adapted from Polzin et al., 2003)

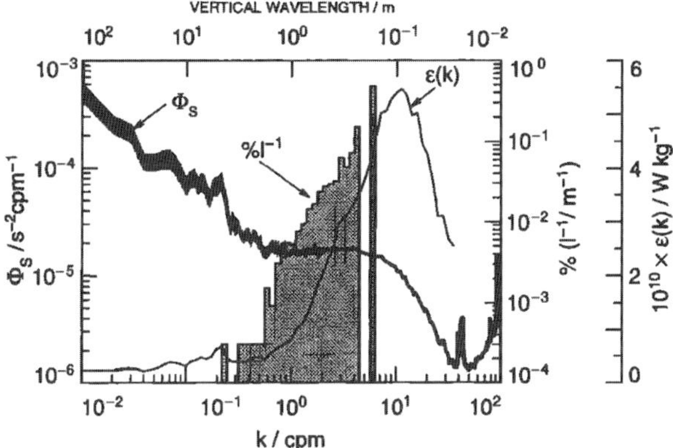

Figure 7.50 Spectrum of vertical shear at 150–250 m on the equator overlaid with histograms of inverse overturn lengths, $\%l^{-1}$, and of $\log_{10} \epsilon$. The extensive saturation range ends near 0.5 cpm. The overturn histogram barely extends into the saturation range. (From Peters et al., 1995)

- *Stable stratification*: The saturation range is bounded at high wavenumbers by the largest overturns (Figure 7.50), demonstrating stable stratification throughout the saturation range.
- *Slopes of horizontal spectra*: High-resolution horizontal tows (Klymak and Moum, 2006, 2007) find spectra of isopycnal slope with $k_h^{1/3}$ slopes spanning saturated and Kolmogorov ranges without changes in slope or offsets in level (Figure 3.15). As discussed in Section 3.8, Riley and Lindborg (2008) interpret these spectra as signatures of strongly stratified turbulence, including vortical motions.

7.8.1 Processes Invoked to Explain the Saturated Range

Because spectral ranges with distinct power laws often indicate uniform dynamics within the range, several processes have been proposed.

Buoyancy Subrange

Extending Kolmogorov's approach to turbulence in stratified fluids led to assumptions of a buoyancy subrange of overturning eddies made highly anisotropic by stratification between the largest overturning scales and the largest isotropic scales (Bolgiano, 1959; Lumley, 1964). Later, Weinstock (1985) derived k_z^{-3} and k_z^{-1} slopes for temperature and strain spectra in the buoyancy subrange. Although these slopes are consistent with observed spectra, the lack of overturns except at the high-wavenumber end of the saturation range (Figure 7.50) demonstrates that the k_z^{-1} range is not a buoyancy subrange of turbulence.

Advective Overturning

As internal waves generated at the top of the troposphere propagate upward into the middle atmosphere, their amplitudes increase as air density decreases exponentially. Wave growth, however, is limited by convective overturning, and spectra have constant amplitudes and slopes of k_z^{-3} for velocity and k_z^{-1} for shear. Assuming that advective overturns dominate the saturation range, Smith et al. (1987) invoke the Orlanski and Bryan (1969) criterion for advective, a.k.a. convective, instability,

$$u' = |c - \overline{u}|. \tag{7.33}$$

Spreading the variance of sinusoidal waves over the local wavenumber bandwidth leads to $\Phi_{vel}^{sat}(k_z) = (1/2)u'^2/|k_z|$. Restricting the waves to the middle of the internal wave range, $f^2 \ll \omega^2 \ll N^2$, reduces the dispersion relation (6.25) to $|k_z| \approx |k_h|N/\omega$. Applying $\omega = k_h(c - \overline{u})$ then yields $|k_z| = N/|c - \overline{u}|$, which combined with (7.33) gives $u' = N/|k_z|$ and

$$\Phi_{vel}^{sat}(k_z) = \frac{N^2}{2|k_z|^3} \quad \left[\frac{(\mathrm{m/s})^2}{\mathrm{m}^{-1}}\right], \quad \Phi_{Fr}^{sat}(k_z) = \frac{1}{2|k_z|} \quad \left[\frac{1}{\mathrm{m}^{-1}}\right]. \tag{7.34}$$

With no adjustable parameters, agreement with oceanic spectra (Figure 7.48) is remarkable. Spectral consistency, however, is not proof, and the lack of overturns throughout the k_z^{-1} range precludes advective overturning from explaining oceanic spectra.

Nonlinear Dynamics and Doppler Spreading

Applying statistical mechanics to the internal wave spectrum, Allen and Joseph (1989) assumed a Maxwell-Boltzman distribution of modes with large- and small-scale cutoffs in a Lagrangian reference frame. When close to statistical equilibrium, the Lagrangian field is entirely wavelike with energy equally distributed among large-scale modes; strong nonlinearity suppresses the small-scale modes. Owing to the absence of the advective nonlinear term, $\boldsymbol{u} \cdot \nabla \boldsymbol{u}$, dynamics differ from those in a Eulerian frame. At low frequency, the Eulerian wavenumber-frequency spectrum nearly matches the Lagrangian spectrum, but at high wavenumber the two differ significantly, owing to the advective nonlinearity. One-dimensional Eulerian temperature spectra have k^{-3} 'tails' at wavenumbers past the small-scale cutoff of energy-containing modes, consistent with similar ranges in spectra of vertical and horizontal oceanic observations (Klymak and Moum, 2006).

Hibiya et al. (1996) addressed the saturated range with a two-dimensional direct numerical simulation (DNS) initialized with a field of randomly phased internal waves having amplitudes matching GM. After five inertial periods, the shear spectrum became quasi-stationary, with a rolloff beginning at 0.04 cpm with a slope close to –1. Because gradient Richardson numbers were too large for either convective or shear instability, the authors concluded that the spectral rolloff was not caused by either form of instability. Rather, the oceanic saturation range appears to result from a mixture of nonlinear decaying internal waves and strongly stratified turbulence or vortical motions.

Horizontal Turbulent Cascade

Noting that the $k_h^{1/3}$ slope of horizontal spectra continues through the Ozmidov wavenumber, Kunze (2019) argued that simultaneous k_z^{1} and $k_h^{1/3}$ spectra are produced by a cascade of stratified turbulence moving energy to high wavenumbers, beginning at the horizontal Coriolis wavenumber, $k_h^f \equiv f/u$, where $\partial u/\partial x \sim f$. The corresponding vertical Coriolis wavenumber, $k_z^f \equiv k_h^f(N/f)$, is identified with the internal wave cutoff wavenumber, k_z^c. In terms of vertical wavenumber, the turbulence is anisotropic between k_z^f, and the Ozmidov wavenumber and is isotropic at higher wavenumbers.

The saturated spectrum/stratified cascade interpretations remain as the best explanations for the saturation range, but detailed horizontal and vertical observations, supplemented with DNS, are needed to distinguish between them or to determine that they are different aspects of the same process.

7.9 Perspectives

Piecing together disparate microstructure observations, process studies, and simulations appears to have produced a zeroth-order understanding of mixing driven by internal waves in the pycnocline. There are, however, many ad hoc elements of this somewhat creaky construct. Replacing them with a solid foundation will require three-dimensional time series of internal wave and vortical mode evolution simultaneous with mixing observations. Rapid improvements in autonomous vehicles suggest that such measurements are within reach. For instance, in 2016 20 EM-APEX floats profiled simultaneously during the Lateral Mixing Experiment (Lien and Sanford, 2019).

8

Mixing in the Stratified Interior

8.1 Overview

Separating the weakly stratified interior from the surface boundary layer, the ocean's pycnocline (Section 8.2) often begins tens of meters below the surface. From peak stratification at the base of the surface layer, the density gradient decreases with increasing depth, until by one kilometer it smoothly merges into the weakly stratified abyss. In some places, strong local processes alter this pattern, particularly during winter at high latitudes, when cold storms homogenize surface layers hundreds or thousands of meters thick. Other special situations include shallow haloclines from intense rain or runoff and double pycnoclines south of the Gulf Stream and Kuroshio. In each case, the shape of the pycnocline offers clues about its formation, and the stratification in turn affects the intensity of the mixing that modifies its shape. Mixing in the homogenous boundary layers atop and below the pycnocline is important, but here attention is focused on the pycnocline.

Finestructure, velocity and scalar variability with vertical scales between one and several hundred meters (Section 8.3), is described by two models. One represents the pycnocline as a sequence of homogenous layers formed by mixing and separated by thin interfaces (sheets). The other approach describes the pycnocline as a sequence of irregular gradients produced by transient internal wave strain. Instabilities arising from these gradients produce mixing patches and in turn modulate evolution of the patches. Finescale variances, such as $\overline{(du/dz)^2}$, are proportional to internal wave intensity and hence to dissipation rates via scaling discussed in the previous chapter. By assuming constant mixing efficiency, global maps of finescale variances scaled as diapycnal diffusivity reveal how mixing varies with depth, latitude, bottom roughness, and season (Section 8.4).

Strong turbulence in the pycnocline occurs in patches a few meters high and up to a few kilometers horizontally (Section 8.5). Some, perhaps most, large patches are associated with near-inertial shear, but observations of two-dimensional structure are few, and there are none of three-dimensional evolution. Horizontal tows find two types of centimeter-scale structures in the patches, broadband and narrowband. The former is consistent with three-dimensional turbulence, while the latter is attributed to salt fingers (Section 8.6). Determining net fluxes from salt fingers too weak to form staircases is one of the major research challenges in the pycnocline.

To supplement the focus on mixing in mid-ocean pycnoclines, three regimes are considered where internal waves, double diffusion, and mixing combine very differently than in open-ocean pycnoclines. With no continental obstructions impeding its circuit of the globe, the Southern Ocean (Section 8.7) exhibits strong mixing where deep mesoscale flows cross rough bottoms. The Arctic (Section 8.8) has very weak mixing, owing to sea ice sheltering the water from wind forcing much of the year and also likely absorbing some of the waves. Diffusive layering is pervasive in shallow haloclines, but diapycnal fluxes are low. As stirring rods for surface tides, mid-ocean ridges (Section 8.9) generate intense local mixing in addition to generating low-mode internal tides that propagate long distances to dissipate on distant continental slopes.

8.2 Vertical Structure of Stratification

Most pycnoclines begin at the base of a surface mixed layer and continue smoothly downward (Figure 8.1), dominated by temperature decreasing with depth, as near

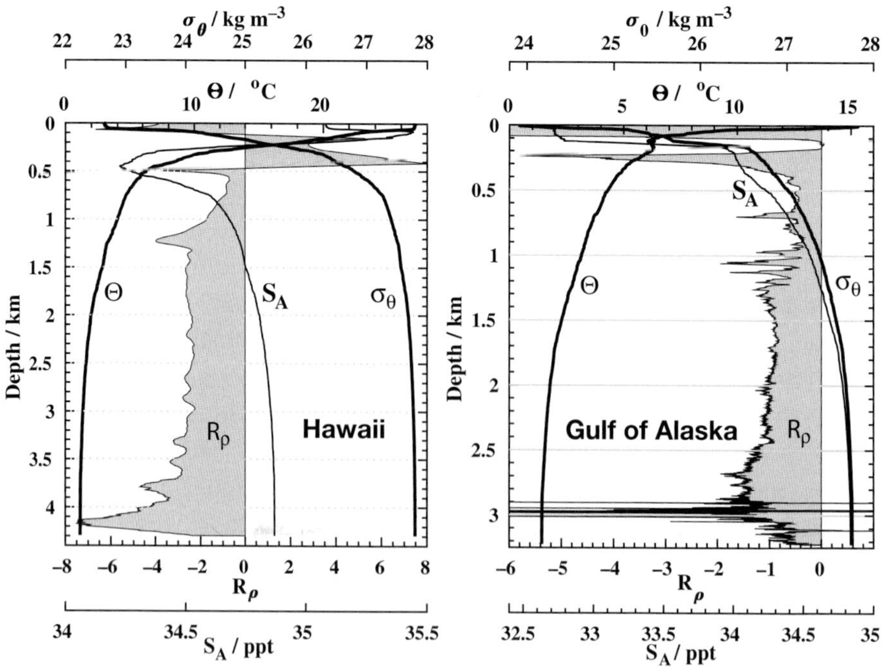

Figure 8.1 (Left) Mid-ocean thermocline (data courtesy of Tom Sanford.) (Right) High-latitude halocline. R_ρ is the density ratio (2.44).

Hawaii. Some are more complex. For example, in the northwest Atlantic and Pacific the principal pycnocline lies between 600 and 900 m, beneath weakly stratified waters formed during winter mixing in the Gulf Stream and Kuroshio. Also, at high latitudes salinity controls shallow stratification, as in the Gulf of Alaska.

8.2.1 Thermoclines and Haloclines

Most of the downward density increase in the water column occurs within a few hundred meters of the surface. At mid- and low latitudes, cooling temperature dominates the downward increase in density, often offsetting salinity decreasing beneath high surface evaporative fluxes. At high latitudes heavy rainfall and melting ice combined with small thermal expansion coefficients produce shallow haloclines dominating stratification. The contrast between thermoclines and haloclines is shown more graphically by plotting temperature versus salinity (Figure 8.2). The 40 m-thick Alaskan halocline is nearly isothermal, in contrast with the 400 m-thick Hawaiian thermocline, which coincides with a destabilizing salinity decrease. Owing to the low molecular diffusivity of salt, differential diffusion should be very important during weak and moderate mixing in the halocline. It would be less important in the thermocline, which is also diffusively unstable to salt fingering. Below the pycnocline, stratification decreases exponentially to reach $N^2 \approx 10^{-7}\ \mathrm{s}^{-2}$ near the bottom (Figure 8.3). In comparison with five-hour buoyancy periods in the abyss, the inertial period is $\approx$ 33 hours at the Hawaiian site and 16 hours in the Gulf of Alaska. How narrowing of the internal wave frequency bandwidth affects mixing is not understood.

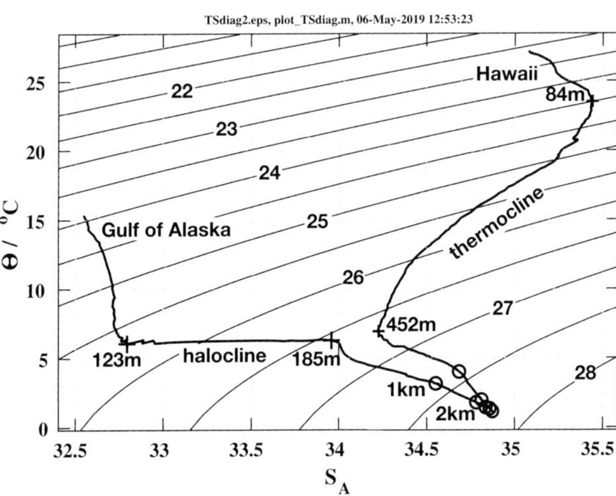

Figure 8.2 Conservative temperature versus salinity for the profiles in Figure 8.1. Contours are potential density referenced to the surface. Salt fingering is possible between 84 m and 452 m in the Hawaii profile. Elsewhere, both profiles are diffusively stable.

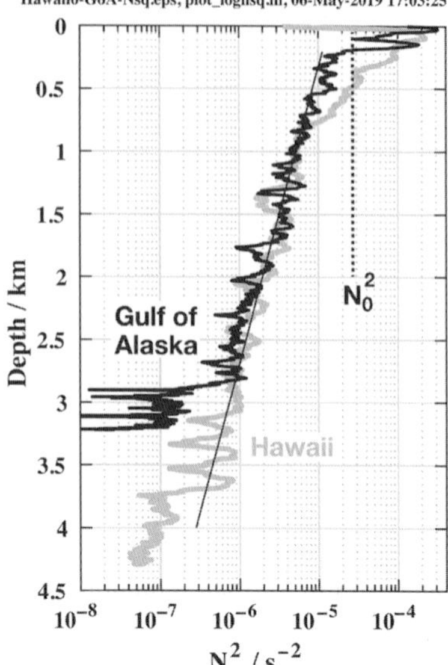

Figure 8.3 Stratification, N^2, for the profiles in Figure 8.1. Between 800 m and the bottom boundary layers, the profiles have the same exponential stratification. $N_0 = 3$ cph is the reference buoyancy frequency for the GM spectrum (Garrett and Munk, 1972) and for Munk (1966), where it was at a depth of 1 km.

8.3 Finestructure

In the 1960s, the application of solid-state electronics to oceanographic profilers increased vertical resolution to reveal temperature and salinity profiles varying in irregular steps rather than smoothly, as previously assumed (Stommel and Federov, 1967; Cox et al., 1969). Having vertical scales from less than a meter to tens of meters, finestructure links the energy-containing scales of internal waves to the microscales of turbulence (Figure 8.4). Finestructure has been characterized as a sequence of 'sheets and layers' formed irreversibly by mixing and, alternatively, as the reversible product of internal wave strain and shear. Strain and shear, in turn, have been modeled with jointly normal Gaussian statistics, going one step beyond the Garrett and Munk (1972) internal wave model. Pinkel and Anderson (1992, 1997) and Pinkel (2020) demonstrate that strain statistics are consistent with Poisson distributions.

8.3.1 Sheets and Layers

Visual observations of dye in a shallow Mediterranean thermocline led Woods (1968) to conclude that finestructure consists of a vertical sequence of well-mixed layers separated by thin interfaces. The basic structure was inferred from

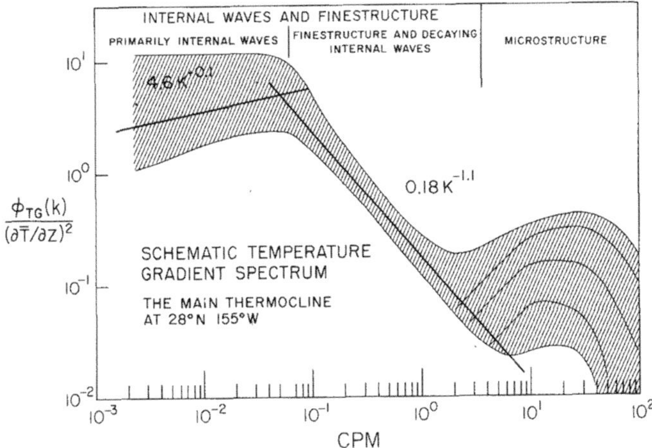

Figure 8.4 Schematic spectrum of vertical temperature gradients, showing three dynamical regimes. (From Gregg, 1977b. © American Meteorological Society. Used with permission)

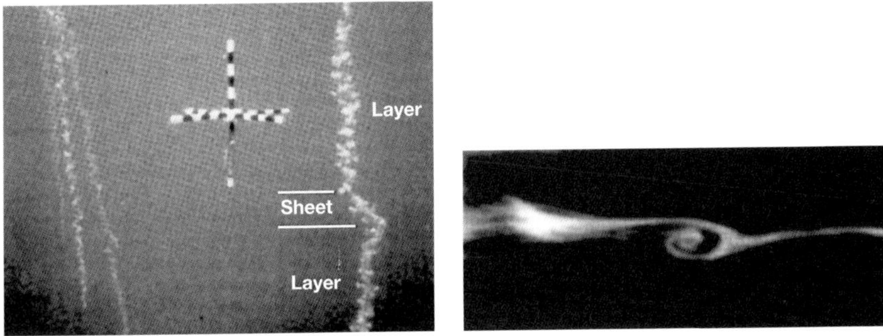

Figure 8.5 Wakes of sinking dye pellets (left) show thick sections moving at uniform velocity (layers) separated by thin regions of high shear (sheets). A dye packet suspended in a sheet (right) spread laterally and was displaced vertically by an internal wave propagating along the sheet. Overturns of the sheet formed ~0.1 m-high billows similar to Kelvin–Helmholtz instabilities in laboratories. (Adapted from Woods, 1968)

horizontal displacements of wakes behind sinking dye pellets (Figure 8.5), showing thin interfaces (sheets) separating much thicker homogenous layers. These observations were supplemented by packets of dye spreading laterally after being suspended in the interfaces. Internal waves often displaced the sheets vertically and sometimes overturned them, apparently as Kelvin–Helmholtz (KH) instabilities, leading Woods (1968) to propose sheets and layers as a model for the thermocline.

To estimate the diapycnal diffusivity of a sheet-and-layer thermocline, Stommel and Federov (1967) assumed that stairstep profiles mix every $2\Delta t$ seconds,

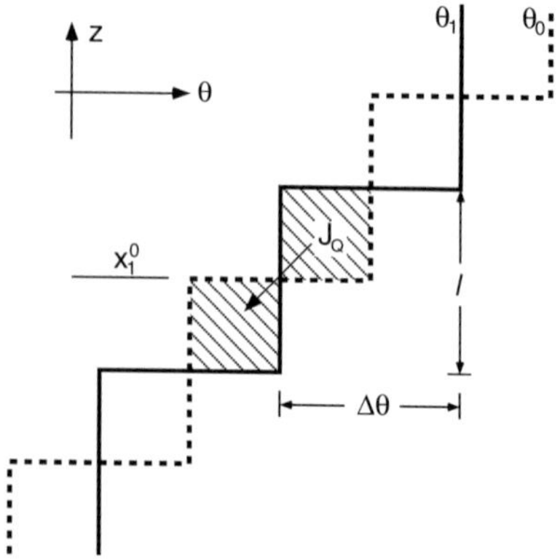

Figure 8.6 Schematic mixing of a sheet-and-layer thermocline with layers l m thick separated by very thin interfaces. Synchronous overturns periodically homogenize water within $\pm l/2$ of the interfaces, causing the water column to alternate between two profiles, solid and dashed. The overturns maintain an average vertical heat flux of J_Q. (Based on Stommel and Federov, 1967)

each time homogenizing layers l meters thick separated by thin temperature steps, $\Delta\theta$. This idealized sequence forms alternating staircases (Figure 8.6) transferring $\Delta Q = -\rho c_p(\Delta\theta/2)(l/2)$ [J m^{-2}] each cycle. The flux, $J_Q = \Delta Q/2\Delta t$, and average gradient, $\partial\overline{\theta}/\partial z = \Delta\theta/l$, produce diffusivity,

$$K_T \equiv -\frac{J_Q}{\rho c_p \partial\overline{\theta}/\partial x_3} = \frac{l^2}{8\Delta t} = \frac{l^2 N}{16\pi n} \quad [\mathrm{m^2\ s^{-1}}], \tag{8.1}$$

where n is the number of buoyancy periods between mixing cycles. Diffusivity increases linearly with N and as the square of overturn height. Achieving $K_T = 1 \times 10^{-4}$ m^2 s^{-1} when $N = N_0$ (3 cycles per hour) requires 1 m overturns every buoyancy period or 4 m overturns every 15 periods (Figure 8.7). In the seasonal thermocline, where stratification may be $3N_0$ (nine cycles per hour), half-meter overturns every buoyancy period produce $K_T = 1 \times 10^{-4}$, while at $N_0/3$ (one cycle per hour) 1.7 m overturns are needed. In view of the inefficiency of mixing, $K_T = 1 \times 10^{-4}$ requires meter-scale overturning continuously. This is not observed in most of the thermocline, where microstructure and tracer observations find $K_T \sim 10^{-5}$ m^2 s^{-1}, requiring 1 m overturns every 12 buoyancy periods, or smaller ones more frequently. Owing to the difficulty of identifying small overturns, statistics are not available for rigorous comparison, but this distribution of size and frequency is reasonable.

8.3.2 Reversible Finestructure Produced by Internal Waves

Using fully resolved temperature profiles, Desaubies and Gregg (1981) found that gradients, $\Delta T/\Delta z$, were randomly distributed in the vertical when Δz was large,

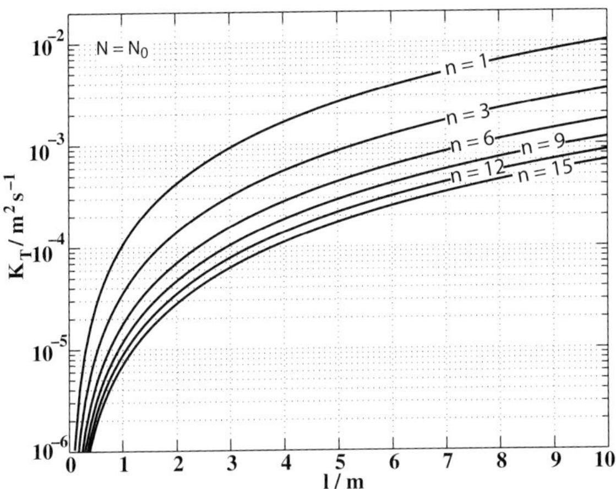

Figure 8.7 Turbulent diffusivity for a sheet-and-layer profile homogenized every n buoyancy periods in an average stratification of $N_0 = 3$ cph. Overturning every buoyancy period, $n = 1$, produces $K_T = 10^{-4}$ m^2 s^{-1} when the layer thickness $l = 1$ m, but one overturn every $n = 15$ buoyancy periods suffices when $l = 4$ m.

but the gradients coalesced into coherent structures as the differencing interval decreased (Figure 8.8). Some intervals had sequences of thin, large gradients and thick, low gradients, e.g. 245–280 m for $\Delta z = 0.5$ m, where a threshold could separate sheets from layers, but those were not the norm, consistent with observations of weak mixing that rarely homogenizes the mean stratification.

A histogram of temperature gradients in a sheet-and-layer profile would be bimodal, with a high peak at zero gradient containing the many samples from thick, homogenous layers, and a small peak at high gradient formed by the small number of samples from thin interfaces. Actual histograms, however, have a large peak at low magnitude and a tail extending to high magnitude (Figure 8.9). There is no objective means based on gradient magnitude to distinguish a sheet from a layer on the histograms.

To estimate vertical temperature gradients produced by internal wave strain, Desaubies and Gregg (1981) modeled the difference in displacement, ζ, for isopycnals having average separation $\overline{\Delta z}$. Displacements of nearby isotherms were assumed to be jointly normal, with individual probability densities consistent with the Garrett and Munk (1975) internal wave model, expressed in terms of root-mean-square (rms) vertical displacement, $< \zeta^2 >^{1/2}$, and wavenumber bandwidth, β_*. For $\overline{\Delta z}$ of a few meters and less, probability densities of temperature gradients produced by strain were highly skewed, with modes less than the mean gradient (Figure 8.9). Fitting the model to observed histograms gave reasonable values of β_* but unrealistically small vertical displacements, $< \zeta^2 >^{1/2}$. As Δz increased to 10 m, distributions became more normal. Although the modeling was unrealistic in assuming jointly normal displacements at all Δz, Desaubies and Gregg (1981)

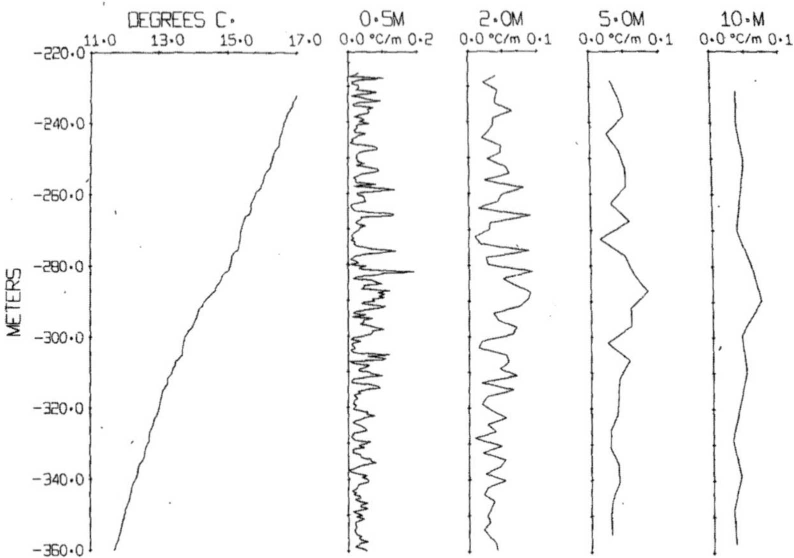

Figure 8.8 Mid-gyre temperature profile and first-difference gradients, $\Delta\theta/\Delta z$, over $\Delta z = 0.5, 2, 5$, and 10 m. (From Desaubies and Gregg, 1981. © American Meteorological Society. Used with permission)

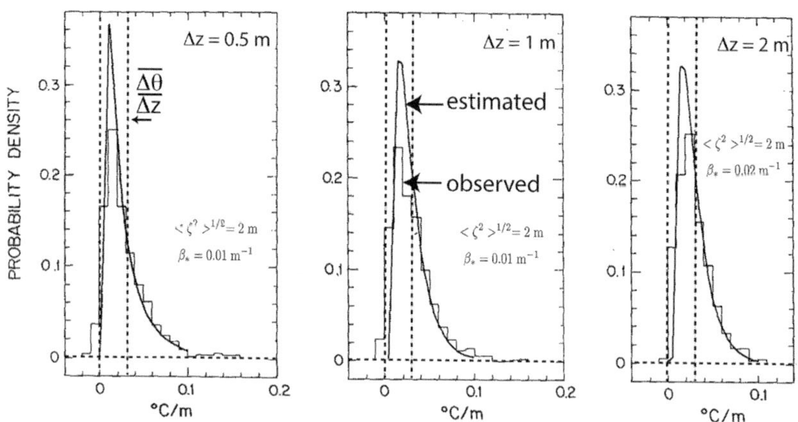

Figure 8.9 Histograms of temperature gradients in the central Pacific pycnocline compared with estimates assuming jointly normal vertical displacements consistent with GM75 and fitted values of wavenumber bandwidth and rms vertical displacement. (Adapted from Desaubies and Gregg, 1981. © American Meteorological Society. Used with permission)

concluded that most finestructure, including sharp kinks, can result from reversible internal wave strain.

Run tests of first-differenced profiles as a function of Δz show random vertical distributions for $\Delta z \geq 2$ m, but not for $\Delta z = 1$ m and smaller (Desaubies and Gregg, 1981). This was interpreted as evidence that strain cuts off between 1 and 2 m.

Desaubies and Smith (1982) applied GM75 to develop probability distributions for the gradient Richardson number, $Ri_g \equiv N^2/Shear^2$, assuming strain, ζ_z, and both shear components, S_x and S_y, are uncorrelated. The probability density of Ri was highly skewed with a mode near 0.6 for an rms strain of 0.5, a realistic value. Numerical simulations found that 6.5% of profiles had $Ri_g < 1/4$. The mean thickness of subcritical regions was $\lesssim 0.5$ m, and the largest was $\sim$3 m.

Strain as a Poisson Process

Following McKean (1974) and Hayes et al. (1975), Pinkel and Anderson (1992) treated finescale strain as a Poisson process, with strain elements randomly distributed in the vertical. Depending on the question, Poisson processes are described by Poisson or gamma forms of the exponential distribution. Expressing strain as

$$\gamma \equiv \Delta z/\overline{\Delta z}, \tag{8.2}$$

Pinkel (2020) gave semi-Langrangian and Eulerian probability densities as

$$Prob_{\mathrm{sL}}\left(\gamma|\overline{\Delta z}\right) = \frac{\kappa(\kappa\gamma)^{\kappa-1}\, e^{-\kappa\gamma}}{\Gamma(\kappa)}, \quad Prob_{\mathrm{Eul}}\left(\gamma|\overline{\Delta z}\right) = \frac{(\kappa\gamma)^{\kappa}\, e^{-\kappa\gamma}}{\Gamma(\kappa)}, \tag{8.3}$$

where $\kappa \equiv \kappa_0\overline{\Delta z}$, $Z_{\mathrm{corr\ sL}}$ is the vertical correlation length of strain in a semi-Lagrangian frame, and $\kappa_0 \equiv Z_{\mathrm{corr\ sL}}/(2 < \zeta^2 >) = 0.5-2$ m. The Eulerian strain covariance has a corresponding spectrum that reproduces the observed change in slope near 0.1 cpm.

8.3.3 Perspectives

Much remains to be learned about finestructure. Further progress will require more detailed measurements distinguishing small-scale signatures of internal waves from those of the vortical mode as a basis for modeling scales of 1–10 meters. This, in turn, is needed to predict the size and intensity of mixing patches.

8.4 Global Mixing Patterns

In spite of 50 years of microstructure profiling, the net results are too sparse to identify most global mixing patterns. Rather, microstructure ensembles have been used to validate mixing estimates from density overturns and from finestructure variances of shear and strain over vertical scales from tens to a few hundred meters (Dillon, 1982; Gregg, 1989b; Polzin et al., 1995; Whalen et al., 2015). This allows the the use of large volumes of strain-based mixing estimates from LADCP/CTD (Lowered Acoustic Doppler Current Profiler/conductivity-temperature-depth)

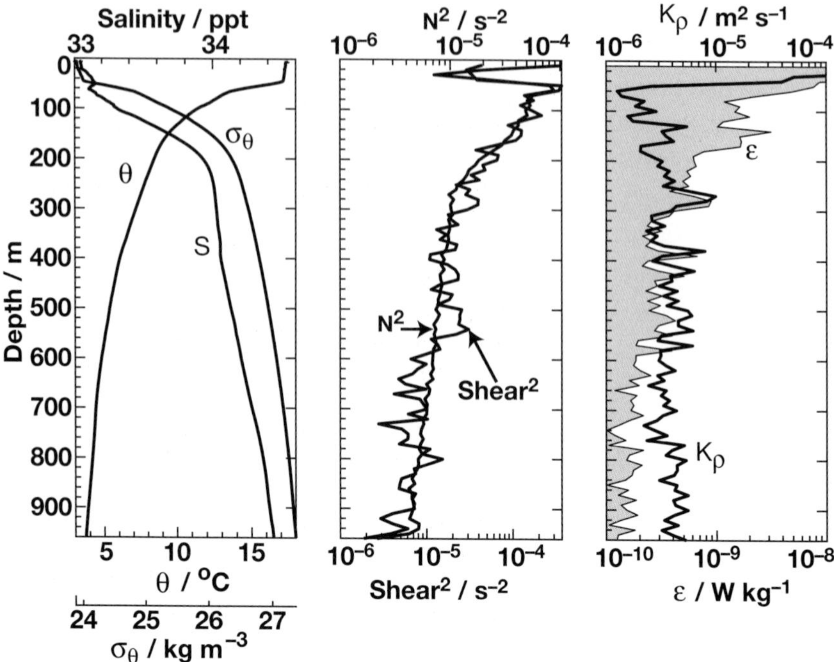

Figure 8.10 Averages over 10 m when internal waves were at GM during PATCHEX in the California Current. Note different N^2 and $Shear^2$ scales.

surveys and Argo profiling floats to reveal basin-scale trends with depth, latitude, and season, as well as regional variations linked to bottom roughness and forcing by tides and winds.

8.4.1 Depth and Latitude Dependence

When the internal wave field is close to background over a smooth seafloor, dissipation rates decrease with depth (Figure 8.10) following the $\epsilon \propto N^2$ dependence of (7.16). Because $K_\rho = \Gamma_{\text{mix}}\epsilon N^{-2}$, diapycnal diffusivity is roughly constant with depth over smooth bottoms, and the buoyancy flux, $J_b = -K_\rho N^2 \approx -0.2\epsilon$ follows the same vertical distribution as dissipation. Consequently, diapycnal velocity (2.66) is upward, with dense water being made lighter, and decreases to near zero as stratification weakens below 1 km.

Internal waves at levels close to GM had 3/4 of the dissipation coming from less than 20% of the samples (Figure 8.11). Below the shallow pycnocline, over half of the samples were below the noise level of 10^{-10} W kg^{-1}, while in the upper 200 m 60–70% were below $16\nu\overline{N^2}$, the threshold for a significant buoyancy flux in laboratory studies of grid turbulence (Rohr et al., 1987). Moreover, these dissipation rates were likely overestimates, as isotropy was assumed but was unlikely

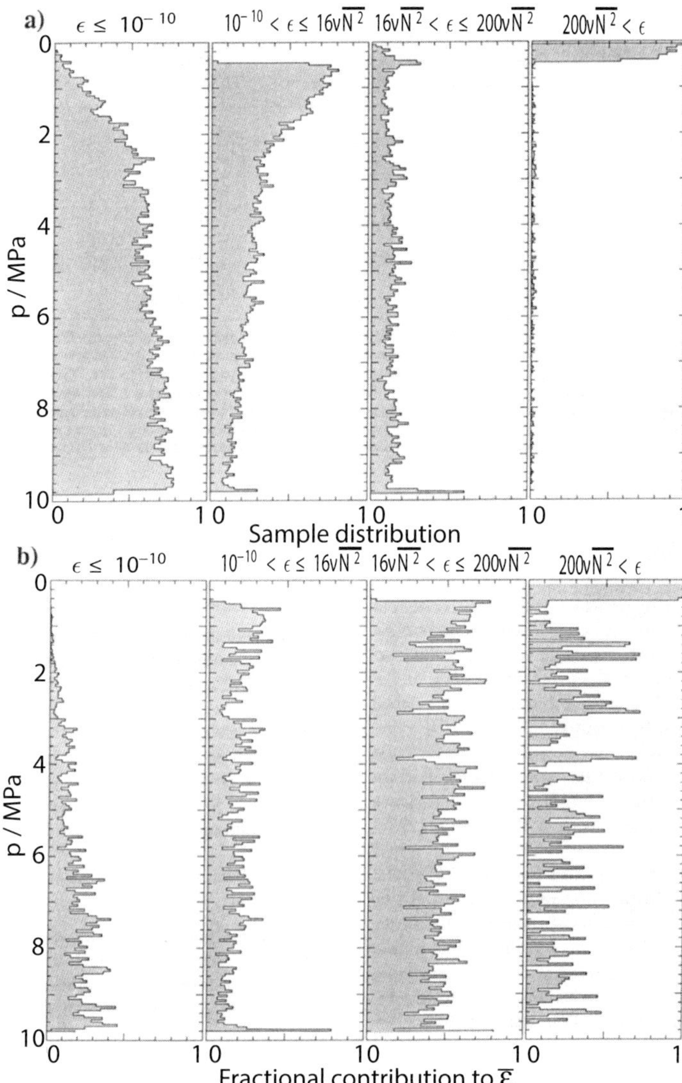

Figure 8.11 (Upper) Dissipation rate by activity class and (lower) contribution of the classes to $\overline{\epsilon}$. The observations were taken with the Multiscale Profiler during PATCHEX. (Adapted from Gregg and Sanford, 1988)

(Section 3.6.2). This class contributed 20–30% of $\overline{\epsilon}$. Roughly half of net dissipation came from a more active class still below the threshold for small-scale isotropy. Only a few percent had $\epsilon > 200\nu\overline{N^2}$, the threshold for full isotropy, but they were about 1/4 of $\overline{\epsilon}$.

Average dissipation and diffusivity profiles approaching the seafloor are strongly affected by bottom roughness. This is demonstrated by separately averaging

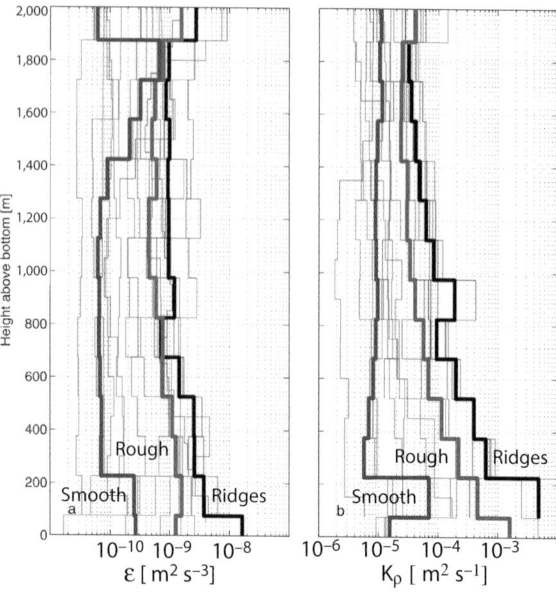

Figure 8.12 Average dissipation rates (a) and diapycnal diffusivities (b) combining ϵ measured directly with microstructure probes and inferred from finescale variability. In the bottom kilometer, dissipation and diffusivity are nearly constant with depth over smooth bottoms, but increase over rough bottoms and even more strongly above ridges. (Adapted from Waterhouse et al., 2014. © American Meteorological Society. Used with permission)

full-depth profiles over smooth bottoms, rough bottoms, and ocean ridges (Figure 8.12). Within the bottom kilometer, the average over smooth bottoms has $\epsilon \lesssim 10^{-10}$ W kg^{-1} and $K_\rho \lesssim 10^{-5}$ m^2 s^{-1}, except for the bottom boundary layer. This average may reflect sensor noise more than signals and should be considered an upper bound. The average dissipation rate over rough bottoms is 10 times larger, and the average over ocean ridges is twice that. Owing to decreased stratification accompanying increased dissipation, increases in K_ρ over roughness are even larger.

Binning K_ρ from LADCP/CTD casts reveals an order-of-magnitude dependence on latitude (Figure 8.13). Below the surface mixed layer and above 1.5 km, turbulent diffusivity is very weak near the equator, roughly 20 times the molecular diffusivity of heat. Diffusivity increases rapidly with latitude to about 30° and then more slowly until low stratification at high latitudes further elevates K_ρ. Other than between 50° and 60°, diffusivity also increases with depth by factors of 2–5.

Averaging by depth bins (Figure 8.14) demonstrates that the latitudinal dependence is consistent with the Henyey et al. (1986) model (7.16), but the rate of change at low latitudes is more gradual than predicted. All averages peak or have

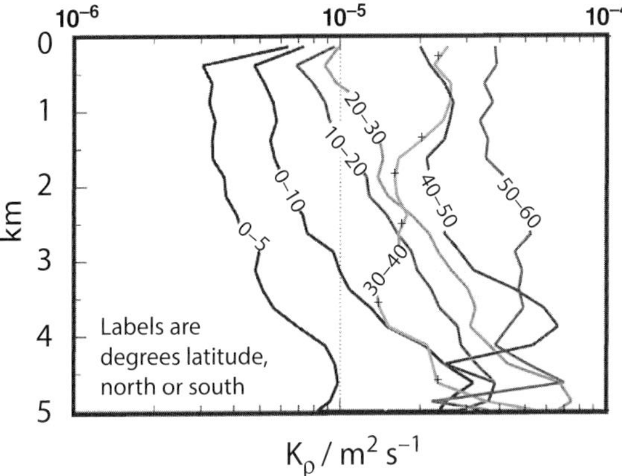

Figure 8.13 Diapycnal diffusivity from finescale shear and strain averaged in latitude bands. Diffusivity increases from the equator and with increasing depth. Ensembles contain 275 to 1,393 profiles. (Adapted from Kunze et al., 2006a. © American Meteorological Society. Used with permission)

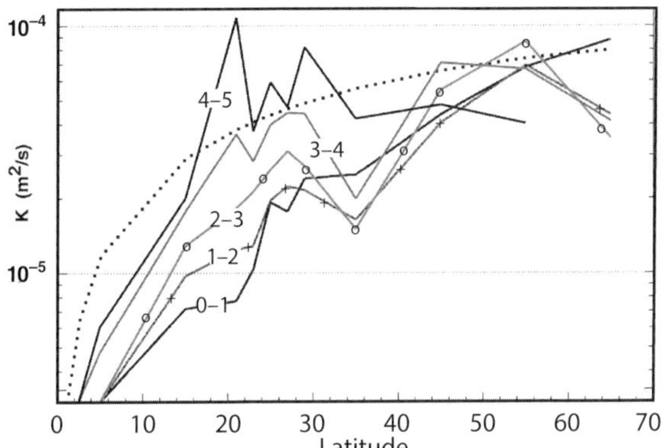

Figure 8.14 K_ρ from finescale shear and strain in depth bands, e.g. 4–5 km. Latitudinal variability roughly follows (7.16), per the dotted line. K_ρ increases with depth at low and mid-latitudes. (Adapted from Kunze et al., 2006a. © American Meteorological Society. Used with permission)

inflections before reaching 30°. MacKinnon et al. (2013) attribute a similar increase along a tidal beam north of Hawaii to parametric subharmonic instability (PSI) acting on the M_2 internal tide, which has a critical latitude of 28.9°. Calculations by Furuichi et al. (2005) find the efficiency of PSI decreasing rapidly equatorward of 25°, resulting in a peak in 30 m shear variance, and presumably mixing, between 25° and 30°. The effect of PSI appears to be superimposed on the broad latitude trend. Diffusivity continues increasing to 45°, but less regularly.

The irregularity results primarily from differences in the latitudinal distribution of bottom roughness.

8.4.2 Geographic Patterns

As found in the Brazil Basin (Polzin et al., 1997), mixing intensity varies by decades across ocean basins in response to variations in bottom roughness and tidal currents. A meridional transect in the North Atlantic (Figure 8.15) shows low diffusivity extending to the bottom at low latitudes, with average levels increasing northward and only a few modest hot spots, e.g. over Rockall Trough. By contrast, a zonal section across the southern Indian Ocean has low diffusivity in the upper half of the water column away from ridges, but nowhere do low levels extend to the bottom. Elevated diffusivities, $\sim 10^{-4}$ m^2 s^{-1}, fill the bottom kilometer and rise to the surface over the Southwest Indian Ridge. Lesser hot spots occur over several other rises.

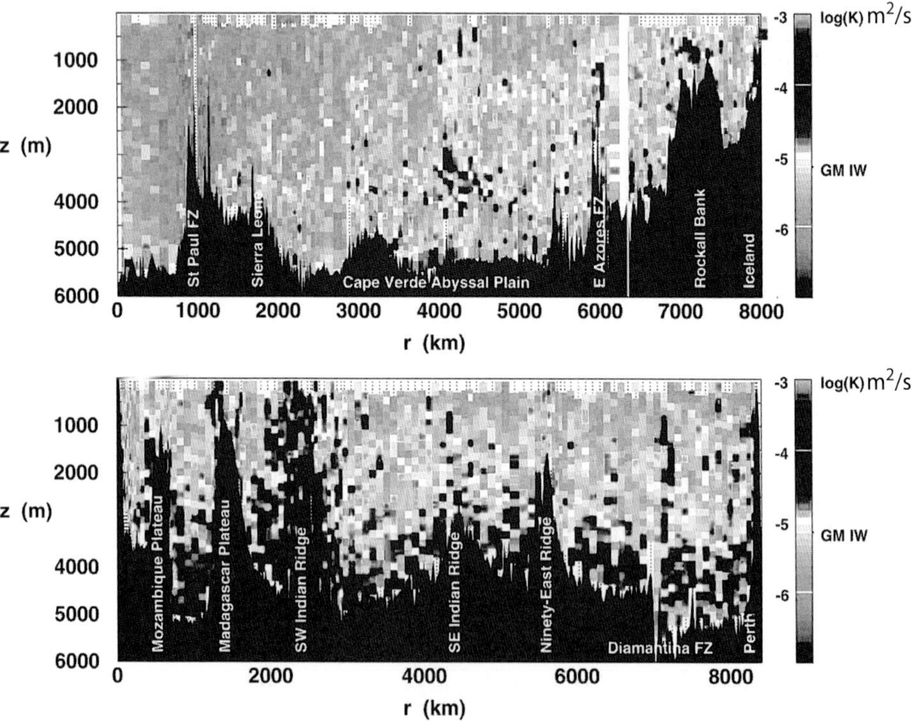

Figure 8.15 Vertically integrated diapycnal diffusivity from finescale shear and strain in the Atlantic from 7°S to Iceland (upper) and in the Indian Ocean along 32°S from Madagascar to Perth (lower). (Adapted from Kunze et al., 2006a. © American Meteorological Society. Used with permission.) (A black and white version of this figure will appear in some formats. For the color version, please refer to the plate section.)

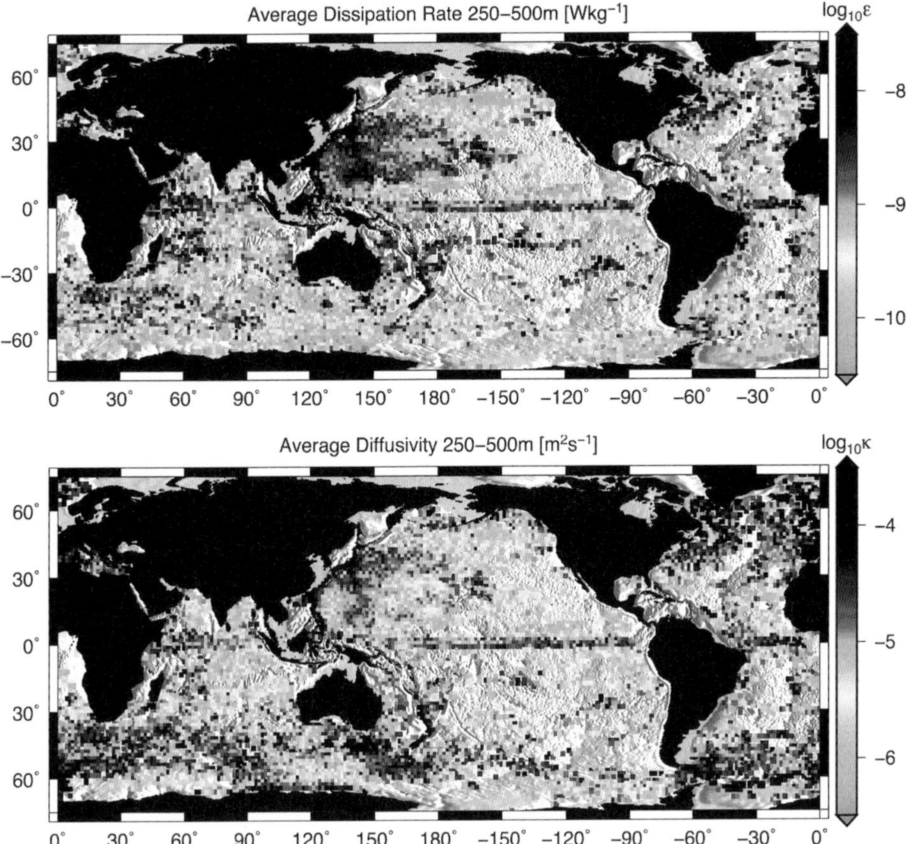

Figure 8.16 Average dissipation rates and diapycnal diffusivities from strain in Argo profiles between 2006 and 2011. (Adapted from Whalen et al., 2012.) (A black and white version of this figure will appear in some formats. For the color version, please refer to the plate section.)

From five years of Argo profiles, pycnocline dissipation rates are highest under storm tracks in the northwestern Pacific and Atlantic (Figure 8.16). Levels were too weak to estimate in the northeastern Pacific and much of the central South Pacific. Dissipation is also surprisingly low in the Southern Ocean, except where the circumpolar flow is constricted poleward of Africa and South America. Owing to weak stratification, however, normalizing dissipation by N^2 to estimate K_ρ shows the Southern Ocean to be the most extensive area of high diffusivity. The North Atlantic is next, also as a consequence of low stratification. Elevated dissipation and diffusivity along the equator are not consistent with Figure 8.14 and may be spurious.

Sufficient data were collected in the northwest Pacific to form time series revealing seasonal cycles of dissipation rates in the upper ocean (Figure 8.17). The largest

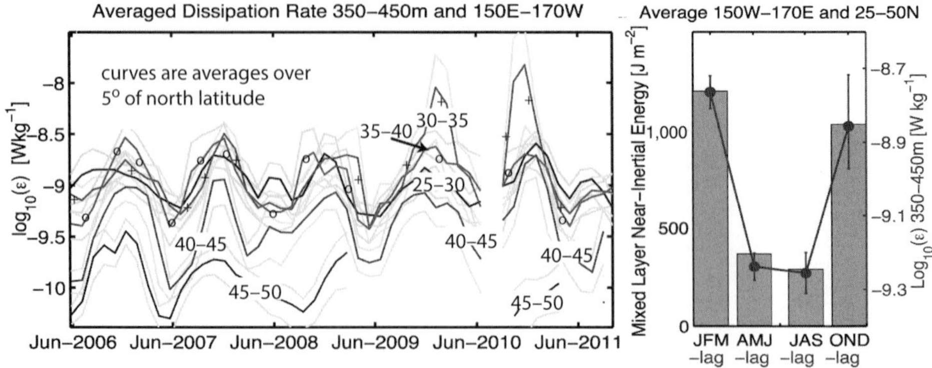

Figure 8.17 (Left) Seasonal variation of ϵ, with confidence levels as light lines. (Right) Mixed-layer near-inertial energy (left axis) and dissipation rate (right panel) in the northwest Pacific. To account for downward propagation, ϵ was lagged 50 days. (Adapted from Whalen et al., 2012)

signals span a decade in amplitude between 30° and 35°. Elsewhere, amplitudes varied by factors of 2–3, peaking in winter and sinking to minima in June. The ϵ cycle at 350–450 m is strongly correlated with near-inertial kinetic energy in surface mixing layers but lags the surface by 50 days, the time required for the waves to propagate to depth.

8.5 Mixing Patches and Overturns from Breaking Internal Waves

In the stratified interior, mixing occurs intermittently in patches (Nasmyth, 1970) characterized by sharply increased amplitudes of the centimeter-scale gradients producing ϵ and χ_T. Some occur as thin 'puffs' (Gregg, 1980a), but most are in distinct overturns. Density overturns contain microstructure patches, but microstructure patches can span multiple overturns (Figure 8.18). The one on the middle interface suggests the early stage of a KH billow. The larger overturns above and below the interface were in relatively uniform gradients, either as late stages of KH or different forms of instability. Sorting out similar events requires that overturns and patches be identified separately and followed in time.

8.5.1 Identifying Mixing Patches

Having well-resolved temperature microstructure but not small-scale velocity, Gregg (1980a) identified microstructure patches as contiguous zero-crossings of fully resolved temperature gradients. Features ranged from isolated negative gradients several centimeters thick to 30 m patches, collectively occupying 7–36% of the water column and distributed independently of finestructure. Adding velocity

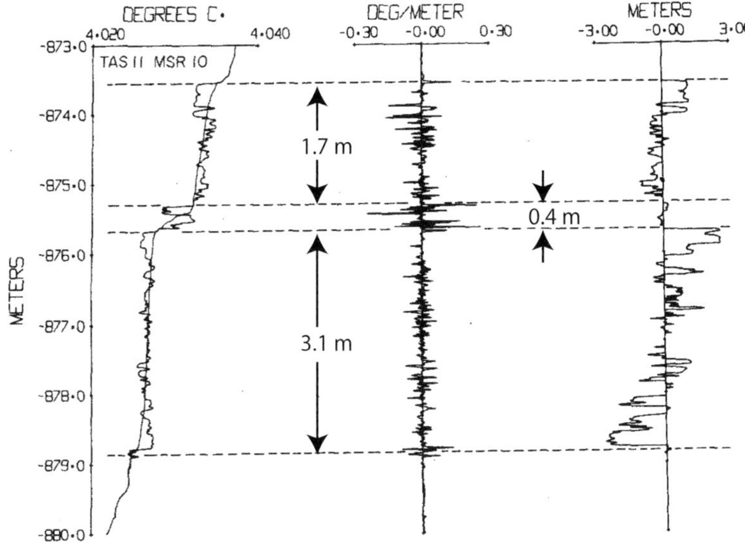

Figure 8.18 (Left) Observed temperature from a profiler falling at 80 mm/s overlaid on the same data sorted to decrease monotonically with depth. (Middle) Microscale temperature gradients. (Right) Vertical displacements of observed temperatures from their depths in the monotonic profile. The 5.2 m patch extends across three overturns. (Adapted from Gregg, 1980a. © American Meteorological Society. Used with permission)

microstructure provides a valuable second measure for identifying patches (Gregg et al., 1986), but the increase in vehicle speed needed to enhance velocity signals decreased the resolution and hence usefulness of the temperature microstructure. Also, identification shifted to ϵ and χ_T computed by integrating spectra over intervals of at least 0.5 m, precluding identification of thinner features.

Thermistor chain tows show the horizontal structure of patches and often reveal the nature of the instability. For example, the patch in Figure 8.19 was 5–10 m high, extended 3 km (Rosenblum and Marmorino, 1990), and contained waves resembling KH instabilities with lengths of 30–50 m. Cox numbers from dual-needle conductivity probes were 10^3-10^4 in the braids separating billows.

8.5.2 Identifying Overturns

Adapting Thorpe (1977), overturns are identified by summing vertical displacements of observed density from their positions after sorting the data to increase monotonically with depth. Owing to noise, summing begins at the first displacement larger than a noise threshold above zero and ends when the cumulative sum returns to that threshold (Figure 8.18). Numerous problems arise, leading Galbraith and Kelley (1996), and many others, to develop tests for validating overturns. When

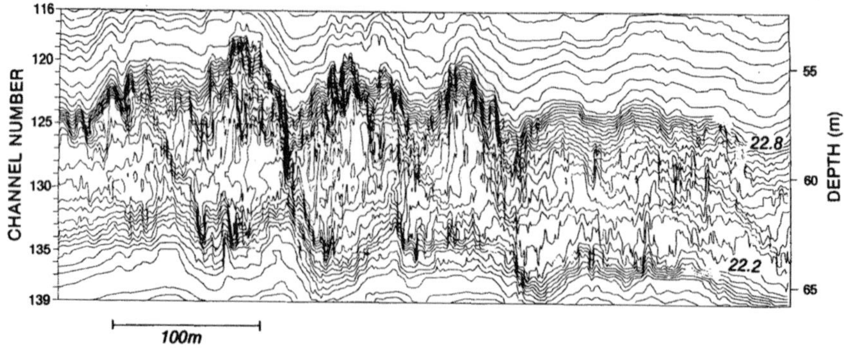

Figure 8.19 Temperature contours at 40 mK intervals from a thermistor chain towed through a mixing patch. Small-scale activity in the billows resembles the acoustic backscatter in Figure 1.4. (From Marmorino, 1987b. © American Meteorological Society. Used with permission)

salinity spiking contaminates density, temperature is used for overturns in regions free of salt-stratified inversions. Noise limits most identifications to strong stratification, but large overturns can rise above noise in weak stratification (Nash et al., 2007), but wavelet denoising (Figure 8.20) can extend identifications to N^2 as low as 2×10^{-6} s^{-2}. (Piera et al., 2002)

8.5.3 Patch Characteristics

In spite of the differences in variables and spatial resolution, profiling and towing demonstrate similar patch characteristics.

From Vertical Profiles

Zero-crossings in temperature gradients analyzed by Gregg (1980a) were thinner than 50 mm, with maximum lifetimes less than N^{-1}. Those in a lake had amplitudes exceeding Rayleigh criteria for instability, further indicating active features. Patches in mid-gyre collectively spanned 7–35% of the water column. Thick patches were more likely where gradient Richardson numbers were low, but otherwise patches were distributed independently of finestructure. Most were thinner than 2 m, but some were 30 m and spanned multiple overturns. Maximum displacements were appreciably less than patch thickness.

Profiling alongside a float attached to a subsurface drogue, Gregg et al. (1986) followed persistent ϵ and χ_T patches in a near-inertia wave. Sustained mixing at the velocity maximum suggested critical layer interactions. Distributions of patch thickness decreased roughly exponentially with increasing thickness (Figure 8.21). Mid-sized patches, 5–10 m, made the largest contributions of the thickness bins, but more than half of net mixing came from patches thinner than 4 m. Dividing the

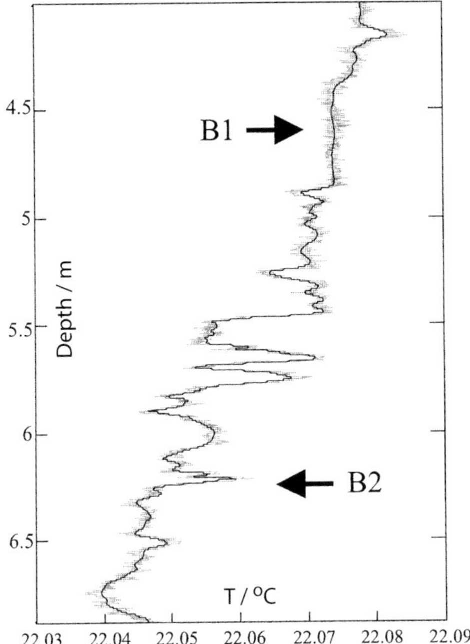

Figure 8.20 Temperature in a weakly stratified surface mixed layer denoised with a discrete wavelet transform and reconstructed with a fast wavelet transform (solid). Comparison with the original profile (dots) shows that, unlike low-pass filtering, denoising smoothes sections with little structure (B1) while retaining thin features (B2). Displacements from denoised data were further screened by dividing the remaining noise by the local gradient. (Adapted from Piera et al., 2002. © American Meteorological Society. Used with permission)

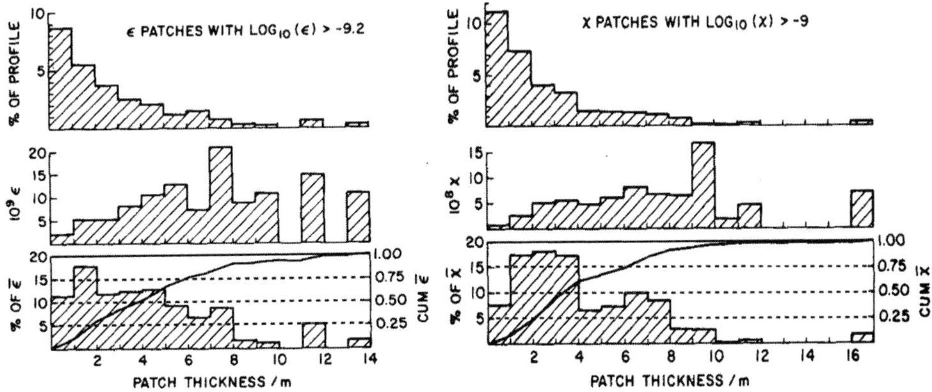

Figure 8.21 From top to bottom, frequency of occurrence, average mixing rate, and fractional cumulative net mixing rate for ϵ (left) and χ_T (right) in patches spanning a near-inertial wave. (Adapted from Gregg et al., 1986. © American Meteorological Society. Used with permission)

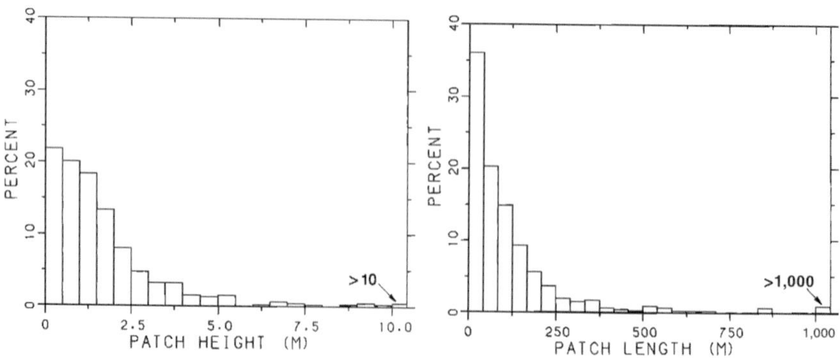

Figure 8.22 Distribution of patch heights and lengths from thermistor-chain tows away from fronts in the seasonal thermocline of the Sargasso Sea. (Adapted from Rosenblum and Marmorino, 1990)

horizontal kinetic energy of the wave by the average dissipation rate gave a decay time of 4–8 days. Within the patches, the largest overturns were 1 m, but most were 0.1–0.2 m.

From Horizontal Tows

Tows in the Sargasso Sea identified patches in 8% of runs adjacent to a warm-core ring. Away from fronts, patches occupied only 2% of the runs, with short, thin patches being most common, and the frequency of occurrence decreasing nearly exponentially in length as well as in thickness (Figure 8.22). Plotting length versus height aligns along an aspect ratio of 100, with values between 20 and 200. Thicker patches were likely to be associated with near-inertial waves (Marmorino et al., 1987; Rosenblum and Marmorino, 1990), but thinner ones roughly followed internal wave potential energy, leading the authors to infer mixing by the internal wave continuum rather than by near-inertial waves. This is similar to a finding from microstructure profiles that the frequency of patches thinner than 2 m was consistent with predictions derived from GM, but the occurrence of thicker patches exceeded predictions (Gregg et al., 1986).

8.6 Double Diffusion in the Pycnocline

Mixing patches produced by salt fingers must be distinguished from those resulting from mechanical turbulence to apply the correct expression for diapycnal diffusivity: (4.25) for fingers or (3.71) for turbulence. Large areas of mid-latitude thermoclines are diffusively unstable to salt fingering (Figure 8.23), but the density ratio is not low enough, $R_\rho \lesssim 1.6$, for fingers to form staircases. Nonetheless, based on the constant R_ρ curvature of temperature–salinity relations in major ocean basins

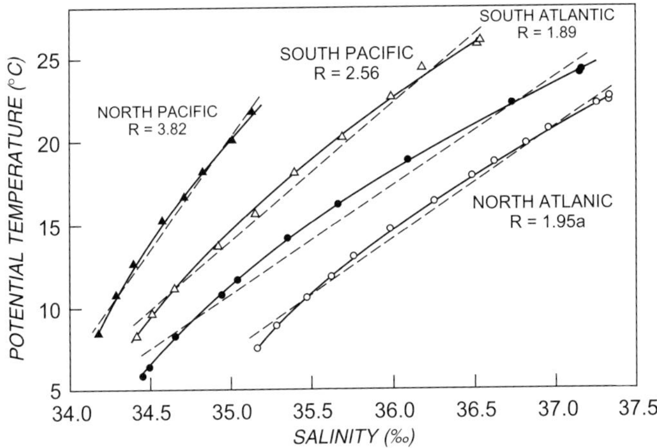

Figure 8.23 Potential temperature versus salinity from hydrographic data (symbols) in Atlantic and Pacific pycnoclines. The observations fit constant-R_ρ curves (solid) better than the linear least-square fits (dashed) expected for mechanical mixing. (From Schmitt, 1981. © American Meteorological Society. Used with permission)

(Ingham, 1966), Schmitt (1981) inferred a strong role for salt fingering, arguing that the inverse dependence of fingering intensity on R_ρ smoothes local divergences and convergences of finger fluxes to produce θS relations with constant R_ρ. By contrast, turbulent mixing produces linear θS relations.

8.6.1 Identifying Salt Fingers with Horizontal Tows

Highly resolved temperature fluctuations measured with a towed cold-film revealed different internal structures within mixing patches several hundred meters apart in the subtropical gyre of the North Pacific (Gargett and Schmitt, 1982). Temperature fluctuations in some patches were more narrowband than others, with lesser ratios of largest to smallest amplitudes (Figure 8.24). The density ratio averaged 1.6 along the tow but rose to 3 in some places, making fingering less likely, although some variation in R_ρ and patch structure may have resulted from differing vertical displacements by internal waves. Rewriting the maximum growth rate for salt fingers as

$$\sigma_{\rm sf} = (\kappa_T/\nu)^{1/2} N \left[(1 - \alpha(T_z/S_z))^{-1/2} - 1 \right] \quad \left[{\rm s}^{-1} \right], \tag{8.4}$$

Gargett and Schmitt (1982) argued that the observations were consistent with fingering being most likely in high-N regions produced by transient internal wave strain.

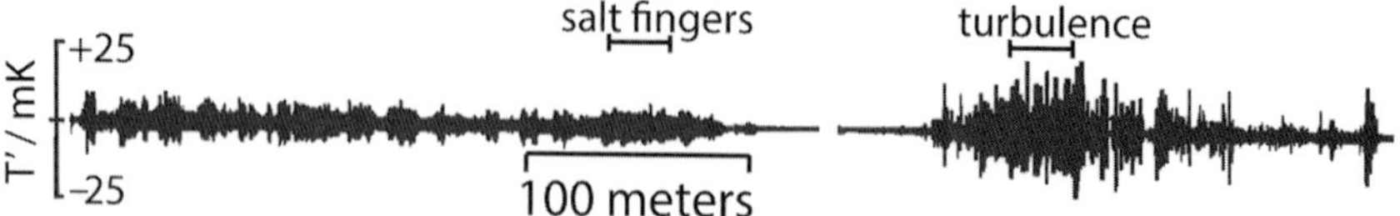

Figure 8.24 Temperature fluctuations from a cold-film towed at 290 m depth in the eastern North Pacific. The mixing patch on the left was attributed to salt fingers, owing to the smaller range of maximum to minimum signals compared to typical turbulent patches. These patches were separated by ~400 m. (Adapted from Gargett and Schmitt, 1982)

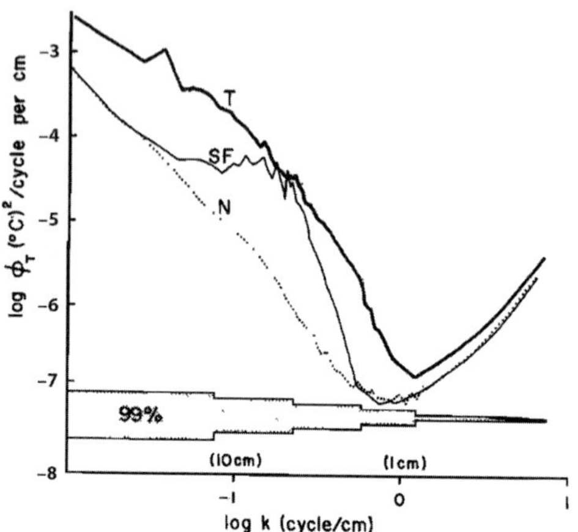

Figure 8.25 Temperature spectra from salt finger, SF, and turbulence, T, tow sections in Figure 8.24. The fingering spectrum has a narrower bandwidth and rises above the noise spectrum, N, over a limited wavenumber range. (From Gargett and Schmitt, 1982)

Spectra confirmed the narrowband nature of fingering patches in contrast to patches with broadband signals (Figure 8.25). The relatively narrow peak at 3–5 cm corresponds to finger diameters predicted for conditions along the tow. Identification in other tow sections was ambiguous, likely owing to varying R_ρ, shear, encounter angles, etc. Marmorino (1987a) found a similar situation along tows in the Sargasso Sea.

Subsequently, Holloway and Gargett (1987) distinguished fingers from turbulence using kurtosis, $K_{T'} \equiv \overline{T'^4}/\left(\overline{T'^2}\right)^2$. A measure of the tails of distributions, kurtosis should be unity for square waves and close to 3/2 for sine waves. For the patches in Figure 8.24, the distinction is large: $K_T = 3.04$ for fingering and 6.75 for turbulence (Figure 8.26). Marmorino and Greenewalt (1988) also found kurtosis a useful discriminant in some situations, but it was not sufficient in the salt finger

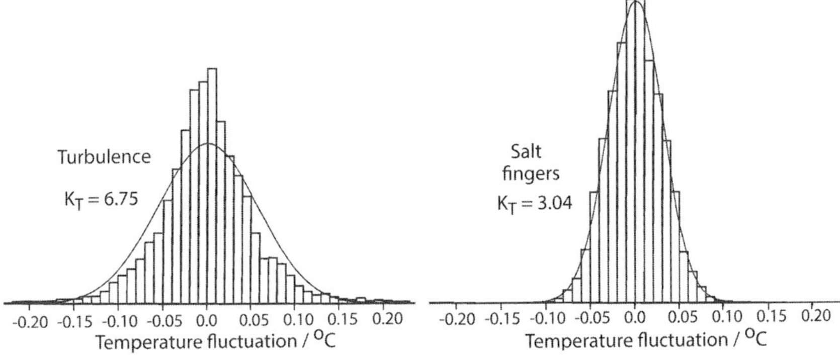

Figure 8.26 Histograms of rms temperature fluctuations between 1 and 90 cpm for the salt finger and turbulence patches in Figure 8.24. Solid curves are normal distributions with the same variance, and K_T is the kurtosis. (Adapted from Holloway and Gargett, 1987)

staircase east of Barbados, where some interface segments had high kurtosis and others had low values.

Zero-crossings of centimeter-scale conductivity gradients versus R_ρ were also useful in the Sargasso Sea (Mack, 1989). With $R_\rho = \pm 12$ along a tow, zero-crossings peaked at $1 < R_\rho < 2$, where finger growth rates are greatest. Salt finger patches were one to several meters vertically and hundreds of meters horizontally, occupying 27% of a tow between 100–115 m depth, 11% at shallower depths, and 35% near a front.

To obtain more robust detections, Mack and Schoerberlein (1993) used joint distributions of kurtosis and slopes of gradient spectra fitted to the low-wavenumber side of the spectral peak. Fingers have higher slopes and lower kurtosis than turbulence. Some sections were characteristic of fingering or of turbulence, while other sections had both characteristics with minimal overlap (Figure 8.27). Formulating the test as a log-likelihood ratio,

$$\lambda = \ln \left[\frac{\text{prob}(Slope, \log(Kurtosis)|\text{salt fingers})}{\text{prob}(Slope, \log(Kurtosis)|\text{turbulence})} \right], \tag{8.5}$$

was more successful than either test alone.

8.6.2 Signatures of Salt Fingers in Profiles

Salt fingering has been inferred from vertical microstructure in monotonic profiles free of thermohaline intrusions based on systematic changes in mixing efficiency with the density ratio. Specifically, $\gamma_{\chi\epsilon}$ is higher where R_ρ is most favorable for fingering and decreases at higher ratios where fingering is not likely (Ruddick

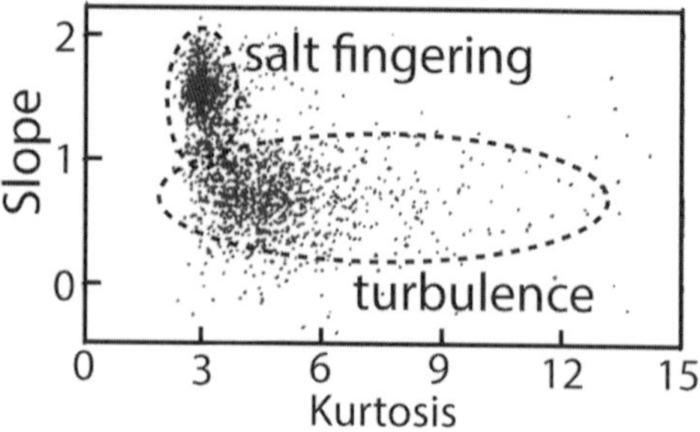

Figure 8.27 Spectral slope at $0.7 \leq k_h \leq 12$ cpm versus kurtosis for horizontal conductivity gradients where salt fingering and turbulence have relatively distinct distributions. (Adapted from Mack and Schoerberlein, 1993. © American Meteorological Society. Used with permission)

et al., 1997; St. Laurent and Schmitt, 1999). The evidence is compelling, but the analysis depends on the implicit assumption by Oakey (1982) that $K_S = K_T$ during turbulent mixing (Section 3.12) and needs to be re-examined for the possibility of differential diffusion.

Fingering is also considered the likely source of microstructure patches beneath salt-stabilized temperature inversions (Figure 8.28). To appear in profiles, fingers must be tilted by background shear, which must be present for the intrusions to form. Washburn and Käse (1987) found similar structures southeast of the Azores with a CTD survey in a broad convergence of several water masses at mid-depths. Mapping R_ρ revealed that 71% of the volume was unstable to fingering, 24% with $R_\rho < 2$, and 5% was unstable to diffusive layering. Also, Mack (1985) traced similar patches for hundreds of meters in the Sargasso Sea.

8.7 The Southern Ocean

Driven eastward by the Roaring Forties, the Antarctic Circumpolar Current (ACC) flows around the globe unobstructed by continents but kept in steady state by the form drag of the Kerguelen Plateau, Campbell Plateau, and Drake Passage (Gille, 1997). Displaced and sheared laterally by meoscale eddies to the north, the ACC contains strong jets centered on narrow fronts: the Subantarctic Front (SAF) to the north, the Polar Front (PF) in the middle, and the Southern ACC Front (SACCF) to the south. To test inferences of strong mixing in the ACC based on water mass analysis and inverse calculations, mixing rates have been estimated using finestructure variances from existing surveys and Argo floats. Focused studies including microstructure profiling have examined the Kerguelen Plateau and the Drake Passage, where the thickening of an artificial tracer was observed.

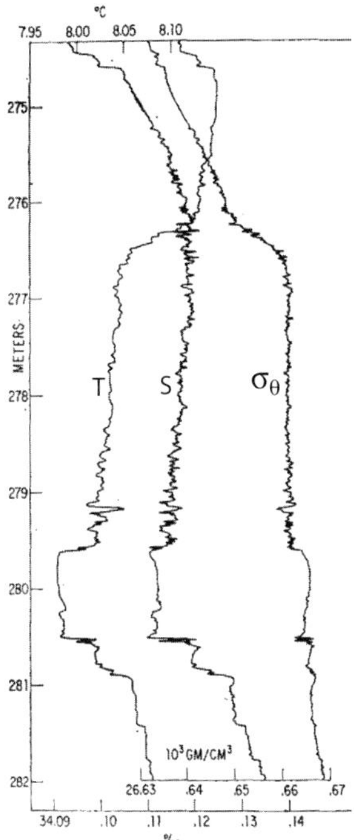

Figure 8.28 Microstructure taken with the probe in Figure 5.19 through a sequence of salt-stabilized temperature inversions in the California Current. The profile is diffusively unstable to salt fingering between the temperature maximum at 276.5 m and the minimum at 279.5 m, a section where potential density, σ_θ, is nearly uniform with $R_\rho \sim 1$. Temperature and salinity fluctuations likely were tilted salt fingers. The homogenous layer between 279.5 and 280.5 is characteristic of profiles with multiple intrusions and was mixed by buoyancy fluxes from fingering above and diffusive layers below. (Adapted from Gregg and Cox, 1972)

Strain in Argo profiles between 300 and 1,800 m reveals elevated mixing intensity over rough bottoms (Wu et al., 2011). For example, $K_\rho \sim 10^{-5}$ m^2 s^{-1} over the abyssal plain upstream (west) of the Drake Passage and a decade higher above roughness in and east of the strait (Figure 8.29). The increase is attributed to deep jets generating upward-propagating internal waves after encountering rough bottoms (Section 7.3.4). Inferences from Argo and CTD profiles agree with microstructure measurements where they have been compared in the Southern Ocean, despite using $R_\omega = 7$ with the strain measurements (Frants et al., 2013). Strain in full-depth CTD casts from global surveys also indicates increased mixing in the bottom 500–1,000 m (Sloyan, 2005), but, owing to low stratification in much of the water column, mixing estimates using density overturns have large errors and were not considered.

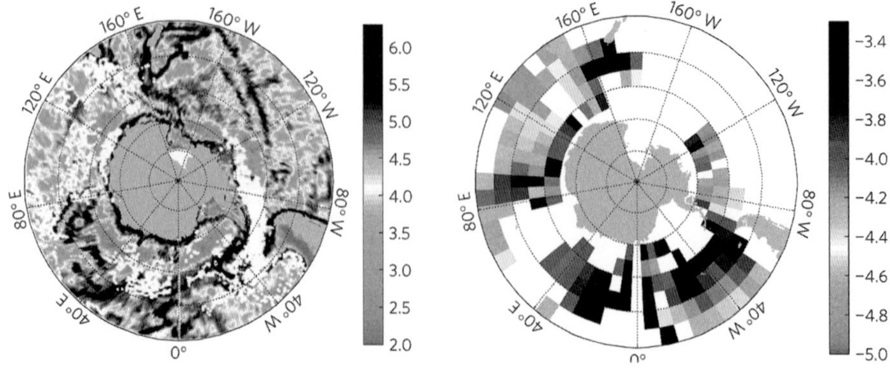

Figure 8.29 (Left) Bottom character as $\log_{10} Roughness(\text{m}^2)$. (Right) $\log_{10}$ $(K_\rho(\text{m}^2\text{s}^{-1})$ over 300–1,800 m. (Adapted from Wu et al., 2011.) (A black and white version of this figure will appear in some formats. For the color version, please refer to the plate section.)

8.7.1 The Kerguelen Plateau

As part of the Southern Ocean Finestructure (SOFine) project to study the momentum balance of the ACC, a survey with microstructure, CTDs, LADCPs, and EM-APEX floats was conducted along the northern side of Kerguelen Plateau, a feature rising 2 km above the surrounding seafloor and so extensive that it is considered a microcontinent. Vertically integrated dissipation rates in the upper 1.5 km of (1–5) mW m^{-2} were similar to estimates of energy fluxes from winds to near-inertial waves and a decade larger than over the abyssal plain to the west (Waterman et al., 2013). Internal waves had vertical and horizontal wavelengths of 0.2 and 15 km and a horizontal group velocity of 30 mm s^{-1} (Meyer et al., 2015b). Varying with current speed, roughness, and wind speed, diapycnal diffusivities, estimated with (7.16), were 10^{-6} to 10^{-3} m^2 s^{-1} (Meyer et al., 2015a). Intense mixing between the Subantarctic Front and the Subtropical Front farther north was carried downstream by the strong currents.

Turbulence was also elevated in the bottom 1–2 km, with integrated dissipation rates of 1 mW m^{-2} where relatively high-frequency internal waves were propagating upward (Waterman et al., 2013). Although associated with fast, deep flows over rough bottoms, observed dissipation rates were about one-tenth of predictions applying finescale parameterizations to observations. Moreover, deep dissipation was only 2–20% of the estimated lee wave flux. Similarly low dissipation-to-production ratios have been found for internal tides generated by barotropic tides where there was independent confirmation of the flux propagating away (e.g. Klymak et al., 2006a). Lacking similar flux measurements for the lee waves, Waterman et al. (2014) explored other possibilities, e.g. dissipation was overestimated because wave fields are not in equilibrium close to generation sites, and/or much of the energy is transferred to the mean flow by wave–mean flow interactions including critical layers (Kunze and Lien, 2019).

8.7.2 The Drake Passage

The Drake Passage (Figure 8.30) constricts the ACC to 500 km between 1 km isobaths, accelerating the flow from 0.05 m s^{-1} just upstream in the southeast Pacific to 0.5 m s^{-1} between South America and Antarctica. A tracer released upstream at 1,500 m depth gave $K_\rho = 1.78 \pm 0.06 \times 10^{-5}$ m^2 s^{-1} over smooth bottoms before entering the constriction, and $3.6 \pm 0.6 \times 10^{-4}$ m^2 s^{-1} in the Drake Passage, a 20-fold increase (Watson et al., 2013).

The Diapycnal and Isopycnal Mixing Experiment (DIMES) ran four survey lines across the strait, taking full-depth CTD/LADCP casts combined with microstructure profiles at some stations (Sheen et al., 2013). In the Southeast Pacific (T1), average dissipation rates gradually decreased downward to 10^{-10} W kg^{-1} at 2 km above the bottom and then rose by half a decade in the bottom kilometer (Figure 8.31).

Owing to decreasing stratification, K_ρ increased from 10^{-5} m^2 s^{-1} near the surface to slightly above 10^{-4} close to the bottom. Bottom-enhanced dissipation under fronts increased in magnitude and thickness over rough bottoms as the flow

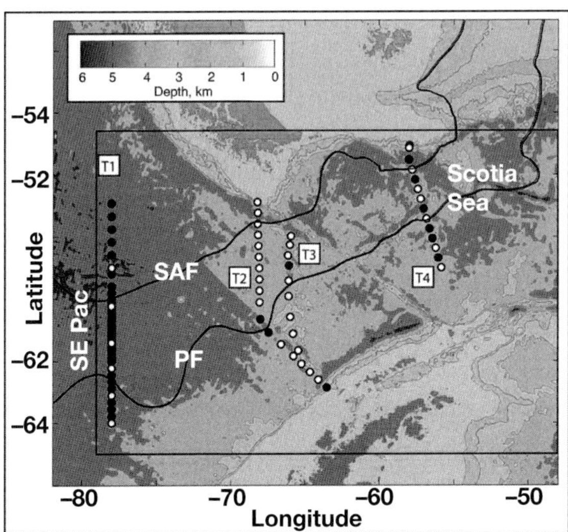

Figure 8.30 Bathymetry of the Drake Passage with DIMES lines T1 through T4 as open circles for combined microstructure and wire-lowered finestructure, and black circles for finestructure only. Dark lines mark the Subantarctic (SAF) and Polar (PF) fronts. (Adapted from Sheen et al., 2013)

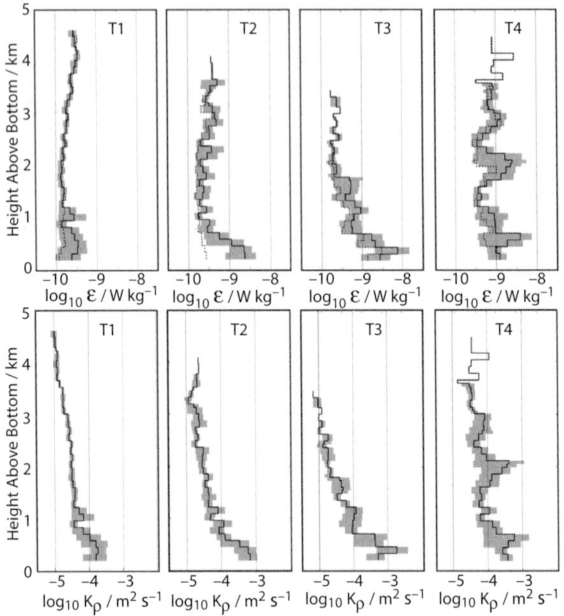

Figure 8.31 Mean (solid) and median (dashed) mixing rates along the DIMES sections. Shading marks 90% confidence limits of the means. Bottom-intensified mixing was strongest in T3 but extended through the water column in T4. (Adapted from Sheen et al., 2013)

passed through the strait until ϵ averaged 10^{-9} W kg^{-1} throughout the water column in the Scotia Sea (T4). The corresponding diffusivity, 10^{-4} m^2 s^{-1} over a vertical span of nearly 5 km, is extraordinary compared with profiles at higher latitudes. At least in the top half of the ocean, these levels continued far downstream of the Drake Passage (Figure 8.29). Away from fronts, dissipation rates were $\sim 10^{-10}$ W kg^{-1}, with no bottom enhancement (St. Laurent et al., 2012). The estimated internal wave energy flux generated at the bottom is comparable to wind work on the surface, and dissipation accounts for only 10–30% of it.

8.8 The Arctic

At shallow depths, the Arctic is strongly stratified by salt, with several haloclines increasing salinity in steps from 25 ppt at the surface to 35 ppt at 280 m (Figure 8.32). The temperature minimum at the bottom of the cold halocline marks the upper boundary of the thick intrusion of Atlantic water that dominates the shallow heat budget. In the Canada Basin, the intrusion core is near 425 m. Owing to the low temperatures, the thermal expansion coefficient is very small, with $\alpha/\beta \approx 0.02$ (Figure 2.15) making temperature a nearly passive tracer. Because heat stored in the Atlantic Water could melt ice if diapycnal fluxes through the halocline

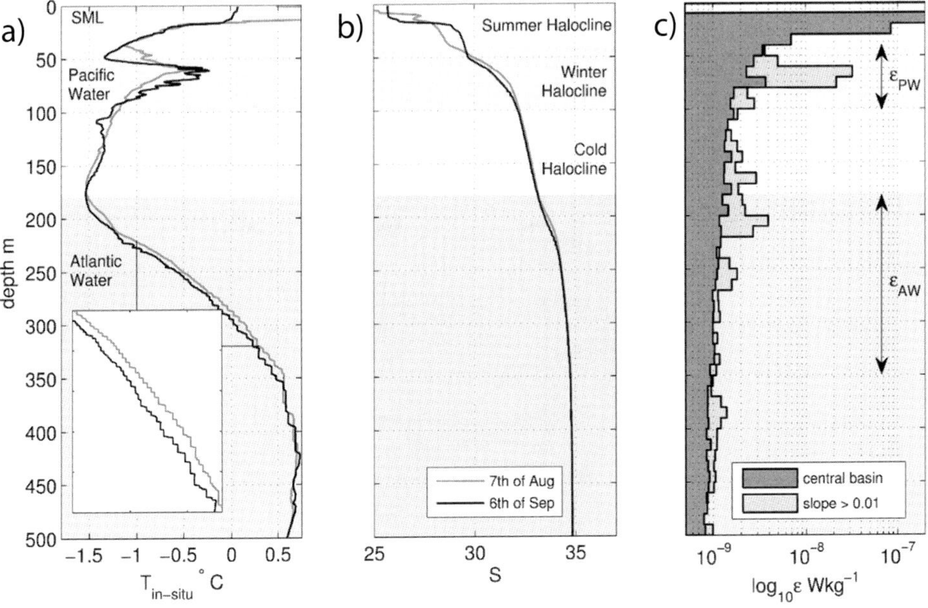

Figure 8.32 Temperature (a) and salinity (b) a month apart in late summer 2007 over the Central Canada Basin. (c) Average dissipation rate in the central basin and over seafloors with slopes > 0.01. The insert in (a) expands the diffusive staircase. (Adapted from Lincoln et al., 2016)

were strong, much work has been done to understand and quantify mixing above the Atlantic core. One aspect involves diffusive layering (Section 4.6).

8.8.1 Internal Waves

Below the ice pack and over abyssal plains of the central Arctic, internal wave energy is a few percent of mid-latitude levels (Figure 8.33). Spectra of vertical displacement have a slope closer to ω^{-1} than to GM's ω^{-2}, and their energy is 3–7% of the model (Levine et al., 1987). Levine (1990) verified that the horizontal coherence of displacement is consistent with a field of random internal waves following Wentzel-Kramers-Brillouin scaling. The low coherence, however, demonstrated that the wavenumber bandwidth was 10 times larger than at lower latitudes. The average spectrum of horizontal velocity versus vertical wavenumber had normalized energy levels 1.3–7.6% of GM and a slope of –0.5 at low wavenumber (Figure 6.26). Above 0.02 cpm, the slope increased and the spectrum merged with those from mid-latitude. Concentrated at high wavenumber, shear variance was 10% of GM. A substantial fraction, 23%, of the horizontal kinetic energy was attributed to the vortical mode.

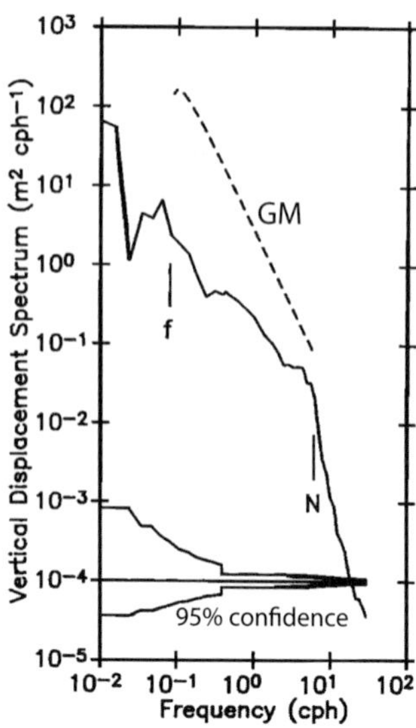

Figure 8.33 Frequency spectrum of internal wave vertical displacements 250 m beneath the ice pack in the Beaufort Sea during March to May 1985. (Adapted from Levine et al., 1987)

The low levels of internal wave energy have been attributed to the ice cover inhibiting generation by wind stress (Levine et al., 1985, 1987) and to the absorption of internal waves by the ice (Morison et al., 1985). In addition, tides are weak in the Arctic, and those generated north of the M_2 critical latitude at 74.5°N must dissipate locally rather than propagate long distances. Consequently, internal tides are not significant in most of the central Arctic (Dosser and Rainville, 2016). Seasonal differences in ice cover and winds over the Beaufort Sea produce annual variations in the amplitude of near-inertial waves, peaking at the end of summer, when ice cover is least, and passing through a sharp minimum at the end of winter (Figure 8.34). Wave amplitude is strongly correlated with the ratio of sea-ice drift speed to wind speed, which is affected by the amount of open water and ice roughness.

Largely owing to topography, internal waves are more variable in the eastern Arctic. Observations from drifting ice camps in late winter and ships in late summer reveal large spatial differences in the wave fields. Levels are so low over the Nansen Abyssal Plain that Expandable Current Profilers (XCPs) can barely detect signals, but over Yermak Plateau amplitudes rise to 10 times GM (D'Asaro and Morison, 1992). The intense internal waves were attributed to generation by strong barotropic tides across rough bottoms (Padman and Dillon, 1991; Wijesekera et al., 1993), and their limited propagation was imputed to absorption in icy surface water (D'Asaro and Morison, 1992).

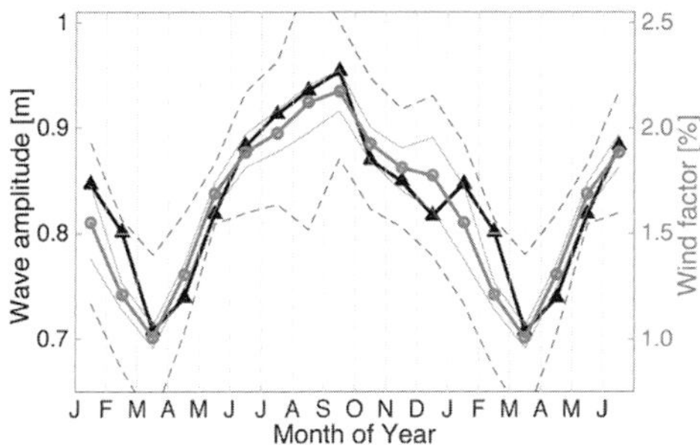

Figure 8.34 Vertical displacement at $\omega = 1.05f$ between the surface layer and the Atlantic Water in the Beaufort Sea from 2005–2014. The gray line is the wind factor, the ratio of sea-ice drift to wind speed. (From Dosser and Rainville, 2016. © American Meteorological Society. Used with permission)

8.8.2 Mixing

Consistent with the distribution of internal wave intensity, large sections of the Arctic have diapycnal diffusivities barely above molecular, with $K_T \sim 10^{-6}$ m^2 s^{-1}. Studying diffusive layers from an ice camp drifting in the Canada Basin, Padman and Dillon (1987) found dissipation rates barely above noise and estimated heat fluxes of $0.01-0.1$ W m^{-2} by applying the 4/3 law (Section 4.2) to $\Delta\theta$ and ΔS across the interfaces, and by calculating molecular fluxes across interfaces that were resolved. Dissipation rates rose to $\mathcal{O}(10^{-9}$ W kg$^{-1})$ in ship-based microstructure profiles during late summer 2012 (Figure 8.32c) after a strong storm left minimal ice cover (Lincoln et al., 2016). Measurements in shallower depths over the basin sides had patches with dissipation rates several times larger. A section of scalar microstructure profiles from the west side of the Canada Basin to the Amundsen Basin in the eastern Arctic showed the same pattern, with $K_T \sim 10^{-6}$ m^2 s^{-1} over basins and 10^{-5} m^2 s^{-1} over ridges (Rainville and Winsor, 2008). Estimates of diapycnal diffusivity from XCP shear in open water near the marginal ice zone of the eastern Arctic found equally low levels over basins and smooth bottoms, but levels of $\mathcal{O}(10^{-4}$ m^2 s$^{-1})$ over the Yermak Plateau and a deeper ridge where surface tides generated internal tides and internal waves (D'Asaro and Morison, 1992). Net mixing is expected to rise as ice-free periods increase.

8.9 Ocean Ridges

Ocean ridges are major stirring rods in ocean basins, where background mixing is usually moderate or weak. Detailed observations have focused on two: the Mid-Atlantic Ridge and the Hawaiian Ridge. The former is deep, rising 2.5 km in a basin 5 km deep, and cut by fracture zones perpendicular to its crest. Tidal and mean currents on the ridge produce strong turbulence that mixes Antarctic Bottom Water into North Atlantic Deep Water. The Hawaiian Ridge is steeper and topped by islands. Channels between the islands rise to within a kilometer of the surface and efficiently generate internal tides as the pycnocline is displaced by surface tides flowing nearly perpendicular to the ridge. The mechanisms and intensity of local mixing and the fate of the internal tides are the principal concerns. Observations are easier than for the Mid-Atlantic Ridge, with most activity in the upper kilometer and surface expressions detectable by satellites.

8.9.1 The Mid-Atlantic Ridge

Microstructure profiles starting at the west side of the Brazil Basin found few signals over the smooth abyssal plain, but encountered increasingly intense mixing going up the flank of the Mid-Atlantic Ridge (Figure 8.35). Diapycnal diffusivities $\leq 10^{-5}$ m^2 s^{-1} throughout the water column over the abyssal plain rose to 2×10^{-3} in the bottom few hundred meters over the crest, and K_ρ was close to 10^{-4} at all depths near the crest. An inverse calculation focused on scalar anomalies over the ridge yielded diapycnal diffusivities of $(1-10) \times 10^{-4}$ m^2 s^{-1}, essentially the same as measured (Mauritzen et al., 2002). A microstructure survey over abyssal hills on the flank found vertically averaged dissipation rates fluctuating with the

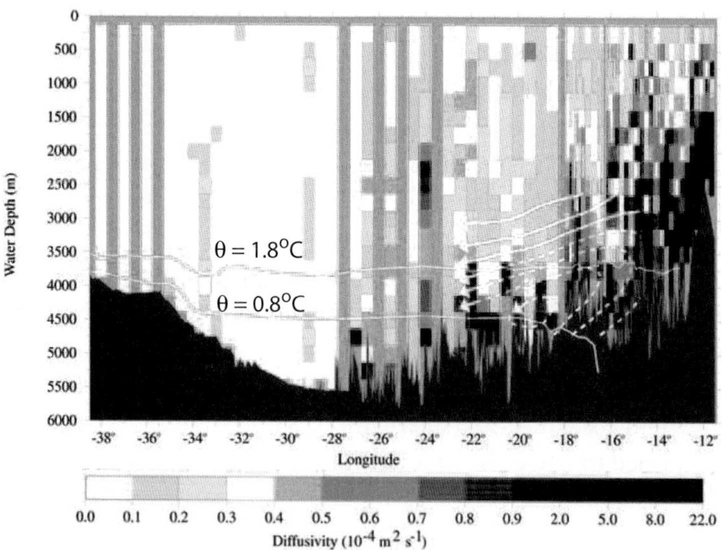

Figure 8.35 Diapycnal diffusivity across the Brazil Basin from velocity microstructure (Polzin et al., 1997; Ledwell et al., 2000). White lines are stream functions inferred from an inverse calculation (St. Laurent et al., 2001). (Adapted from Mauritzen et al., 2002.) (A black and white version of this figure will appear in some formats. For the color version, please refer to the plate section.)

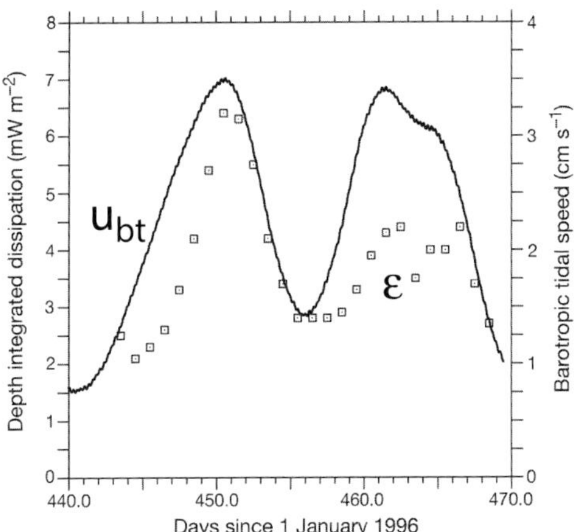

Figure 8.36 Depth-integrated dissipation and model tidal speeds during the microstructure survey following the tracer release in the Brazil Basin. The short time lag is consistent with propagation from the bottom up to where waves broke. (Adapted from Ledwell et al., 2000)

barotropic tide, leading Polzin et al. (1997) and Ledwell et al. (2000) to infer that the mixing resulted from internal waves breaking after propagating up from the bottom (Figure 8.36). Tracer released 500 m above abyssal hills on the flank diffused at $(2–4) \times 10^{-4}$ m^2 s^{-1}, but diffusion was more intense, $\sim 10^{-2}$ m^2 s^{-1}, when the tracer was closer to the bottom (Ledwell et al., 2000).

Mixing by internal wave breaking is not the whole story, however, at least over the ridge flanks. The original microstructure section was in a 1–1.5 km-deep fracture zone perpendicular to the ridge axis, and most of the turbulence was deeper than the canyon rim (St. Laurent et al., 2001). Subsequent surveys in other fracture zones with microstructure, moorings, and hydrography found strong mixing and density gradients associated with mean along-canyon flow over cross-canyon sills (Thurnherr et al., 2005). The net effect of fracture zone canyons every 50 km along two-thirds of ocean ridges is substantial. Although only 15% of the interfacial area between North Atlantic Deep Water and Antarctic Bottom Water occurs in canyons, Thurnherr et al. (2005) estimate that 53% of the diapycnal buoyancy flux mixing them together occurs there. In contrast, 44% of the interfacial area is over rough flanks, but only 33% of the buoyancy flux is there.

8.9.2 The Hawaiian Ridge

Internal Tide Generation and Interaction

Acoustic detection of an internal tide propagating northward from Hawaii (Dushaw et al., 1995) was confirmed when Ray and Mitchum (1996) extracted horizontally

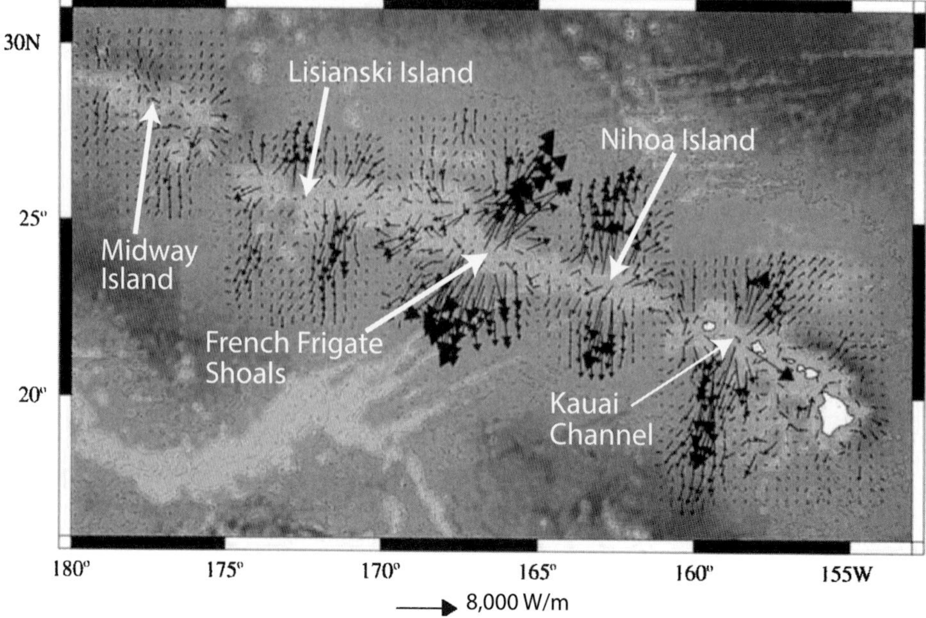

Figure 8.37 Vertically integrated M_2 baroclinic energy propagating from the Hawaiian Ridge, computed with the Princeton Ocean Model in five adjacent domains. (Adapted from Merrifield et al., 2001)

coherent 50 mm surface displacements from satellite altimetry. This was possible because a significant fraction of the baroclinic energy is phase-locked to lunar and solar forcing, allowing extensive averaging along repeat satellite tracks. Applying shallow-water wave equations to the global displacement field yielded 0.7 TW of global tidal dissipation, with (18 ± 6) GW from the Hawaiian Ridge (Egbert and Ray, 2001).

Primitive-equation simulations using bathymetry of the Hawaiian Ridge showed that most tidal conversion occurs at three sites (Figure 8.37). The calculation, however, accounted for only half of the conversion estimated from altimetry (Holloway and Merrifield, 2003). Two-thirds of the 10 GW propagated away as mode-1 internal tides. The remainder, in mode-2 and higher, lost all of its energy within a few hundred kilometers of the ridge. Either the model underestimated conversion to internal tides, or half the energy lost from the surface tide is dissipated at the ridge.

Guided by the simulations, the Hawaii Ocean Mixing Experiment (HOME) examined conversion and mixing along the ridge, concentrating on the three major sites (Rudnick et al., 2003). Direct observations of energy flux vectors with Sanford's Absolute Velocity Profiler (AVP) were within a factor of 2 of model

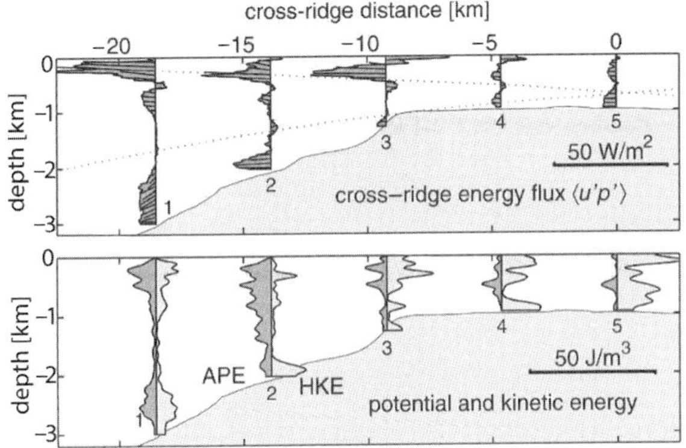

Figure 8.38 Cross-ridge energy flux and tidal characteristics (upper) and energy densities (lower) on the southern flank of Kaena Ridge, Hawaii. Within 21 hours, seven profiles were taken at each station. (Adapted from Nash et al., 2006. © American Meteorological Society. Used with permission)

predictions (Lee et al., 2006), with magnitudes of (9–33) kW/m at the French Frigate Shoals, (4–18) kW/m at the Kauai Channel, and (0.2–23) kW/m west of Nihoa Island. Figure 8.38 shows the structure of the internal tide originating on the south side of the Kauai Channel, obtained by profiles every three hours with XCPs supplemented by an AVP at the deep station. (Figure 7.6 shows barotropic forcing for this section.) On the ridge, most energy is kinetic and potential energy is small, but the ratio reverses along the slope. The cross-ridge energy flux, negligible on the ridge, focuses into two beams over the slope, one upward near the surface and the other downward close to the bottom. The surface beam is the more intense and is concentrated closer to the surface than in the model. Fitting the measurements onto the first 10 flat-bottom baroclinic modes shows mode-1 dominant at the deeper station and mode-2 carrying most of the remaining energy.

Acoustic Doppler Current Profiler (ADCP) observations over Kaena Ridge, in the Kauai Channel, revealed a striking pattern of horizontally coherent internal waves with near-diurnal frequency (Carter and Gregg, 2006). A period seven hours less than the local inertial period ruled out wind generation, and the frequency of $\omega_{M_2}/2$ suggested formation by PSI of the internal tide beam crossing the ridge at 525–595 m depth. Bicoherence analysis confirmed nonlinear coupling between the two frequencies in the beam.

Local Dissipation and Mixing

Extensive microstructure measurements, dropped and towed, are summarized by an empirical function with separable cross-ridge and depth variability, $K_\rho = 10^{-5} H(y - y_0) V(z/z_{\text{bot}})$, where y_0 is on the ridge axis and z_{bot} is the bottom depth

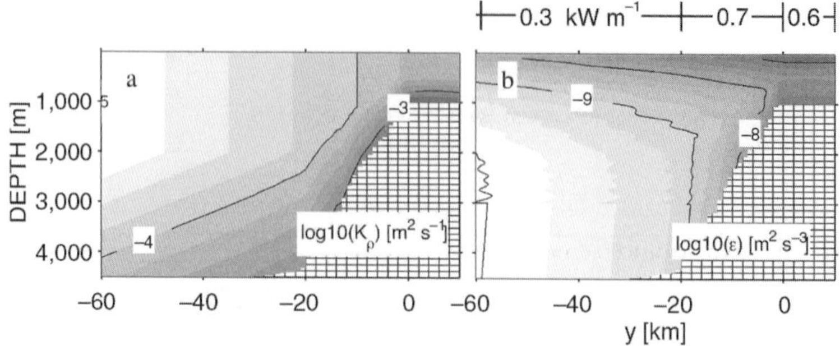

Figure 8.39 Structure of diapycnal diffusivity (a) and dissipation rate (b) for the bathymetry of the Kauai Channel. Net dissipations offshore, on the flanks, and on the crest are at the top of panel (b). (Adapted from Klymak et al., 2006a. © American Meteorological Society. Used with permission)

(Klymak et al., 2006a). Dissipation rates exceed 10^{-8} W kg^{-1} on top and gradually decrease by a decade along the flank (Figure 8.39). Close to the bottom, diffusivity exceeds 10^{-3} m^2 s^{-1}.

Mixing mechanisms include internal hydraulic jumps just off the edge of the crest, where $dz_{\rm bot}/dx > \omega/N$ (Legg and Klymak, 2008). In addition to the PSI over the crest already mentioned, 100 m overturns were observed deep on the flank during peak downslope flow, when stratification was least, and again when the flow reversed to upslope (Aucan et al., 2006). Time-averaged dissipation rates of 2×10^{-8} W kg^{-1} were estimated in large overturns.

Net dissipation on the ridge was (3 ± 1.5) GW, about one-sixth of the energy lost from the surface tide and perhaps one-third of the energy radiating away as internal tide. These estimates, however, do not include systematic measurements in thin bottom boundary layers on the crests. A microstructure profile that inadvertently hit bottom had dissipation rates so high that its inclusion changed the average for the crest. It is likely that thin bottom boundary layers make significant, but overlooked, contributions to the energy budgets of ridges and other shallow tidal flows.

Propagation from the Ridge

To examine the fate of internal tides propagating from the ridge (Figs. 8.40, 8.41), Alford et al. (2007) measured velocity along the northward beam from the French Frigate Shoals with ship transects and profiling moorings. Although simulations (MacKinnon and Winters, 2005) predicted intense mixing induced by PSI at the critical latitude, the observations showed that at most 12% of the energy was lost from the beam to PSI (Alford et al., 2007). Nevertheless, diapycnal diffusivity

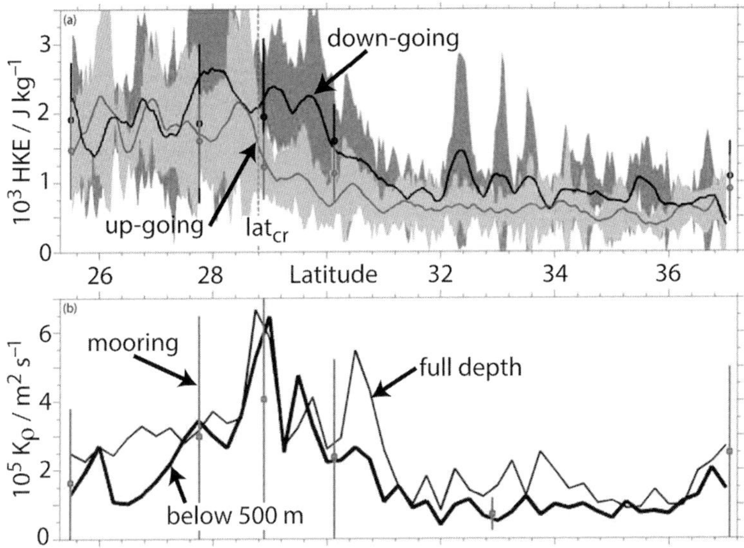

Figure 8.40 Horizontal kinetic energy (upper) and diapycnal diffusivity (lower) along the northward internal tide beam from the French Frigate Shoals. Diffusivity peaked at the M_2 critical latitude, and the ratio of upgoing to downgoing kinetic energy increased to the south. K_ρ was estimated from ADCP shear using (7.16). (Adapted from Alford et al., 2007)

peaked at the critical latitude, particularly between depths of 0.5 and 1 km (Figure 8.40b), and PSI is the most likely source. More direct evidence for PSI came from unique signatures in meridional shear. At the critical latitude, near-inertial shear appeared as vertically standing waves, unlike observations a few degrees north and south (Figure 7.16). Waves formed at the surface propagated downward and those generated at the bottom propagated upward, but those created at mid-depth by PSI can propagate in either direction. Hence, vertical standing waves suggest PSI. In addition, downgoing wave energy dominated upward energy north of the critical latitude, but up- and downgoing energy were comparable to the south, where PSI is possible.

Internal tide beams propagating much farther from Hawaii (Figure 8.41) than previously realized have been identified using TOPEX/POSEIDON tandem satellite tracks from the extended mission (Zhao and Alford, 2009). In particular, reanalysis revealed a strong beam from Lisianski Island (LL), west of the French Frigate Shoals, that can be tracked to the Aleutians. Narrowing of this and the other beams implies energy loss, but simulations by Rainville et al. (2010) demonstrate that this may not be correct. Rather, the narrowing may result from interference between the multiple sources along the ridge. For example, the energy may become incoherent

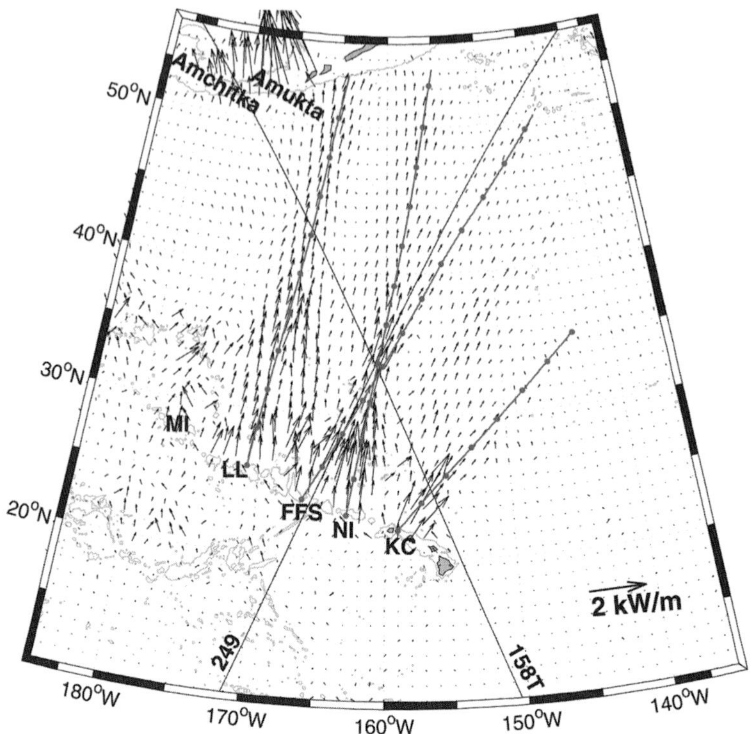

Figure 8.41 Northward energy fluxes of the coherent part of the mode-1 M_2 internal tide derived from altimetry of the extended TOPEX/POSEIDON mission. Shipboard and mooring measurements were along mission track 249. (Adapted from Zhao and Alford, 2009. © American Meteorological Society. Used with permission)

with its sources rather than being dissipated. This behavior is consistent with inferences by St. Laurent and Garrett (2002) that, rather than enhancing mixing in the near offshore, most internal tide energy may contribute to mixing at distant sources, on the Aleutian continental shelf in this case.

Appendix A

Glossary

A.1 English Symbols

$\mathcal{A}$: Wave action density (Section 6.3.2)
AABW: Antarctic Bottom Water (Section 1.3)
ACC: Antarctic Circumpolar Current (Section 1.2.5)
aie: Available internal energy (Section 2.7.5)
ape: Available potential energy (Sections 2.7.3, 2.7.5)
ake: Available kinetic energy (Section 7.7.1)
b: Specific buoyancy (Section 2.6.1)
c: Electrical conductivity of seawater (Section 2.2.1) and sound speed (Section 2.6)
c^{dr}: Flux coefficient for diffusive regime of double diffusion (Section 4.2)
c_g: Group velocity (Section 6.3.1)
$c_{g\mathrm{E}}$: Eulerian group velocity (Section 6.3.1)
$c_{g\mathrm{L}}$: Lagrangian group velocity (Section 6.3.1)
c_{gz}: Vertical group velocity (Section 6.6.2)
c^{sf}: Flux coefficient for salt fingers and salt sheets (Section 4.2)
c_p: Specific heat of seawater at constant pressure (Section 2.2.1)
c_{phase}: Phase velocity (Section 6.3.1)
dz_{wkb}: WKB depth increment (Section 6.3.4)
E_{iw}: Internal wave energy density (Section 6.3.2)
e: Specific internal energy (Section 2.2.2)
f_{eff}: Effective Coriolis frequency (Section 6.1)
Fr: Froude number (Section 3.2)
Fr_h: Horizontal Froude number (Section 3.8)
$Fr_N \equiv u/Nl_z$: Stratification Froude number (Section 3.2)
$Fr_S \equiv u/Sl_z$: Shear Froude number (Section 3.2)
gpe: Gravitational potential energy (Section 2.7.2)
gpe_r: Reference gravitational potential energy (Section 2.7.2)
g: Gravitational acceleration (Section 2.6.1)
$\tilde{g}$: Specific Gibbs function (Section 2.2.2)
g_{ss}: Grams of sea salt (Section 2.2)
hke: Horizontal kinetic energy (Section 6.3.5)
h: Specific enthalpy (Section 2.2.2)
h_{S_A}: Specific enthalpy of sea salt (Section 2.2.2)
h^o: Potential enthalpy (Section 2.3.1)

h_Θ: Partial specific enthalpy with respect to conservative temperature (Section 2.3.1)
IDW: Indian Deep Water (Section 1.3)
J_b: Buoyancy flux (Section 2.7.1)
J_b^{dr}: Buoyancy flux due to diffusive regime of double diffusion (Section 4.2)
J_b^{sf}: Buoyancy flux due to salt fingers/salt sheets (Section 4.2)
J_b^{turb}: Turbulent buoyancy flux (Section 2.53)
J_h, J_z: Horizontal and vertical fluxes (Section 8.2.1)
J_{iso}, J_{dia}: Isopycnal and diapycnal fluxes (Section 1.3.4)
J_{iw}: Internal wave energy flux (Section 6.3.2)
$\boldsymbol{J}_Q$: Heat flux vector (Section 2.4)
$\boldsymbol{J}'_Q$: Reduced heat flux vector (Section 2.4)
$\boldsymbol{J}_S$: Salt flux vector (Section 2.4)
J_z: Vertical flux (Section 8.2.1)
$\boldsymbol{J}_\eta$: Vector flux of specific entropy (Section 2.3.2)
K: Absoute temperature in degrees Kelvin (Section 3.7.1)
k: Three-dimensional wavenumber (Section 3.7.1)
K_T: Turbulent diffusivity of temperature (Section 1.1)
K_{dia}: General diapycnal turbulent diffusivity, an alternate expression for K_ρ (Section 1.3.4)
K_ρ: Diapycnal turbulent diffusivity of density (Section 1.1)
K_{iso}: Isopycnal turbulent diffusivity (Section 1.3.4)
kg_{sw}: Kilogram of seawater (Section 2.2)
k_h: Horizontal wavenumber (Section 3.8.2)
k_v: von Karman's constant (Section 3.5.1)
k_z: Vertical wavenumber (Section 3.6.1)
l_O: Ozmidov length (Section 3.2)
l_κ: Diffusive, or Batchelor, length scale for scalar diffusivity κ (Section 3.7.2)
N: Buoyancy frequency (Section 2.6.1)
N_0: Reference buoyancy frequency used by Garrett and Munk (1972) (Table B.1)
N_{ref}: Reference buoyancy frequency (Section 6.3.4)
NADW: North Atlantic Deep Water (Section 1.3)
Nu: Nusselt number (Section 4.2)
PDW: Pacific Deep Water (Section 1.3)
Pr: Prandtl number (Section 2.4.5)
p_{ref}: Reference pressure used for potential temperature and density (Section 2.2.1)
PSI: Parametric subharmonic instability (Section 7.5.1)
PV: Potential vorticity (Section 6.3.3)
Ra_S, Ra_T: Saline and thermal Rayleigh numbers (Section 4.2)
Re: Reynolds number (Section 3.2)
Re_b: Buoyancy Reynolds number (Section 3.2)
Re_h: Horizontal Reynolds number (Section 3.8)
Re_S: Shear Reynolds number (Section 3.3.3)
Re_ϵ: Reynolds number using Taylor scaling for ϵ (Section 3.2)
Ri_{flux}: Flux Richardson number (Section 3.2)
Ri_g: Gradient Richardson number (Section 3.2)
$R_{\text{flux}}^{\text{dr}}$: Diffusive regime flux ratio (Section 4.2)
$R_{\text{flux}}^{\text{sf}}$: Salt finger/salt sheet flux ratio (Section 4.2)
Ro: Rossby number (Section 6.1)
R_ω: Internal wave shear-to-strain ratio (Section 6.6.2)
R_ρ: Density ratio using practical salinity and temperature (Section 2.6.1)

R_ρ^Θ: Density ratio using absolute salinity and conservative temperature (Section 2.6.1)
S: Salinity in grams of sea salt per kilogram of seawater, as either practical or absolute salinity
S_A: Absolute salinity (Section 2.2.2)
s: Salinity in concentration units, i.e. kilograms of sea salt per kilogram of seawater (Section 2.3.2)
S_c: Schmidt number (Section 2.4.5)
S_P: Practical salinity (Section 2.2.1)
S_T: Soret coefficient (Section 2.4.1)
$T_{\rm abs}$: Absolute temperature (Section 2.2.2)
tke: Turbulent kinetic energy (Section 3.2)
Tu: Turner angle (Section 4.2.2)
u_b: Buoyancy velocity (Section 3.2)
$u_{\rm cw}, u_{\rm acw}$: Clockwise and anticlockwise velocity components (Section 6.3.6)
u_s, v_s, w_s: Stretched horizontal velocities (Section 6.3.4)
w_ρ: Diapycnal velocity (Section 1.3.4)
$w_{\rm n}$: Dianeutral velocity (Section 2.9.3)
WKB: Wentzel-Krammers-Brillouin scaling for internal wave changes in response to variations in stratification (Section 6.3.4)
z: Vertical coordinate (Section 3.5.1)
$z_{\rm wkb}$: Depth normalized to the reference used for WKB scaling (Section 6.3.4)

A.2 Greek Symbols

α: Horizontal wavenumber in Appendix B
α_T: Coefficient of thermal expansion in terms of practical variables (Section 2.2.1)
α^Θ: Coefficient of thermal expansion in terms of conservative temperature and absolute salinity (Section 2.6)
β: Radian vertical wavenumber in Appendix B, and coefficient of haline contraction in terms of practical variables (Section 2.43)
β_*: Vertical mode number in Garrett and Munk internal wave spectrum (Table B.1)
β^Θ: Coefficient of haline contraction in terms of conservative temperature and absolute salinity (Section 2.6)
β_u: Upper cutoff of GM, the internal wave spectrum given by Garrett and Munk (1972) and amended in several subsequent papers (Appendix B)
$\Gamma_{\rm adiabatic}$: Adiabatic lapse rate (Section 2.2.1)
$\Gamma_{\rm mix}$: Mixing coefficient (Section 3.12)
Γ_S: The equilibrium salt gradient (Section 2.4.2)
$\Gamma_{\chi\epsilon}$: Mixing coefficient calculated from observed χ_T and ϵ (Section 3.12)
Γ_S: The equilibrium salt gradient (Section 2.4.2)
γ^Θ: Isothermal compressibility (Section 2.6)
$\epsilon_{\rm iw}^{\rm HWF}$: Turbulent dissipation rate produced by internal waves according to Henyey et al. (1986) (Section 7.5.4)
$\epsilon_{\rm iw}^{\rm MM}$: Turbulent dissipation rate produced by internal waves according to McComas and Müller (1981b) (Section 7.5.2)
$\epsilon_{\rm iw}^{\rm shear}$: Turbulent dissipation rate produced by internal waves estimated primarily from shear scaling (Section 7.6.2)
$\epsilon_{\rm iw}^{\rm strain}$: Turbulent dissipation rate produced by internal waves estimated primarily from strain scaling (Section 7.6.4)

ζ: Vertical displacement from equilibrium depth (Section 6.2.1)
ζ_z: Vertical strain (Section 6.2.1)
$\zeta_{\rm wkb}$: Vertical displacement after WKB normalization (Section 6.3.4)
η: Specific entropy (Section 2.2.2)
θ: Potential temperature (Section 2.2.2)
Θ: Conservative temperature (Section 2.3.1)
$\theta_{\rm topo}^{\rm critical}$: Critical inclination of topography for critical reflection (Section 6.3.7)
κ_S: Molecular diffusivity of salt in seawater (Section 2.4)
κ_T: Molecular diffusivity of heat in seawater (Section 2.4)
μ: Relative chemical potential of seawater (Section 2.4)
μ_S: Partial chemical potential of sea salt in seawater (Section 2.3.2)
μ_W: Partial chemical potential of pure water in seawater (Section 2.3.2)
ν: Kinematic viscosity of seawater (Section 2.4.4)
ρ: Density of water (Section 1.1)
σ: Growth rate for infinitesimal disturbance (Section 4.2) and electrical conductivity (Section 5.3)
σ_η: Rate of entropy generation (Section 2.3.2)
σ_θ: Potential density (Section 2.5.1)
$\tau_{\rm spice}$: A measure of variability along isopycnals (Section 2.8.1)
υ: Specific volume (Section 2.7.3)
$\Phi_{hke}, \Phi{vke}, \Phi_{pe}$: Spectra of horizontal and vertical kinetic energy and of potential energy (Section 6.6)
Φ_u, Φ_v: Spectra of u and v velocity components in Eulerian frame (Section 6.6)
Φ_{xy}: Cross-spectrum between two variables, e.g. x and y (Section 6.6)
$\Phi_{shear}, \Phi_{strain}$: Spectra of vertical shear and strain (Section 6.6)
Φ_{vel}: Spectrum of horizontal velocity in Eulerian frame (Section 6.6)
$\Phi_{vel}^{cw}, \Phi_{vel}^{acw}$: Clockwise and anticlockwise spectra of horizontal velocity (Section 6.6)
$\chi_{\rm ape}$: Dissipation rate of available potential energy (Section 3.4.1)
χ_T: Thermal dissipation rate (Section 1.1)
χ_S: Saline dissipation rate (Section 1.1)
$\omega^{\rm critical}$: Critical frequency for internal wave reflection (Section 6.3.7)
ω_E, ω_L: Internal wave frequency in Eulerian and Lagrangian frames (Section 6.5)

Appendix B

The GM79 Internal Wave Spectrum, Prepared with R.-C. Lien

GM79 refers to the final update to the internal wave spectra initiated by Garrett and Munk (1972). First circulated in 1979, GM79 appeared in Munk (1981). The principal change was representing spatial structure as modes rather than analytic functions of wavenumber. That is useful for modeling, but here we follow the standard observational practice of converting modes to wavenumbers with the mode number j as a parameter.

$$\beta = \frac{j\pi}{b}\frac{N}{N_0} \quad \left[\mathrm{m}^{-1}\right], \quad \alpha = \frac{j\pi}{b}\frac{(\omega^2 - f^2)^{1/2}}{(N^2 - \omega^2)^{1/2}}\frac{N}{N_0} \quad \left[\mathrm{m}^{-1}\right], \tag{B.1}$$

where β and α are vertical and horizontal wavenumbers, with $\alpha = (\alpha_x^2 + \alpha_y^2)^{1/2}$. (Here, we follow the GM practice of using Greek symbols for wavenumber.) Because studies of diapycnal mixing are based on β and the frequency dependence of α introduces complications, we consider only vertical spectra. Frequencies and wavenumbers are in radians and are converted to cyclic versions by dividing by 2π. Accordingly, one- and two-dimensional spectra must be multiplied by 2π and $(2\pi)^2$, respectively. N_0 is the reference stratification, and b is the scale-depth of abyssal pycnoclines (Table B.1). Wavenumber bandwidths and corresponding low wavenumber spectral rolloffs are specified by β_* and α_* evaluated with j_*, e.g. $\beta_* \equiv (j_*\pi/b)(N/N_0)$.

Because GM spectral forms are empirical approximations to observations rather than fitted curves, there is little difference in shape between GM79 and the previous version GM76 (Gregg and Kunze, 1991). An important difference for mixing applications, however, is that GM79 shear and strain variances are $2/\pi$ of those in GM76, as described in Cairns and Williams (1976). Reflecting subsequent observations and usage, GM79 is here extended by allowing for the observed dimensionless energy density, E, to exceed E_0, the GM reference.

B.1 Energy Density and Two-Dimensional Spectra

Internal wave intensity is specified by the energy density,

$$E_{\mathrm{iw}} = b^2 N_0^2 E_0 \left(\frac{N}{N_0}\right)\left(\frac{E}{E_0}\right) \quad \left[\frac{\mathrm{J}}{\mathrm{kg}}\right], \tag{B.2}$$

which varies with the observed dimensionless energy density, E_0, and follows Wentzel-Kramers-Brillouin (WKB) scaling, e.g. E_{iw} at stratification $N = N_0/2$ is reduced by half

Table B.1 *Parameters of the Garrett and Munk internal wave model.*

Symbol	Expression	Name
N_0	5.2×10^{-3} rad s^{-1}	Reference buoyancy frequency
b	1,300 m	Scale-depth of the thermocline
j_*	3	Reference vertical mode number
E_0	6.3×10^{-5}	Dimensionless reference energy density

from the reference level. The distribution of energy is assumed to have the same frequency dependence at all wavenumbers and the same wavenumber distribution at all frequencies. Consequently, the two-dimensional energy spectrum is separable, i.e. the product of two functions, one varying with frequency and the other with wavenumber. Choosing vertical wavenumber,

$$\Phi_E(\beta,\omega) = \frac{2j_* b N_0^2 E_0}{\beta_*^2 + \beta^2}\left(\frac{N}{N_0}\right)^2 \left(\frac{E}{E_0}\right) \frac{2f/\pi}{\omega(\omega^2 - f^2)^{1/2}} \left[\frac{\text{J kg}^{-1}}{\text{m}^{-1}\ \text{s}^{-1}}\right] \tag{B.3}$$

$$\text{for} \quad f \leq \omega \leq N \quad \text{and} \quad \beta \leq \beta_u \quad \text{with} \quad \beta_u = 0.6(E_0/E) \quad [\text{m}^{-1}], \tag{B.4}$$

where β_u corresponds to k_z^c in the text.

The internal wave field described by (B.3) is presumed to be dominated by linear interactions between waves at $\beta < \beta_u$, where the slope steepens from β^{-2} to β^{-3}, i.e.

$$\Phi_E(\beta,\omega) = \Phi_E(\beta_u,\omega)\frac{\beta_u}{\beta} \quad \text{for} \quad \beta > \beta_u \tag{B.5}$$

(Gregg et al., 1973; Gregg, 1977b; Gargett et al., 1981). When the wave field is at the GM reference level, the transition occurs at $\beta_u = 0.6$ m^{-1}. As discussed below, β_u shifts to lower wavenumber as energy increases, consistent with saturation of the wave field when the variance reaches the level corresponding to 0.6 m^{-1} (0.1 cpm) at background intensity. Turbulence dominates the spectrum at wavenumbers exceeding the Ozmidov wavenumber,

$$\beta_{\text{Oz}} \equiv (N^3/\epsilon)^{1/2} \quad \left[\text{m}^{-1}\right]. \tag{B.6}$$

By construction, integrating the two-dimensional spectrum returns the energy density,

$$\int_0^{\beta_u}\int_f^N \Phi_E(\beta,\omega)\, d\beta\, d\omega = E_{\text{iw}}, \tag{B.7}$$

with components from vertical wavenumber and frequency,

$$\int_0^{\beta_u} \frac{d\beta}{\beta_*^2 + \beta^2} = \frac{1}{\beta_*}\tan^{-1}\left(\frac{\beta_u}{\beta_*}\right) \approx \frac{\pi}{2\beta_*} \quad \text{when} \quad \beta_u \gg \beta_*, \tag{B.8}$$

$$\int_f^N \frac{d\omega}{\omega(\omega^2 - f^2)^{1/2}} = \frac{1}{f}\sec^{-1}\left(\frac{N}{f}\right) \approx \frac{\pi}{2f} \quad \text{when} \quad N \gg f. \tag{B.9}$$

Very good in most of the ocean, the approximations are marginal in the abyssal Arctic, where N/f may be as low as 5, and in strong stratification at shallow depths, where β_* can

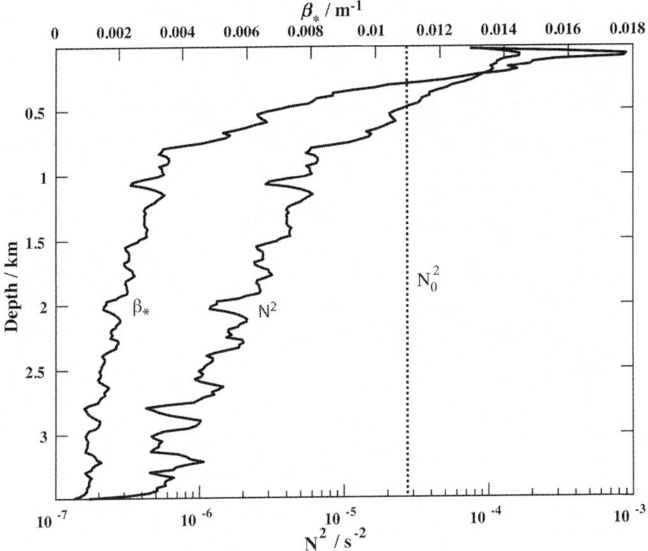

Figure B.1 The reference vertical wavenumber, $\beta_* = j_*\pi(N/N_0)/b$, varied 10-fold in a profile near Hawaii (Figure 8.1). For reference, the high-wavenumber cutoff is $\beta_u = 0.6$ m^{-1} when internal wave energy is at the GM background level.

increase to 0.02 m^{-1} (Figure B.1) and β_u decrease to 0.1 m^{-1} when $E = 6E_0$. With these approximations, one-dimensional forms are

$$\Phi_E(\beta) = \frac{2j_*bN_0^2E_0}{\beta_*^2 + \beta^2}\left(\frac{N}{N_0}\right)^2\left(\frac{E}{E_0}\right) \quad \left[\frac{\text{J kg}^{-1}}{\text{m}^{-1}}\right] \tag{B.10}$$

and

$$\Phi_E(\omega) = b^2N_0^2E_0\left(\frac{N}{N_0}\right)\left(\frac{E}{E_0}\right)\frac{2f/\pi}{\omega(\omega^2 - f^2)^{1/2}} \quad \left[\frac{\text{J kg}^{-1}}{\text{s}^{-1}}\right]. \tag{B.11}$$

B.2 Wave Functions

Spectra of energy components, e.g. horizontal kinetic energy (hke),

$$E_{iw} = E_{hke} + E_{vke} + E_{pe}, \tag{B.12}$$

are obtained from the energy spectrum using relations for single internal waves developed by Fofonoff (1969),

$$\frac{E_{pe}}{E_{vke}} = \frac{N^2}{\omega^2}, \quad \frac{E_{hke}}{E_{vke}} = \frac{(\omega^2 + f^2)(N^2 - \omega^2)}{\omega^2(\omega^2 - f^2)}, \quad \frac{E_{iw}}{E_{vke}} = 2\frac{N^2 - f^2}{\omega^2 - f^2}. \tag{B.13}$$

Defining $\Phi_\xi(\beta,\omega) = \Phi_x(\beta,\omega) + \Phi_y(\beta,\omega)$ $\left[\text{m}^2/(\text{m}^{-1}\ \text{s}^{-1})\right]$ as the spectrum of horizontal displacement, the horizontal wave function is

$$X^2(\omega) \equiv 2\frac{\Phi_{hke}}{\omega^2 \Phi_E} = \frac{\Phi_\xi(\beta,\omega)}{\Phi_{E_0}(\beta,\omega)} = \frac{\omega^2 + f^2}{\omega^4}\frac{N^2-\omega^2}{N^2-f^2} \quad \left[\mathrm{s}^2\right]. \tag{B.14}$$

Likewise, taking Φ_ζ as the spectrum of vertical displacement, the vertical wave function, $Z^2(\omega)$, is the ratio of twice the potential energy spectrum to the energy spectrum,

$$Z^2(\omega) \equiv \frac{2\,\Phi_{pe}(\beta,\omega)}{\Phi_E(\beta,\omega)} = \frac{N^2\,\Phi_\zeta(\beta,\omega)}{\Phi_E(\beta,\omega)} = \frac{\omega^2 - f^2}{\omega^2}\frac{N^2}{N^2-f^2} \text{ [dimensionless]}. \tag{B.15}$$

Garrett and Munk (1975) used only the left halves of the wave functions, implicitly assuming hydrostatic pressure ($N^2 \gg \omega^2$) and a wide frequency bandwidth ($N^2 \gg f^2$). Here, the full expression is used and approximated as needed.

B.3 Horizontal Velocity

The spectrum of horizontal velocity, $\Phi_{vel} = \phi_u + \phi_v = 2\phi_{hke}$, is obtained by multiplying ω^2 by Φ_ξ, the spectrum of horizontal displacement, which in turn is $X^2(\omega)$ times the energy spectrum,

$$\Phi_{vel}(\beta,\omega) = \omega^2\Phi_\xi(\beta,\omega) = \omega^2 X^2(\omega)\Phi_E(\beta,\omega) \quad \left[\frac{(\mathrm{m/s})^2}{\mathrm{m}^{-1}\,\mathrm{s}^{-1}}\right] \tag{B.16}$$

$$= \frac{2j_* b N_0^2 E_0}{\beta_*^2+\beta^2}\left(\frac{N}{N_0}\right)^2\left(\frac{E}{E_0}\right)\frac{(2f/\pi)(\omega^2+f^2)(N^2-\omega^2)}{\omega^3(\omega^2-f^2)^{1/2}(N^2-f^2)}.$$

One-dimensional spectra are obtained by integrating with ω or β, e.g.

$$\int_f^N \frac{(2f/\pi)(\omega^2+f^2)(N^2-\omega^2)}{\omega^3(\omega^2-f^2)^{1/2}(N^2-f^2)}\,d\omega \tag{B.17}$$

$$= \left(\frac{2}{\pi} + \frac{(N/f)^2}{\pi((N/f)^2-1)}\right)\sec^{-1}\left(\frac{N}{f}\right) + \frac{1}{\pi((N/f)^2-1)^{1/2}},$$

which asymptotes to 1.5 when $N/f \gg 1$, 1.44 when $N/f = 10$, and 1.39 when $N/f = 5$. Then,

$$\Phi_{vel}(\beta) = \int_f^N \Phi_{vel}(\beta,\omega)\,d\omega \approx \frac{3j_* b N_0^2 E_0}{\beta_*^2+\beta^2}\left(\frac{N}{N_0}\right)^2\left(\frac{E}{E_0}\right)\left[\frac{(\mathrm{m/s})^2}{\mathrm{m}^{-1}}\right]. \tag{B.18}$$

By comparision with (B.10), $\Phi_{hke}(\beta) = 1/2\;\Phi_{vel}(\beta) = 3/2\;\Phi_E(\beta)$. Likewise, applying (B.8) yields

$$\Phi_{vel}(\omega) = \int_0^{\beta_u} \Phi_{vel}(\beta,\omega)\,d\beta \quad \left[\frac{(\mathrm{m/s})^2}{\mathrm{s}^{-1}}\right]$$

$$\approx b^2 N_0^2 E_0\left(\frac{N}{N_0}\right)\left(\frac{E}{E_0}\right)\frac{(2f/\pi)(\omega^2+f^2)}{\omega^3(\omega^2-f^2)^{1/2}}. \tag{B.19}$$

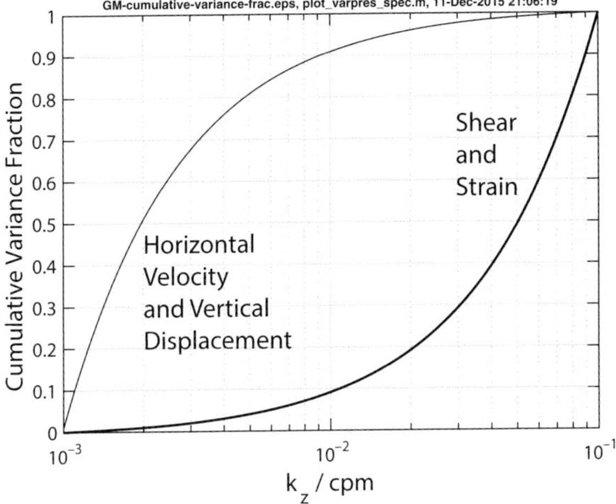

Figure B.2 Cumulative variances for horizontal velocity and vertical displacement (thin line) and for vertical shear and strain (thick line). For a 1,000 m record, 90% of velocity and displacement variances are at $k_z < 0.01$ cpm, and 90% of shear and strain variance are at $k_z > 0.01$ cpm. Half of the shear and strain variance is between 0.05 and 0.1 cpm.

Velocity variance follows as

$$
\begin{aligned}
< vel^2 > &= \int_0^{\beta_u} \Phi_{vel}(\beta)\, d\beta = \int_f^N \Phi_{vel}(\omega)\, d\omega \quad \left[(\mathrm{m/s})^2\right] \qquad \text{(B.20)} \\
&\approx \frac{3}{2} E_{\mathrm{iw}} = \frac{3b^2 N_0^2 E_0}{2} \left(\frac{N}{N_0}\right)\left(\frac{E}{E_0}\right) = (0.066)^2 \left(\frac{N}{N_0}\right)\left(\frac{E}{E_0}\right).
\end{aligned}
$$

The variance is concentrated at low wavenumbers (Figure B.2).

For typical stratification and GM energy levels, root-mean-square (rms) horizontal velocities vary from ~ 0.1 m s^{-1} in the shallow pycnocline to ~ 20 mm s^{-1} near the bottom (Figure B.3).

B.4 Vertical Shear of Horizontal Velocity

The spectrum of vertical shear is β^2 times the velocity spectrum,

$$
\begin{aligned}
\Phi_{vshear}(\beta,\omega) &= \beta^2 \Phi_{vel}(\beta,\omega) \quad \left[\frac{\mathrm{s}^{-2}}{\mathrm{m}^{-1}\ \mathrm{s}^{-1}}\right] \qquad \text{(B.21)} \\
&= 2 j_* b N_0^2 E_0 \frac{\beta^2}{\beta_*^2 + \beta^2} \left(\frac{N}{N_0}\right)^2 \left(\frac{E}{E_0}\right) \frac{(2f/\pi)(\omega^2 + f^2)(N^2 - \omega^2)}{\omega^3(\omega^2 - f^2)^{1/2}(N^2 - f^2)}.
\end{aligned}
$$

Integrating the frequency dependence with (B.17) gives the shear spectrum as a function of vertical wavenumber,

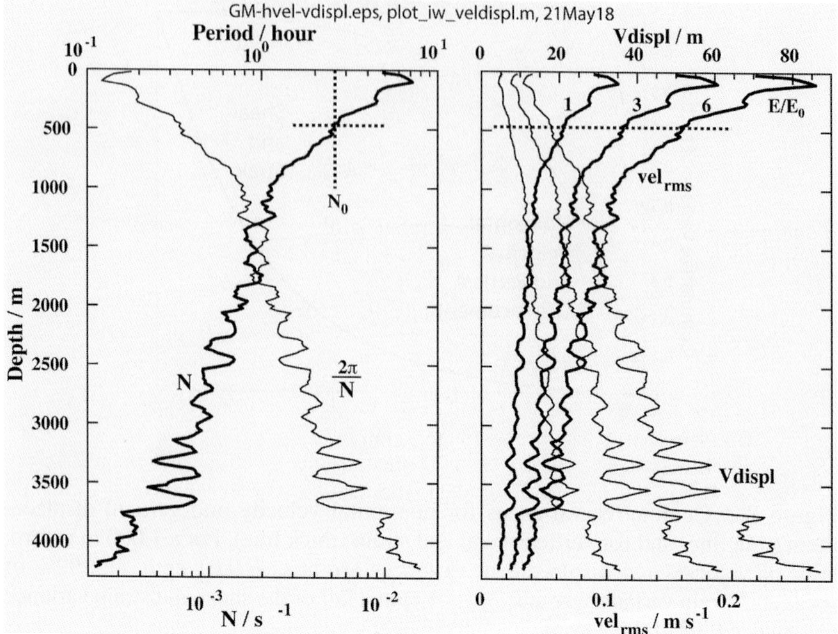

Figure B.3 (Left) Stratification as buoyancy frequency (bottom axis) and as buoyancy period (top axis). (Right) Horizontal velocity (B.20) and vertical displacement (B.31) for internal waves with one, three, and six times the GM energy level.

$$\Phi_{vshear}(\beta) = \int_f^N \Phi_{vshear}(\omega, \beta)\, d\omega \qquad \left[\frac{\mathrm{s}^{-2}}{\mathrm{m}^{-1}}\right] \tag{B.22}$$

$$\approx 3 j_* b N_0^2 E_0 \,\frac{\beta^2}{\beta_*^2 + \beta^2} \left(\frac{N}{N_0}\right)^2 \left(\frac{E}{E_0}\right),$$

and integrating the wavenumber dependence with

$$\int_0^{\beta_u} \frac{\beta^2}{\beta_*^2 + \beta^2}\, d\beta = \beta_u - \beta_* \tan^{-1}(\beta_u/\beta_*) \approx \beta_u \tag{B.23}$$

gives the shear spectrum as a function of frequency,

$$\Phi_{vshear}(\omega) = \int_0^{\beta_u} \Phi_{vshear}(\beta, \omega)\, d\omega \qquad \left[\frac{\mathrm{s}^{-2}}{\mathrm{s}^{-1}}\right] \tag{B.24}$$

$$\approx 2 j_* b \beta_u N_0^2 E_0 \left(\frac{N}{N_0}\right)^2 \left(\frac{E}{E_0}\right) \frac{(2f/\pi)(\omega^2 + f^2)}{\omega^3(\omega^2 - f^2)^{1/2}}.$$

Concentrated at high wavenumbers (Figure B.2), shear variance follows as

$$< Shear^2 > = \int_0^{\beta_u} \Phi_{vshear}(\beta)\, d\beta = \int_f^N \Phi_{vshear}(\omega)\, d\omega \quad [\mathrm{s}^{-2}] \tag{B.25}$$
$$\approx \frac{3}{\pi}\beta_* \beta_u E_{\mathrm{iw}} = 3 j_* b \beta_u N_0^2 E_0 \left(\frac{N}{N_0}\right)^2 \left(\frac{E}{E_0}\right).$$

When energy is elevated, the cutoff wavenumber is obtained by integrating $\Phi_{shear}(\beta)$ to produce the same variance as integrating the GM background spectrum to 0.6 m^{-1},

$$\frac{\int_0^{\beta_u} \Phi_{vshear}(\beta)\, d\beta}{N^2} = \frac{\int_0^{0.6} \Phi_{vshear}^{\mathrm{GM79}}(\beta)\, d\beta}{N_0^2} = 3 j_* b E_0\, 0.6 = 0.44, \tag{B.26}$$

where β_u was replaced with (B.4). For GM76, the background rolloff occurs when the variance is 0.66 instead of 0.44 in GM79.

Variance of linear internal waves is concentrated at frequencies just above f (Figure B.4). At background (N_0 and E_0), the wavenumber spectrum is flat within a

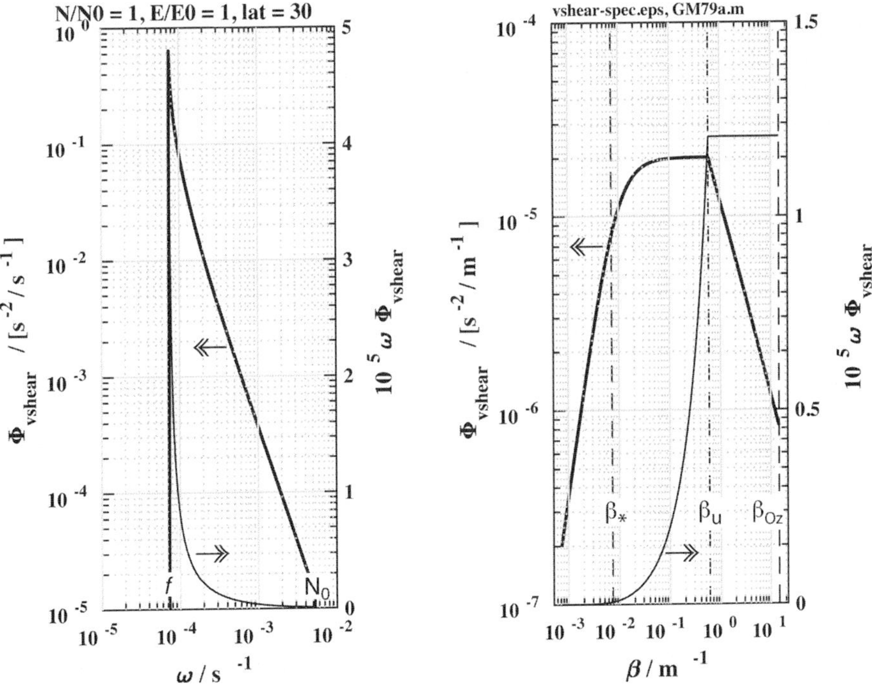

Figure B.4 Shear spectra as functions of frequency (left) and vertical wavenumber (right) in log-log (thick) and variance-preserving (thin) forms. As shown by the variance-preserving forms, most shear variance is at frequencies between f and $3f$ and at vertical wavenumbers close to β_u. Variance in the saturated range considerably exceeds that in the linear range.

decade of β_u, and that is where the variance is concentrated. Increasing β_* and decreasing β_u compresses this part of the wavenumber spectrum, eventually eliminating the flat portion when $N \gg N_0$ and $E \gg E_0$.

B.5 Vertical Displacement

The two-dimensional spectrum of vertical displacement, $\Phi_\zeta(\beta,\omega)$, is obtained by multiplying the normalized vertical wave function (B.15) by the energy spectrum,

$$\begin{aligned}\Phi_\zeta(\beta,\omega) &= \left(\frac{Z^2(\omega)}{N^2}\right)\Phi_E(\beta,\omega) \quad \left[\frac{\mathrm{m}^2}{\mathrm{m}^{-1}\ \mathrm{s}^{-1}}\right] \\ &= \frac{2j_* b E_0}{\beta_*^2+\beta^2}\left(\frac{E}{E_0}\right)\frac{(2f/\pi)(\omega^2-f^2)^{1/2}(N^2-f^2)}{\omega^3 N^2}.\end{aligned} \tag{B.27}$$

The integral of the frequency term,

$$\begin{aligned}&\int_f^N \frac{(2f/\pi)(\omega^2-f^2)^{1/2}(N^2-f^2)}{\omega^3 N^2}\,d\omega \\ &= \frac{((N/f)^2-1)}{\pi(N/f)^2}\left[\sec^{-1}\left(\frac{N}{f}\right)-\left(\left(\frac{N}{f}\right)^2-1\right)^{1/2}\right],\end{aligned} \tag{B.28}$$

is 0.5 when $N \gg f$, 0.45 when $N/f = 10$, and 0.40 when $N/f = 5$. Therefore,

$$\Phi_\zeta(\beta) = \int_f^N \Phi_\zeta(\beta,\omega)\,d\omega \approx \frac{j_* b E_0}{\beta_*^2+\beta^2}\left(\frac{E}{E_0}\right) \quad \left[\frac{\mathrm{m}^2}{\mathrm{m}^{-1}}\right]. \tag{B.29}$$

The integral in (B.8) gives the frequency spectrum of displacement as

$$\begin{aligned}\Phi_\zeta(\omega) &= \int_0^{\beta_u} \Phi_\zeta(\beta,\omega)\,d\beta \quad \left[\frac{\mathrm{m}^2}{\mathrm{s}^{-1}}\right] \\ &= b^2 E_0\left(\frac{N}{N_0}\right)\left(\frac{E}{E_0}\right)\frac{(2f/\pi)(\omega^2-f^2)^{1/2}(N^2-f^2)}{\omega^3 N^2}.\end{aligned} \tag{B.30}$$

Integrating again yields displacement variance,

$$\begin{aligned}<\zeta^2> &= \int_0^{\beta_u}\Phi_\zeta(\beta)\,d\beta = \int_f^N \Phi_\zeta(\omega)\,d\omega \quad [\mathrm{m}^2] \\ &\approx \frac{b^2 E_0}{2}\left(\frac{N_0}{N}\right)\left(\frac{E}{E_0}\right) = (7.3)^2\left(\frac{N_0}{N}\right)\left(\frac{E}{E_0}\right).\end{aligned} \tag{B.31}$$

For typical stratification and the GM energy level, rms displacements vary from ~2 m in the shallow pycnocline to ~20 m in the abyss (Figure B.3). Potential energy is 1/4 of total energy,

$$PE = \frac{N^2 < \zeta^2 >}{2} \approx \frac{b^2 N_0^2 E_0}{4} \left(\frac{N}{N_0}\right)\left(\frac{E}{E_0}\right) = \frac{E_{iw}}{4} \quad \left[\frac{\text{J}}{\text{kg}}\right]. \tag{B.32}$$

B.6 Vertical Velocity

The spectrum of vertical velocity is ω^2 times the displacement spectrum,

$$\Phi_w(\beta,\omega) = \omega^2 \Phi_\zeta(\beta,\omega) \quad \left[\frac{(\text{m/s})^2}{\text{m}^{-1}\ \text{s}^{-1}}\right] \tag{B.33}$$

$$= \frac{2 j_* b E_0}{\beta_*^2 + \beta^2} \left(\frac{E}{E_0}\right) \frac{(2f/\pi)(\omega^2 - f^2)^{1/2}(N^2 - f^2)}{\omega N^2}.$$

The integral of the frequency term is

$$\int_f^N \frac{(2f/\pi)(\omega^2 - f^2)^{1/2}(N^2 - f^2)}{\omega N^2}\, d\omega \tag{B.34}$$

$$= \frac{2fN}{\pi}\left(1 - \left(\frac{f}{N}\right)^2\right)\left[\left(1 - \left(\frac{f}{N}\right)^2\right)^{1/2} - \left(\frac{f}{N}\right)\cos^{-1}\left(\frac{f}{N}\right)\right],$$

where the product of the f/N terms equals 0.98 for $N/f = 100$, 0.84 for $N/f = 10$, and 0.68 for $N/f = 5$. Approximating the integral as 1,

$$\Phi_w(\beta) = \int_f^N \Phi_w(\beta,\omega)\, d\omega \quad \left[\frac{(\text{m/s})^2}{\text{m}^{-1}}\right] \tag{B.35}$$

$$\approx \frac{4 j_* b N_0 f_{30} E_0}{\pi(\beta_*^2 + \beta^2)} \left(\frac{N}{N_0}\right)\left(\frac{f}{f_{30}}\right)\left(\frac{E}{E_0}\right).$$

Using the approximation to (B.8), the frequency spectrum becomes

$$\Phi_w(\omega) = \int_{\beta_*}^{\beta_u} \Phi_w(\beta,\omega)\, d\beta \quad \left[\frac{(\text{m/s})^2}{\text{s}^{-1}}\right] \tag{B.36}$$

$$= 2b^2 E_0 \left(\frac{N_0}{N}\right)\left(\frac{E}{E_0}\right)\left(\frac{\beta_u}{\beta_*}\right) \frac{(2f/\pi)(\omega^2 - f^2)^{1/2}(N^2 - f^2)}{\omega N^2}.$$

The variance follows as

$$< w^2 > = \int_{\beta_*}^{\beta_u} \Phi_w(\beta)\, d\beta = \int_f^N \Phi_w(\omega)\, d\omega \quad \left[(\mathrm{m/s})^2\right] \tag{B.37}$$
$$\approx \frac{2b^2 N_0 f_{30} E_0}{\pi}\left(\frac{f}{f_{30}}\right)\left(\frac{E}{E_0}\right) = (5.1\times 10^{-3})^2\left(\frac{f}{f_{30}}\right)\left(\frac{E}{E_0}\right).$$

B.7 Vertical Strain

The two-dimensional strain spectrum has the same β dependence but not the same frequency function as the shear spectrum,

$$\Phi_{strain}(\omega,\beta) = \beta^2 \Phi_\zeta(\beta,\omega) \quad \left[\frac{1}{\mathrm{m}^{-1}\ \mathrm{s}^{-1}}\right] \tag{B.38}$$
$$= 2 j_* b E_0 \frac{\beta^2}{\beta_*^2+\beta^2}\left(\frac{E}{E_0}\right)\frac{(2f/\pi)(\omega^2-f^2)^{1/2}(N^2-f^2)}{\omega^3 N^2}.$$

Integrating with (B.28), the vertical wavenumber spectrum is

$$\Phi_{strain}(\beta) = \int_f^N \Phi_{Strain}(\omega,\beta)\, d\omega \quad \left[\frac{1}{\mathrm{m}^{-1}}\right] \tag{B.39}$$
$$\approx j_* b E_0 \frac{\beta^2}{\beta_*^2+\beta^2}\left(\frac{E}{E_0}\right).$$

Applying (B.23) to the two-dimensional spectrum gives

$$\Phi_{strain}(\omega) = \int_0^{\beta_u} \Phi_{strain}(\omega,\beta)\, d\beta \quad \left[\frac{1}{\mathrm{s}^{-1}}\right] \tag{B.40}$$
$$\approx 2 j_* b \beta_u E_0 \left(\frac{E}{E_0}\right)\frac{(2f/\pi)(\omega^2-f^2)^{1/2}(N^2-f^2)}{\omega^3 N^2}.$$

The variance is then independent of N,

$$< strain^2 > = \int_0^{\beta_u} \Phi_{strain}(\beta)\, d\beta = \int_f^N \Phi_{strain}(\omega)\, d\omega \tag{B.41}$$
$$\approx j_* b \beta_u E_0 \left(\frac{E}{E_0}\right) = (0.496)^2\left(\frac{E}{E_0}\right) \quad \text{[dimensionless]}.$$

B.8 Shear-to-Strain Ratio

Consistent with horizontal kinetic energy being three times potential energy, the shear-to-strain ratio is

$$R_\omega(\beta) \equiv \frac{\Phi_{shear}(\beta)/N^2}{\Phi_{strain}(\beta)} = \frac{< shear^2 > /N^2}{< strain^2 >} = 3. \tag{B.42}$$

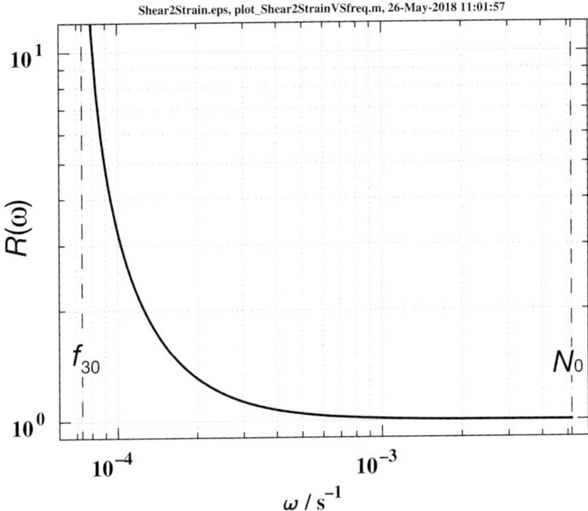

Figure B.5 Shear-to-strain ratio, R_ω, at N_0 and f_{30} and versus ω from (B.43).

As a function of frequency, the shear-to-strain ratio,

$$R_\omega(\omega) \equiv \frac{\Phi_{shear}(\omega)/N^2}{\Phi_{strain}(\omega)} = \frac{\omega^2 + f^2}{\omega^2 - f^2}, \tag{B.43}$$

is large near f and ≈ 1 within a decade of N (Figure B.5).

Bibliography

Agee, E.M. 1987. Mesoscale cellular convection over the oceans. *Dyn. Atmos. Oceans*, **10**, 317–341.

Agrawal, Y.C., and Belting, C.J. 1988. Laser velocimetry for benthic sediment transport. *Deep-Sea Res.*, **35**, 1047–1067.

Alford, M.H. 2008. Observations of parametric subharmonic instability of the diurnal internal tide in the South China Sea. *Geophys. Res. Lett.*, **35**(L15602).

Alford, M.H. 2010. Sustained, full-water-column observations of internal waves and mixing near Mendocino Escarpment. *J. Phys. Oceanogr.*, **40**, 2643–2660.

Alford, M.H., Gerdt, D.W., and Adkins, C.M. 2006a. An ocean refractometer: Resolving millimeter-scale turbulent density fluctuations via the refractive index. *J. Atmos. Ocean. Tech.*, **23**, 121–137.

Alford, M.H., Gregg, M.C., and D'Asaro, E.A. 2005. Mixing, 3-D mapping and Lagrangian evolution of a thermohaline intrusion. *J. Phys. Oceanogr.*, **35**, 1689–1711.

Alford, M.H., Gregg, M.C., and Merrifield, M.A. 2006b. Structure, propagation and mixing of energetic baroclinic tides in Mamala Bay, Oahu, Hawaii. *J. Phys. Oceanogr.*, **36**, 997–1018.

Alford, M.H., Gregg, M.C., Zervakis, V., and Kontoyiannis, H. 2012. Internal wave measurements on the Cycladic Plateau of the Aegean Sea. *J. Geophys. Res.*, **117**(C01015).

Alford, M.H., MacKinnon, J.A., Pinkel, R., and Klymak, J. 2017. Space-time scales of shear in the North Pacific. *J. Phys. Oceanogr.*, **47**, 2455–2478.

Alford, M.H., MacKinnon, J.A., Simmons, H.L., and Nash, J.D. 2016. Near-inertial internal gravity waves in the ocean. *Annu. Rev. Mar. Sci.*, **8**, 95–123.

Alford, M.H., MacKinnon, J.A., Zhao, Z., Pinkel, R., Klymak, J., and Peacock, T. 2007. Internal waves across the Pacific. *Geophys. Res. Lett.*, **34**(L24601).

Alford, M.H., and Pinkel, R. 2000. Patterns of turbulent and double-diffusive phenomena: Rapid-profiling microconductivity probe. *J. Phys. Oceanogr.*, **30**, 833–854.

Alford, M.H., Shcherbina, A.Y., and Gregg, M.C. 2013. Observations of near-inertial internal gravity waves radiating from a frontal jet. *J. Phys. Oceanogr.*, **43**, 1225–1239.

Alford, M.H., and Whitmont, M. 2007. Seasonal and spatial variability of near-inertial kinetic energy from historical moored velocity records. *J. Phys. Oceanogr.*, **37**, 2022–2037.

Allen, K.R., and Joseph, R.I. 1989. A cannonical statistical theory of oceanic internal waves. *J. Fluid Mech.*, **204**, 185–228.

Annis, A., and Moum, J.N. 1995. Surface wave-turbulence interactions: Scaling $\epsilon(z)$ near the sea surface. *J. Phys. Oceanogr.*, **25**, 2025–2045.

Apel, J.R. 2002. Oceanic internal waves and solitons. In: *An Atlas of Oceanic Internal Solitary Waves (May 2002)*. Global Ocean Associates.

Armi, L., and Farmer, D. 2002. Stratified flow over topography: Bifurcation fronts and transition to the uncontrolled state. *Proc. Roy. Soc. Lond. A*, **458**, 513–538.

Armi, L., Hebert, D., Oakey, N., and Price, J.F. 1989. Two years in the life of a Mediterranean salt lens. *J. Phys. Oceanogr.*, **19**, 354–370.

Arons, A.B. 1981. The scientific work of Henry Stommel. Pages xiv–xviii of: Warren, B.A., and Wunsch, C. (eds), *Evolution of Physical Oceanography*. Boston: MIT Press.

Aucan, J., Merrifield, M.A., and Luther, D.S. 2006. Tidal mixing events on the deep flanks of Kaena Ridge, Hawaii. *J. Phys. Oceanogr.*, **36**, 1202–1219.

Baines, P.G., and Gill, A.E. 1969. On thermohaline convection with linear gradients. *J. Fluid Mech.*, **37**, 289–306.

Batchelor, G.K. 1946. The theory of axisymmetric turbulence. *Proc. Roy. Soc. London Ser. A*, **186**, 480–502.

Batchelor, G.K. 1959. Small-scale variation of convected quantitites like temperature in turbulent fluid. Part 1. General discussion and the case of small conductivity. *J. Fluid Mech.*, **5**, 113–139.

Beaird, N., Fer, I., Rhines, P., and Eriksen, C. 2012. Dissipation of turbulent kinetic energy inferred from Seagliders: An application to the eastern Nordic Seas overflows. *J. Phys. Oceanogr.*, **42**, 2268–2282.

Bebieva, Y., and Timmermans, M.-L. 2017. The relationship between double diffusive intrusions and staircases in the Arctic Ocean. *J. Phys. Oceanogr.*, **47**(4), 867–878.

Bell, T.H., 1975a. Lee waves in stratified flow with simple harmonic time dependence. *J. Fluid Mech.*, **67**, 705–722.

Bell, T.H. 1975b. Topographically generated internal waves in the open ocean. *J. Geophys. Res.*, **80**, 320–337.

Bender, C.M., and Orszag, S.A. 1978. *Advanced Mathematical Methods for Scientists and Engineers*. New York, NY: McGraw-Hill.

Bertuccioli, L., Roth, G.I., Katz, J., and Osborn, T.R. 1999. A submersible particle image velocimetry system for turbulence measurements in the bottom boundary layer. *J. Atmos. Ocean. Tech.*, **16**, 1635–1646.

Biescas, B., Armi, L., Sallarès, V., and Gràcia, E. 2010. Seismic imaging of staircase layers below the Mediterranean undercurrent. *Deep-Sea Res. I*, **57**, 1345–1353.

Billant, P., and Chomaz, J.-M. 2000a. Experimental evidence for a new instability of a vertical columnar vortex pair in a strongly stratified fluid. *J. Fluid Mech*, **418**, 167–188.

Billant, P., and Chomaz, J.-M. 2000b. Three-dimensional stability of a vertical columnar vortex pair in a stratified fluid. *J. Fluid Mech.*, **419**, 65–91B.

Billant, P., and Chomaz, J.-M. 2001. Self-similarity of strongly stratified inviscid flows. *Phys. Fluids*, **13**(6), 1645–1651.

Bogucki, D., and Garrett, C. 1993. A simple model for the shear-induced decay of an internal solitary wave. *J. Phys. Oceanogr.*, **23**, 1767–1776.

Bogucki, D., Dickey, T., and Redekop, L.G. 1997. Sediment resuspension and mixing by resonantly generated internal solitary waves. *J. Phys. Oceanogr.*, **27**, 1181–1196.

Bolgiano, R. 1959. Turbulent spectra in a stably stratified atmosphere. *J. Geophys. Res.*, **64**, 2226–2229.

Bouruet-Aubertot, P., Koudella, C., Staquet, C., and Winters, K.B. 2001. Particle dispersion and mixing by breaking internal gravity waves. *Dyn. Atmos. Ocean*, **33**, 95–134.

Brainerd, K.E., and Gregg, M.C. 1992. Turbulent decay during diurnal restratification of the oceanic mixed layer. Pages (J8)203–(J8)206 of: *AMS 10th Symposium on Turbulence and Diffusion, September 29–October 2, 1992, Portland, Oregon*.

Brandt, P., Rubino, A., Alpers, W., and Backhaus, J.O. 1997. Internal waves in the Strait of Messina studied by a numerical model and synthetic aperature radar images from the ERS 1/2 satellites. *J. Phys. Oceanogr.*, **27**, 648–663.

Bretherton, F.P. 1966. The propagation of groups of internal gravity waves in a shear flow. *Quart. J. Roy. Meteor. Soc.*, **92**, 466–480.

Bretherton, F.P., and Garrett, C.J.R. 1968. Wavetrains in inhomogeneous moving media. *Proc. Roy. Soc. Lond. A*, **302**, 529–554.

Brethouwer, G., Billiant, P., Lindborg, E., and Chomaz, J.-M. 2007. Scaling analysis and simulation of strongly stratified turbulent flows. *J. Fluid Mech.*, **585**, 343–368.

Briscoe, M.G. 1975. Preliminary results from the trimoored Internal Waves Experiment. *J. Geophys. Res.*, **80**(27), 3872–3884.

Briscoe, M.G., and Weller, R.A. 1984. Preliminary results from the Long-term Upper-Ocean Study (LOTUS). *Dyn. Atmos. Oceans*, **8**, 243–265.

Brown, N. 1974. A precision CTD microprofiler. Pages 270–278 of: *Engr. in the Ocean Envir., Ocean '74*, vol. 2.

Browne, L.W.B., Antonia, R.A., and Shad, D.A. 1987. Turbulent energy dissipation in a wake. *J. Fluid Mech.*, **179**, 307–326.

Bryden, H.L., and Nurser, A.J.G. 2003. Effect of strait mixing on ocean stratification. *J. Phys. Oceanogr.*, **33**, 1870–1872.

Cacchione, D.A., Pratson, L.F., and Ogston, A.S. 2002. The shaping of continental slopes by internal tides. *Science*, **296**, 724–727.

Cairns, J.L., and Williams, G.O. 1976. Internal wave observations from a midwater float, 2. *J. Geophys. Res.*, **81**, 1943–1950.

Caldwell, D.R. 1973. Thermal and Fickian diffusion of sodium chloride in a solution of oceanic concentration. *Deep-Sea Res.*, **20**, 1029–1039.

Caldwell, D.R. 1974a. The effect of pressure on thermal and Fickian diffusion of sodium chloride. *Deep-Sea Res.*, **21**, 369–375.

Caldwell, D.R. 1974b. Experimental studies on the onset of thermohaline convection. *J. Fluid Mech.*, **64**, 347–367.

Caldwell, D.R., Wilcox, S.D., and Matsier, M. 1975. A relatively simple freely-falling probe for small-scale temperature gradients. *Limnol. Oceanogr.*, **20**, 1035–1047.

Carmack, E.C., Williams, W.J., Zimmerman, S.L., and McLaughlin, F.A. 2012. The Arctic Ocean warms from below. *Geophys. Res. Lett.*, **39**(L07604).

Carpenter, J.R., Sommer, T., and Wüest, A. 2012a. Simulations of a double-diffusive interface in the diffusive convection regime. *J. Fluid Mech.*, **711**, 411–436.

Carpenter, J.R., Sommer, T., and Wüest, A. 2012b. Stability of a double-diffusive interface in the diffusive convection regime. *J. Phys. Oceanogr.*, **42**, 840–854.

Carter, G.S., and Gregg, M.C. 2006. Persistent near-diurnal internal waves observed above a site of M_2 barotropic-to-baroclinic conversion. *J. Phys. Oceanogr.*, **36**, 1136–1147.

Carter, G.S., Gregg, M.C., and Lien, R.-C. 2005. Internal waves, solitary-like waves, and mixing on the Monterey Bay shelf. *Cont. Shelf Res.*, **25**, 1499–1520.

Chang, M.-H., Lien, R.-C., Yang, Y.J., Tang, T.Y., and Wang, J. 2008. A composite view of surface signatures and interior properties of nonlinear internal waves: Observations and applications. *J. Atmos. Ocean. Tech.*, **25**, 1218–1227.

Corrsin, S. 1963. Estimates of the relations between Eulerian and Lagrangian scales in large Reynolds number turbulence. *J. Atmos. Sci.*, **20**, 115–119.

Cox, C.S., Nagata, Y., and Osborn, T. 1969. Oceanic fine structure and internal waves. *Bull. Japanese Soc. Fisheries Oceanography*, Papers in Dedication to Prof. Michitaka Uda(Nov.), 67–71.

Crapper, P.F. 1975. Measurements across a diffusive interface. *Deep-Sea Res.*, **22**, 537–545.

Curry, J.A., and Webster, P.J. 1999. *Thermodynamics of Atmospheres and Oceans*. Intl. Geophys. Ser., vol. 65. 24–28 Oval Rd., London, NW1 7DX, UK: Academic Press.

D'Asaro, E.A. 1989. The decay of wind-forced mixed layer inertial oscillations due to the β effect. *J. Geophys. Res.*, **94**, 2045–2056.

D'Asaro, E.A. 2001. Turbulent vertical kinetic energy in the ocean mixed layer. *J. Phys. Oceanogr.*, **31**, 3530–3537.

D'Asaro, E.A., Farmer, D.M., Osse, J.T., and Dairiki, G.T. 1996. A Lagrangian float. *J. Atmos. Ocean. Tech.*, **13**(6), 1230–1246.

D'Asaro, E.A., and Lien, R.-C. 2000a. Lagrangian measurements of waves and turbulence in stratified flows. *J. Phys. Oceanogr.*, **30**, 641–655.

D'Asaro, E.A., and Lien, R.-C. 2000b. The wave-turblence transition for stratified flows. *J. Phys. Oceanogr.*, **30**, 1669–1678.

D'Asaro, E.A., and Lien, R.-C. 2007. Measurement of scalar variance dissipation from Lagrangian floats. *J. Atmos. Ocean. Tech.*, **24**, 1066–1077.

D'Asaro, E.A., and Morehead, M.D. 1991. Internal waves and velocity fine structure in the Arctic Ocean. *J. Geophys. Res.*, **96**, 12,725–12.738.

D'Asaro, E.A., and Morison, J.H. 1992. Internal waves and mixing in the Arctic Ocean. *Deep-Sea Res A*, **39**(2), S459–S484.

Davis, R.E. 1994a. Diapycnal mixing in the ocean: Equations for large-scale budgets. *J. Phys. Oceanogr.*, **24**, 777–800.

Davis, R.E. 1994b. Diapycnal mixing in the ocean: The Osborn-Cox model. *J. Phys. Oceanogr.*, **24**, 2560–2576.

Davis, R.E., Regier, L.A., Dufour, J., and Webb, D.C. 1992. The Autonomous Lagrangian Circulation Explorer (ALACE). *J. Atmos. Ocean. Tech.*, **9**, 264–285.

Davis, R.E., Sherman, J.T., and Dufour, J. 2001. Profiling ALACEs and other advances in autonomous subsurface floats. *J. Atmos. Ocean. Tech.*, **18**, 982–993.

de Groot, S.R., and Mazur, P. 1969. *Non-Equilibrium Thermodynamics*. Amsterdam: North-Holland.

Desaubies, Y.J.F., and Gregg, M.C. 1981. Reversible and irreversible finestructure. *J. Phys. Oceanogr.*, **11**, 541–556.

Desaubies, Y.J.F., and Smith, W.K. 1982. Statistics of Richardson number and instability in oceanic internal waves. *J. Phys. Oceanogr.*, **12**, 1245–1259.

Dickson, R.R., and Brown, J. 1994. The production of North Atlantic Deep Water: Sources, rates, and pathways. *J. Geophys. Res.*, **99**, 12,319–12,341.

Dillon, T.M. 1982. Vertical overturns: A comparison of Thorpe and Ozmidov length scales. *J. Geophys. Res.*, **87**, 9601–9613.

Dillon, T.M., and Caldwell, D.R. 1980. The Batchelor spectrum and dissipation in the upper ocean. *J. Geophys. Res.*, **85**(C4), 1910–1916.

Döös, K, and Coward, A. 1997. The Southern Ocean as the major upwelling zone of North Atlantic Deep Water. *WOCE newsletter*, **27**, 3–4.

Doron, P., Bertuccioli, L., Katz, J., and Osborn, T.R. 2001. Turbulence characteristics and dissipation estimates in the coastal ocean bottom boundary layer using PIV data. *J. Phys. Oceanogr.*, **31**(8), 2108–2134.

Dosser, H.V., and Rainville, L. 2016. Dynamics of the changing near-inertial internal wave field in the Arctic Ocean. *J. Phys. Oceanogr.*, **46**(2), 395–415.

Dunkerton, T.J. 1997. Shear instability of internal inertial-gravity waves. *J. Atmos. Sci.*, **54**, 1628–1641.

Dushaw, B.D., Cornuelle, B.D., Worcester, P.F., Howe, B.M., and Luther, D.S. 1995. Barotropic and baroclinic tides in the central North Pacific Ocean determined from long range reciprocal acoustic transmissions. *J. Phys. Oceanogr.*, **25**, 631–647.

Efron, B., and Gong, G. 1983. A leisurely look at at the bootstrap, the jackknife, and cross-validation. *Amer. Stat.*, **37**, 36–48.

Egbert, G.D., and Ray, R.D. 2000. Significant dissipation of tidal energy in the deep ocean inferred from satellite altimeter data. *Nature*, **405**(15 June 2000), 775–778.

Egbert, G.D., and Ray, R.D. 2001. Estimates of M_2 tidal energy dissipation from TOPEX/Poseidon altimeter data. *Geophys. Res. Ltrs.*, **106**(C10), 22,475–22,502.

Ekman, V.W. 1905. On the influence of the Earth's rotation on ocean-currents. *Ark. Mat. Astron. Fys.*, **2**, 1–52.

Elliott, J.A., and Oakey, N.S. 1975. Horizontal coherence of temperature microstructure. *J. Phys. Oceanogr.*, **5**(7), 506–515.

Elliott, J.A., and Oakey, N.S. 1976. Spectrum of small-scale oceanic temperature gradient. *J. Fish. Res. Bd. Canada*, **33**, 2296–2306.

Ellison, T.H. 1957. Turbulent transport of heat and momentum from an infinite rough plane. *J. Fluid Mech*, **2**, 456–466.

Eriksen, C.C. 1982. Observations of internal wave reflection off sloping bottoms. *J. Geophys. Res.*, **87**, 525–538.

Eriksen, C.C., Osse, T.J., Light, R.D., Wen, T., Lehman, T.W., Sabin, P.L., Ballard, J.W., and Chiodi, A.M. 2001. Seaglider: A long-range autonomous underwater vehicle for oceanographic research. *IEEE J. Ocean Engr.*, **26**(4), 424–436.

Evans, D.L., Rossby, H.T., Mork, M., and Gytre, T. 1979. YVETTE – a free-fall shear profiler. *Deep-Sea Res A*, **26**, 703–718.

Ewart, T.E., and Bendiner, W.P. 1981. An observation of the horizontal and vertical diffusion of a passive tracer in the deep ocean. *J. Geophys. Res.*, **86**(C11), 10,974–10,982.

Fabula, A.G. 1968a. The dynamic response of towed thermometers. *J. Fluid Mech.*, **34**, 449–464.

Fabula, A.G. 1968b. The dynamic response of towed thermometers. *J. Fluid Mech.*, **34**(3), 449–464.

Farmer, D.M., and Smith, J.D. 1980. Tidal interaction of stratified flow with a sill in Knight Inlet. *Deep-Sea Res.*, **27A**, 239–254.

Feistel, R. 2008. A Gibbs function for seawater thermodynamics for -6° to 80^oC and salinity up to 120 gkg^{-1}. *Deep-Sea Res. I*, **55**, 1639–1671.

Fer, I., Peterson, A.K., and Ullgren, J.E. 2014. Microstructure measurements from an underwater glider in the turbulent Faroe Bank Channel overflow. *J. Atmos. Ocean. Tech.*, **31**, 1128–1150.

Fernando, H.J.S. 1987. The formation of a layered structure when a stable salinity gradient is heated from below. *J. Fluid Mech.*, **182**, 525–541.

Fernando, H.J.S. 1989. Oceanographic implications of laboratory experiments on diffusive interfaces. *J. Phys. Oceanogr.*, **19**, 1707–1715.

Ferrari, R., Mashayek, A., McDougall, T.J., Nukurashin, M., and Campin, J.-M. 2016. Turning ocean mixing upside down. *J. Phys. Oceanogr.*, **46**, 2239–2260.

Ferrari, R., and Rudnick, D.L. 2000. Thermohaline variability in the upper ocean. *J. Geophys. Res.*, **105**, 16,857–16,883.

Ferron, B., Mercier, H., Speer, K., Gargett, A., and Polzin, K. 2003. Mixing in the Romanche Fracture Zone. *J. Phys. Oceanogr.*, **28**, 1929–1945.

Flanagan, J.D., Lefler, A.S., and Radko, R. 2013. Heat transport through diffusive interfaces. *Geophys. Res. Lett.*, **40**, 2466–2470.

Flanagan, J.D., Radko, T., Shaw, W.J., and Stanton, T.P. 2014. Dynamic and double-diffusive instabilities in a weak pycnocline. Part II: Direct numerical simulations and flux laws. *J. Phys. Oceanogr.*, **44**, 1992–2012.

Flatté, S.M., Henyey, F.S., and Wright, J.A. 1985. Eikonal calculations of short-wavelength internal wave spectra. *J. Geophys. Res.*, **90**, 7265–7272.

Fleury, M., and Lueck, R.G. 1994. Direct heat flux estimates using a towed vehicle. *J. Phys. Oceanogr.*, **24**(4), 801–818.

Fofonoff, N.P. 1962. Physical properties of seawater. Pages 3–30 of: Hill, M.N. (ed.), *The Sea*, vol. 1. New York, NY: Wiley-Interscience.

Fofonoff, N.P. 1969. Spectral characteristics of internal waves in the ocean. *Deep-Sea Res*, **16, Suppl.**, 58–71.

Fofonoff, N.P. 1998. Nonlinear limits to ocean thermal structure. *J. Mar. Res.*, **56**(4), 783–811.

Fofonoff, N.P. 2001. Thermal stability of the world ocean thermocline. *J. Phys. Oceanogr.*, **31**(8), 2169–2177.

Foster, T.D. 1972. An analysis of the cabbeling instability in sea water. *J. Phys. Oceanogr.*, **2**, 294–301.

Foster, T.D., and Carmack, E.C. 1976. Temperature and salinity structure in the Weddell Sea. *J. Phys. Oceanogr.*, **6**(1), 36–44.

Frants, M., Damerell, G.M., Gille, S.T., Heywood, K.J., MacKinnon, J., and Sprintall, J. 2013. An assessment of density-based finescale methods for estimating diapycnal diffusivity in the Southern Ocean. *J. Atmos. Ocean. Tech.*, **30**(11), 2647–2661.

Fringer, O.B., and Street, R.L. 2003. The dynamics of breaking progressive interfacial waves. *J. Fluid Mech.*, **494**, 319–353.

Fu, L.-L. 1981. Observations and models of inertial waves in the deep ocean. *Rev. Geophys. and Space Phys.*, **19**(1), 141–170.

Furuichi, N., Hibiya, T., and Niwa, Y. 2005. Bispectral analysis of energy transfer within the two-dimensional oceanic internal wave field. *J. Phys. Oceanogr.*, **35**, 2104–2109.

Galbraith, P.S., and Kelley, D.E. 1996. Identifying overturns in CTD profiles. *J. Atmos. Ocean. Tech.*, **13**, 688–702.

Gargett, A.E. 1985. Evolution of scalar spectra with the decay of turbulence in a stratified fluid. *J. Fluid Mech.*, **159**, 379–407.

Gargett, A.E. 1990. Do we really know how to scale the turbulent kinetic energy dissipation rate ϵ due to breaking of oceanic internal waves? *J. Geophys. Res.*, **95**, 15,971–15,974.

Gargett, A.E. 2003. Differential diffusion: an oceanographic primer. *Progr. Oceanogr.*, **56**, 559–570.

Gargett, A.E., Hendricks, P.J., Sanford, T.B., Osborn, T.R., and Williams, III, A.J. 1981. A composite spectrum of vertical shear in the upper ocean. *J. Phys. Oceanogr.*, **11**, 1258–1271.

Gargett, A.E., Merryfield, W., and Holloway, G. 2003. Direct numerical simulation of differential scalar diffusion in three-dimensional stratified turbulence. *J. Phys. Oceanogr.*, **33**, 1758–1782.

Gargett, A.E., Osborn, T.R., and Nasmyth, P.W. 1984. Local isotropy and the decay of turbulence in a stratified fluid. *J. Fluid Mech.*, **144**, 231–280.

Gargett, A.E., and Schmitt, R.W. 1982. Observations of salt fingers in the central waters of the eastern North Pacific. *J. Geophys. Res.*, **87**, 8017–8029.

Garrett, C. 2001. What is the 'near-inertial' band and why is it different from the rest of the internal wave spectrum? *J. Phys. Oceanogr.*, **31**, 962–971.

Garrett, C., and Horne, E. 1978. Frontal circulation due to cabbeling and double diffusion. *J. Geophys. Res.*, **83**, 4651–4656.

Garrett, C.J.R, and Munk, W.H., 1972. Space-time scales of internal waves. *Geophys. Fluid Dyn.*, **2**, 225–264.

Garrett, C.J.R., and Munk, W.H. 1975. Space-time scales of internal waves: A progress report. *J. Geophys. Res.*, **80**, 291–297.

Gibson, C.H., and Schwarz, W.H. 1963. The universal equilibrium spectra of turbulent velocity and scalar fields. *J. Fluid Mech.*, **16**, 365–384.

Giles, A.B., and McDougall, T.J. 1986. Two methods for the reduction of salinity spiking of CTDs. *Deep-Sea Res.*, **33**, 1253–1274.

Gill, A.E. 1982. *Atmosphere-Ocean Dynamics*. New York, NY: Academic.

Gill, A.E. 1984. On the behavior of internal waves in the wake of a storm. *J. Phys. Oceanogr.*, **14**, 1129–1151.

Gille, S.T. 1997. The Southern Ocean momentum balance: Evidence for topographic effects from numerical model output and altimeter data. *J. Phys. Oceanogr.*, **27**, 2219–2232.

Gille, S.T. 2004. How nonlinearities in the equation of state of seawater can confound estimates of steric sea level change. *J. Geophys. Res. Oceans*, **109**(C03005).

Goodman, L. 1990. Acoustic scattering from ocean microstructure. *J. Geophys. Res.*, **95**(C7), 11,557–11,573.

Goodman, L., Levine, E.R., and Lueck, R.G. 2006. On measuring the terms of the turbulent kinetic energy budget from an AUV. *J. Atmos. Ocean. Tech.*, **23**(7), 977–990.

Goodman, L., and Wang, Z. 2009. Turbulence observations in the northern bight of Monterey Bay from a small AUV. *J. Mar. Syst.*, **77**(4), 441–458.

Gordon, A.L. 2013. Bottom water formation. In: *Reference Module in Earth Systems and Environmental Sciences*. Elsevier.

Graham, F.S., and McDougall, T.J. 2013. Quantifying the nonconservative production of conservative temperature, potential temperature, and entropy. *J. Phys. Oceanogr.*, **43**, 838–862.

Grant, H.L., Hughes, B., Vogel, W., and Moillet, A. 1968b. The spectrum of temperature fluctuations in turbulent flow. *J. Fluid Mech.*, **34**, 423–441.

Grant, H.L., Moilliet, A., and Stewart, R.W. 1959. A spectrum of turbulence at very high Reynolds number. *Nature*, **184**, 808–810.

Grant, H.L., Moilliet, A., and Vogel, W.M. 1968a. Some observations of the occurrence of turbulence in and above the thermocline. *J. Fluid Mech.*, **33**, 443–448.

Grant, H.L., Stewart, R.W., and Moilliet, A. 1962. Turbulence spectra from a tidal channel. *J. Fluid Mech.*, **12**, 241–268.

Gregg, M.C. 1968. Mechanical stirring and salt fingers. Pages 42–51 of: *WHOI Summer Institute of Geophysical Fluid Dynamics Participant Reports*, vol. II. Woods Hole, MA: Woods Hole Oceanographic Institution.

Gregg, M.C. 1975a. Microstructure and intrusions in the California Current. *J. Phys. Oceanogr.*, **5**, 253–278.

Gregg, M.C. 1975b. Oceanic fine and microstructure. *Rev. Geophys. and Space Phys.*, **13**, 586–591 and 635–636.

Gregg, M.C. 1977a. *Cruise report of the Mixed Layer Experiment (MILE) on the U.S.N.S. DE STEIGUER, 18 August – 8 September 1977*. Tech. rept. APL-UW Technical Note 5–77. Applied Physics Laboratory, University of Washington, Seattle, WA.

Gregg, M.C. 1977b. Variations in the intensity of small-scale mixing in the main thermocline. *J. Phys. Oceanogr.*, **7**, 436–454.

Gregg, M.C. 1980a. Microstructure patches in the thermocline. *J. Phys. Oceanogr.*, **10**, 915–943.

Gregg, M.C. 1980b. The three-dimensional mapping of a small thermohaline intrusion. *J. Phys. Oceanogr.*, **10**, 1468–1492.

Gregg, M.C. 1984. Entropy generation in the ocean by small-scale mixing. *J. Phys. Oceanogr.*, **14**, 688–711.

Gregg, M.C. 1987. Structures and fluxes in a deep convecting mixed layer. Pages 1–23 of: Muller, P., and Henderson, D. (eds.), *Dynamics of the Oceanic Surface Mixed Layer, Proceedings, 'Aha Huliko'a Hawaiian Winter Workshop, University of Hawaii at Manoa, January 14–16, 1987*. Hawaii Institute of Geophysics.

Gregg, M.C. 1989a. Scaling turbulent dissipation in the thermocline. *J. Geophys. Res.*, **94**, 9686–9698.

Gregg, M.C. 1989b. Small-scale mixing: A first-order process? Pages 117–126 of: Müller, P., and Henderson, D. (eds.), *Parameterization of Small-Scale Processes: Proceedings, 'Aha Huliko'a Hawaiian Winter Workshop, University of Hawaii at Manoa, January 17–20, 1989*. Hawaii Institute of Geophysics. Special Publication.

Gregg, M.C. 1991. The study of mixing in the ocean: A brief history. *Oceanography*, **4**, 39–45.

Gregg, M.C. 1999. Uncertainities and limitations in measuring ϵ and χ_T. *J. Atmos. Ocean. Tech.*, **16**, 1483–1490.

Gregg, M.C., Alford, M.H., Kontoyiannis, H., Zervakis, V., and Winkel, D. 2012. Mixing over the steep side of the Cycladic Plateau in the Aegean Sea. *J. Mar. Syst*, **89**, 30–47.

Gregg, M.C., and Cox, C.S. 1971. Measurements of the oceanic microstructure of temperature and electrical conductivity. *Deep-Sea Res.*, **18**, 925–934.

Gregg, M.C., and Cox, C.S. 1972. The vertical microstructure of temperature and salinity. *Deep-Sea Res.*, **19**, 355–376.

Gregg, M.C., Cox, C.S., and Hacker, P.W. 1973. Vertical microstructure measurements in the Central North Pacific. *J. Phys. Oceanogr.*, **3**, 458–469.

Gregg, M.C., D'Asaro, E.A., Shay, T.J., and Larson, N. 1986. Observations of persistent mixing and near-inertial internal waves. *J. Phys. Oceanogr.*, **16**, 856–885.

Gregg, M.C., D'Asaro, E.A., Riley, J.J., and Kunze, E. 2018. Mixing efficiency in the ocean. *Annu. Rev. Mar. Sci.*, **10**, 443–473.

Gregg, M.C., and Hess, W.C. 1985. Dynamic response calibration of Sea-Bird temperature and conductivity probes. *J. Atmos. Ocean. Tech.*, **2**, 304–313.

Gregg, M.C., and Kunze, E. 1991. Shear and strain in Santa Monica Basin. *J. Geophys. Res.*, **96**, 16,709–16,719.

Gregg, M.C., and Meagher, T.B. 1980. The dynamic response of glass-rod thermistors. *J. Geophys. Res.*, **85**, 2779–2786.

Gregg, M.C., Meagher, T.B., Pederson, A.M., and Aagaard, E.A. 1978. Low noise temperature microstructure measurements with thermistors. *Deep-Sea Res.*, **25**, 843–856.

Gregg, M.C., Meagher, T.B., Aagaard, E.E., and Hess, W.C. 1981. A salt-stratified tank for measuring the dynamic response of conductivity probes. *IEEE J. of Oceanic Engr.*, **OE-6**, 113–118.

Gregg, M.C., Nodland, W.E., Aagaard, E.E., and Hirt, D.H. 1982. Use of a fiber-optic cable with a free-fall microstructure profiler. Pages 260–265 of: *Oceans '82: Conference Record, Sept. 20–22, 1982*. Washington, DC: Marine Technology Soc.

Gregg, M.C., and Özsoy, E. 2002. Flow, water mass changes and hydraulics in the Bosphorus. *J. Geophys. Res.*, **107**(C3), 2000JC000485.

Gregg, M.C., and Sanford, T.B. 1980. Signatures of mixing from the Bermuda Slope, the Sargasso Sea and the Gulf Stream. *J. Phys. Oceanogr.*, **10**, 105–127.

Gregg, M.C., and Sanford, T.B. 1981. Reply to Dillon and Caldwell. *J. Phys. Oceanogr.*, **11**, 1438–1439.

Gregg, M.C., and Sanford, T.B. 1987. Shear and turbulence in thermohaline staircases. *Deep-Sea Res.*, **34**, 1689–1696.

Gregg, M.C., and Sanford, T.B. 1988. The dependence of turbulent dissipation on stratification in a diffusively stable thermocline. *J. Geophys. Res.*, **93**, 12,381–12,392.

Gregg, M.C., Sanford, T.B., and Winkel, D.P. 2003. Reduced mixing from the breaking of internal waves in equatorial ocean waters. *Nature*, **422**, 513–515.

Gregg, M.C., Seim, H.E., and Percival, D.B. 1993. Statistics of shear and turbulent dissipation profiles in random internal wave fields. *J. Phys. Oceanogr.*, **23**, 1777–1799.

Gregg, M.C., Winkel, D.P., Sanford, T.B., and Peters, H. 1996. Turbulence produced by internal waves in the oceanic thermocline at mid and low latitudes. *Dyn. Atmos. Oceans*, **24**, 1–14.

Gross, T.F., Williams, A.J., and Terray, E.A. 1984. Bottom boundary layer spectral dissipation estimates in the presence of wave motions. *Cont. Shelf Res.*, **14**(10–11), 1239–1256.

Groves, G.W. 1959. Flow estimate for the perpetual salt fountain. *Deep-Sea Res.*, **5**, 209–214.

Gurevich, A.V., and Pitaevskii, L.P. 1973. Nonstationary structure of a collisionless shock wave. *Sov. Phys JETP*, **38**, 291–297.

Gurvich, A.S., and Yaglom, A.M. 1993. Breakdown of eddies and probability distributions for small scale turbulence. *Phys. Fluids*, **10**, 59–65.

Guthrie, J.D., Fer, I., and Morison, J. 2015. Observational validation of the diffusive flux laws in the Amundsen Basin. *J. Geophys. Res.*, **120**, 7880–7896.

Hasselman, K. 1966. Feynman diagrams and interaction rules of wave-wave scattering processes. *Rev. Geophys. and Space Phys.*, **4**, 1–32.

Hayes, S.P., Joyce, T.M., and Millard, R.C. 1975. Measurements of vertical finestructure. *J. Geophys. Res.*, **80**, 314–320.

Hayes, S.P., Milburn, H.B., and Ford, E.F. 1984. TOPS: A free-fall velocity and CTD profiler. *J. Atmos. Ocean. Tech.*, **1**, 220–236.

Hazel, P. 1972. Numerical studies of the stability of inviscid stratified shear flows. *J. Fluid Mech.*, **51**, 39–61.

Head, M.J. 1983. *The Use of Miniature Four-Electrode Conductivity Probes for High Resolution Measurement of Turbulent Density or Temperature Variations in Salt-Stratified Water Flows*. Ph.D. thesis, University of California, San Diego.

Hebert, D. 1988a. The available potential energy of an isolated feature. *J. Geophys. Res.*, **93**(C1), 556–564.

Hebert, D. 1988b. Estimates of salt finger fluxes. *Deep-Sea Res.*, **35**(12), 1887–1901.

Henyey, F.S., and Hoering, A. 1997. Energetics of borelike internal waves. *J. Geophys. Res.*, **102**, 3323–3330.

Henyey, F.S., and Pomphrey, N. 1983. Eikonal description of internal-wave interactions. *Dyn. Atmos. Oceans*, **7**, 189–208.

Henyey, F.S., Wright, J., and Flatté, S.M. 1986. Energy and action flow through the internal wave field: An eikonal approach. *J. Geophys. Res.*, **91**, 8487–8495.

Hesselberg, T., and Sverdrup, H.U. 1914. Die stabilitätsverhältnisse des seewassers bei vertikalen verschiebungen. *Bergens Mus. Aarb.*, **15**, 1–16.

Hibiya, T., Niwa, Y., Nakajima, K., and Suginohara, N. 1996. Direct numerical simulation of the roll-off range of internal wave shear spectra in the ocean. *J. Geophys. Res.*, **101**(C6), 14,123–14,129.

Hieronymus, M., and Carpenter, J.R. 2016. Energy and variance budgets of a diffusive staircase with implications for heat flux scaling. *J. Phys. Oceanogr.*, **46**, 2553–2569.

Hill, K.D., and Woods, D.J. 1988. The dynamic response of the two-electrode conductivity cell. *IEEE J. of Oceanic Engr.*, **13**(3), 118–123.

Hinze, J.O. 1975. *Turbulence*. 2nd ed. New York, NY: Mc-Graw Hill.

Hirst, E. 1991. Internal wave-wave resonance theory: Fundamentals and limitations. Pages 211–226 of: Müller, P., and Henderson, D. (eds.), *Dynamics of Oceanic Internal Gravity Waves: Proceedings, 'Aha Huliko'a Hawaiian Winter Workshop, University of Hawaii at Manoa, January 11–13, 1991*. Hawaii Institute of Geophysics.

Ho, D.T., Ledwell, J.R., and Jr., W.M. Smethie. 2008. Use of SF_5CF_3 for ocean tracer release experiments. *Geophys. Res. Lett.*, **35**(L04702).

Holbrook, W.S., and Fer, I. 2005. Oceanic internal wave spectra inferred from seismic reflection transects. *Geophys. Res. Lett.*, **32**(L15604).

Holbrook, W.S., Páramo, P., Pearse, S., and Schmitt, R.W. 2003. Thermohaline fine structure in a oceanographic front from seismic reflection profiling. *Science*, **301**(5634), 821–824.

Holleman, R.C., Geyer, W.R., and Ralston, D.K. 2016. Stratified turbulence and mixing efficiency in a salt wedge estuary. *J. Phys. Oceanogr.*, **46**, 1769–1783.

Holloway, G. 1980. Oceanic internal waves are not weak waves. *J. Phys. Oceanogr.*, **10**, 906–914.

Holloway, G. 1982. On interaction time scales of oceanic internal waves. *J. Phys. Oceanogr.*, **12**, 293–296.

Holloway, G., and Gargett, A.E. 1987. The inference of salt fingering from towed microstructure observations. *J. Geophys. Res.*, **92**, 1963–1966.

Holloway, P.E., and Merrifield, M.A. 2003. On the spring-neap variability and age of the internal tide at the Hawaiian Ridge. *J. Geophys. Res.*, **108**(C4), 3126, doi:10.1029/2002JC001486.

Holtermann, P.L., Umlauf, L., Tanhua, T., Schmale, O., Rehder, G., and Waniek, J.J. 2012. The Baltic Sea tracer release experiment. 1. Mixing rates. *J. Geophys. Res.*, **117**(C01021).

Horne, E.P.W., and Toole, J.M. 1980. Sensor response mismatches and lag correction techniques for temperature-salinity profilers. *J. Phys. Oceanogr.*, **10**(7), 1122–1130.

Howard, L.N. 1961. Note on a paper by John W. Miles. *J. Fluid Mech.*, **10**, 509–512.

Huang, R.X. 2005. Available potential energy in the world's oceans. *J. Mar. Res.*, **63**(1), 141–158.

Huppert, H.E. 1971. On the stability of a series of double-diffusivity layers. *Deep-Sea Res.*, **18**, 1005–1021.

Huppert, H.E., and Linden, P.F. 1979. On heating a stable salinity gradient from below. *J. Fluid Mech.*, **95**, 431–464.

Ijichi, T., and Hibiya, T. 2015. Frequency-based correction of finescale parameterization of turbulent dissipation in the deep ocean. *J. Atmos. Ocean. Tech.*, **32**, 1526–1535.

Ingham, M.C. 1966 (June). *The Salinity Extrema of the World Ocean*. PhD thesis, Oregon State University, Corvalis, OR.

Inoue, R., Kunze, E., Laurent, L. St., Schmitt, R.W., and Toole, J.M. 2008. Evaluating salt-fingering theories. *J. Mar. Res.*, **66**(4), 413–440.

IOC, SCOR, and IAPSO. 2010. *The International Thermodynamic Equation of Seawater – 2010: Calculation and Use of Thermodynamic Properties*. Manuals and Guides 56. Intergovernmental Oceanographic Commission, UNESCO.

Irish, J.D., and Nodland, W.E. 1978. Evaluation of metal-film temperature and velocity sensors and the stability of a self-propelled research vehicle for making measurements of ocean turbulence. Pages 180–187 of: *IEEE/MTS Proceeding of OCEANS-78, September 1978, Washington, D.C.*

Iselin, C. O'D. 1939. The influence of vertical and lateral turbulence on the characteristics of the waters at mid-depths. *Trans., American Geophys. Union*, **20**(3), 414–417.

Ivers, W.D. 1975. *The Deep Circulation in the Northern North Atlantic, with Especial Reference to the Labrador Sea*. PhD thesis, University of California San Diego, San Diego, CA.

Ivey, G.N., and Imberger, J. 1991. On the nature of turbulence in a stratified fluid. Part I: The energetics of mixing. *J. Phys. Oceanogr.*, **21**, 650–658.

Jackett, D.R., and McDougall, T.J. 1985. An oceanographic variable for the characterization of intrusions and water masses. *Deep-Sea Res., Part A*, **32**, 1195–1207.

Jackett, D.R., and McDougall, T.J. 1997. A neutral density surface for the world's oceans. *J. Phys. Oceanogr.*, **27**(2), 237–263.

Jackson, P.R., and Rehmann, C.R. 2003. Laboratory measurements of differential diffusion in a diffusively stable, turbulent flow. *J. Phys. Oceanogr.*, **33**(8), 1592–1603.

Jones, W.P., and Musonge, P. 1988. Closure of the Reynolds stress and scalar flux equations. *Phys. Fluids*, **31**.

Joyce, T.M. 1977. A note on the lateral mixing of water masses. *J. Phys. Oceanogr.*, **7**, 626–629.

Joyce, T.M. 1980. On production and dissipation of thermal variance in the ocean. *J. Phys. Oceanogr.*, **10**, 460–463.

Kao, T.W., and Pao, H.-P. 1978. Note on the flow of a stratified fluid over a stationary obstacle in a channel. *Geophys. Astrophys. Fluid Dyn.*, **10**(1), 109–114.

Kelley, D.E. 1984. Effective diffusivities within oceanic thermohaline staircases. *J. Geophys. Res.*, **89**, 10,484–10,488.

Kelley, D.E. 1989. Explaining effective diffusivities within diffusive oceanic staircases. Pages 481–502 of: Nihoul, J.C.J., and Jamart, B.M. (eds.), *Small-Scale Turbulence and Mixing in the Ocean*. Amsterdam: Elsevier.

Kelley, D.E. 1990. Fluxes through diffusive staircases, a new formulation. *J. Geophys. Res.*, **95**, 3365–3371.

Kelley, D.E., Fernando, H.J.S., Gargett, A.E., Tanny, J., and Özsoy, E. 2003. The diffusive regime of double diffusion. *Prog. Oceanogr.*, **56**(3–4), 461–481.

Kennelly, M.A., McKeown, P.A., and Sanford, T.B. 1986. *XCP Performances Near the Geomagnetic Equator*. Report APL-UW Tech. Rept. 8607. University of Washington, Applied Physics Laboratory, Seattle, WA.

Killworth, P.D. 1977. Mixing on the Weddell Sea Continental Slope. *Deep-Sea Res.*, **24**, 427–448.

Kimura, S., and Smyth, W. 2007. Direct numerical simulation of salt sheets and turbulence in a double-diffusive shear layer. *Geophys. Res. Lett.*, **34**(L21610).

Kimura, S., and Smyth, W. 2011. Turbulence in a sheared, salt-fingering-favorable environment: Anisotropy and effective diffusivities. *J. Phys. Oceanogr.*, **41**(6), 1141–1159.

Klaassen, G.P., and Peltier, W.R. 1985. The onset of turbulence in finite-amplitude Kelvin–Helmholtz billows. *J. Fluid Mech.*, **155**, 1–35.

Klocker, A., and McDougall, T.J. 2010. Influence of the nonlinear equation of state on global estimates of dianeutral advection and diffusion. *J. Phys. Oceanogr.*, **40**, 1690–1709.

Klymak, J.M., and Gregg, M.C. 2004. Tidally generated turbulence over the Knight Inlet sill. *J. Phys. Oceanogr.*, **34**, 1135–1151.

Klymak, J.M., Legg, S.M., and Pinkel, R. 2010. High-mode stationary waves in stratified flow over large obstacles. *J. Fluid Mech.*, **664**(321–336).

Klymak, J.M., and Moum, J.N. 2006. Oceanic isopycnal slope spectra: Part I: Internal waves. *J. Phys. Oceanogr.*, **36**(12), 1–16.

Klymak, J.M., and Moum, J.N. 2007. Oceanic isopycnal slope spectra. Part II: Turbulence. *J. Phys. Oceanogr.*, **37**, 1232–1245.

Klymak, J.M., Moum, J.N., Nash, J.D., Kunze, E., Girton, J.B., Carter, G.S., Lee, C.M., Sanford, T.B., and Gregg, M.C. 2006a. An estimate of tidal energy lost to turbulence at the Hawaiian Ridge. *J. Phys. Oceanogr.*, **36**, 1148–1164.

Klymak, J.M., Pinkel, R., Liu, C.-T., Liu, A.K., and David, L. 2006b. Prototypical solitons in the South China Sea. *Geophys. Res. Lett.*, **33**(L11607).

Kolmogorov, A.N. 1941. The local structure of turbulence in incompressible viscous fluid for very large Reynolds' numbers. *Dokl. Akad. Nauk SSSR*, **30**, 299–303.

Kolmogorov, A.N. 1962. A refinement of previous hypotheses concerning the local structure of turbulence in a viscous incompressible fluid at high Reynolds number. *J. Fluid Mech.*, **13**, 82–85.

Kraichnan, R.H. 1968. Small-scale structure of a scalar field convected by turbulence. *Phys. Fluids*, **11**, 945–953.

Kukuruznyak, D.A., Bulkey, S.A., Omland, K.A., Ohuchi, F.S., and Gregg, M.C. 2001. Preparation and properties of thermistor-thin-films by metal organic decomposition. *Thin Solid Films*, **385**, 89–95.

Kundu, P.K., and Cohen, I.M. 2004. *Fluid Mechanics*. 3rd ed. Academic Press.

Kunze, E. 1985. Near-inertial propagation in geostrophic shear. *J. Phys. Oceanogr.*, **15**, 544–565.

Kunze, E. 1987. Limits on growing, finite-length salt fingers, a Richardson number constraint. *J. Mar. Res.*, **45**, 533–556.

Kunze, E. 1993. Submesoscale dynamics near a seamount. Part II: The partition of energy between internal waves and geostrophy. *J. Phys. Oceanogr.*, **23**(12), 2589–2601.

Kunze, E. 2003. A review of oceanic salt-fingering theory. *Prog. Oceanogr.*, **56**, 399–417.

Kunze, E. 2019. Biologically generated mixing in the ocean. *Annu. Rev. Mar. Sci.*, **11**.

Kunze, E., and Lien, R.-C. 2019. Energy sinks for lee waves in shear flow. *J. Phys. Oceanogr.*, **49**(3), 2851–2865.

Kunze, E., Dower, J.F., Beveridge, I., Dewey, R., and Bartlett, K.P. 2006b. Observations of biologically generated mixing in a coastal inlet. *Science*, **313**, 1768–1770.

Kunze, E., Firing, E., Hummon, J.M., Chereskin, T.K., and Thurnherr, A.M. 2006a. Global abyssal mixing inferred from lowered ADCP shear and CTD strain profiles. *J. Phys. Oceanogr.*, **36**, 1553–1576.

Kunze, E., and Lueck, R. 1986. Velocity profiles in a warm-core ring. *J. Phys. Oceanogr.*, **16**(5), 991–995.

Kunze, E., Rosenfeld, L.K., Carter, G.S., and Gregg, M.C. 2002. Internal waves in Monterey Submarine Canyon. *J. Phys. Oceanogr.*, **32**, 1890–1913.

Kunze, E., and Sanford, T.B. 1984. Observations of near-inertial waves in a front. *J. Phys. Oceanogr.*, **14**, 566–581.

Kunze, E., Schmitt, R.W., and Toole, J.M. 1995. The energy balance in a warm-core ring's near-inertial critical layer. *J. Phys. Oceanogr.*, **25**, 942–957.

Kunze, E., Williams, III, A.J., and Briscoe, M.G. 1990. Observations of shear and vertical stabiliity from a neutrally buoyant float. *J. Geophys. Res.*, **95**, 18,127–18,142.

Kunze, E., Williams, III, A.J., and Schmitt, R.W. 1987. Optical microstructure in the thermocline staircase east of Barbados. *Deep-Sea Res.*, **34**, 1697–1704.

Lai, D.Y., Paka, V.T., Delisi, D.P., Arjannikov, A.V., and Khanaev, S.A. 2000. An intercomparison study using electromagnetic three-component turbulent velocity probes. *J. Atmos. Ocean. Tech.*, **17**(7), 980–994.

Lamb, K.G., Lien, R.-C., and Diamessis, P. 2019. *Encyclopedia of Ocean Sciences*. Elsevier Ltd. Chap. Internal solitary waves and mixing.

Landau, L.D., and Lifshitz, E.M. 1959. *Fluid Mechanics*. New York, NY: Addison-Wesley.

Larson, N.G., and Gregg, M.C. 1983. Turbulent dissipation and shear in thermohaline intrusions. *Nature*, **306**, 26–32.

Lazier, J.R.N. 1973. Temporal changes in some fresh water temperature structures. *J. Phys. Oceanogr.*, **3**, 226–229.

Ledwell, J.R., and Bratkovich, A. 1995. A tracer study of mixing in the Santa Cruz Basin. *J. Geophys. Res.*, **100**(C10), 20,681–20,704.

Ledwell, J.R., Duda, T.F., Sundermeyer, M.A., and Seim, H.E. 2004. Mixing in a coastal environment: 1. A view from dye dispersion. *J. Geophys. Res.*, **109**(C10013).

Ledwell, J.R., He, R., Xue, Z., DiMarco, S.F., Spencer, L.J., and Chapman, P. 2016. Dispersion of a tracer in the deep Gulf of Mexico. *J. Geophys. Res. Oceans*, **121**, 1110–1132.

Ledwell, J.R., and Hickey, B.M. 1995. Evidence for enhanced boundary mixing in the Santa Monica Basin. *J. Geophys. Res.*, **100**(C10), 20,665–20,679.

Ledwell, J.R., Laurent, L.C. St., Girton, J.B., and Toole, J.M. 2011. Diapycnal mixing in the Antarctic Circumpolar Current. *J. Phys. Oceanogr.*, **41**, 241–246.

Ledwell, J.R., Montgomery, E.T., Polzin, K.L., Laurent, L. C. St., Schmitt, R.W., and Toole, J.M. 2000. Evidence for enhanced mixing over rough topography in the abyssal ocean. *Nature*, **403**, 179–182.

Ledwell, J.R., and Watson, A.J. 1991. The Santa Monica Basin Tracer Experiment: A study of diapycnal and isopycnal mixing. *J. Geophys. Res.*, **96**(C5), 8695–8718.

Ledwell, J.R., Watson, A.J., and Broecker, W.S. 1986. A deliberate tracer experiment in Santa Monica Basin. *Nature*, **323**, 322–324.

Ledwell, J.R., Watson, A.J., and Law, C.S. 1993. Evidence for slow mixing across the pycnocline from an open-ocen tracer-release experiment. *Nature*, **364**(6439), 701–703.

Ledwell, J.R., Watson, A.J., and Law, C.S. 1998. Mixing of a tracer in the pycnocline. *J. Geophys. Res.*, **103**(C10), 21,499–21,529.

Lee, C.M., Kunze, E., Sanford, T.B., Nash, J.D., Merrifield, M.A., and Holloway, P.E. 2006. Internal tides and turbulence along the 3000-m isobath of the Hawaiian Ridge. *J. Phys. Oceanogr.*, **36**, 1165–1183.

Lee, I-H., Lien, R.-C., Liu, J.T., and Chuang, W. 2009. Turbulence mixing and internal tides in Gaoping (Kaoping) submarine canyon, Taiwan. *J. Mar. Syst.*, **76**(4), 383–396.

Legg, S., and Huijts, K.M.H. 2006. Preliminary simulations of internal waves and mixing generated by finite amplitude tidal flow over finite topography. *Deep-Sea Res II*, **53**(1–2), 140–156.

Legg, S., and Klymak, J. 2008. Internal hydraulic jumps and overturning generated by tidal flow over a tall steep ridge. *J. Phys. Oceanogr.*, **38**(9), 1949–1964.

Lelong, M.-P., and Dunkerton, T.J. 1998a. Inertia-gravity wave breaking in three dimensions. I. Convectively unstable waves. *J. Atmos. Sci.*, **55**(15), 2489–2501.

Lelong, M.-P., and Dunkerton, T.J. 1998b. Inertia-gravity wave breaking in three dimensions. II. Convectively stable waves. *J. Atmos. Sci.*, **55**(15), 2473–2488.

Lelong, M.-P., and Sundermeyer, M.A. 2005. Geostrophic adjustment of an isolated diapycnal mixing event and its implications for small scale lateral dispersion. *J. Phys. Oceanogr.*, **35**(12), 2352–2367.

Levine, E.R., and Lueck, R.G. 1999. Turbulence measurements from an autonomous underwater vehicle. *J. Atmos. Ocean. Tech.*, **16**, 1533–1544.

Levine, M.D. 1990. Internal waves under the Arctic ice pack during the Arctic Internal Waves Experiment: The coherence structure. *J. Geophys. Res.*, **95**, 7347–7357.

Levine, M.D. 2002. A modification of the Garrett-Munk internal wave spectrum. *J. Phys. Oceanogr.*, **32**, 3166–3181.

Levine, M.D., Paulson, C.A., and Morison, J.H. 1985. Internal waves in the Arctic Ocean: Comparison with lower-latitude observations. *J. Phys. Oceanogr.*, **15**, 800–809.

Levine, M.D., Paulson, C.A., and Morison, J.H. 1987. Observations of internal gravity waves under the Arctic ice pack. *J. Geophys. Res.*, **92**(C1), 779–782.

Libby, P.A. 1996. *Introduction to Turbulence*. Washington, D.C.: Taylor & Francis.

Lien, R.-C., and D'Asaro, E.A. 2002. The Kolmogorov constant for the Lagrangian velocity spectrum and structure function. *Phys. Fluids*, **14**, 4456–4459.

Lien, R.-C., and D'Asaro, E.A. 2006. Measurement of turbulent kinetic energy dissipation rate with a Lagrangian float. *J. Atmos. Ocean. Tech.*, **23**, 964–976.

Lien, R.-C., and Müller, P. 1992. Consistency relations for gravity and vortical modes in the ocean. *Deep-Sea Res A*, **39**(9), 1595–1612.

Lien, R.-C., and Sanford, T.B. 2019. Small-scale potential vorticity in the upper ocean thermocline. *J. Phys. Oceanogr.*, **49**, 1845–1872.

Lighthill, J. 1979. *Waves in Fluids*. Cambridge University Press.

Lincoln, B.J., Rippeth, R.P., Lenn, Y.-D., Timmermans, M.L., Williams, W.J., and Bacon, S. 2016. Wind-driven mixing at intermediate depths in an ice-free Arctic. *Geophys. Res. Lett.*, **43**, 9749–9756.

Lindborg, E. 2006. The energy cascade in a strongly stratified fluid. *J. Fluid Mech.*, **550**, 207–242.

Linden, P.F. 1971. Salt fingers in the presence of grid-generated turbulence. *J. Fluid Mech.*, **49**, 611–624.

Linden, P.F. 1973. On the structure of salt fingers. *Deep-Sea Res.*, **20**, 325–340.

Linden, P.F. 1974. Salt fingers in a steady shear flow. *Geophys. Fluid Dyn.*, **6**, 1–27.

Liu, A.K. 1988. Analysis of nonlinear internal waves in the New York Bight. *J. Geophys. Res.*, **93**, 12,317–12,329.

Liu, A.K., Chang, Y.S., Hsu, M.-K., and Liang, N.K. 1998. Evolution of nonlinear internal waves in the East and South China Seas. *J. Geophys. Res.*, **103**, 7995–8008.

Liu, H.T. 1995. Energetics of grid turbulence in a stably stratified fluid. *J. Fluid Mech.*, **296**, 127–157.

Llewellyn Smith, S., and Young, W. 2002. Conversion of the barotropic tide. *J. Phys. Oceanogr.*, **32**, 1554–1566.

Lombardo, C.P., and Gregg, M.C. 1989. Similarity scaling of viscous and thermal dissipation in a convecting surface boundary layer. *J. Geophys. Res.*, **94**, 6273–6284.

Lorenz, E. 1955. Available potential energy and maintenance of the general circulation. *Tellus*, **7**, 157–167.

Lorke, A., and Wuest, A. 2005. Application of coherent ADCP for turbulence measurements on the bottom boundary layer. *J. Atmos. Ocean. Tech.*, **22**, 1821–1828.

Lueck, R.G. 1987. Microstructure measurements in a thermohaline staircase. *Deep-Sea Res.*, **34**, 1677–1688.

Lueck, R.G., Huang, D., Newman, D., and Box, J. 1997. Turbulence measurement with a moored instrument. *J. Atmos. Ocean. Tech.*, **14**(2), 143–161.

Lueck, R.G., and Osborn, T.R. 1986. The dissipation of kinetic energy in a warm-core ring. *J. Geophys. Res.*, **4**, 681–698.

Lumley, J.L. 1964. The spectrum of nearly inertial turbulence in a stably stratified fluid. *J. Atmos. Sci.*, **21**, 99–102.

Lynn, R.J., and Reid, J.L. 1968. Characteristics and circulation of deep and abyssal waters. *Deep-Sea Res*, **15**, 577–598.

Mack, S.A. 1985. Two-dimensional measurements of ocean microstructure: The role of double diffusion. *J. Phys. Oceanogr.*, **15**, 1581–1604.

Mack, S.A. 1989. Towed-chain measurements of ocean microstructure. *J. Phys. Oceanogr.*, **19**, 1108–1129.

Mack, S.A., and Schoerberlein, H.C. 1993. Discriminating salt fingering from turbulence-induced microstructure: Analysis of towed temperature-conductivity chain data. *J. Phys. Oceanogr.*, **23**(9), 2073–2106.

MacKinnon, J.A., and Gregg, M.C. 2003. Mixing on the late-summer New England shelf – solibores, shear and stratification. *J. Phys. Oceanogr.*, **33**, 1476–1492.

MacKinnon, J.A., Johnston, T.M.S., and Pinkel, R. 2008. Strong transport and mixing of deep water through the Southwest Indian Ridge. *Nat. Geosci.*, **38**, 1943–1950.

MacKinnon, J.A., Laurent, L. St., and Garbato, A.C.N. 2013. Diapycnal mixing processes in the ocean interior. Chap. 7, pages 159–183 of: Siedler, G., Griffies, S.M., Gould, J., and Church, J.A. (eds.), *Ocean Circulation and Climate: A 21st Century Perspective*. International Geophysics, vol. 103. Amsterdam: Elsevier.

MacKinnon, J.A., and Winters, K.B. 2005. Subtropical catastrophe: Significant loss of low-mode tidal energy at 28.9°. *Geophys. Res. Lett.*, **32**, doi:10.1029/2005GL023376.

MacKinnon, J.A., Zhao, Z., Whalen, C.B., et al. 2017. Climate process team on internal wave-driven ocean mixing. *Bull. Am. Met. Soc.*, Nov., 2429–2454.

Macoun, P., and Lueck, R. 2004. Modelling the spatial response of the airfoil shear probe using different sized probes. *J. Atmos. Ocean. Tech.*, **21**, 284–297.

Marmorino, G.O. 1987a. Observations of small-scale mixing in the thermocline. Part I: Salt fingering. *J. Phys. Oceanogr.*, **17**(9), 1339–1347.

Marmorino, G.O. 1987b. Observations of small-scale mixing processes in the seasonal thermocline: Part II: Wave breaking. *J. Phys. Oceanogr.*, **17**, 1348–1355.

Marmorino, G.O. 1991. Intrusions and diffusive interfaces in a salt-finger staircase. *Deep-Sea Res. A*, **38**(11), 1431–1454.

Marmorino, G.O., and Caldwell, D. 1976. Heat and salt transport through a diffusive thermohaline interface. *Deep-Sea Res.*, **23**, 59–67.

Marmorino, G.O., and Greenewalt, D. 1988. Inferring the nature of microstructure signals. *J. Geophys. Res.*, **93**, 1219–1225.

Marmorino, G.O., Rosenblum, L.J., and Trump, C.L. 1987. Finescale temperature variability: The influence of near-inertial waves. *J. Geophys. Res.*, **92**, 13,049–13,062.

Marshall, J., and Speer, K. 2012. Closure of the meridional overturning circulation through Southern Ocean upwelling. *Nature-Geo.*, **5**(March).

Marshall, J., Jamous, D., and Nilsson, J. 1999. Reconciling thermodynamic and dynamic methods of computation of water-mass transformation rates. *Deep-Sea Res. I*, **46**, 545–572.

Mashayek, A., and Peltier, W.R. 2011. Three-dimensionalization of the stratified mixing layer at high Reynolds number. *Phys. Fluids*, **23**(111701).

Mater, B.D., and Venayagamoorthy, S.K. 2014. A unifying framework for parameterizing stably stratified shear-flow turbulence. *Phys. Fluids*, **26**(036601).

Mauritzen, C., Polzin, K.L., McCartney, M.S., Millard, R.C., and West-Mack, D.E. 2002. Evidence in hydrography and density fine structure for enhanced vertical mixing over the Mid-Atlantic Ridge in the western Atlantic. *J. Geophys. Res.*, **107**(C10).

May, B.D., and Kelley, D.E. 1997. Effect of baroclinicity on double-diffusive interleaving. *J. Phys. Oceanogr.*, **27**, 1997–2008.

McComas, C.H., and Bretherton, F.P. 1977. Resonant interaction of oceanic internal waves. *J. Geophys. Res.*, **82**, 1397–1412.

McComas, C.H., and Müller, P. 1981a. The dynamic balance of internal waves. *J. Phys. Oceanogr.*, **11**(July), 970–986.

McComas, C.H., and Müller, P. 1981b. Time scales of resonant interactions among oceanic internal waves. *J. Phys. Oceanogr.*, **11**(Feb.), 139–147.

McDougall, T.J. 1981. Fluxes of properties through a series of double-diffusive interfaces with a non-linear equation of state. *J. Phys. Oceanogr.*, **11**, 1294–1299.

McDougall, T.J. 1984. The relative roles of diapycnal and isopycnal mixing on subsurface water mass conversion. *J. Phys. Oceanogr.*, **14**, 1577–1589.

McDougall, T.J. 1987a. Neutral surfaces. *J. Phys. Oceanogr.*, **17**(12), 1950–1964.

McDougall, T.J. 1987b. Thermobaricity, cabbeling and water-mass conversion. *J. Geophys. Res.*, **92**(C5), 5448–5464.

McDougall, T.J. 1988. Some implications of ocean mixing for ocean modelling. Pages 21–36 of: Nihoul, J.C.J., and Jamart, B.M. (eds.), *Small-Scale Turbulence and Mixing in the Ocean*. Amsterdam: Elsevier.

McDougall, T.J. 1991. Interfacial advection in the thermocline staircase east of Barbados. *Deep-Sea Res.*, **38**(3), 357–370.

McDougall, T.J. 2003. Potential enthalpy: A conservative oceanic variable for evaluating heat content and heat fluxes. *J. Phys. Oceanogr.*, **33**, 945–963.

McDougall, T.J., and Barker, P.M. 2012. Comment on 'Buoyancy frequency profiles and internal semidiurnal tide turning depths in the oceans' by B. King et al. *J. Geophys. Res. Oceans*, **119**, 1–7.

McDougall, T.J., and Feistel, R. 2003. What causes the adiabatic lapse rate? *Deep-Sea Res I*, **50**, 1523–1535.

McDougall, T.J., Feistel, R., and Pawlowicz, R. 2013. Thermodynamics of seawater. Chap. 6, pages 141–158 of: Siedler, G., Griffies, S.M., Gould, J., and Church, J.A. (eds.), *Ocean Circulation and Climate*. International Geophysics, vol. 103. Amsterdam: Elsevier.

McDougall, T.J., and Ferrari, R. 2017. Abyssal upwelling and downwelling driven by near-boundary mixing. *J. Phys. Oceanogr.*, **47**(2), 261–283.

McDougall, T.J., and Jackett, D.R. 1988. On the helical nature of neutral surfaces. *Progr. Oceanogr.*, **20**, 153–183.

McDougall, T.J., and Krzysik, O.A. 2015. Spiciness. *J. Mar. Res.*, **73**, 141–152.

McDougall, T.J., and Taylor, J.R. 1984. Flux measurements across a finger interface at low values of the stability ratio. *J. Mar. Res.*, **42**, 1–14.

McKean, R.S. 1974. Interpretation of internal wave measurements in the presence of finestructure. *J. Phys. Oceanogr.*, **4**, 200–213.

McPhee, M.G. 1992. Turbulent heat flux in the upper ocean under sea ice. *J. Geophys. Res.*, **97**, 5365–5379.

Meagher, T.B., Pederson, A.M., and Gregg, M.C. 1982. A low-noise conductivity microstructure instrument. Pages 283–290 of: *Oceans '82: Conference Record, Sept. 20–22, 1982*. Washington, D.C.: Marine Technology Society.

Merrifield, M.A., Holloway, P.E., and Johnston, T.M.S. 2001. The generation of internal tides at the Hawaiian Ridge. *Geophys. Res. Lett.*, **28**(4), 559–562.

Merryfield, W.J. 2000. Origin of thermohaline staircases. *J. Phys. Oceanogr.*, **30**, 1046–1068.

Merryfield, W.J., and Grinder, M. 1999. Salt fingering fluxes from numerical simulations. Unpublished manuscript.

Meyer, A., Polzin, K.L., Sloyan, B.M., and Phillips, H.E. 2015a. Internal waves and mixing near the Kerguelen Plateau. *J. Phys. Oceanogr.*, **46**(2), 417–437.

Meyer, A., Sloyan, B.M., Polzin, K.L., Phillips, H.E., and Bindoff, N.L. 2015b. Mixing variability in the Southern Ocean. *J. Phys. Oceanogr.*, **45**(4), 966–987.

Miles, J. 1961. On the stability of heterogenous shear flows. *J. Fluid Mech.*, **10**, 496–508.

Miller, J.B., Gregg, M.C., Miller, V.W., and Welsh, G.L. 1989. Vibration of tethered microstructure profilers. *J. Atmos. Ocean. Technol.*, **6**, 980–984.

Millero, F., Chen, C.T., Bradshaw, A., and Schleicher, K. 1980. A new high pressure equation of state for seawater. *Deep-Sea Res.*, **27A**, 255–264.

Miyake, Y., and Koizumi, M. 1948. The measurement of the viscosity coefficient of sea water. *J. Mar. Res.*, **7**, 63–66.

Monin, A.S., and Yaglom, A.M. 1971. *Statistical Fluid Mechanics: Mechanics of Turbulence.* Vol. 1. Cambridge, MA: The MIT Press.

Monin, A.S., and Yaglom, A.M. 1975. *Statistical Fluid Mechanics: Mechanics of Turbulence.* Vol. 2. Cambridge, MA: The MIT Press.

Montgomery, R.B. 1938. Circulation in the upper layers of the southern North Atlantic, deduced with the use of isentropic analysis. *Pap. Phys. Oceanogr. Meteor*, **6**(2), 55.

Montroll, E.W., and Shlesinger, M.F. 1982. On 1/*f* and other distributions with long tails. *Proc. Natl. Acad. Sci. USA*, **79**, 3380–3383.

Morison, J.H., Long, C.E., and Levine, M.D. 1985. Internal wave dissipation under sea ice. *J. Geophys. Res.*, **90**, 11,959–11,966.

Morrison, A.T., Billings, J.D., and Doherty, K.W. 2000. The McLane moored profiler: An autonomous platform for oceanographic measurements. In: *Oceans 2000 MTS/IEEE Conf. & Exhib.*

Moum, J.N. 1990. Profiler measurements of vertical velocity microstructure in the ocean. *J. Atmos. Ocean. Tech.*, **7**, 323–333.

Moum, J.N. 1996. Energy-containing scales of turbulence in the ocean thermocline. *J. Geophys. Res.*, **101**(C6), 14,095–14,109.

Moum, J.N. 2015. Ocean speed and turbulence measurements using pitot–static tubes on moorings. *J. Atmos. Ocean. Tech.*, **32**, 1400–1413.

Moum, J.N., and Nash, J.D. 2009. Mixing measurements on an equatorial ocean mooring. *J. Atmos. Ocean. Tech.*, **26**, 317–336.

Moum, J.N., Gregg, M.C., Lien, R.C., and Carr, M.E. 1995. Comparison of turbulent kinetic energy dissipation rate estimates from two ocean microstructure profilers. *J. Atmos. Ocean. Tech.*, **12**, 346–366.

Moum, J.N., Caldwell, D.R., Nash, J.D., and Gunderson, G.D. 2002. Observations of boundary mixing over the continental slope. *J. Phys. Oceanogr.*, **32**, 2113–2130.

Moum, J.N., Farmer, D.M., Smyth, W.D., Armi, L., and Vagle, S. 2003. Structure and generation of turbulence at interfaces strained by internal solitary waves propagating shoreward over the continental shelf. *J. Phys. Oceanogr.*, **33**(10), 2093–2112.

Moum, J.N., Farmer, D.M., Shroyer, E.L., Smyth, W.D., and Armi, L. 2007a. Dissipative losses in nonlinear internal waves propagating across the continental shelf. *J. Phys. Oceanogr.*, **37**(7), 1989–1995.

Moum, J.N., Klymak, J.M., Nash, J.D., Perlin, A., and Smyth, W.D. 2007b. Energy transport by nonlinear internal waves. *J. Phys. Oceanogr.*, **37**(7), 1968–1988.

Müller, P., and Siedler, G. 1976. Consistency relations for internal waves. *Deep-Sea Res*, **23**, 613–628.

Müller, P., Olbers, D.J., and Willebrand, J. 1978. The IWEX spectrum. *J. Geophys. Res.*, **83**, 479–500.

Munk, W.H. 1966. Abyssal recipes. *Deep-Sea Res.*, **13**, 707–730.

Munk, W.H. 1981. Internal waves and small-scale processes. Pages 264–291 of: Warren, B.A., and Wunsch, C. (eds.), *Evolution of Physical Oceanography: Scientific Surveys in Honor of Henry Stommel*. Cambridge, MA: MIT Press.

Munk, W.H., and Wunsch, C. 1998. Abyssal recipes II: Energetics of tidal and wind mixing. *Deep-Sea Res. I*, **45**(12), 1977–2010.

Nagai, T., Tandon, A., Kunze, E., and Mahadevan, A. 2015. Spontaneous generation of near-inertial waves by the Kuroshio Front. *J. Phys. Oceanogr.*, **45**, 2381–2406.

Nagasawa, M., Hibiya, T., Niwa, Y., Watanabe, M., Isoda, Y., Takagi, S., and Kamei, Y. 2002. Distribution of fine-scale shear in the deep waters of the North Pacific obtained using expendable current profilers. *J. Geophys. Res.*, **107**(C2).

Nandi, P., Holbrook, W.S., Pearse, S., Páramo, P., and Schmitt, R.W. 2004. Seismic reflection profiling of water mass boundaries in the Norwegian Sea. *Geophys. Res. Lett.*, **31**(L23311).

Nash, J.D., Alford, M.H., Kunze, E., Martini, K., and Kelly, S. 2007. Hotspots of deep ocean mixing on the Oregon continental slope. *Geophys. Res. Lett.*, **34**(L01605).

Nash, J.D., and Moum, J.N. 1999. Estimating salinity variance dissipation rate from conductivity microstructure measurements. *J. Atmos. Ocean. Tech.*, **16**, 263–274.

Nash, J.D., and Moum, J.N. 2002. Microstructure estimates of turbulent salinity flux and the dissipation spectrum of salinity. *J. Phys. Oceanogr.*, **32**(8), 2312–2334.

Nash, J.D., Caldwell, D.R., Zeiman, M.J., and Moum, J.N. 1999. A thermocouple probe for high speed temperature measurement in the ocean. *J. Atmos. Ocean. Tech.*, **16**, 1474–1482.

Nash, J.D., Kunze, E., Lee, C.M., and Sanford, T.B. 2006. Structure of the baroclinic tide generated at Kaena Ridge, Hawaii. *J. Phys. Oceanogr.*, **36**, 1123–1135.

Nasmyth, P.W. 1970. *Oceanic Turbulence*. Ph.D. thesis, University of British Columbia, Vancouver, Canada.

Naveira Garabato, A.C., Oliver, K.I.C., Watson, A.J., and Messias, M.-J. 2004. Turbulent diapycnal mixing in the Nordic seas. *J. Geophys. Res.*, **109**(C12010).

Neal, V.T., Neshbya, S., and Denner, W. 1969. Thermal stratification in the Arctic. *Science*, **166**(3903), 373–374.

Newman, F.C. 1976. Temperature steps in Lake Kivu: A bottom heated saline lake. *J. Phys. Oceanogr.*, **6**, 157–163.

Nikurashin, M., and Ferrari, R. 2010a. Radiation and dissipation of internal waves generated by geostrophic flows impinging on small-scale topography: Application to the Southern Ocean. *J. Phys. Oceanogr.*, **40**(5), 2025–2042.

Nikurashin, M., and Ferrari, R. 2010b. Radiation and dissipation of internal waves generated by geostrophic motions impinging on small-scale topography: Theory. *J. Phys. Oceanogr.*, **40**(5), 1055–1074.

Nikurashin, M., and Ferrari, R. 2011. Global energy conversion rate from geostrophic flows into internal lee waves in the deep ocean. *Geophys. Res. Lett.*, **38**(L08610).

Nimmo Smith, W.A.M. 2008. A submersible three-dimensional particle tracking velocimetry system for flow visualization in the coastal ocean. *Limnol. Oceanogr.*, **6**, 96–104.

Nimmo Smith, W.A.M., Katz, J., and Osborn, T.R. 2005. On the structure of turbulence in the bottom boundary layer of the coastal ocean. *J. Phys. Oceanogr.*, **35**(1), 72–93.

Niwa, Y., and Hibiya, T. 2014. Generation of baroclinic tide energy in a global three-dimensional numerical model with different spatial grid resolutions. *Ocean Modelling*, **80**, 59–73.

Oakey, N.S. 1977. *OCTUPROBE III: An Instrument to Measure Oceanic Turbulence and Microstructure*. Rept. Series BI-R-77-3. Bedford Inst. Oceanogr., Dartmouth, N.S., Canada.

Oakey, N.S. 1982. Determination of the rate of dissipation of turbulent energy from simultaneous temperature and velocity shear microstructure measurements. *J. Phys. Oceanogr.*, **12**, 256–271.

Oakey, N.S., and Greenan, B.J.W. 2004. Mixing in a coastal environment: 1. A view from microstructure measurements. *J. Geophys. Res.*, **109**(C10014).

Oort, A.H., Anderson, L.A., and Peixoto, J.P. 1994. Estimates of the energy cycle of the oceans. *J. Geophys. Res.*, **99**(C4), 7665–7688.

Orlanski, I., and Bryan, K. 1969. Formation of the thermocline step structure of large-amplitude internal gravity waves. *J. Geophys. Res.*, **74**, 6975–6983.

Orr, M.H. 1981. Remote, acoustic detection of zooplankton response to fluid processes, oceanographic instrumentation, and predators. *Can. J. Fish. Aquat. Sci.*, **38**, 1096–1105.

Orr, M.H., and Mignerey, P.C. 2003. Nonlinear internal waves in the South China Sea: Observation of the conversion of depression internal waves to elevation internal waves. *J. Geophys. Res.*, **108**(3064).

Osborn, T.R. 1974. Vertical profiling of velocity microstructure. *J. Phys. Oceanogr.*, **4**, 109–115.

Osborn, T.R. 1980. Estimates of the local rate of vertical diffusion from dissipation measurements. *J. Phys. Oceanogr.*, **10**, 83–89.

Osborn, T.R., and Cox, C.S. 1972. Oceanic fine structure. *Geophys. Fluid Dyn.*, **3**, 321–345.

Osborn, T.R., and Crawford, W.R. 1980. An airfoil probe for measuring turbulent velocity fluctuations in water. Pages 369–386 of: Dobson, F., Hasse, L., and Davis, R. (eds.), *Air-Sea Interactions: Instruments and Methods*. New York: Plenum.

Osborn, T.R., and Lueck, R.G. 1985. Turbulence measurement with a submarine. *J. Phys. Oceanogr.*, **15**, 1502–1520.

Ozmidov, R.V. 1965. On the turbulent exchange in a stably stratified ocean. *Izv., Atmos. Oceanic Phys*, **1**, 853–860.

Padman, L., and Dillon, T.M. 1987. Vertical heat fluxes through the Beaufort Sea thermohaline staircases. *J. Geophys. Res.*, **92**(C10), 10,799–10,806.

Padman, L., and Dillon, T.M. 1991. Turbulent mixing near the Yermak Plateau during the Coordinated Eastern Arctic Experiment. *J. Geophys. Res.*, **96**, 4769–4782.

Palmer, M.R., Stephenson, G.R., Inall, M.E., Balfour, C., Düsterhus, A., and Green, J.A.M. 2015. Turbulence and mixing by internal waves in the Celtic Sea determined from ocean glider microstructure measurements. *J. Mar. Syst*, **144**, 57–69.

Panchev, S., and Kesich, D. 1969. Energy spectrum of isotropic turbulence at large wavenumbers. *Comptes rendus de l'Academie bulgare des Sciences*, **22**, 627–630.

Pawlowicz, R.T., McDougall, T., Feistel, R., and Tailleux, R. 2012. An historical perspective on the development of the development of the Thermodynamic Equation of Seawater–2010. *Ocean Sci.*, **8**, 161–174.

Pederson, A.M. 1973. A small in-situ conductivity instrument. Pages 68–75 of: *IEEE Int. Conf. Eng. in the Ocean*.

Pederson, A.M., and Gregg, M.C. 1979. Development of a small in-situ conductivity instrument. *IEEE J. Ocean Engr.*, **OE-4**, 69–75.

Perlin, A., and Moum, J.N. 2012. Comparison of thermal variance dissipation rates from moored and profiling instruments at the equator. *J. Atmos. Ocean. Tech.*, **29**, 1347–1362.

Peters, H., Gregg, M.C., and Sanford, T.B. 1995. Detail and scaling of turbulent overturns in the Pacific Equatorial Undercurrent. *J. Geophys. Res.*, **100**, 18,349–18,368.

Phillips, O.M. 1966. *Dynamics of the Upper Ocean*. Cambridge University Press.

Piera, J., Roget, E., and Catalan, J. 2002. Turbulent patch identification in microstructure profiles: A method based on wavelet denoising and Thorpe displacement analysis. *J. Atmos. Ocean. Tech.*, **19**, 1390–1402.

Pingree, R.D. 1972. Mixing in the deep stratified ocean. *Deep-Sea Res*, **19**, 549–561.

Pinkel, R. 1981. On the use of Doppler sonar for internal wave measurements. *Deep-Sea Res*, **28A**(3), 269–289.

Pinkel, R. 2000. Internal solitary waves in the warm pool of the western equatorial Pacific. *J. Phys. Oceanogr.*, **30**, 2906–2926.

Pinkel, R. 2008a. Advection, phase distortion, and the frequency spectrum of finescale fields in the sea. *J. Phys. Oceanogr.*, **38**, 291–313.

Pinkel, R. 2008b. The wavenumber-frequency spectrum of vortical and internal-wave shear in the western Arctic Ocean. *J. Phys. Oceanogr.*, **38**(Feb.), 277–290.

Pinkel, R. 2014. Vortical and internal wave shear and strain. *J. Phys. Oceanogr.*, **44**, 2070–2092.

Pinkel, R. 2020. The Poisson link between internal wave and dissipation scales in the thermocline. Part I. Probability density functions and the Poisson modeling of vertical strain. *J. Phys. Oceanogr.*, submitted.

Pinkel, R., and Anderson, S. 1992. Toward a statistical description of finescale strain in the thermocline. *J. Phys. Oceanogr.*, **22**, 773–795.

Pinkel, R., and Anderson, S. 1997. Shear, strain, and Richardson number variations in the thermocline. Part I: Statistical description. *J. Phys. Oceanogr.*, **27**(2), 264–281.

Pinkel, R., Golden, M.A., Smith, J.A., Sun, O.M., Aja, A.A., Bui, M.N., and Hughen, T. 2011. The wirewalker: A vertically profiling instrument carrier powered by ocean waves. *J. Atmos. Ocean. Tech.*, **28**(3), 426–435.

Pinkel, R., Rainville, L., and Klymak, J. 2012. Semidiurnal baroclinic wave momentum fluxes at Kaena Ridge, Hawaii. *J. Phys. Oceanogr.*, **42**(8), 1249–1269.

Pollard, R.T., and Millard, R.C. 1970. Comparison between observed and simulated wind-generated intertial currents. *Deep-Sea Res.*, **17**, 813–821.

Pollman, F., Eden, C., and Olbers, D. 2017. Evaluating the global internal wave model IDEMIX using finestructure methods. *J. Phys. Oceanogr.*, **47**(9), 2267–2289.

Polyakov, I.V., Padman, L., Lenn, Y.-D., Pnyushkov, A., Rember, R., and Ivanov, V.V. 2019. Eastern Arctic Ocean diapycnal heat fluxes through large double-diffusive steps. *J. Phys. Oceanogr.*, **49**(1), 227–246.

Polzin, K.L. 1996. Statistics of Richardson number: Mixing models and finestructure. *J. Phys. Oceanogr.*, **26**, 1409–1425.

Polzin, K.L., and Ferrari, R. 2004. Isopycnal dispersion in NATRE. *J. Phys. Oceanogr.*, **34**, 247–257.

Polzin, K.L., Garabato, A.C.N., Hussen, T.N., Sloyan, B.M., and Waterman, S. 2014. Finescale parameterizations of turbulent dissipation. *JGR*, **119**, 1–29.

Polzin, K.L., Kunze, E., Hummon, J., and Firing, E. 2002. The finescale response of lowered ADCP velocity profiles. *J. Atmos. Ocean. Tech.*, **19**(2), 205–224.

Polzin, K.L., Kunze, E., Toole, J.M., and Schmitt, R.W. 2003. The partition of finescale energy into internal waves and subinertial motions. *J. Phys. Oceanogr.*, **33**, 234–248.

Polzin, K.L., Toole, J.M., Ledwell, J.R., and Schmitt, R.W. 1997. Spatial variability of turbulent mixing in the abyssal ocean. *Science*, **276**(5309), 93–96.

Polzin, K.L., and Lvov, Y.L. 2011. Toward regional characterizations of the oceanic internal wavefield. *Rev. Geophys.*, **49**(RG4003).

Polzin, K.L., Speer, K.G., Toole, J.M., and Schmitt, R.W. 1996. Intense mixing of Antarctic Bottom Water in the equatorial Atlantic Ocean. *Nature*, **380**, 54–57.

Polzin, K.L., Toole, J.M., and Schmidt, R.W. 1995. Finescale parameterization of turbulent dissipation. *J. Phys. Oceanogr.*, **25**, 306–328.

Pope, S.B. 2000. *Turbulent Flows*. Cambridge, UK: Cambridge University Press.

Proni, J.R., and Apel, J.R. 1975. On the use of high-frequency acoustics for the study of internal waves and microstructure. *J. Geophys. Res.*, **80**, 1147–1151.

Pytkowicz, R.M. 1963. Gravity and the properties of sea water. Pages 286–287 of: *The Sea*, vol. 8. Wiley-Interscience.

Radko, T. 2014. Applicability and failure of the flux-gradient laws in double-diffusive convection. *J. Fluid Mech.*, **750**, 33–72.

Rainville, L., and Pinkel, R. 2004. Observations of energetic high-wavenumber internal waves in the Kuroshio. *J. Phys. Oceanogr.*, **34**, 1495–1505.

Rainville, L., and Winsor, P. 2008. Mixing across the Arctic Ocean: Microstructure observations during the Beringia 2005 Expedition. *Geophys. Res. Lett.*, **35**(8).

Rainville, L., Johnston, T.M.S., Carter, G.S., Merrifield, M.A., Pinkel, R., Worcester, P.F., and Dushaw, B.D. 2010. Interference pattern and propagation of the M_2 internal tide south of the Hawaiian Ridge. *J. Phys. Oceanogr.*, **40**, 311–325.

Ray, G.T, and Mitchum, R.D. 1996. Surface generation of internal tides generated near Hawaii. *Geophys. Res. Lett.*, **23**, 2101–2104.

Reynolds, O. 1895. On the dynamical theory of incompressible viscous fluids and the determination of the criterion. *Proc. Roy. Soc. Lond. A*, **A186**, 123–164.

Rice, J.A. 1988. *Mathematical Statistics and Data Analysis*. Wadsworth and Brooks/Cole.

Riley, J.J., and Lelong, M.-P. 2000. Fluid motions in the presence of strong stable stratification. *Annu. Rev. Fluid Mech.*, **32**, 613–657.

Riley, J.J., and Lindborg, E. 2008. Stratified turbulence: A possible interpretation of some geophysical turbulence measurements. *J. Atmos. Sci.*, **65**(7), 2416–2424.

Riley, J.J., Metcalfe, R.W., and Weissman, M.A. 1981. Direct numerical simulations of homogenous turbulence in density-stratified fluids. Pages 79–112 of: West, B.J. (ed.), *Proc. AIP Conf. Nonlinear Properties of Internal Waves*. American Inst. Physics.

Rimac, A., and von Storch, J.-S. 2016. The total energy flux leaving the ocean's mixed layer. *J. Phys. Oceanogr.*, **46**(6), 1885–1900.

Roden, G.I. 1964. Shallow temperature inversions in the Pacific Ocean. *J. Geophys. Res.*, **61**, 255–263.

Roemmich, D., Hautala, S., and Rudnick, D. 1996. Northward abyssal transport through the Samoan Passage and adjacent regions. *J. Geophys. Res.*, **101**(C6), 14,039–14,066.

Rohr, J.J., Helland, K.N., Itsweire, E.C., and Atta, C.W. Van. 1987. Turbulence in a stably stratified shear flow: A progress report. In: *Turbulent Shear Flows*, vol. 5. New York, NY: Springer-Verlag.

Rosenblum, L.J., and Marmorino, G. 1990. Statistics of mixing patches observed in the Sargasso Sea. *J. Geophys. Res.*, **95**, 5349–5357.

Ross, C.K. 1984. Temperature-salinity characteristics of the 'overflow' water in Denmark Strait during 'Overflow 73'. *Rapp. P.-V. Reun. Cons. Int. Explor. Mer*, **185**, 111–119.

Rossby, H.T. 1969. A vertical profile of currents near Plantagenet Bank. *Deep-Sea Res.*, **16**, 377–385.

Ruddick, B. 1983. A practical indicator of the stability of the water column to double-diffusive activity. *Deep-Sea Res.*, **3**, 1105–1107.

Ruddick, B., Walsh, D., and Oakey, N. 1997. Variations in apparent mixing efficiency in the North Atlantic central water. *J. Phys. Oceanogr.*, **27**(12), 2589–2605.

Rudnick, D. 1997. Direct velocity measurements in the Samoan Passage. *J. Geophys. Res.*, **102**(C2), 3293–3302.

Rudnick, D.L., Boyd, T.J., Brainerd, R.E., Carter, G.S., Egbert, G.D., Gregg, M.C., Holloway, P.E., Klymak, J.M., Kunze, E., Lee, C.M., Levine, M.D., Luther, D.S., Martin, J.P., Merrifield, M.A., Moum, J.N., Nash, J.D., Pinkel, R., Rainville, L., and Sanford, T.B. 2003. From tides to mixing along the Hawaiian Ridge. *Science*, **301**, 355–357.

Salehipour, H., Peltier, W.R., and Mashayek, A. 2015. Turbulent diapycnal mixing in stratified shear flows: The influence of Prandtl number on mixing efficiency and transition at high Reynolds number. *J. Fluid Mech.*, **773**, 178–223.

Sanchez, X., Roget, E., Planella, J., and Forcat, F. 2011. Small-scale spectrum of a scalar field in water: The Batchelor and Kraichnan models. *J. Phys. Oceanogr.*, **41**, 2155–2167.

Sanford, T.B. 1975. Observations of the vertical structure of internal waves. *J. Geophys. Res.*, **80**(27), 3861–3871.

Sanford, T.B., Carlson, J.A., Dunlap, J.H., Prater, M.D., and Lien, R.-C. 1999. An electromagnetic vorticity and velocity sensor for observing finescale kinetic fluctuations in the ocean. *J. Atmos. Ocean. Tech.*, **16**, 1647–1667.

Sanford, T.B., Drever, R.G., and Dunlap, J.H. 1978. A velocity profiler based on the principles of geomagnetic induction. *Deep-Sea Res*, **25**, 183–210.

Sanford, T.B., Drever, R.G., and Dunlap, J.H. 1985. An acoustic Doppler and electromagnetic profiler. *J. Atmos. Ocean. Tech.*, **2**(6), 110–124.

Sanford, T.B., Drever, R.G., Dunlap, J.H., and D'Asaro, E.A. 1982. *Design, Operation and Performance of an Expendable Temperature and Velocity Profiler*. Tech. Report 8110. Applied Physics Laboratory, University of Washington, 1013 NE 40th St., Seattle, WA 98105-6698.

Sanford, T.B., Dunlap, J.H., Carlson, J.A., Webb, D.C., and Girton, J.B. 2005. Autonomous velocity and density profiler: EM-APEX. Pages 152–156 of: White, J.R., and Anderson, S. (eds.), *Proceedings of the IEEE/OES/CMTC Eighth Working Conference on Current Measurement Technology*. The Printing House, 445 Hoes Lane, Piscataway, NJ, 08854: IEEE.

Sanford, T.B., and Lien, R.-C. 1999. Turbulent properties in a homogenous tidal bottom boundary layer. *J. Geophys. Res.*, **104**(C1), 1245–1257.

Sanford, T.B., Price, J.F., and Girton, J.B. 2011. Upper-ocean response to Hurricane Frances (2004) observed by profiling EM-APEX floats. *J. Phys. Oceanogr.*, **41**, 1041–1055.

Schanze, J.J., and Schmitt, R.W. 2013. Estimates of cabbeling in the global ocean. *J. Phys. Oceanogr.*, **43**(4), 698–705.

Schmitt, R.W. 1979. Flux measurements on salt fingers at an interface. *J. Mar. Res.*, **37**, 419–436.

Schmitt, R.W. 1981. Form of the temperature-salinity relationship in the central water: Evidence for double-diffusive mixing. *J. Phys. Oceanogr.*, **11**(7), 1015–1026.

Schmitt, R.W., Ledwell, J.R., Montromery, E.T., Polzin, K.L., and Toole, J.M. 2005. Enhanced diapycnal mixing by salt fingers in the thermocline of the tropical Atlantic. *Science*, **308**(5722), 685–688.

Schmitt, R.W., Perkins, H., Boyd, J.D., and Stalcup, M.C. 1987. C-SALT: An investigation of the thermohaline staircase in the western tropical North Atlantic. *Deep-Sea Res.*, **34**(10), 1655–1665.

Schmitt, R.W., Toole, J.M., Koehler, R.L., Mellinger, E.C., and Doherty, K.W. 1988. The development of a fine- and microstructure profiler. *J. Atmos. Ocean. Tech.*, **5**, 484–500.

Seim, H.E. 1999. Acoustic backscatter from salinity microstructure. *J. Atmos. Ocean. Tech.*, **16**, 1491–1498.

Seim, H.E., and Gregg, M.C. 1994. Detailed observations of a naturally occurring shear instability. *J. Geophys. Res.*, **99**, 10,049–10,073.

Seim, H.E., Gregg, M.C., and Miyamoto, R.T. 1995. Acoustic backscatter from turbulent microstructure. *J. Atmos. Ocean. Tech.*, **12**, 367–380.

Shaw, W.J., and Stanton, T.P. 2014. Dynamic and double-diffusive instabilities in a weak pycnocline. Part I: Observations of heat flux and diffusivity in the vicinity of Maud Rise, Weddell Sea. *J. Phys. Oceanogr.*, **44**, 1973–1991.

Shay, T.J., and Gregg, M.C. 1986. Convectively driven turbulent mixing in the upper ocean. *J. Phys. Oceanogr.*, **16**, 1777–1798.

Shcherbina, A.Y., Gregg, M.C., Alford, M.H., and Harcourt, R.R. 2009. Characterizing thermohaline intrusions in the North Pacific subtropical frontal zone. *J. Phys. Oceanogr.*, **39**(11), 2735–2756.

Shcherbina, A.Y., Gregg, M.C., Alford, M.H., and Harcourt, R.R. 2010. Three-dimensional structure and temporal evolution of submesoscale thermohaline intrusions in the north Pacific subtropical frontal zone. *J. Phys. Oceanogr.*, **40**(8), 1669–1689.

Shen, C.Y. 1993. Heat-salt finger fluxes across a density interface. *Phys. Fluids A*, **5**, 2633–2643.

Shen, C.Y. 1995. Equilibrium salt-fingering convection. *Phys. Fluids*, **7**, 706–717.

Sheen, K.L., Brearley, J.A., Garabato, A.C.N., Smeed, D.A., Waterman, S., Ledwell, J.R., Meredith, M.P., Laurent, L. St., Thurnherr, A.M., Toole, J.M., and Watson, A.J. 2013. Rates and mechanisms of turbulent dissipation and mixing in the Southern Ocean: Results from the Diapycnal and Isopycnal Mixing Experiment in the Southern Ocean (DIMES). *J. Geophys. Res.*, **118**, 2774–2792.

Sherman, J.T., and Davis, R.E. 1995. Observations of temperature microstructure in NATRE. *J. Phys. Oceanogr.*, **25**, 1913–1929.

Sherman, J.T., and Pinkel, R. 1991. Estimates of the vertical wavenumber-frequency spectra of vertical shear and strain. *J. Phys. Oceanogr.*, **21**, 292–303.

Shirtcliffe, T.G.L. 1967. Thermosolutal convection: Observation of an overstable mode. *J. Fluid Mech.*, **213**, 480–490.

Shroyer, E.L., Moum, J.N., and Nash, J.D. 2010. Energy transformations and dissipation of nonlinear internal waves over New Jersey's continental shelf. *Nonlin. Process Geophys.*, **17**, 345–360.

Simmons, H.L. 2008. Spectral modification and geographic redistribution of the semi-diurnal internal tide. *Ocean Modell.*, **21**, 126–138.

Simmons, H.L., and Alford, M.H. 2012. Simulating the long range swell of internal waves generated by ocean storms. *Oceanography*, **25**(2), 126–138.

Simpson, J.H., Howe, M.R., Morris, N.C.G., and Stratford, J. 1979. Velocity shear in the steps below the Mediterranean outflow. *Deep-Sea Res.*, **26A**, 1381–1386.

Sloyan, B.M. 2005. Spatial variability of mixing in the Southern Ocean. *Geophys. Res. Lett.*, **32**(L18603).

Smith, P.C. 1976. Baroclinic instability in the Denmark Strait overflow. *J. Phys. Oceanogr.*, **6**, 355–371.

Smith, S.A., Fritts, D.C., and VanZandt, T.E. 1987. Evidence for a saturated spectrum of atmospheric gravity waves. *J. Atmos. Sci.*, **44**(10), 1404–1410.

Smyth, W.D. 1999. Dissipation-range geometry and scalar spectra in sheared stratified turbulence. *J. Fluid Mech.*, **401**, 209–242.

Smyth, W.D., and Kimura, S. 2011. Mixing in a moderately sheared salt-fingering layer. *J. Phys. Oceanogr.*, **41**(7), 1364–1384.

Smyth, W.D., and Moum, J.N. 2000. Anisotropy of turbulence in stably stratified layers. *Phys. Fluids*, **12**(6), 1343–1362.

Smyth, W.D., Moum, J.N., and Caldwell, D.R. 2001. The efficiency of mixing in turbulent patches: Inferences from direct simulations and microstructure observations. *J. Phys. Oceanogr.*, **31**, 1969–1992.

Smyth, W.D., and Moum, J.N. 2012. Ocean mixing by Kelvin–Helmholtz instability. *Oceanography*, **25**(2), 140–149.

Smyth, W.D., Nash, J.D., and Moum, J.N. 2005. Differential diffusion in breaking Kelvin–Helmholtz billows. *J. Phys. Oceanogr.*, **35**(6), 1004–1020.

Smyth, W.D., and Thorpe, S.A. 2012. Glider measurements of overturning in a Kelvin–Helmholtz billow train. *J. Mar. Res.*, **70**, 119–140.

Smyth, W.D., and Winters, K.B. 2003. Turbulence and mixing in Holmboe waves. *J. Phys. Oceanogr.*, **33**, 694–711.

Smyth, W.D., Carpenter, J.R., and Lawrence, G. 2007. Mixing in symmetric Holmboe waves. *J. Phys. Oceanogr.*, **37**, 1566–1583.

Solomon, H. 1971. On the representation of isentropic mixing in ocean circulation models. *J. Phys. Oceanogr.*, **1**, 233–234.

Soloviev, A., Lukas, R., Hacker, P., Schoerberlein, H., Baker, M., and Arjannikov, A. 1999. A near-surface microstructure sensor system used during TOGA COARE. Part II: Turbulence measurements. *J. Atmos. Ocean. Tech.*, **16**(11), 1598–1618.

Sommer, T., Carpenter, J.R., Schmid, M., Lueck, R.G., Schurter, M., and Wüest, A. 2013a. Interface structure and flux laws in a natural double-diffusive layering. *J. Geophys. Res.*, **118**, 6092–6106.

Sommer, T., Carpenter, J.R., Schmid, M., Lueck, R.G., and Wüest, A. 2013b. Revisiting microstructure sensor responses with implications for double-diffusive fluxes. *J. Atmos. Ocean. Tech.*, **30**, 1907–1923.

Sommer, T., Schmid, M., and Wüest, A. 2019. The role of double diffusion for heat and salt balances in Lake Kivu. *Limnol. Oceanogr.*, **64**(2019), 650–660.

Spilhaus, A.F. 1939. A detailed study of the surface layers of the ocean in the neighborhood of the Gulf Stream with the aid of rapid measuring hydrographic instruments. *J. Mar. Res.*, **3**, 51–75.

Sreenivasan, K. 1995. On the universality of the Kolmogorov constant. *Phys. Fluids*, **7**, 2778–2784.

Sreenivasan, K. 1996. The passive scalar spectrum and the Obukhov-Corrsin constant. *Phys. Fluids*, **8**(189), 189–196.

St. Laurent, L. 2008. Turbulent dissipation on the margins of the South China Sea. *Geophys. Res. Lett.*, **35**(L23615).

St. Laurent, L., Garabato, A.C.N., Ledwell, J.R., Thurnherr, A.M., Toole, J.M., and Watson, A.J. 2012. Turbulence and diapycnal mixing in Drake Passage. *J. Phys. Oceanogr.*, **42**(12), 2143–2152.

St. Laurent, L., and Garrett, C. 2002. The role of internal tides in mixing the deep ocean. *J. Phys. Oceanogr.*, **32**(10), 2882–2899.

St. Laurent, L., and Schmitt, R.W. 1999. The contribution of salt fingers to vertical mixing in the North Atlantic Tracer Release Experiment. *J. Phys. Oceanogr.*, **29**, 1404–1424.

St. Laurent, L., Stringer, S., Garrett, C., and Perrault-Joncas, D. 2003. The generation of internal tides at abrupt topography. *Deep-Sea Res. I*, **50**, 987–1003.

St. Laurent, L., Toole, J.M., and Schmitt, R.W. 2001. Buoyancy forcing by turbulence above rough topography in the abyssal Brazil Basin. *JPO*, **31**, 3476–3485.

Stacey, M.T., Monismith, S.G., and Burau, J.R. 1999. Measurement of Reynolds stress profiles in unstratified profiles. *J. Geophys. Res.*, **104**, 10,933–10,949.

Steele, E., Nimmo-Smith, A., Vlasenko, A., Vlasenko, V., and Hosegood, P. 2013. Examination of turbulence structures in the bottom boundary layer of the coastal ocean by submersible 3D-PTV. In: *10th International Symposium of Particle Image Velocimetry – PIV13, Delft, The Netherlands, July 1–3, 2013*.

Steffen, E.L., and D'Asaro, E.A. 2002. Deep-convection in the Labrador Sea observed by Lagrangian floats. *J. Phys. Oceanogr.*, **32**(2), 475–492.

Stern, M.E. 1960. The 'salt-fountain' and thermohaline convection. *Tellus*, **12**, 172–175.

Stern, M.E. 1967. Lateral mixing of water masses. *Deep-Sea Res.*, **14**, 747–753.

Stern, M.E. 1968. T-S gradients on the micro scale. *Deep-Sea Res.*, **15**, 245–250.

Stern, M.E. 1969. Collective instability of salt fingers. *J. Fluid Mech.*, **35**, 209–218.

Stern, M.E. 1975. *Ocean Circulation Physics*. New York: Academic.

Stern, M.E., Radko, T., and Simeonov, J. 2001. Salt fingers in an unbounded thermocline. *J. Mar. Res.*, **59**, 355–390.

Stewart, R.W. 1959. The problem of diffusion in a stratified fluid. *Annu. Rev. Geophys.*, **6**, 303–311.

Stillinger, D.C., Helland, K.N., and Atta, C.W. Van. 1983. Experiments on the transition of homogeneous turbulence to internal waves in a stratified fluid. *J. Fl.*, **131**, 91–122.

Stommel, H. 1962. On the cause of the temperature-salinity curve in the ocean. *Nat. Acad. Sci.*, **48**, 764–766.

Stommel, H., Arons, A. B., and Blanchard, D. 1956. An oceanographical curiosity, the perpetual salt fountain. *Deep-Sea Res.*, **3**, 152–153.

Stommel, H., and Federov, K.N. 1967. Small scale structure in temperature and salinity near Timor and Mindanao. *Tellus*, **19**, 306–326.

Sundermeyer, M.A., and Lelong, M.-P. 2005. Numerical simulations of lateral dispersion by the relaxation of diapycnal mixing events. *J. Phys. Oceanogr.*, **35**(12), 2368–2386.

Sundermeyer, M.A., and Price, J.F. 1998. Lateral mixing and the North Atlantic Tracer Release Eexperiment: Observations and numerical simulations of Lagrangian particles and a passive tracer. *J. Geophys. Res.*, **103**, 481–497.

Sutherland, B.R. 2010. *Internal Gravity Waves*. Cambridge, UK: Cambridge University Press.

Swift, S.A., Bower, A.S., and Schmitt, R.W. 2012. Vertical, horizontal, and temporal changes in temperature in the Atlantis II and Discovery hot brine pools, Red Sea. *Deep-Sea Res. I*, **64**, 118–128.

Tailleux, R. 2009. On the energetics of stratified turbulent mixing, irreversible thermodynamics, Boussinesq models and the ocean heat engine controversy. *J. Fluid Mech.*, **638**, 339–382.

Tailleux, R. 2013. Available potential energy and energy in stratified fluids. *Ann. Rev. Fluid Mech.*, **45**, 35–58.

Tait, R.I., and Howe, M.R. 1968. Some observations of thermohaline stratification in the deep ocean. *Deep-Sea Res.*, **15**, 275–280.

Takahashi, A., and Hibiya, T. 2019. Assessment of finescale parameterizations of deep ocean mixing in the presence of geostrophic current shear: Results of microstructure measurements in the Antarctic Circumpolar Current region. *J. Geophys. Res: Oceans*, **124**, 135–153.

Talley, L.D. 2013. Closure of the global overturning circulation through the Indian, Pacific, and Southern Oceans: Schematics and transport. *Oceanography*, **26**, 80–97.

Talley, L.D., Pickard, G.L., Emery, W.J., and Swift, J.H. 2011. *Descriptive Physical Oceanography: An Introduction*. 6th ed. Elsevier Ltd.

Talley, L.D., and Yun, J.-Y. 2001. The role of cabbeling and double diffusion in setting the density of the North Pacific Intermediate Water salinity minimum. *J. Phys. Oceanogr.*, **31**(6), 1538–1549.

Taylor, G.I. 1935. Statistical theory of turbulence. *Proc. R. Soc. Lond. A*, **151**, 421–444.

Taylor, J., and Bucens, P. 1989. Laboratory experiments on the structure of salt fingers. *Deep-Sea Res*, **36**, 1675–1704.

Tennekes, H., and Lumley, J.L. 1972. *A First Course in Turbulence*. Cambridge, MA: MIT Press.

Thorpe, S.A. 1971. Experiments on the instability of stratified shear flows: Miscible fluids. *J. Fluid Mech.*, **46**, 299–319.

Thorpe, S.A. 1977. Turbulence and mixing in a Scottish loch. *Proc. Roy. Soc. Lond. A*, **286**, 125–181.

Thorpe, S.A. 1987. On the reflection of a strain of finite-amplitude internal waves from a uniform slope. *J. Fluid Mech.*, **178**, 299–302.

Thorpe, S.A. 2005. *The Turbulent Ocean*. Cambridge University Press.

Thorpe, S.A., and Brubaker, J.M. 1983. Observation of sound reflection by temperature microstructure. *Limnol. Oceanogr.*, **28**, 601–613.

Thorpe, S.A., Osborn, T.R., Jackson, J.F.E., Hall, A.J., and Lueck, R.G. 2002. Measurements of turbulence in the upper ocean mixing layer using Autosub. *J. Phys. Oceanogr.*, **32**(1), 122–145.

Thurnherr, A.M., St. Laurent, L.C., Speer, K.G., Toole, J.M., and Ledwell, J.R. 2005. Mixing associated with sills in a canyon on the Midocean Ridge flank. *J. Phys. Oceanogr.*, **35**, 1370–1381.

Thwaites, F.T., Krishfield, R., Timmermans, M.-L., Toole, J.M., and III, A.J. Williams. 2011. Noise in ice-tethered profiler and McLane Moored Profiler velocity measurements. In: *Proc. IEEE/OES/CWTH Tenth Working Conf. on Current Measurement Technology*.

Thwaites, F.T., Williams, III, A.J., Terray, E.A., and Trowbridge, J.H. 1995. A family of acoustic vorticity meters to measure ocean boundary layer shear. Pages 193–198 of: *Proc. IEEE Fifth Working Conf. of Current Measurement*.

Timmermans, M.-L., Garrett, C., and Carmack, E. 2003. The thermohaline structure and evolution of deep waters in the Canada Basin, Arctic Ocean. *Deep-Sea Res I*, **50**, 1305–1321.

Timmermans, M.-L., Toole, J., Krishfield, R., and Winsor, P. 2008. Ice-tethered profiler observations of the double-diffusive staircases in the Canada Basin thermocline. *J. Geophys. Res.*, **113**(C00A02).

Toggweiler, J.R., and Samuels, B.J. 1998. On the ocean's large-scale circulation near the limit of no vertical mixing. *J. Phys. Oceanogr.*, **28**, 1832–1852.

Toole, J.M., Polzin, K.L., and Schmitt, R.W. 1994. Estimates of diapycnal mixing in the abyssal ocean. *Science*, **264**, 1120–1123.

Turner, J.S. 1965. The coupled transports of salt and heat across a sharp density interface. *Intl. J. Heat Mass Transfer*, **8**, 759–767.

Turner, J.S. 1968a. The behavior of a stable salinity gradient heated from below. *J. Fluid Mech.*, **33**, 183–200.

Turner, J.S. 1968b. The influence of molecular diffusivity on turbulent entrainment across a density interface. *J. Fluid Mech.*, **33**, 639–656.

Turner, J.S. 1973. *Buoyancy Effects in Fluids*. Cambridge University Press.

Turner, J.S. 1978. Double-diffusive intrusions into a density gradient. *J. Geophys. Res.*, **83**(C6), 2887–2901.

Turner, J.S., and Stommel, H. 1964. A new case of convection in the presence of combined vertical salinity and temperature gradients. *Proc. Natl. Acad. Sci. USA*, **52**, 49–53.

UNESCO. 1980. *The Practical Salinity Scale 1978 and the International Equation of State of Seawater 1980*. Technical Papers 36. UNESCO.

Vanneste, J. 2012. Balance and spontaneous wave generation in geophysical flows. *Annu. Rev. Fluid Mech.*, **45**, 147–172.

Veronis, G. 1965. On finite amplitude instability in thermohaline convection. *J. Mar. Res.*, **13**, 1–17.

Veronis, G. 1968. Effect of a stabilizing gradient of solute on thermal convection. *Tellus*, **34**, 315–336.

Veronis, G. 1972. On properties of seawater defined by temperature, salinity, and pressure. *J. Mar. Res.*, **30**, 227–255.

Visbeck, M., and Rhein, M. 2000. Is boundary mixing slowly ventilating Greenland Sea Deep Water? *J. Phys. Oceanogr.*, **30**, 215–224.

Voulgaris, G., and Trowbridge, J.H. 1998. Evaluation of the acoustic Doppler velocimeter (ADV) for turbulence measurements. *J. Atmos. Ocean. Tech.*, **15**, 272–289.

Waite, M.L., and Bartello, P. 2004. Stratified turbulence dominated by vortical motion. *J. Fluid Mech.*, **517**, 281–308.

Wang, W., and Huang, R.X. 2004. Wind energy input to the surface waves. *J. Phys. Oceanogr.*, **34**, 1276–1280.

Washburn, L., and Käse, R.H. 1987. Double diffusion and the distribution of the density ratio in the Mediterranean waterfront south of the Azores. *J. Phys. Oceanogr.*, **17**(1), 12–25.

Washburn, L., Duda, T.F., and Jacobs, D.C. 1996. Interpreting conductivity microstructure: Estimating the temperature variance dissipation rate. *J. Atmos. Ocean. Tech.*, **13**, 1166–1188.

Waterhouse, A.F., MacKinnon, J.A., Nash, J.D., Alford, M.H., Kunze, E., Simmons, H.I., Polzin, K.L., Laurent, L.C. St., Sun, O.M., Pinkel, R., Talley, L.D., Whalen, C.B., Huussen, T.N., Carter, G.S., Fer, I., Waterman, S., Garabato, A.C.N., Sanford, T.B., and Lee, C.M. 2014. Global patterns of diapycnal mixing from measurements of the turbulent dissipation rate. *J. Phys. Oceanogr.*, **44**(7), 1854–1872.

Waterman, S., Garabato, A.C. Naveira, and Polzin, K. 2013. Internal waves and turbulence in the Antarctic Circumpolar Current. *J. Phys. Oceanogr.*, **43**(2), 259–282.

Waterman, S., Polzin, K.L., Garabato, A.C.N., Sheen, K.L., and Forryan, A. 2014. Suppression of internal wave breaking in the Antarctic Circumpolar Current near topography. *J. Phys. Oceanogr.*, **44**, 1466–1492.

Watson, A.J., and Ledwell, J.R. 2000. Oceanographic tracer release experiments using sulphur hexafluoride. *J. Geophys. Res. Oceans*, **105**(C6), 14,325–14,337.

Watson, A.J., Messias, M.-J., Fogelqvist, E., Scoy, K.A. Van, Johannessen, T., Oliver, K.I.C., Stevens, D.P., Rey, F., Tanhua, T., Olsson, K.A., Carse, F., Simonsen, K., Ledwell, J.R., Jansen, E., Cooper, D.J., Kreupke, J.A., and Guilyardi, E. 1999. Mixing and convection in the Greenland Sea from a tracer-release experiment. *Nature*, **401**, 902–904.

Watson, A.J., Ledwell, J.R., Messias, M.-J., King, B.A., Mackay, N., Meredith, M.P., Mills, B., and Garabatoo, A.C. Naveira. 2013. Rapid cross-density ocean mixing at mid-depths in the Drake Passage measured by tracer release. *Nature*, **501**, 408–411.

Webb, D.J., and Suginohara, N. 2001. Vertical mixing in the ocean. *Nature*, **406**, 37.

Weinstock, J. 1985. On the theory of temperature spectra in a stably stratified fluid. *J. Phys. Oceanogr.*, **15**(4), 475–477.

Welander, P. 1959. An advective model of the ocean thermocline. *Tellus*, **11**, 309–318.

Wesson, J.C., and Gregg, M.C. 1988. Turbulent dissipation in the Strait of Gibraltar and associated mixing. Pages 201–212 of: Nihoul, J.C.J., and Jamart, B.M. (eds.), *Small-Scale Turbulence and Mixing in the Ocean, Proceedings of the 19th International Liege Colloquium on Ocean Hydrodynamics*. Amsterdam: Elsevier.

Wesson, J.C., and Gregg, M.C. 1994. Mixing at Camarinal Sill in the Strait of Gibraltar. *J. Geophys. Res.*, **99**, 9847–9878.

Whalen, C.B., Talley, L.D., and MacKinnon, J.A. 2012. Spatial and temporal variability of global ocean mixing inferred from Argo profiles. *Geophys. Res. Lett.*, **39**(L18612).

Whalen, C.B., MacKinnon, J.A., Talley, L.D., and Waterhouse, A.F. 2015. Estimating the mean diapycnal mixing using a finescale strain parameterization. *J. Phys. Oceanogr.*, **45**(4), 1174–1188.

White, W., and Bernstein, R. 1981. Large-scale vertical eddy diffusion in the main pycnocline of the central North Pacific. *J. Phys. Oceanogr.*, **11**(4), 434–441.

Wijesekera, H., Padman, L., Dillon, T., Levine, M., Paulson, C., and Pinkel, R. 1993. The application of internal-wave dissipation models to a region of strong mixing. *J. Phys. Oceanogr.*, **23**, 269–286.

Willebrand, J., Müller, P., and Olbers, D.J. 1977. *Inverse Analysis of the Trimoored Internal Wave Experiment (IWEX)*. Berichte 20a. Inst. für Meereskunde, Christian-Albrechts-Universität, Kiel, Kiel, Germany.

Williams, A.J. 2014. Current measurement by differential acoustic travel-time reviewed. Pages 1–5 of: *Baltic International Symposium (BALTIC)*. Xplore, vol. IEEE/OES. IEEE.

Winkel, D.P., Gregg, M.C., and Sanford, T.B. 1996. Resolving oceanic shear and velocity with the Multi-Scale Profiler. *J. Atmos. Ocean. Tech.*, **13**, 1046–1072.

Winkel, D.P., Gregg, M.C., and Sanford, T.B. 2002. Patterns of shear and turbulence across the Florida Current. *J. Phys. Oceanogr.*, **32**, 3269–3285.

Winters, K.B., and D'Asaro, E.A. 1994. Three-dimensional wave instability near a critical level. *J. Fluid Mech.*, **272**(August), 255–284.

Winters, K.B., and D'Asaro, E.A. 1996. Diascalar flux and the rate of fluid mixing. *J. Fluid Mech.*, **317**(June), 179–193.

Winters, K.D., and D'Asaro, E.A. 1997. Direct simulation of internal wave energy transfer. *J. Phys. Oceanogr.*, **27**(9), 1937–1945.

Woods, J.D. 1968. Wave-induced shear instability in the summer thermocline. *J. Fluid Mech.*, **32**(4), 791–800.

Woods, J.D., Onken, R., and Fischer, J. 1986. Thermohaline intrusions created isopycnically at oceanic fronts are inclined to isopycnals. *Nature*, **322**(31 July 1986), 446–449.

Wu, L.X., Jing, Z., Riser, S., and Visbeck, M. 2011. Seasonal and spatial variations of Southern Ocean diapycnal mixing from Argo profiling floats. *Nat. Geosci.*, **4**, 363–366.

Wunsch, C. 2006. *Discrete Inverse and State Estimation Problems: With Geophysical Fluid Applications*. Cambridge University Press.

Wunsch, C., and Ferrari, R. 2004. Vertical mixing, energy, and the general circulation of the oceans. *Annu. Rev. Fluid Mech.*, **36**, 281–314.

Wunsch, C., and Webb, S. 1979. The climatology of deep ocean internal waves. *J. Phys. Oceanogr.*, **9**, 235–243.

Würger, A. 2010. Thermal non-equilibrium transport in colloids. *Rep. Progr. Phys.*, **73**.

Wust, G. 1933. Das bodenwasser und die Gliederung der Atlantischen Tiefsee. *Wiss. Ergebn. Dtsch. Atlant. Exped. 'Meteor'*, **6**(1), 1–107.

Yamazaki, H., and Lueck, R.G. 1990. Why oceanic dissipation rates are not lognormal. *J. Phys. Oceanogr.*, **20**, 1907–1908.

Yamazaki, H., and Osborn, T. 1990. Dissipation estimates for stratified turbulence. *J. Geophys. Res.*, **95**(C6), 9739–9744.

You, Y. 2002. A global ocean climatological atlas of the Turner angle: Implications for double-diffusion and water-mass structure. *Deep-Sea Res. I*, **49**, 2075–2093.

Zhao, Z. 2016. Internal tide oceanic tomography. *Geophys. Res. Lett.*, **43**, 9157–9164.

Zhao, Z. 2018. The global mode-2 M_2 internal tide. *J. Geophys. Res.: Oceans*, **123**, 7725–7746.

Zhao, Z., and Alford, M.H. 2009. New altimetric estimates of Mode-1 M_2 internal tides in the central North Pacific Ocean. *J. Phys. Oceanogr.*, **39**(7), 1669–1684.

Zhao, Z., Alford, M.H., Girton, J.B., Rainville, L., and Simmons, H.L. 2016. Global observations of open-ocean mode-1 M2 internal tides. *J. Phys. Oceanogr.*, **46**(6), 1657–1684.

Zodiatis, G., and Gasparini, G.P. 1996. Thermohaline staircase formations in the Tyrrhenian Sea. *Deep-Sea Res. I*, **43**(5), 655–678.

Index